Fundamental Bacterial Genetics

Nancy Trun and Janine Trempy

350 Main Street, Malden, MA 02148-5020, USA
108 Cowley Road, Oxford OX4 1JF, UK
550 Swanston Street, Carlton, Victoria 3053, Australia

First published 2004 by Blackwell Science Ltd

Library of Congress Cataloging-in-Publication Data

Trun, Nancy Jo.
Fundamental bacterial genetics / Nancy Trun and Janine Trempy.
p. ; cm.
Includes bibliographical references and index.
ISBN 0-632-04448-9 (pbk. : alk. paper)
1. Bacterial genetics. 2. Escherichia coli—Genetics.
[DNLM: 1. Bacteria—genetics. 2. DNA, Bacterial. 3. Escherichia coli—genetics. QW 51 T871f 2004] I. Trempy, J. E. II. Title.

QH434.T78 2004
579.3'135—dc21

2003000141

A catalogue record for this title is available from the British Library.

Set in 10/12½ pt Stone Serif
by SNP Best-set Typesetter Ltd., Hong Kong
Printed and bound in the United Kingdom
by William Clowes Ltd, Beccles, Suffolk

For further information on
Blackwell Publishing, visit our website:
http://www.blackwellpublishing.com

Fundamental Bacterial Genetics

Nancy dedicates this book to all the really interesting teachers that she had the pleasure of learning from.

Janine dedicates this book to her child, Kieley, for keeping herself delightfully yet independently busy on those nights and weekends that writing had to get done, and to her parents, for emphasizing the importance of being a life-long learner.

Brief contents

Full contents	vi
Preface	xii
Acknowledgments	xiv
Chapter 1 Introduction to the cell	1
Chapter 2 The bacterial DNA molecule	17
Chapter 3 Mutations	38
Chapter 4 DNA repair	58
Chapter 5 Recombination	74
Chapter 6 Transposition	89
Chapter 7 Bacteriophage	105
Chapter 8 Transduction	126
Chapter 9 Natural plasmids	142
Chapter 10 Conjugation	156
Chapter 11 Transformation	175
Chapter 12 Gene expression and regulation	191
Chapter 13 Plasmids, bacteriophage, and transposons as tools	213
Chapter 14 DNA cloning	234
Chapter 15 Bioinformatics and proteomics	256
Glossary	267
Further reading	278
Index	281

Full contents

Preface xii
Acknowledgments xiv

Chapter 1 Introduction to the cell 1
The bacterial cell: a quick overview 5
How do cells grow? 13
What is genetics? 15
Summary 16

Chapter 2 The bacterial DNA molecule 17
The structure of DNA and RNA 17
Deoxyribonucleosides and deoxyribonucleotides 17
DNA is only polymerized 5′ to 3′ 19
Double-stranded DNA 21
Supercoiling double-stranded DNA 23
Replication of the *Escherichia coli* chromosome 26
Constraints that influence DNA replication 28
The replication machinery 28
  DNA polymerases 28
  DnaG primase 30
  DnaA, DnaB, and DnaC 31
  Replication of both strands 32
  Theta mode replication 34
  Minimizing mistakes in DNA replication 34
The DNA replication machinery as molecular tools 36
Summary 37

Chapter 3 Mutations 38
Phenotype and genotype 39
Classes of mutations 40
  Point mutations and their consequences 42
Measuring mutations: rate and frequency 44
Spontaneous and induced mutations 44
  Errors during DNA replication: incorporation errors 44
  Errors due to tautomerism 45

Spontaneous alteration by depurination 45
Spontaneous alteration by deamination 46
Alterations by spontaneous genetic rearrangement 48
Alterations caused by transposition 48
Induced mutations 48
Chemicals that mimic normal DNA bases: base analogs 49
Chemicals that react with DNA bases: base modifiers 49
Chemicals that bind DNA bases: intercalators 51
Mutagens that physically damage the DNA: ultraviolet light and ionizing radiation 51
Mutator strains 52
Reverting mutations 52
Suppression 53
Ames test 54
How have we exploited bacterial mutants? 56
Summary 56

Chapter 4 DNA repair 58
Lesions that constitute DNA damage 58
Reverse, excise or tolerate? 60
Mechanisms that reverse DNA damage 60
Photoreactivation 60
O6-methylguanine or O4-methylthymine methyltransferase 61
Mechanisms that excise DNA damage 62
UvrABC directed nucleotide excision repair 62
MutHLS methyl directed mismatch repair 65
Very short patch repair 65
Glycosylases 67
Uracil-N-glycosylase coupled with AP excision repair 67
Deaminated bases removed by DNA glycosylase 68
Alkylated bases removed by DNA glycosylase 68
MutM/MutY: oxidative damage 68
N-glycosylases specific for pyrimidine dimers 68
Mechanisms that tolerate DNA damage 69
Transdimer synthesis 69
Post replication/recombinational repair (PRR) 69
Introduction to the SOS regulon 71
Summary 72

Chapter 5 Recombination 74
Homologous recombination 74
Models for homologous recombination 78
The Holliday or double-strand invasion model of recombination 79
An alternative to the Holliday model: the single-strand invasion model of Meselson and Radding 81
Further enzymatic considerations 82
Site-specific recombination 83
A typical site-specific recombinational event 83
Bacteriophage λ: a model for site-specific recombination 84
Other microbial examples of site-specific recombination 85

Illegitimate recombination 86
Summary 87

Chapter 6 Transposition 89
The structure of transposons 89
The frequency of transposition 91
The two types of transposition reactions 93
The transposition machinery 93
  Accessory proteins encoded by the transposon 94
  Accessory proteins encoded by the host 94
Non-replicative transposition 95
Replicative transposition 95
Does the formation of a cointegrate predict the transposition mechanism? 97
The fate of the donor site 97
Target immunity 99
Transposons as molecular tools 100
Summary 104

Chapter 7 Bacteriophage 105
The structure of phage 105
The lifecycle of a bacteriophage 106
Lytic–lysogenic options 107
The λ lifecycle 107
  λ adsorption 107
  λ DNA injection 107
  Protecting the λ genome in the bacterial cytoplasm 109
  What happens to the λ genome after it is stabilized? 109
  λ and the lytic–lysogenic decision 110
  The λ lysogenic pathway 112
  The λ lytic pathway 113
  DNA replication during the λ lytic pathway 114
  Making λ phage 114
  Getting out of the cell – the λ S and R proteins 116
  Induction of λ by the SOS system 117
  Superinfection 117
  Restriction and modification of DNA 117
The lifecycle of M13 118
  M13 adsorption and injection 118
  Protection of the M13 genome 118
  M13 DNA replication 118
  M13 phage production and release from the cell 119
The lifecycle of P1 119
  Adsorption, injection, and protection of the genome 119
  P1 DNA replication and phage assembly 119
  The location of the P1 prophage in a lysogen 120
  P1 transducing particles 120
The lifecycle of T4 121
  T4 adsorption and injection 121
  T4rII mutations and the nature of the genetic code 123
Summary 125

Chapter 8 Transduction 126
Generalized transduction vs. specialized transduction 126
P1 as a model for generalized transducing phage 126
Packaging the chromosome 127
Moving pieces of the chromosome from one cell to another 128
Identifying transduced bacteria: selection vs. screening 129
Carrying out a transduction 129
Uses for transduction 130
Two-factor crosses to determine gene linkage 131
Mapping the order of genes – three-factor crosses 132
Strain construction 133
Localized mutagenesis 133
Specialized transducing phage 134
Making merodiploids with specialized transducing phage 135
Moving mutations from plasmids to specialized transducing phage to the chromosome 138
Summary 140

Chapter 9 Natural plasmids 142
Origins of replication 142
Plasmid copy number 145
Setting the copy number 148
Plasmid incompatibility 148
Plasmid amplification 149
Other genes that can be carried by plasmids 149
Plasmids can be circular or linear DNA 152
Broad host range plasmids 152
Moving plasmids from cell to cell 153
Summary 154

Chapter 10 Conjugation 156
The F factor 156
The R factors 156
The conjugation machinery 158
Transfer of the DNA 158
Surface exclusion 159
F, Hfr, or F-prime 160
Formation of the Hfr 161
Transfer of DNA from an Hfr to another cell 162
Formation of F-primes 163
Transfer of F-primes from one cell to another 164
Genetic uses of F-primes 165
Genetic uses of Hfr strains—mapping genes on the *E. coli* chromosome using Hfr crosses 167
The 50% rule 171
Using several Hfr strains to cover the chromosome 171
Mobilization of non-conjugatible plasmids by R and F 172
Conjugation from prokaryotes to eukaryotes 173
Summary 173

Chapter 11 Transformation **175**
Natural competency 176
The process of natural transformation 183
The machinery of naturally transformable cells 184
Artificial transformation 187
Transformation as a genetic tool: gene mapping 188
Transformation as a molecular tool 188
Summary 189

Chapter 12 Gene expression and regulation **191**
The players in the regulation game 192
Operons and regulons 196
Repression of the *lac* operon 197
Activation of the *lac* operon by cyclic AMP and the CAP protein 200
Regulation of the tryptophan biosynthesis operon by attenuation 200
Regulation of the heat-shock regulon by an alternate sigma factor, mRNA stability, and proteolysis 205
Regulation of the SOS regulon by proteolytic cleavage of the repressor 208
Two-component regulatory systems: signal transduction and the *cps* regulon 209
Summary 211

Chapter 13 Plasmids, bacteriophage, and transposons as tools **213**
What is a cloning vector? 213
Why not use naturally occurring plasmids as vectors? 215
The importance of copy number 215
An example of how a cloning vector works—pBR322 215
Multiple cloning sites 216
Determining which plasmids contain an insert 217
Expression vectors 217
Vectors for purifying the cloned gene product 219
Vectors for localizing the gene product 221
Vectors for studying gene expression 221
Shuttle vectors 224
Artificial chromosomes 225
Constructing phage vectors 225
Suicide vectors 227
Phage display vectors 228
Combining phage vectors and transposons 229
Summary 232

Chapter 14 DNA cloning **234**
Isolating DNA from cells 234
Plasmid DNA isolation 234
Chromosomal DNA isolation 235
Cutting DNA molecules 235
Type I restriction–modification systems 236
Type II restriction–modification systems 236
Type III restriction–modification systems 237

Restriction–modification as a molecular tool 237
Generate double-stranded breaks in DNA by shearing the DNA 239
Joining DNA molecules 239
Manipulating the ends of molecules 240
Visualizing the cloning process 241
Constructing libraries of clones 242
DNA detection—Southern blotting 243
DNA amplification—polymerase chain reaction 245
Adding novel DNA sequences to the ends of a PCR amplified sequence 248
Site-directed mutagenesis using PCR 248
Cloning and expressing a gene 249
DNA sequencing using dideoxy sequencing 251
DNA sequence searches 253
Summary 254

Chapter 15 Bioinformatics and proteomics 256
Bioinformatics 256
Strategies for sequencing genomes 257
Bacterial genomes 259
Analyzing genomes 260
The *E. coli* K12 genome 261
Proteomics 262
Techniques for examining the proteome—SDS-PAGE and 2-D PAGE 262
Techniques for examining the proteome—microarray technology 264
Summary 266

Glossary 267
Further reading 278

Index 281

Preface

Neither one of us had planned on writing a textbook at this point in our careers. As we reflected on this past decade of teaching bacterial genetics to undergraduate students, we came to realize that even though the amount of knowledge in this field had exploded, our students still had to start their learning at the beginning. We could not teach the cool stuff, such as gene regulation, bioinformatics, and recombinant DNA manipulations without first making sure they understood DNA structure, its function, and the events that impacted this structure and function in a bacterial cell. Very few bacterial molecular genetic textbooks were available that started at the beginning, using a format and language easy to digest. This textbook does start at the beginning and makes no assumptions. Students first learn about the bacterial cell and its contents (Chapter 1), the chemistry, structure, and function of the DNA molecule (Chapter 2), and then move on to learn about all the things that can happen to the DNA molecule, both inside and outside the bacterial cell (Chapters 3–11). The book ends with chapters on gene expression and regulation (Chapter 12), molecular tools and DNA cloning (Chapters 13–14), and bioinformatics and proteomics (Chapter 15). We used *Escherichia coli* as our "gold standard" to explain concepts. When available, we included examples of other prokaryotic and eukaryotic systems. Furthermore, we provided new options, along with traditional options, for teaching concepts such as operon and regulon regulation.

The chapters are descriptive, and hopefully organized in such a way that captures the attention of today's click and flick society. Our goal was to get to the point as soon as possible, sometimes borrowing language and phrases from this current generation. The language of bacterial genetics is difficult for the novice. Students cannot be expected to understand complex experimental results and corresponding interpretations without first having an understanding of the applicable language. In that respect, we envision this textbook to be used as a fundamental tool for learning the applicable language. This learning process would most likely take place in a beginning bacterial genetic course for the microbiology major, or in a general bacterial genetic course for general science, biology, or engineering majors who need an introductory understanding of the content in this field for application purposes. This textbook can be used by courses structured either for a quarter or semester system. We also felt it was important to emphasize relevance of material, because relevance increases retention. With that in mind, most sections contain information on the real-life impacts of a particular molecular mechanism or process.

We have included a number of features to facilitate the use of this textbook to teach bacterial genetics:

- Each chapter is organized in a similar fashion, starting with an introduction and concluding with a summary, study questions, and suggestions for further reading.
- Chapters build on concepts learned in previous chapters.
- Key words are identified in bold, and are defined in a definitions section located at the end of the textbook.
- Each chapter contains FYIs, "for your information" sections that provide additional and relevant information as it relates to a described concept.
- An accompanying website is accessible through www.blackwellpublishing.com/trun

Acknowledgments

An enormous number of talented scientists contributed to the knowledge described in this textbook. We apologize if we failed to adequately identify contributions in the reference section at the end of this text. This oversight was not intentional, but rather a reflection of the overwhelming number of contributors to this field.

We were fortunate to receive our scientific training during a time when the technological advances in this field rapidly pushed our understanding of complex molecular genetic mechanisms in bacteria. The newly developed molecular approaches combined with traditional genetic approaches have allowed scientists to extract previously elusive answers from bacteria. Nancy acknowledges the contributions of John Reeve, Tom Silhavy, Susan Gottesman, and Sue Wickner to her scientific training. Janine acknowledges the contributions of Ken Bingman, Richard Consigli, Joe Bolen, William Haldenwang, and Susan Gottesman to her scientific training. These individuals imparted to us the skills needed to be both a successful researcher and science educator. And for this we are forever grateful.

We thank our editor at Blackwell Publishers, Nancy Whilton, for her extraordinary patience. We are not sure this project would have been completed without her gentle pushing and unwavering belief that this textbook was needed. We also thank Nathan Brown, assistant editor, for communicating "motivational" thoughts that even Janine could not ignore, thus keeping us on track to finish this textbook before the content became outdated or he retired.

We acknowledge the contributions of our outside reviewers:

Laurie Achenbach, S. Illinois University, Carbondale
Paul Gulig, University of Florida
Jim Hu, Texas A & M University
Philip Lessard, MIT
Amy Medlock, University of Georgia
Leland Pierson, University of Arizona
Paul Rainey, University of Oxford
Dave Westenberg, University of Missouri, Rolla

Reviewing is an enormous and time-consuming activity. We greatly appreciate the time spent, by our reviewers, generating insightful and helpful comments.

Finally, we thank our students! We have listened! We have acknowledged your needs! And, we have finally done something about it! So now it is your turn to go learn this stuff and carry on where we left off!

Chapter 1

Introduction to the cell

Living cells are formed from a small number of the different types of molecules that make up the earth. Most biomolecules contain carbon and many contain nitrogen. Both carbon and nitrogen are very scarce in non-living entities. While cells may not always utilize the most abundant molecules, they do use molecules whose unique chemistry is capable of carrying out the reactions necessary for life.

There are three levels of organization to describe the molecules that make up living organisms (Fig. 1.1).

1 The simplest level is the individual elements such as carbon, nitrogen, or oxygen.

2 The basic elements can be arranged into a series of small molecules known as **building blocks**. Building blocks include compounds such as amino acids and nucleic acids.

3 The building blocks are organized into larger compounds, known as **macromolecules**. Macromolecules comprise the different structures that are found in cells.

Four different types of macromolecules are used to construct a cell:

- nucleic acids;
- proteins;
- lipids;
- carbohydrates.

Each type of macromolecule is used for a specific purpose in the cell. There are many examples of macromolecules being combined in different configurations to form larger cell structures.

The **nucleic acids** can be subdivided into **DNA** and **RNA**. DNA is composed of two kinds of building blocks, the bases (adenine, guanine, cytosine, and thymine) and a sugar–phosphate backbone. DNA is used by the cell as a repository for all of the information necessary to direct synthesis of the

FYI 1.1

Elements, ions, and trace minerals that make up living systems

Elements	Ions	Trace minerals
Oxygen	Sodium	Manganese
Carbon	Potassium	Iron
Nitrogen	Magnesium	Cobalt
Hydrogen	Calcium	Copper
Phosphorus	Chloride	Zinc
Sulfur		Aluminum
		Iodine
		Nickel
		Chromium
		Selenium
		Boron
		Vanadium
		Molybdenum
		Silicon
		Tin
		Fluorine

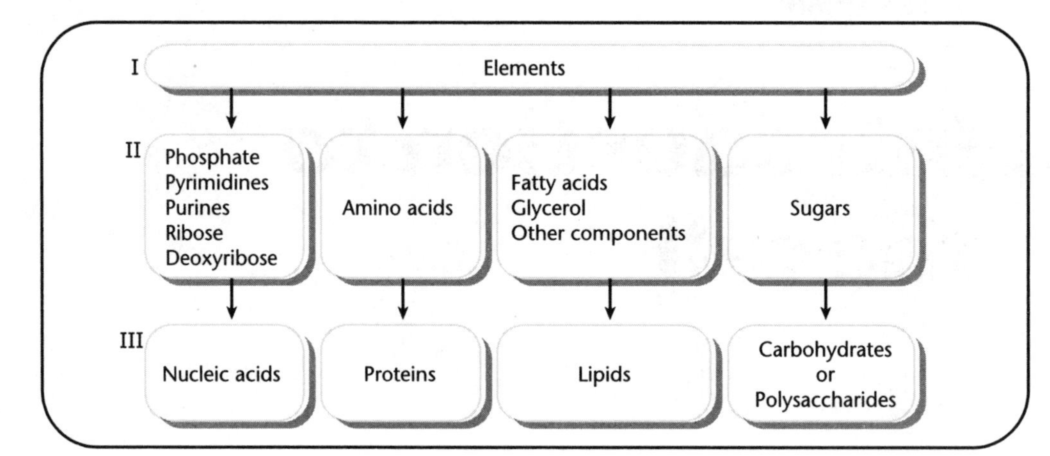

Fig. 1.1 Three levels of organization describe the compounds that make up living organisms.

macromolecules and to produce energy for this synthesis. DNA is also used to transmit information from one generation to the next. RNA has a very similar composition to DNA. The two major differences between DNA and RNA are in the sugar used in the sugar–phosphate backbone (ribose for RNA and deoxyribose for DNA) and in one of the bases (uracil for RNA and thymine for DNA). In *Escherichia coli*, DNA exists as a double-stranded molecule. Chapter 2 describes in detail both the structure and replication of DNA.

The RNA in the cell has at least four different functions.

1 **Messenger RNA** (mRNA) is used to direct the synthesis of specific proteins.
2 **Transfer RNA** (tRNA) is used as an adapter molecule between the mRNA and the amino acids in the process of making the proteins.
3 **Ribosomal RNA** (rRNA) is a structural component of a large complex of proteins and RNA known as the ribosome. The ribosome is responsible for binding to the mRNA and directing the synthesis of proteins.
4 The fourth class of RNA is a catch-all class. There are small, stable RNAs whose functions remain a mystery. Some small, stable RNAs have been shown to be involved in regulating expression of specific regions of the DNA. Other small, stable RNAs have been shown to be part of large complexes that play a specific role in the cell. In general, RNA is used to convey information from the DNA into proteins.

Proteins are composed of amino acids. Most proteins are made from a unique combination of 20 different amino acids (Fig. 1.2). The order in which amino acids appear in a protein are specified by the mRNA used to direct synthesis of the protein. All amino acids have a common core of repeating amino–carbon–carboxyl groups, with varying side chains on the central carbon. Proteins, therefore, have a repeating backbone with an amino terminus and carboxyl terminus. The amino acids can be grouped together and described by physical properties such as charge (acid or basic), size, interactions with water (hydrophobic—water "hating" or hydrophilic—water "loving"), a specific element (sulfur containing) or structure they contain (aromatic rings). The types of amino acids used to make up a protein specify what the protein is capable of. Proteins perform many duties in the cell, including functioning as structural and motor components, enzymes, signaling molecules, and regulatory molecules. Some proteins perform only one function while others are multifunctional.

Lipids are an unusual group of molecules that, in bacteria, are used to make the membranes that surround a cell. One type of lipid, known as a fatty acid, is composed

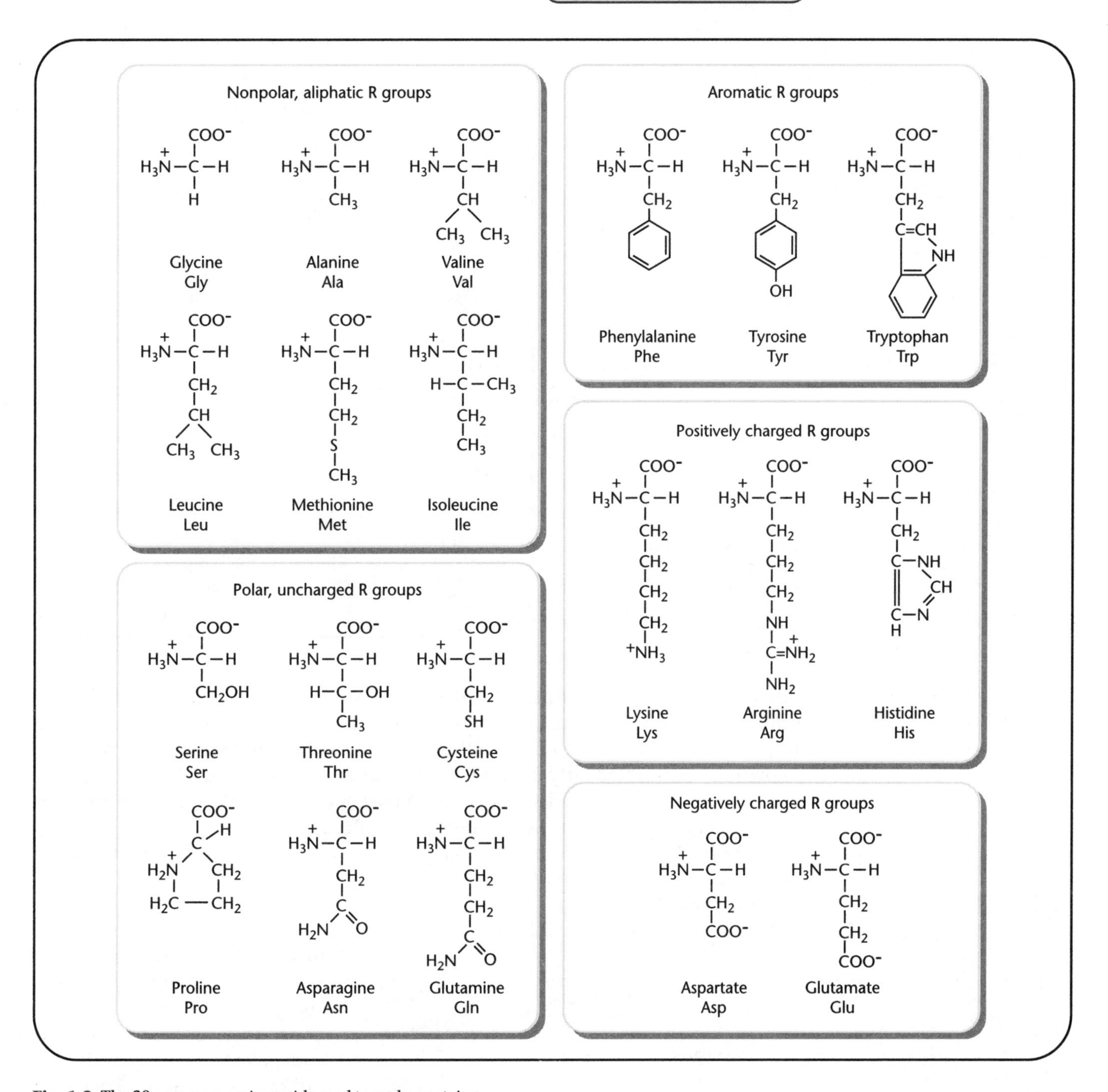

Fig. 1.2 The 20 common amino acids used to make proteins.

of long chains of carbon molecules attached to a smaller head group (Fig. 1.3a). The small head group is known as the polar head group. The fatty acid chains are extremely hydrophobic and line up with the long chains of carbons near each other. Because of the properties of the fatty acid chains, membranes are double-sided (Fig. 1.3b). One side of a membrane is known as a leaflet. The polar head groups face the outside surfaces of the membranes because the polar head groups are water soluble. Other chemical groups can be added to the head groups of the lipids but, in general,

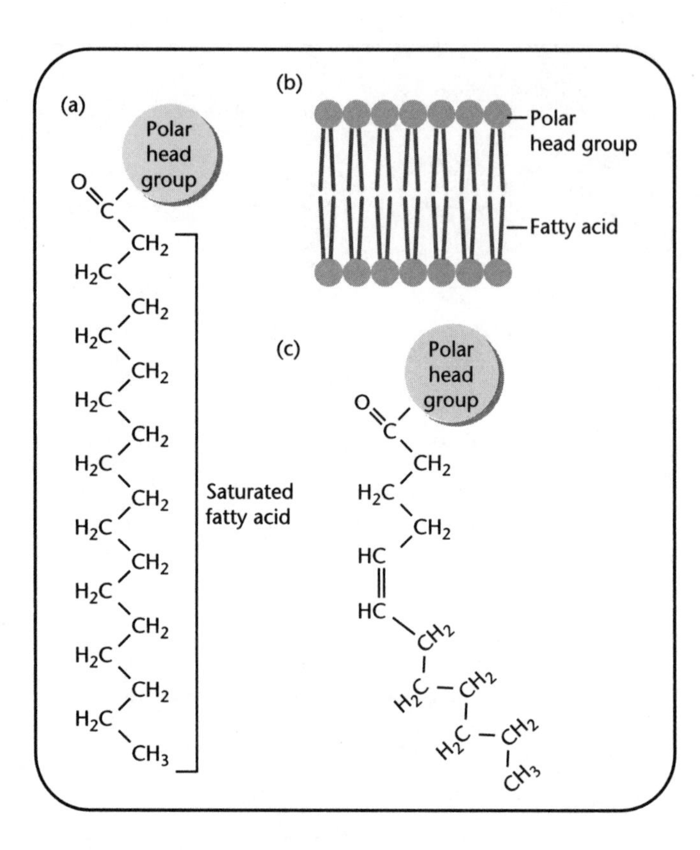

Fig. 1.3 Lipid molecules. (a) The general structure of a saturated fatty acid. (b) The structure of membranes with two leaflets. (c) The general structure of an unsaturated fatty acid.

other chemical groups cannot be added to the fatty acids. If the fatty acids contain only single bonds, they are known as saturated fatty acids. Saturated fatty acids are flexible and can be tightly packed. If the fatty acids contain any double bonds they are known as unsaturated fatty acids (Fig. 1.3c). Unsaturated fatty acids have a kink in them and cannot be packed as closely. Membranes usually contain a mixture of fatty acids to maintain the right packing density and fluidity.

Carbohydrates are composed of simple sugars (Fig. 1.4). They can be used as:

1 an immediate source of energy;
2 a stored source of energy;
3 structural components of the cell.

In bacteria, carbohydrates that are used as immediate sources of energy are the simple carbohydrates such as glucose, lactose, and galactose. Frequently, the carbohydrates that a bacterium utilizes as energy sources can be used to distinguish one species of bacteria from another. Energy is usually stored in a carbohydrate that is a long polymer of glucose known as glycogen.

Structurally, carbohydrates are used to make a protective covering for the outside of the

FYI 1.2

The composition of an average *E. coli* cell

Component	No. of different species	% of cell
Water	1	70
Proteins	3000 or more	16.5
Nucleic acids		
DNA	1	0.9
RNA	1000 or more	6.2
Lipids	40	2.7
Lipopolysaccharide	1	1.0
Peptidoglycan	1	0.8
Glycogen	1	0.8
Polyamines	2	0.1
Metabolites, cofactors, and ions	800 or more	1.0

Note: Although a nearly infinite number of different species of proteins, nucleic acids, polysaccharides, etc. are possible, only a tiny fraction of them are found in a given cell type.

cell called the **capsule** or capsular polysaccharide. Carbohydrates are also added to the polar head groups of the fatty acids on the face of the membrane that is exposed to the outside of the cell. The fatty acid attached carbohydrates help protect the cell from detergents and antibiotics. A complex mixture of carbohydrates is used to make the cell wall. Cell walls maintain the shape of the cell. In some species of bacteria, the cell wall is located outside the membrane. In other species, the cell wall is located underneath the outer membrane.

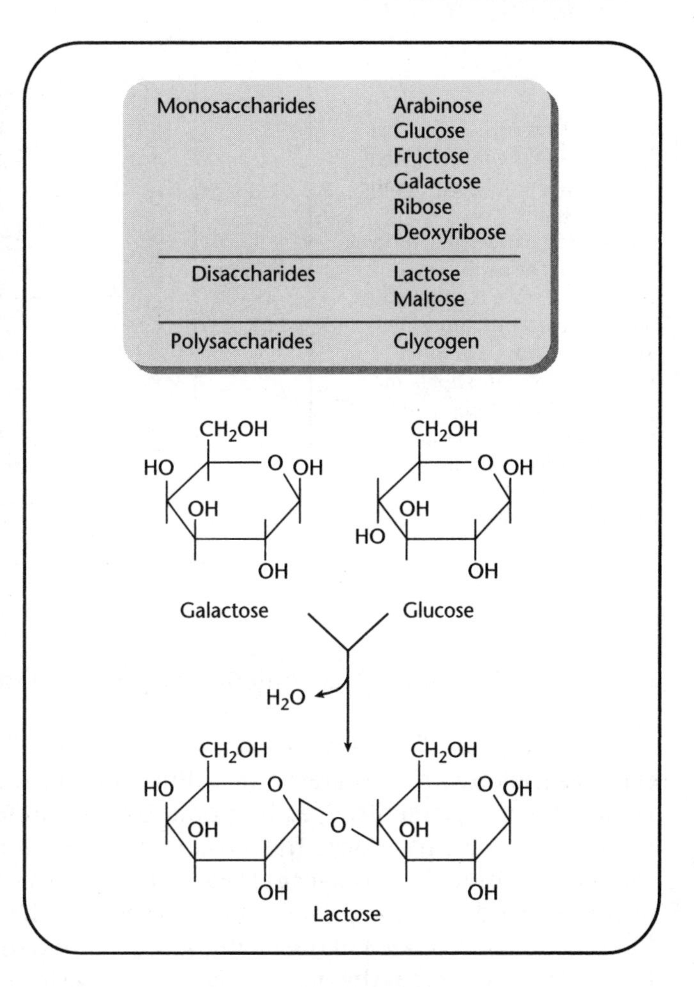

Fig. 1.4 Types and general structures of carbohydrates. The polymerizing of galactose with glucose results in the formation of lactose upon liberation of one H_2O molecule.

Each of the four types of macromolecules provides unique functions to the cell. For some of the cell's requirements, a single type of macromolecule suffices. In other situations, a mixture of macromolecules is required. It is interesting that beginning with a limited number of elements and ending with only four major classes of large molecules, the 5000–6000 different compounds needed to make a cell can be constructed.

The bacterial cell: a quick overview

Escherichia coli is a simple, single-celled organism. It is most commonly found in one of two places, either as a normal, though minor, component of the human intestinal tract or outside the body in areas contaminated by feces. *E. coli* can grow in either the absence or presence of oxygen. It is easy to grow in the laboratory and grows in the absence of other organisms. It can be grown on chemically defined media and divides as quickly as every 20 to 30 minutes. Many of its genetic variants are non-disease causing. These attributes are the main reasons that *E. coli* has been so extensively and intensively studied. The last 50 years of research have made it one of the best-understood life forms.

E. coli is a Gram-negative, rod-shaped bacteria. It is 1 to 2 microns in length and 0.5 to 1 micron in diameter. The cell is surrounded by two membranes (the inner and outer membranes) and as such has four cellular locations, the **outer membrane**, the **inner membrane**, the space between the two membranes or the **periplasm**, and the space surrounded by the inner membrane called the **cytoplasm** (Fig. 1.5). Each

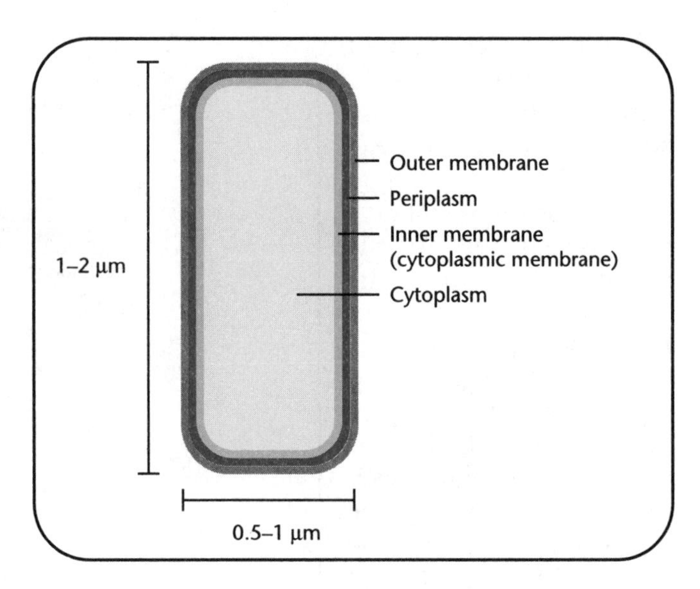

Fig. 1.5 The four cellular locations of the Gram-negative *E. coli* cell. The outer membrane contains lipids, lipopolysaccharides, and protein molecules. The periplasm or periplasmic space is an aqueous environment containing many kinds of proteins. The inner membrane contains lipids and proteins. The proteins used to generate energy are located here. The cytoplasm contains the DNA and is the site of synthesis for most of the macromolecules in the cell.

of these locations has unique contents and properties that provide specific functions to the cell.

The **outer membrane** is the main barrier of the cell to certain kinds of toxic substances. The outer membrane is the cellular surface that interacts with the outside environment. Detergents, dyes, hydrophobic antibiotics, and bile salts from the intestines that could be toxic to the cell cannot cross this membrane. It is composed of the very common double layer of lipids with the polar head groups of the fatty acids facing outward (Fig. 1.6a). One unusual aspect of the outer membrane is that the lipids on the outer leaflet of the membrane are different from the lipids on the inner leaflet. This means that the cell can distinguish the inside of the cell from the outside of the cell by the composition of the membrane leaflets.

Attached to the outer membrane and facing away from the cell are specialized lipids, known as **lipopolysaccharides** or LPS, that contain large carbohydrate side chains (Fig. 1.6b). LPS is essential for the cell to prevent toxic compounds from passing through the outer membrane. Because the LPS is very dense, with six to seven fatty acid chains per molecule, it is very hard for any compound to get through. LPS contains predominantly saturated fatty acids, which favor tight packing of the chains and a greater impermeability of the membrane. In mutants that are partially defective in LPS, the outer membrane is much more permeable and the cells become very fragile. The carbohydrate side chains of LPS are usually negatively charged. The side chains allow the cells growing in the intestinal tract to avoid being engulfed by intestinal cells, attacked by the immune system, or degraded by digestive enzymes. The side chains may also allow the bacterial cells to colonize the surface of human cells in the intestinal tract.

Anchored in the outer membrane and facing the outside of the cell are a number of specialized structures. Many different types of hair-like projections, known as **fimbriae** (singular = fimbria), can cover the cell surface (Fig. 1.7). A single cell can express several types of fimbriae at the same time and there can be between 100 and 1000 fimbriae per cell. Fimbriae are made of protein and are not generally associated with helping the cell move. Rather, fimbriae usually bind to a specific component of eukaryotic cells and allow the bacteria to adhere to the eukaryotic cell. Some of the more

FYI 1.3

The Gram stain

The Gram stain was discovered in 1884 by Christian Gram. It distinguishes two major groups of bacteria, Gram-positive and Gram-negative. The major difference in these groups of bacteria is the nature of their cell wall. In a Gram stain, actively growing cells are heat-fixed, stained with the basic dye, crystal violet, and then with iodine. They are briefly decolorized by treatment with a solvent such as acetone or alcohol. Gram-positive cells are not decolorized and remain stained a deep blue-black. Gram-negative cells are completely decolorized and appear red. The Gram stain remains one of the fixtures for classifying bacteria.

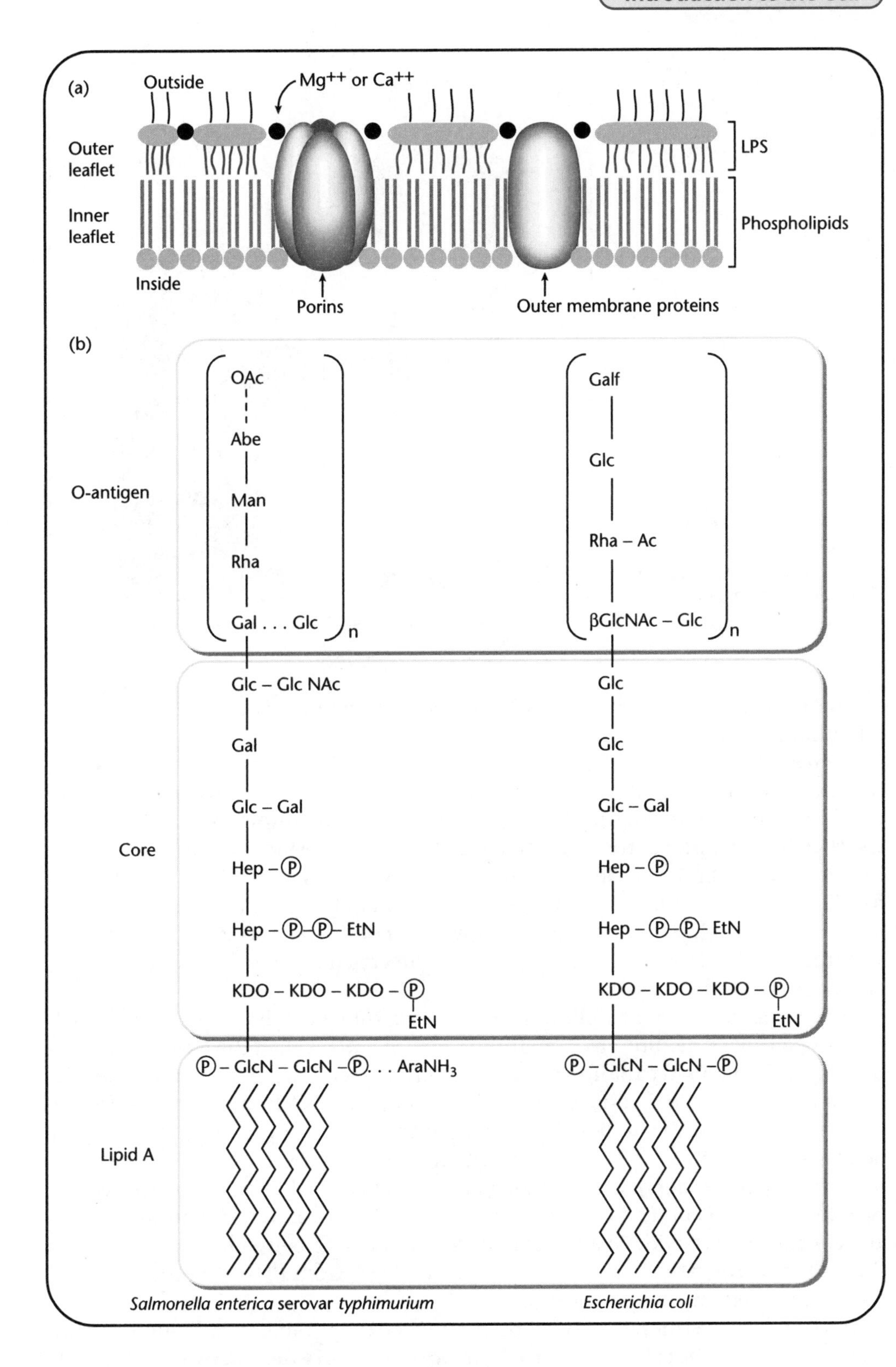

Fig. 1.6 (a) Structure of *E. coli*'s outer membrane. (b) The lipopolysaccharide of *E. coli* and *Salmonella*. Mutants defective in the O antigen lose their virulence. Mutants defective in the parts of the core closest to the lipid A are hypersensitive to dyes, bile salts, antibiotics, detergents, and mutagens. Mutants defective in lipid A are dead. OAc, O-acetyl; Abe, abequose; Man, D-mannose; Rha, L-rhamnose; Gal, D-galactose; GlcNAc, N-acetyl-D-glucosamine; Hep, L-glycero-D-manno-heptose; KDO, 2-keto-3-deoxy-octonic acid; EtN, ethanolamine; P, phosphate; GlcN, D-glucosamine; $AraNH_3$, 4-aminoarabinose.

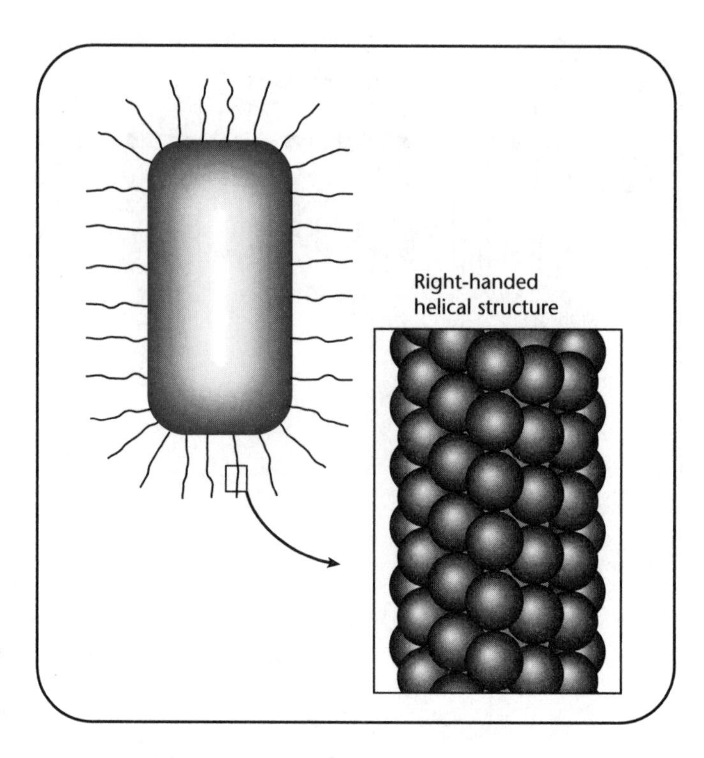

Fig. 1.7 Structure of *E. coli* fimbriae. Note the right-handed helical structure. Each subunit of the fimbriae is a single identical protein.

FYI 1.4

The colonization of human cell surfaces by bacteria

Virtually every exposed surface of the body is inhabited by some type of bacteria. Usually each exposed surface has a mixed population of many different species of bacteria. Different bacteria live on different cell surfaces: for example, the skin is widely inhabited by *Staphylococcus epidermis* while *E. coli* is a natural but minor resident of the colon. Problems for humans arise when either the number of a certain type of bacteria increases dramatically or the bacteria end up in a part of the body where bacteria do not normally reside. Some parts of the body contain extremely large numbers of bacteria. The colon contains ~10^9 bacteria per gram. A few simple calculations lead to the conclusion that there are about 10 times more bacterial cells than there are human cells in the human body!

common receptors for fimbriae are blood group antigens, collagens, and eukaryotic cell surface sugars.

The **flagella** (singular = flagellum) are anchored in the inner membrane and face the outside of the cell. They differ from the fimbriae in that they are corkscrew-like and 5 to 10 microns in length. Flagella are present in fewer numbers (1 to 10 per cell) and they have a motor at their base that is imbedded in the membrane (Fig. 1.8). The flagella rotate in both the clockwise and counter-clockwise directions. The result of flagella rotation is to move the cell. Flagella movement is coupled to a sensing system in the cytoplasm that results in the net movement of the cell towards a favorable environment and away from a harmful one. This directed movement of a bacterial cell is known as **chemotaxis**.

Some *E. coli* cells have an additional surface structure that has a unique function. It is also a hair-like structure 2 to 3 microns in length and is called the **F pilus** (plural = pili) (Fig. 1.9). It is used to build a bridge between two cells, the male that contains the F pilus and the female that does not. A single strand of DNA from the male chromosome is transferred to the female cell. The transferred DNA is replicated and the double-stranded DNA can be incorporated into the chromosome of the female cell. F pilus mediated transfer of genetic information between cells is called **conjugation** (see Chapter 10). Conjugation has been instrumental in our understanding of *E. coli* and continues to provide insights into exchange of DNA between cells. It has been shown that conjugation can be used to transfer DNA from *E. coli* to other bacterial species and even into eukaryotic cells.

A second surface structure that is loosely attached to the outer membrane and completely surrounds the cell is the **capsule** or **capsular polysaccharide**. The capsule is a thick slime layer that is sometimes present on the cell. Cells use the capsule to protect themselves from dehydration or osmotic shock when they are outside the intestines and to avoid being eaten by macrophages when they are inside the body.

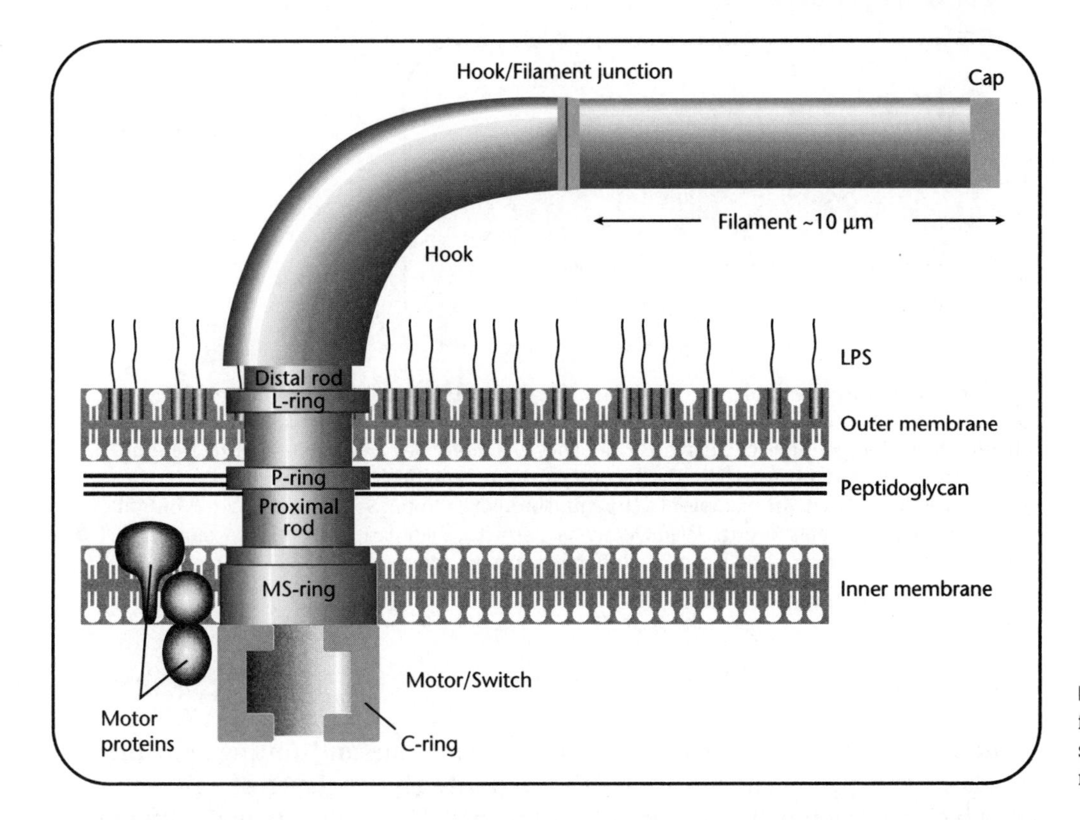

Fig. 1.8 The base of the *E. coli* flagella. Note the motor structure anchored in the membrane.

Imbedded in the outer membrane are a number of different types of proteins (Fig. 1.10). Some act as receptors for specific substances and are used to transport these substances through the membrane. Some of the proteins form gated holes or pores through the membrane. The pores allow passive diffusion across the outer membrane of chemicals required by the cell for growth. Pores that go all the way through the membrane must be gated or closed off when not in use. The pores must be gated so that cell contents do not leak out of the cell or toxic compounds do not leak into the cell. Other proteins have no transport function and appear to play a structural role. It is generally thought that the outer membrane must contain a certain amount of protein. If one of the protein species is missing from the membrane, the cell responds by increasing the amounts of the other proteins.

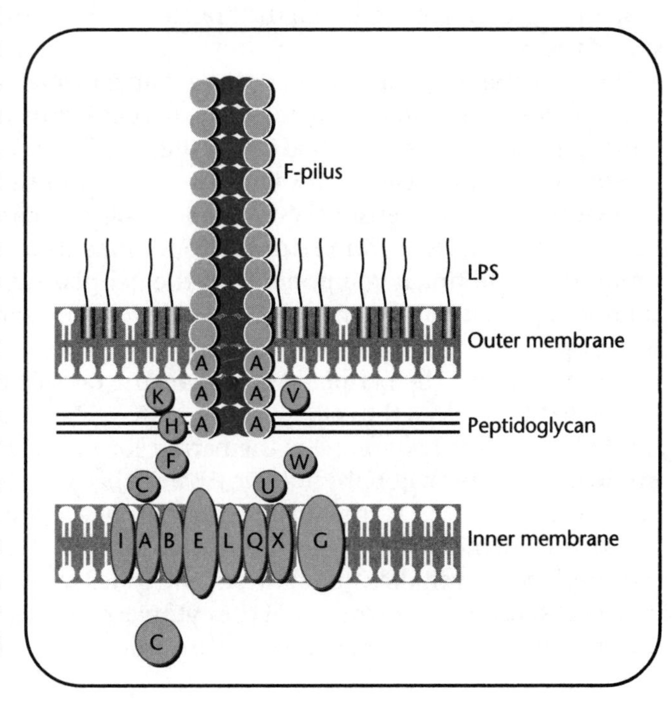

Fig. 1.9 Assembly of the F pilus. The proteins A, B, C, E, F, G, H, I, L, Q, U, V, and W are needed for F pilus biogenesis. Additional F encoded proteins are needed for mating pair stabilization (two proteins), surface exclusion (two proteins), and DNA synthesis (four proteins). There are 13 F encoded proteins with no known function.

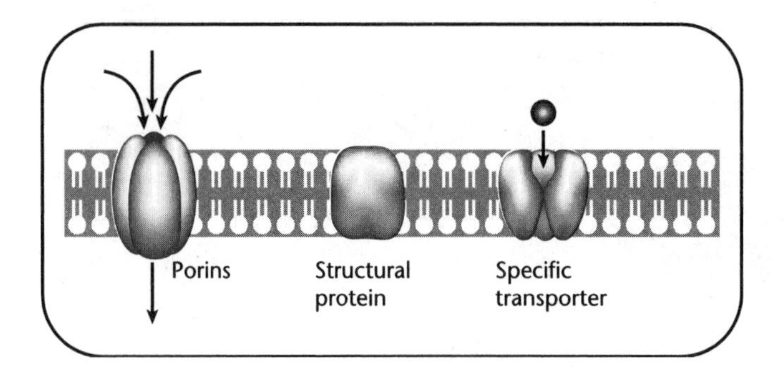

Fig. 1.10 The porins, structural proteins, and specific transport receptors spanning *E. coli*'s outer membrane. Porins are composed of three identical subunits with three channels, one through each subunit. They are responsible for passive diffusion of small molecules. Examples of porins are OmpF and OmpC. Structural proteins are required in the outer membrane. When they are missing, the outer membrane is destabilized. An example of a structural protein is OmpA. Specific transporters only allow specific compounds to pass through them. Many have structural similarities to porins. Examples include LamB that transports maltodextrin sugars, BtuB that transports vitamin B, and FadL that transports fatty acids.

FYI 1.5

KHO antigens

Many different subspecies of *E. coli* can be isolated. More than 50 years ago, Kauffman and coworkers described a system to classify *E. coli* based on their outer membrane surface structures. The system uses three major surface structures that are capable of eliciting an immune response in mammals: the lipopolysaccharide or O antigen, the flagella or H antigen, and the capsule or K antigen. To date, 173 different O groups, 56 H groups, and 80 K groups have been found. The most widely used laboratory strain is *E. coli* K12. In contrast to the innocuous, non-pathogenic *E. coli* K12 strain, outbreaks of *E. coli* O157:H7, a strain with very distinct O and H antigens, continue to arise in the United States each year, usually from contaminated food. Infection by *E. coli* O157:H7 can be lethal in very young, very old, or immunocompromised individuals.

Anchored to the inner face of the outer membrane and jutting out into the periplasm is a rigid structure called the **peptidoglycan** layer or cell wall (Fig. 1.11a,b,c). The peptidoglycan surrounds the cell and is required to maintain the rod shape of the cell. The peptidoglycan is arranged in rings that go around the short axis of the cell. The penicillin antibiotics interfere with the assembly of the peptidoglycan rings into the completed structure. When a cell is growing in the presence of penicillin, the rings are not connected to one another so that when the cell grows, the rings come apart and the cell explodes. Thus, the peptidoglycan maintains the cell shape and protects the cell from the pressure differences between the inside and outside of the cell.

The **periplasmic space** or periplasm is an aqueous compartment that resembles the aqueous environment outside the cell. It contains proteins to help concentrate nutrients from outside the cell to inside the cell. Other periplasmic proteins sense the external environment and transduce this information across the inner membrane to the cytoplasm. These sensing systems allow the cell to respond to the environment, either protecting itself from harmful conditions or taking advantage of beneficial conditions. Additional components of the periplasm include proteases to degrade abnormal proteins and a system to provide for the formation of disulfide bonds in specific proteins. Disulfide bonds are only formed in the periplasm; they cannot form in the cytoplasm. The periplasm monitors the outside of the cell and concentrates solutes to the inside of the cell.

While the outer membrane is the barrier for many toxic compounds, the **inner membrane** is the major physiological barrier between the outside of the cell and the cytoplasm (Fig. 1.12). It contains the proteins needed to generate energy and has an electrical gradient, called the proton motive force, across it. If the proton motive force is disrupted, the cell dies. In addition, the inner membrane contains proteins that transport substances into or out of the cytoplasm, proteins that transmit signals into the cytoplasm, and systems to transport proteins out of the cytoplasm. The inner

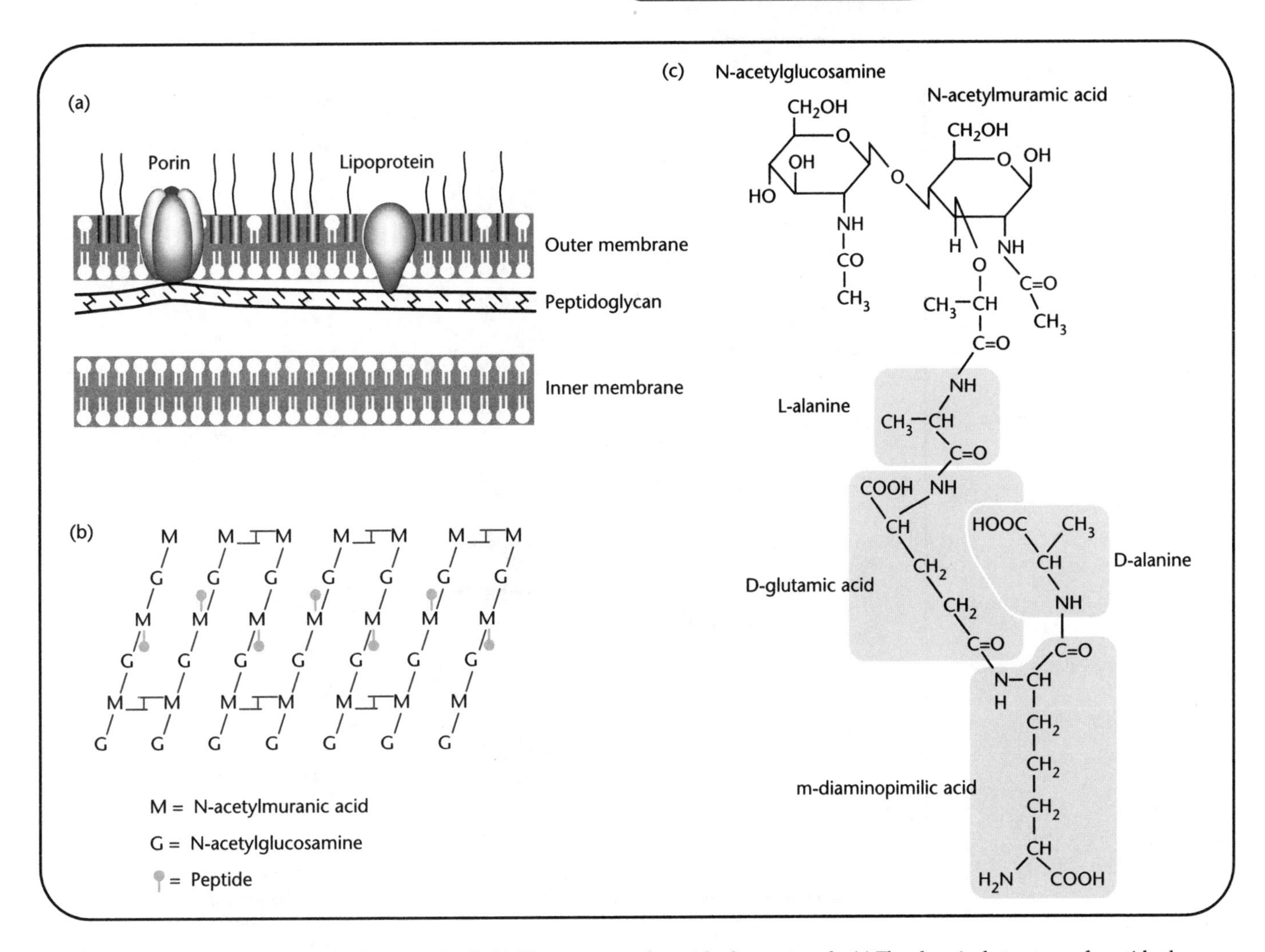

Fig. 1.11 (a) The location of peptidoglycan in *E. coli*. (b) The structure of peptidoglycan strands. (c) The chemical structure of peptidoglycan.

membrane is the location of many proteins involved in cell division. Some of these division proteins are only found in limited places in the inner membrane, such as in a ring around the center of the short axis of the cell. The inner membrane is the site of many and varied biochemical reactions, signaling reactions, and transport activities.

The **cytoplasm** is the hub of cellular activities. It is the location of the single *E. coli* chromosome, a circular double-stranded molecule of 4,639,221 base pairs. All DNA replication and repair of damaged DNA takes place here. **Transcription** of the DNA into messenger RNA (mRNA) and **translation** of the mRNA into protein occur in the cytoplasm. Decisions about which gene products to make are made in the cytoplasm. Signals from the environment are received and acted upon in the cytoplasm. In the cytoplasmic compartment, food sources are broken down or stored and energy is generated. Because the cytoplasm contains the DNA, it is the site of synthesis of the majority of the molecules that the cell makes.

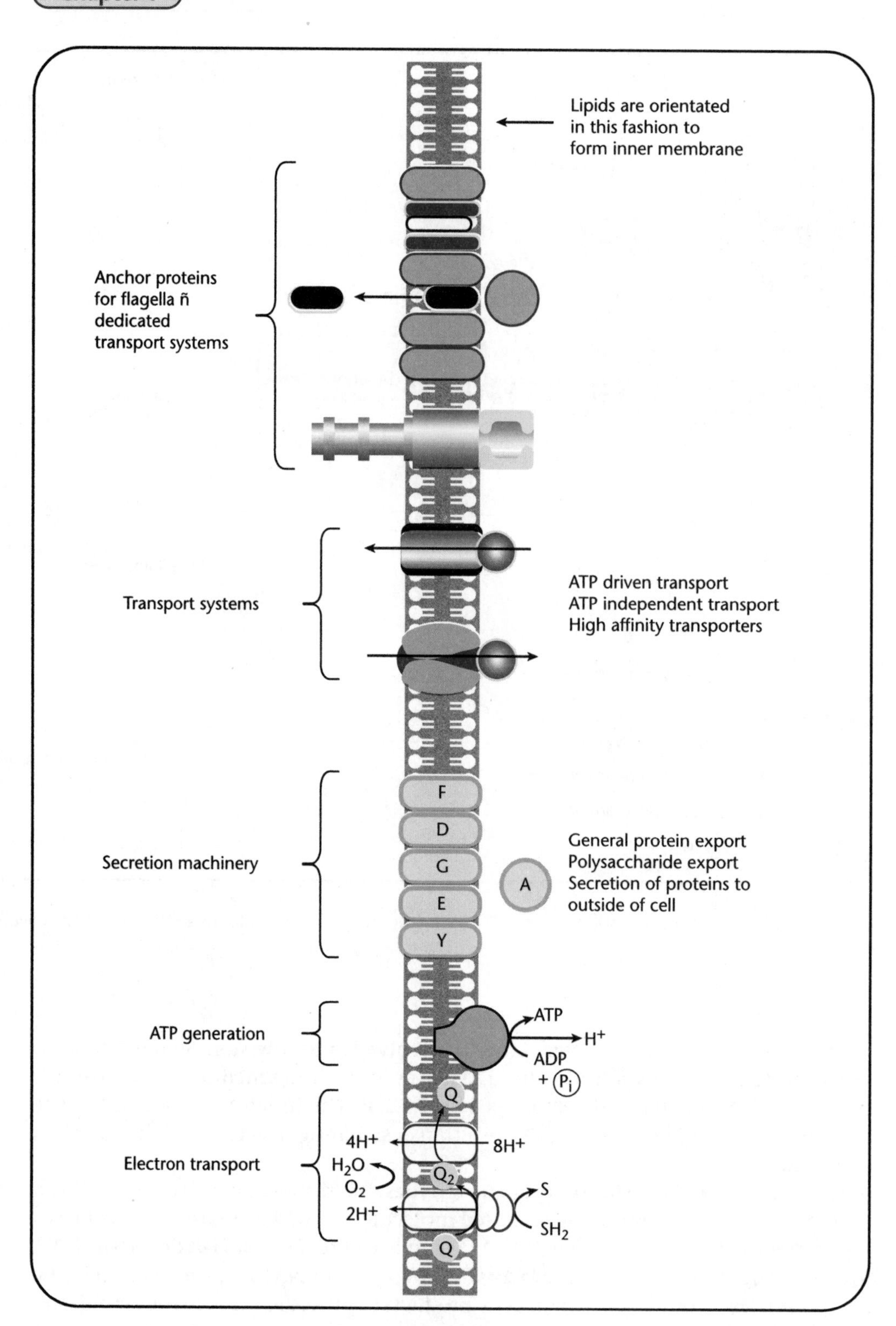

Fig. 1.12 The inner membrane of *E. coli* is composed of approximately 50% protein and 50% phospholipid. The major types of phospholipid are phosphotidylethanolamine, phosphotidylglycerol, and cardiolipin. The inner membrane contains the proteins for transport systems, the protein export machinery, ATP generation machinery, and electron transport, to name a few.

How do cells grow?

In order to reproduce, cells must grow and divide. Growth is not random, but rather follows a very regular pattern (Fig. 1.13). The diameter of most rod-shaped bacteria does not change during growth. It is the length of the cylinder that increases until the cell is approximately double in size. At this point, *E. coli* builds a physical barrier, called the **septum**, between the two daughter cells. Subsequently, the daughter cells are physically separated from one another.

During growth, all components of the cell must be

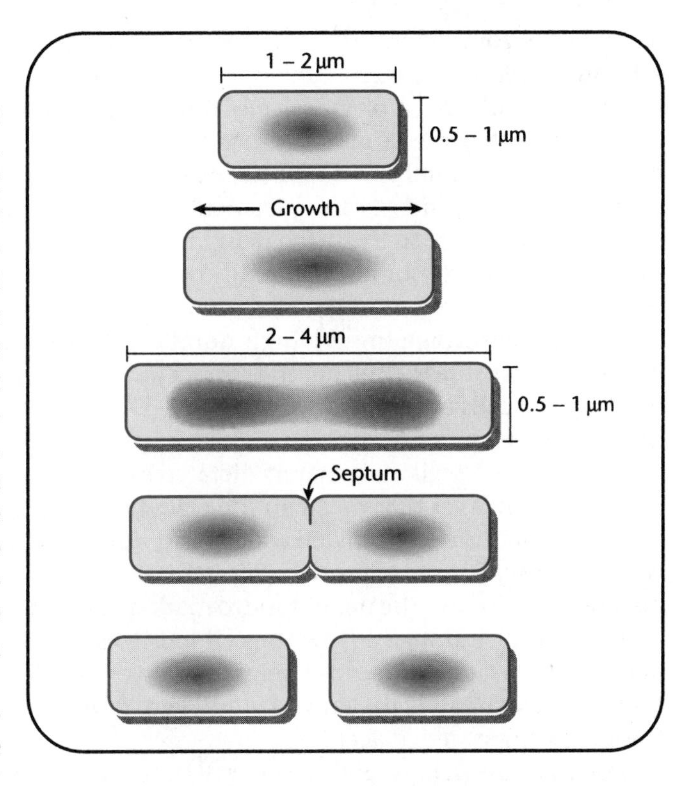

Fig. 1.13 Cell growth. During growth, all components of the cell must be doubled. The resulting daughter cells are physically separated from one another by a septum.

FYI 1.6

Measuring cell growth

Measuring cells as they grow and divide provides a reliable and quantitative indication of the health of the population. Cell growth is directly related to the nutrients the cells are consuming. Measurements of cell growth on different energy sources or in mutant cells can be directly compared. For energy sources, this has allowed the determination of which combinations of nutrients are most or least beneficial to the cells. For different mutant cells, it indicates how important the gene is to the cell and if there are conditions where it is more or less important.

Cell growth can be measured in three different ways. First, the **optical density (OD)** of cells growing in liquid media can be measured in a spectrophotometer. This technique is based on the fact that small particles (cells, in this case) scatter light approximately proportionally to their number. A suspension of cells is put into a holder or cuvette in the spectrophotometer and a beam of 600 nm wavelength light is passed through the suspension. The amount of light that comes out of the suspension is measured. The amount of light coming out is either the same as the amount that went in or it is less. If it is the same, there are not enough cells in the suspension to measure. If it is less it is because the cells have scattered the light. How much less is proportional to the number of cells. For example, if the OD600 is 0.25 in one culture and 0.5 in a second culture, the second culture has twice as many cells. OD600 is used to make measurements on populations of cells and to compare the relative numbers of cells in these populations.

The main drawback of this method is that it does not distinguish between live and dead cells or between cells and large particles of debris. It also relies on the cells being approximately the same size. If a mutant cell forms long filaments or other odd-shaped cells, the OD will not give a reading that is proportional to the number of cells. The strengths of this method are that it is rapid and easy so that many samples can be measured in a short amount of time. OD measurements do not kill cells so after an OD measurement is made, the cells can be used for further experimentation.

The second way to measure bacterial growth is by a **viable cell count (VCC)**. A sample is taken, diluted and spread on solid media, and incubated overnight. The next day, each colony, or group of cells, that form on the agar represents the growth from a single cell. The colonies are counted and the number of bacteria in each sample is determined. Frequently, the OD600 measurements and the VCC are both carried out. This allows determination of the correlation of an OD value with an exact count of the number of viable cells. The combination of OD and VCC measurements eliminates the drawbacks of only taking OD readings.

Both the OD600 and the VCC take measurements on populations of cells. If a change occurred at the level of a single cell, these measurements would miss it. A third method, called **flow cytometry**, was devised to measure the light scatter of single cells. A stream of media with bacteria in it is passed across a microscope slide field. Aimed at the field is a beam of visible light and opposite the beam is a detection system that is attached to a computer. As individual cells pass through the beam, the amount of light they scatter is recorded. Several thousand cells can be measured per second. This is a very powerful technique but the main drawback is the cost of the flow cytometer. For certain types of experiments where measurements on individual cells are required, it is the only way the data can be obtained.

doubled. For components that are present in very large numbers such as ribosomes (2×10^4 molecules/cell), membrane lipids (2×10^7 molecules/cell), or outer membrane pore proteins (2×10^5 molecules/cell), doubling is approximate. Because of the molecules' abundance, the numbers do not have to be exact. For cell components that are present in small numbers, doubling must be exact. In an *E. coli* cell that is dividing every 30 minutes, there is a single circular chromosome. The chromosome must be replicated only once so that each daughter cell can inherit one molecule. The duplication of the chromosome is a carefully controlled and regulated event. Different cell components are present in specific numbers with some being rare, some at intermediate levels, and some in very high numbers. Many different mechanisms are employed to make sure the cell has the correct number of any given component and that the daughter cells inherit the right amount. The rarer a component is, the more effort the cell invests to ensure that it is divided evenly.

The growth of a cell is ultimately dictated by the growth of the membranes and the peptidoglycan layer that make up the cell's boundaries. For both the inner and the outer membrane, new lipids are added randomly into the existing membranes, there is no one specific zone of growth (Fig. 1.14). Practically, this means that the cell cannot use the growth of the membrane to mark specific places along the membrane or to move molecules along it. How the cell localizes specific structures along the membrane and how it measures its length are not known. The fact that daughter cells are born at approximately the same size and very rarely are mistakes made indicates that the measuring system is accurate.

Growth of the peptidoglycan cell wall occurs randomly around the peptidoglycan along the long axis of the cell. The peptidoglycan that forms the caps of the cells and the septum is not exactly the same as that on the long axis of the cell and is synthesized by different enzymes. For cells dividing every 48 minutes, 50,000 **muropeptides** or strands of peptidoglycan per minute are added to the growing cell wall. The new strands are inserted next to pre-existing muropeptides at ~200 different sites in the cell wall (Fig. 1.15). The enzymes that synthesize the peptidoglycan reside in the inner membrane and move in a unidirectional fashion around the cell. They can circumnavigate the cell in about 8 minutes. Excluding the poles, a cell has ~1100 rings of muropeptides that maintain its shape.

As the membranes grow, the volume of the cytoplasm also increases. The cytoplasmic contents including the DNA and the

FYI 1.7

The growth curve

When *E. coli*, or any bacteria, is placed in a nutrient-rich broth, they begin to grow. Invariably, they follow a simple pattern of growth. First, the cells must prepare for growth and make sure they have all the necessary proteins required for growth, this is called the lag phase. Next, cells begin growing logarithmically and double at a constant rate, this is log or exponential growth. Once cells have used all the available nutrients, they begin to slow down and prepare for a period of inactivity known as the stationary phase. If the nutrient deprivation lasts long enough, the cells will begin to die. Historically, most experiments have been carried out on exponentially growing cells at 37°C. In the past few years, the genes that respond to different growth conditions have been investigated. New families of genes have been discovered that are required for stationary phase, cold shock, and heat shock, to name a few.

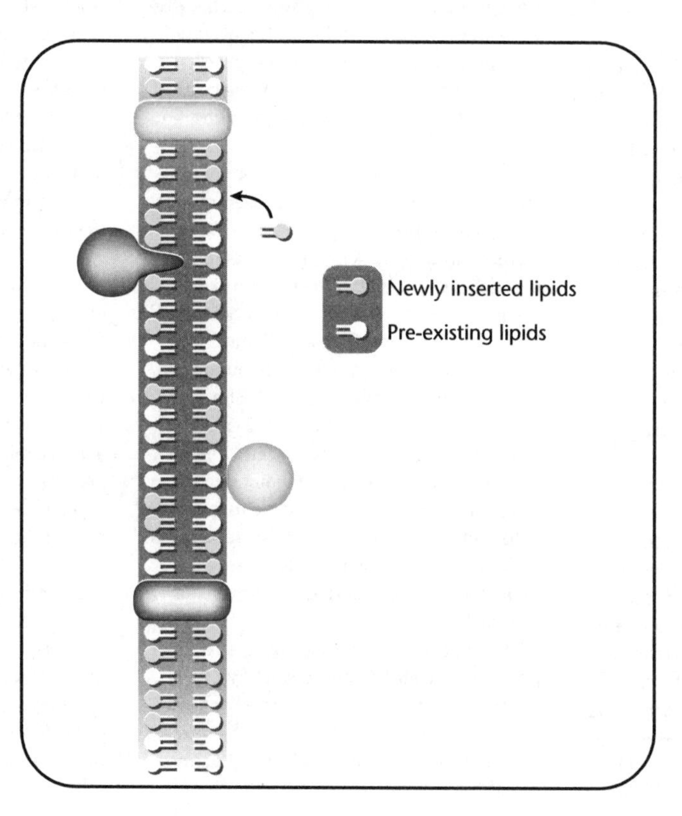

Fig. 1.14 Growth of membranes. Lipid insertion into growing membranes does not occur at a specific place on the membrane, rather it is random.

several thousand different protein species must double in number. Because proteins are regulated independently of each other, there is no singular mechanism for their doubling. The vast majority of the proteins in the cytoplasm freely diffuse throughout the boundaries of the cytoplasm. When the division septum is laid down, approximately half of the contents of the cytoplasm are trapped on one side of the septum and half on the other side. There are a few known cytoplasmic proteins that are not freely diffusable but they are the exception rather than the rule. Aside from the DNA, the components of the cytoplasm are numerous enough to be divided approximately in half.

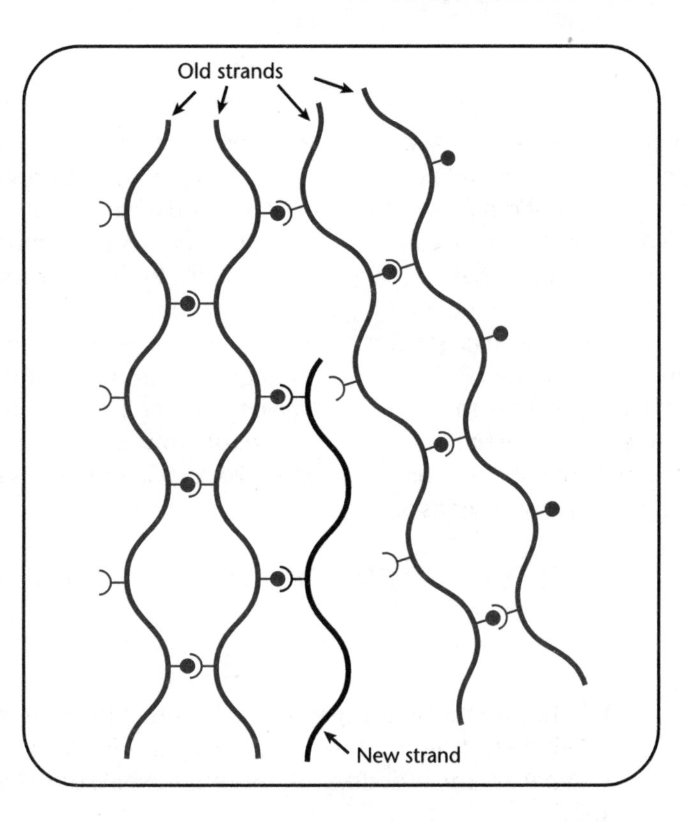

Fig. 1.15 Growth of peptidoglycan. Each strand is actually a ribbon that is in a helical conformation.

What is genetics?

The main goal of biology is to understand how a living cell works. Genetics is specifically concerned with the characteristics of the cell (size, shape, ability to grow on a certain sugar, specific internal structures, etc.) and how they are passed from one generation to the next. Early geneticists described physical characteristics or **phenotypes** of the cell and followed their inheritance without knowing that they were encoded by the DNA. After the discovery of DNA and the proof that it is the molecule used to transmit information from generation to generation, geneticists began to correlate the phenotype with the site in the DNA (or gene) that is responsible for it. The information in the DNA is referred to as the **genotype** and the characteristic it specifies is described as the phenotype.

One of the main characteristics defining genetics is that genetics studies the cell when the cell is intact and growing. Genetics does not attempt to separate each molecule and study the behavior of an individual molecule in a test tube. Rather, genetics focuses on a specific molecule and isolates that molecule in the cell by making a mutation in the DNA or gene that specifies the molecule. The behavior of the mutant cell is then compared to a **wild-type**, or normal cell. From the behavior patterns of the two cells, the function of the molecule is inferred. The greater the number of specific tests that can be devised for the wild-type and mutant cells, the more reliable is the assignment of the function for the gene being studied. Genetics always tries to find the simplest, logical argument to describe what has been observed.

Summary

E. coli is one of the best-understood life forms. As a simple, single-celled organism with a completely sequenced genome, much is known about *E. coli*. Our understanding of *E. coli*'s lifestyle and the molecular structures needed to support that lifestyle has contributed to *E. coli*'s widespread use as a model organism for techniques involving molecular genetic analysis. When undertaking a genetic analysis, the first goal is to make mutants that affect the specific pathway or molecule of interest. To make mutants, it is necessary to understand DNA metabolism, how mutants arise or can be induced, and what kinds of alterations are possible. Each of these topics can dramatically influence how to go about isolating mutants and what interpretations can be drawn from the mutants. Next, it is important to know what tools are available to study the mutants. How can they be moved from cell to cell? How can the mutants be cloned and subjected to the tools of molecular biology? In the chapters that follow, these subjects will be examined in more detail. Each of the chapters contributes one more piece to the puzzle of how we study biological systems using a genetic point of view.

Study questions

1 What are the building blocks for each of the four macromolecules?
2 What are differences between DNA and RNA? Similarities?
3 What are the functions of messenger RNA, transfer RNA, and ribosomal RNA?
4 What two features of a fatty acid molecule dictate how it is oriented in a membrane?
5 What are the differences between carbohydrates used as energy sources and carbohydrates used as structural components of the cell?
6 What is the function of the outer membrane? The inner membrane?
7 How many different cell surface structures can be present on the outer surface of a bacterial cell? What is the function of each?
8 In what cellular compartment are disulfide bonds formed?
9 What is the function of the septum and where is the septum located?

References

Gerhardt, P., Murray, M.G.E., Wood, W.A., and Krieg, N.R. eds. 1994. *Methods for General and Molecular Bacteriology*. Washington, DC: ASM Press.

Inouye, M. ed. 1986. *Bacterial Outer Membranes as Model Systems*. New York: Wiley.

Joklik, W., Willett, H., Amos, D., and Wilfert, C. eds. 1992. *Zinsser Microbiology*. Norwalk, CN: Appleton and Lange. Chapter 3: Bacterial morphology and ultrastructure.

Neidhardt, F.C., Curtiss III, R., Ingraham, J.L., Lin, E.C.C., Low, K.B., Hagasanik, B., Rexnikoff, W.S., Riley, M., Schaechter, M., and Umbarger, H.E. eds. 1996. *Escherichia coli* and *Salmonella typhimurium: Cellular and Molecular Biology*, 2nd edn., Washington, DC: ASM Press. Part One, Chapters 3 through 13. Each of these chapters explores a part of the cell in detail. Each chapter is fully referenced.

Chapter 2

The bacterial DNA molecule

The structure of DNA and RNA

Deoxyribonucleic acid or **DNA** serves as an organism's genetic material. It is divided into functional units called **genes**. Many, but not all genes, encode **polypeptides** or **proteins**. Other genes encode ribonucleic acid (RNA) molecules that are not made into proteins. The base sequence of a gene dictates the structure of the polypeptide it encodes. Faithful replication of the base sequence of DNA ensures that the inherent features of an organism are passed to its progeny. Alterations in the base sequence fuel the evolution of a species, impacting its ability to survive. This apparent dichotomy, faithful replication, but not 100% accuracy in replicating the base sequence, requires an understanding of how the structure of DNA defines yet places restrictions on the machinery involved in DNA replication and repair.

RNA is structurally similar, but functionally different than DNA. Some RNA is used as a messenger molecule to allow transfer of information from DNA to proteins (mRNA). Some RNA is a structural component of large protein—RNA complexes used to translate mRNA into protein (rRNA). Some RNA is used as an adapter molecule in the process of making polypeptides (tRNA).

DNA is composed of four different **nucleic acid** bases: two purines, adenine and guanine; and two pyrimidines, cytosine and thymine (Fig. 2.1a). RNA uses adenine, guanine, and cytosine but substitutes uracil for thymine (Fig. 2.1b). 2-Deoxyribose is the sugar used in DNA, whereas ribose is the sugar used in RNA (Fig. 2.2). The structures of ribose and 2-deoxyribose are similar. They differ in that 2-deoxyribose does not have an oxygen attached to the C2 position. In DNA, the nucleic acid bases are anchored to 2-deoxyribose by a N,N-glycosidic bond between the C1 of the sugar and the N1 of a pyrimidine (Fig. 2.3a) or the N9 of a purine (Fig. 2.3b).

Deoxyribonucleosides and deoxyribonucleotides

A base bonded to the sugar, 2-deoxyribose, is called a **deoxyribonucleoside,** a deoxynucleoside, or a nucleoside (Fig. 2.3a,b). Nucleosides are linked to each other by phosphodiester bonds. A **phosphodiester bond** is a single phosphate connected by ester linkages to two sugars (Fig. 2.4a). Another way to view this is that the DNA molecule consists of alternating units of phosphate and 2-deoxyribose. The phosphate covalently connects two sugars by bonding with the 3′ carbon of one sugar and

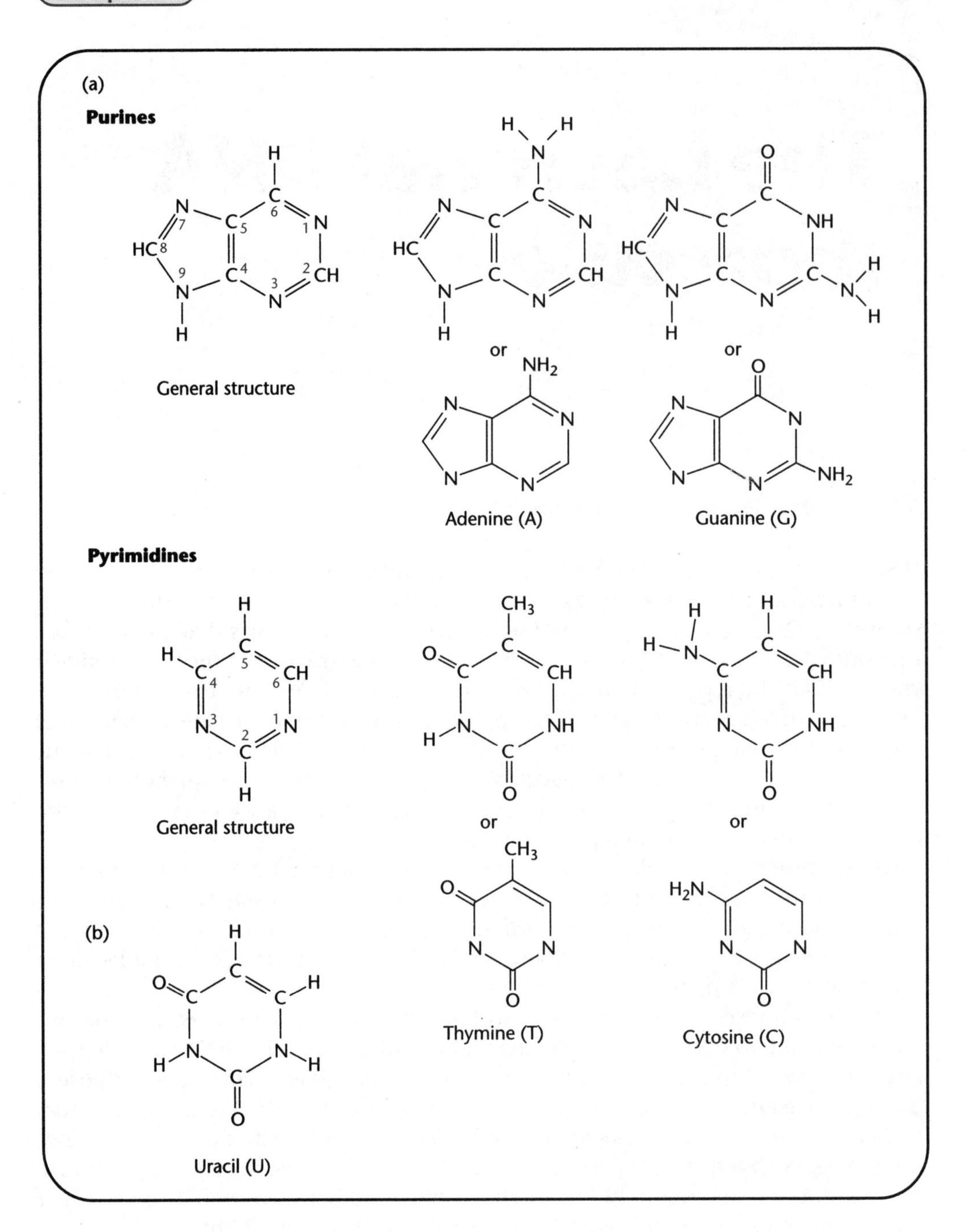

Fig. 2.1 (a) The nucleic acid bases used to construct DNA. (b) Uracil is found in place of thymine in RNA.

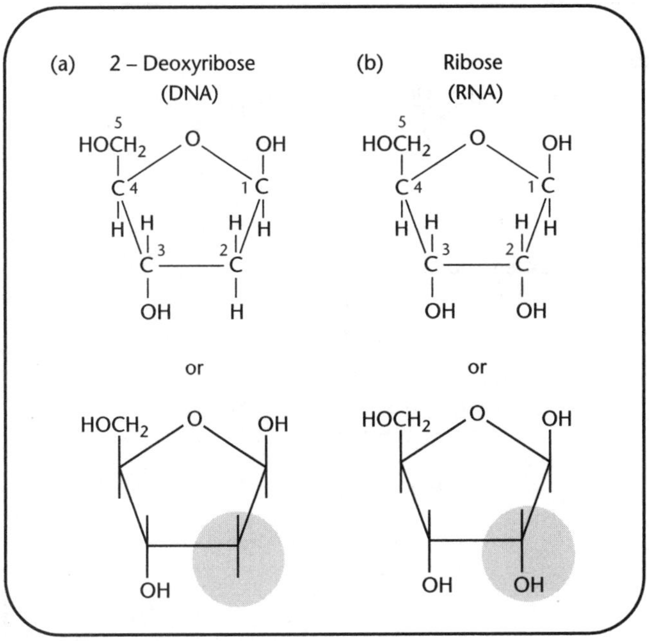

Fig. 2.2 The sugars found in nucleic acids. (a) 2-Deoxyribose is found in DNA. (b) Ribose is found in RNA. 2-Deoxyribose and ribose differ in that 2-deoxyribose does not have an oxygen attached to the C2 position.

with the 5′ carbon of the next sugar. A base attached to the phosphate—2-deoxyribose backbone is referred to as a **deoxyribonucleotide**, a nucleotide, or a nucleoside 5′ monophosphate. Table 2.1 summarizes nucleoside and nucleotide nomenclature.

DNA is only polymerized 5′ to 3′

The precursor to the nucleotide is the **nucleoside 5′ triphosphate**, which in the course of being polymerized to a growing DNA chain, will lose two phosphates and become a nucleotide (Fig. 2.4b). The phosphate closest to the nucleoside in a

Table 2.1 Nucleoside and nucleotide nomenclature.

Base	**Nucleoside base + sugar**	**Nucleoside 5′ triphosphate (precursor to nucleotide) base + sugar + 3 phosphates**	**Nucleotide (nucleoside 5′ monophosphate) base + sugar + 1 phosphate**
Adenine A	Adenosine	Adenosine 5′ triphosphate ATP	Adenosine 5′ monophosphate AMP
Guanine G	Guanosine	Guanosine 5′ triphosphate GTP	Guanosine 5′ monophosphate GMP
Thymine T	Thymidine	Thymidine 5′ triphosphate TTP	Thymidine 5′ monophosphate TMP
Cytosine C	Cytidine	Cytidine 5′ triphosphate CTP	Cytidine 5′ monophosphate CMP
Uracil U	Uridine	Uridine 5′ triphosphate UTP	Uridine 5′ monophosphate UMP

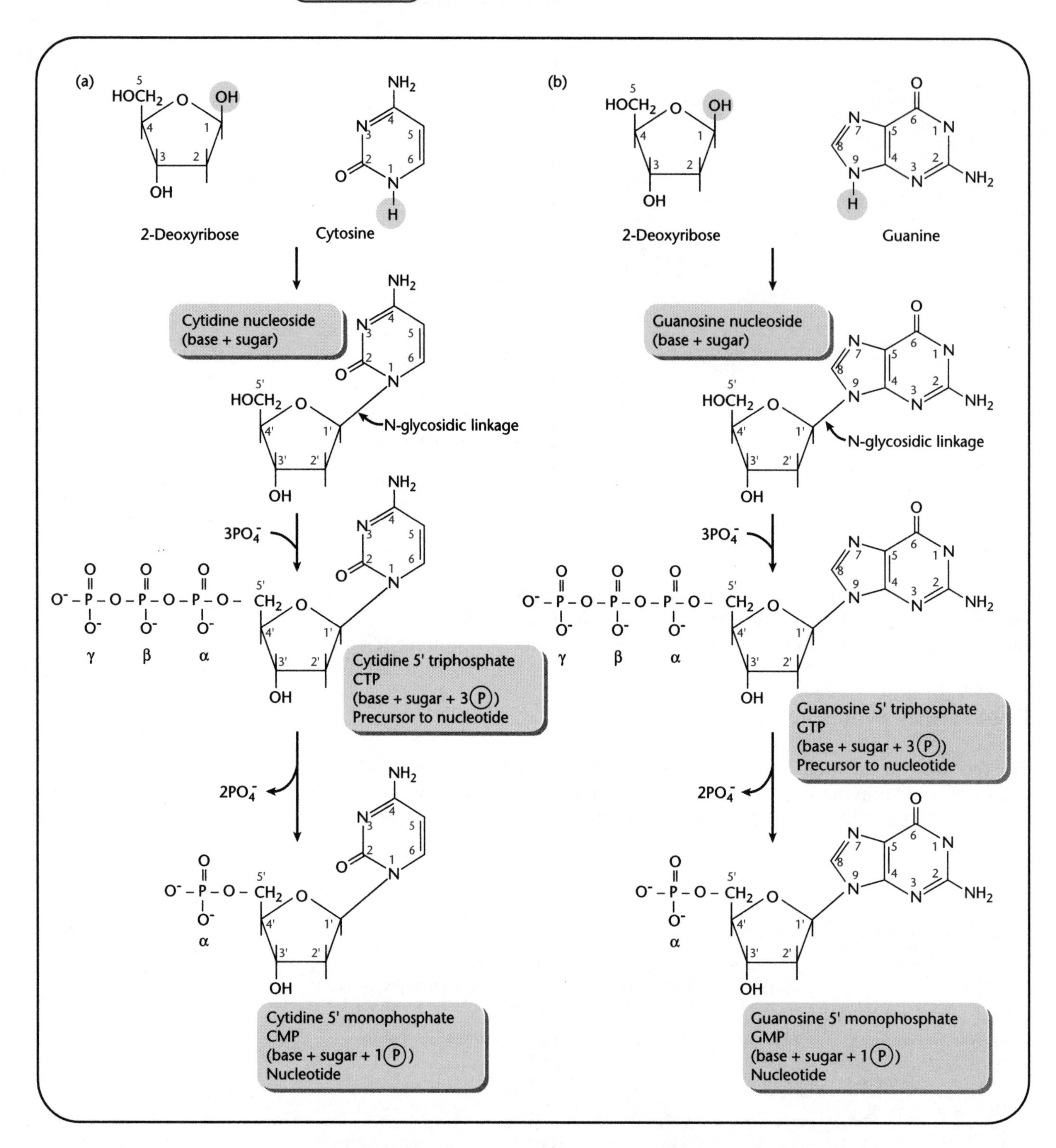

Fig. 2.3 The synthesis of a nucleoside and nucleotide. (a) The synthesis of cytidine (the nucleoside of cytosine) and its corresponding nucleotide, cytidine 5′ monophosphate. (b) The synthesis of guanosine and its corresponding nucleotide, guanosine 5′ monophosphate. 2-Deoxyribose is attached at the C1 position to the N1 of pyrimidines and to the N9 of purines. In both examples, the sites of reaction, the N,N-glycosidic bond, and the addition and removal of phosphate groups are noted.

nucleoside 5′ triphosphate molecule is known as the α phosphate. The middle phosphate is the β phosphate and the outermost phosphate is the γ phosphate. The two phosphates that are lost during polymerization of the nucleoside 5′ triphosphate are the β and γ phosphates. Polymerizing a nucleoside 5′ triphosphate to a growing DNA chain can only occur where there is a free hydroxyl (OH) group. A free hydroxyl group exists only at the 3′ end of a chain of nucleotides. At the 5′ end there is a phosphate group. Thus, growth of a DNA chain can only occur in a 5′ to 3′ direction (Fig. 2.4b). The enzyme that catalyzes the polymerization of a nucleoside 5′ triphosphate to the free hydroxyl group at the 3′ end of a growing DNA chain is a DNA polymerase. *E. coli* has five DNA polymerases: DNA Pol I, DNA Pol II, DNA Pol III, DNA Pol IV (or DinB), and DNA Pol V (or UmuD UmuC). DNA Pol III is the polymerase responsible for the DNA polymerization of a growing DNA chain.

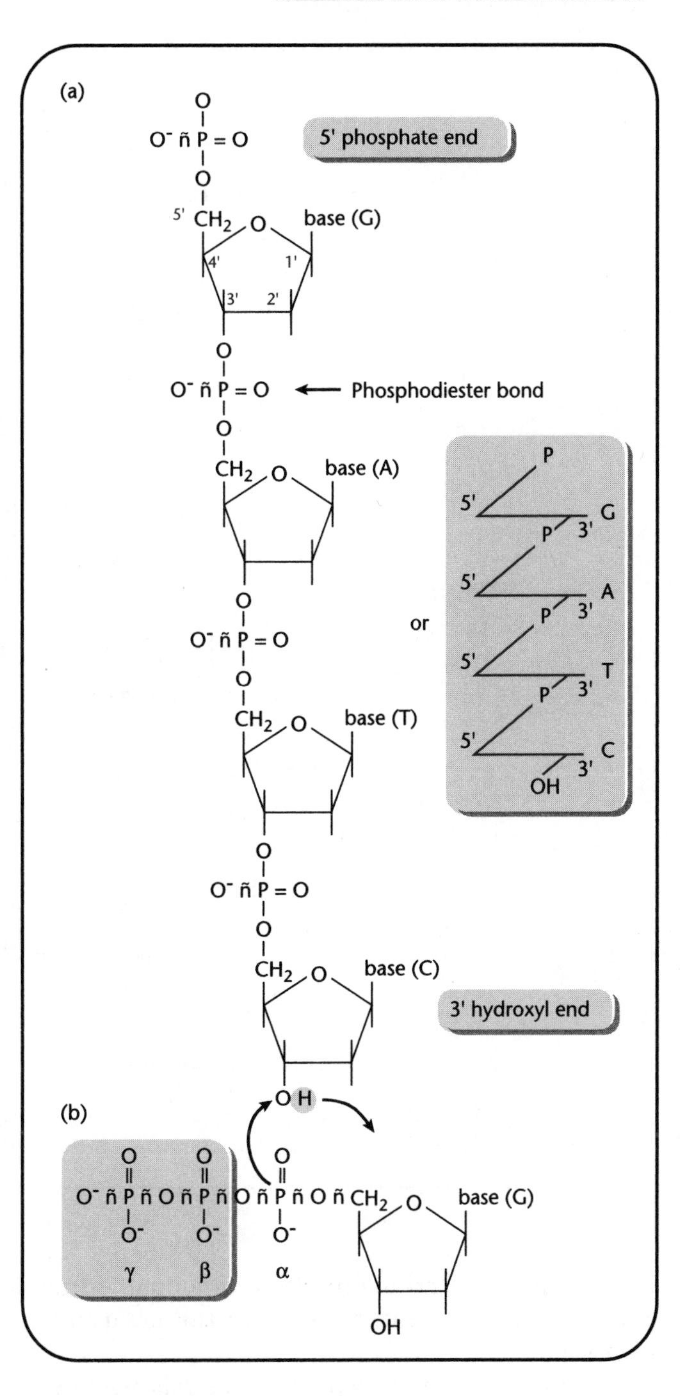

Fig. 2.4 (a) Two schematics illustrating a chain of nucleotides, also referred to as an oligonucleotide. The 5′ phosphate end (5′ P), the 3′ hydroxyl end (3′ OH), and the phosphodiester bonds are noted. (b) Addition of a nucleotide can only occur at the free 3′ hydroxyl end, thus growth occurs in a 5′ to 3′ direction. The β and γ phosphates are cleaved from the precursor nucleoside 5′ triphosphate to drive the addition of this precursor via its α phosphate group.

Double-stranded DNA

With few exceptions, the DNA molecule comprising an organism's chromosome is double-stranded. It consists of non-identical, yet complementary base sequences. The two strands are held together by hydrogen bonds between the purines and pyrimidines. The most stable hydrogen bonds form between an adenine and thymine or a guanine and cytosine. Two hydrogen bonds form between adenine and thymine, while three hydrogen bonds form between guanine and cytosine (Fig. 2.5). The complementary nature of the base sequence is dictated by the specific pairing

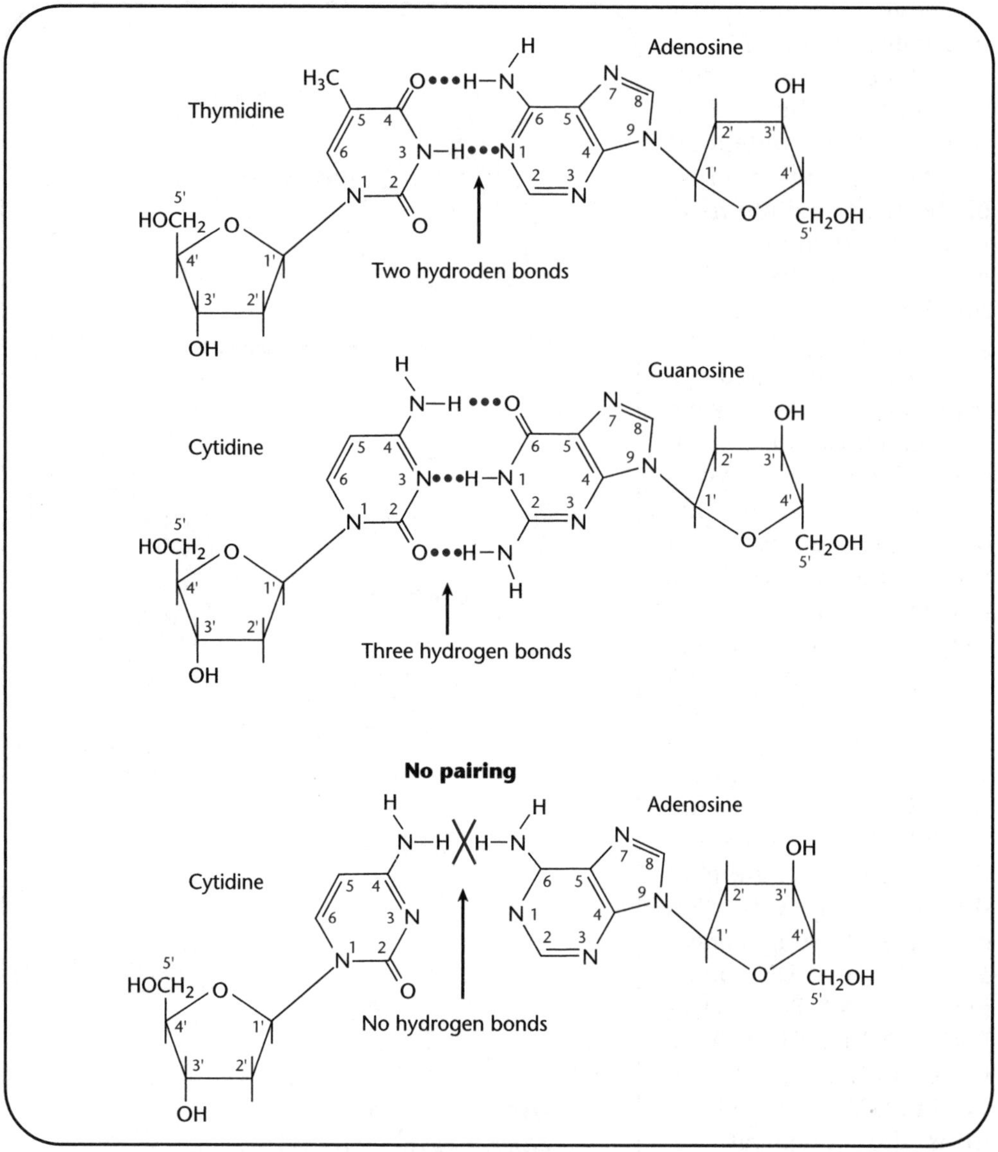

Fig. 2.5 Hydrogen bonding between bases. Each base pair is held together by hydrogen bonds with two hydrogen bonds holding an A–T pair together and three hydrogen bonds holding a G–C pair together. Hydrogen bonds are weak attractions that occur between a hydrogen atom on one molecule and a non-hydrogen atom on another molecule (illustrated by blue dotted line). Based on this definition, it is clear why C cannot hydrogen bond with A and G cannot hydrogen bond with T.

between purines and pyrimidines. In a double-stranded DNA molecule, the molar amounts of guanine and cytosine and the molar amounts of adenine and thymine are identical.

The orientation of the two DNA strands is described as **antiparallel** (Fig. 2.6). One strand is chemically oriented in a 5′ to 3′ direction, while its complementary strand runs 3′ to 5′. The two DNA strands are not actually found side by side, but rather are woven around each other to form a characteristic double helix. This structure has one major and one minor groove per helical turn (Fig. 2.7). Many proteins with the capacity to bind DNA and regulate expression of the genes in the DNA, interact predominantly with the major groove. In the major groove, the atoms comprising the bases are more exposed. The atoms of the bases found in these interactions are situated farthest from the atoms involved in the N,N-glycosidic bond linking the base to the sugar (Fig. 2.3).

Supercoiling double-stranded DNA

Many DNA molecules, such as the *E. coli* chromosome, are circular. Circular DNA molecules are twisted and compacted through a process called **supercoiling** (Fig. 2.8). The form predominantly found in *E. coli* is negative supercoiling. Negative supercoiling occurs when the DNA is twisted about its axis in the direction opposite to the direction of the right-handed double helix. Supercoiling is used to bring into close proximity pieces of DNA that are widely separated on the chromosome or to compact the DNA molecule so that it fits inside the bacterial cell. Relaxing supercoiled DNA is necessary so that large molecular machines needed for DNA replication, RNA transcription, or DNA repair can have access to the base sequence of the DNA molecule. **Topoisomerases** are responsible for either creating or relaxing supercoiled DNA.

Topoisomerases interconvert various topological forms of DNA. Topoisomerase can catalyze the interconversion of relaxed and supercoiled DNA, knotted and

FYI 2.1

Chargaff's rule

Erwin Chargaff measured the relative amounts of A, T, G, and C in the DNA from many organisms, including bacteriophages ϕX174 and T7, *Escherichia coli* B, *Neurospora*, maize, *Drosophila*, salmon, rat, calf, and human. The composition of these bases varied from one species to another. What was intriguing, yet at the time unexplainable, was that in all organisms examined except for bacteriophage ϕX174, the molar amounts of A were equal to T and the molar amounts of G were equal to C. This equal distribution of A to T and G to C became known as Chargaff's rule. We now understand that the reason for these ratios are that A pairs with T and G pairs with C. Chargaff's rule defines the principle of complementarity: each base can only pair with one other base, its complement. Single-stranded DNA molecules such as the chromosome of bacteriophage ϕX174 do not follow Chargaff's rule.

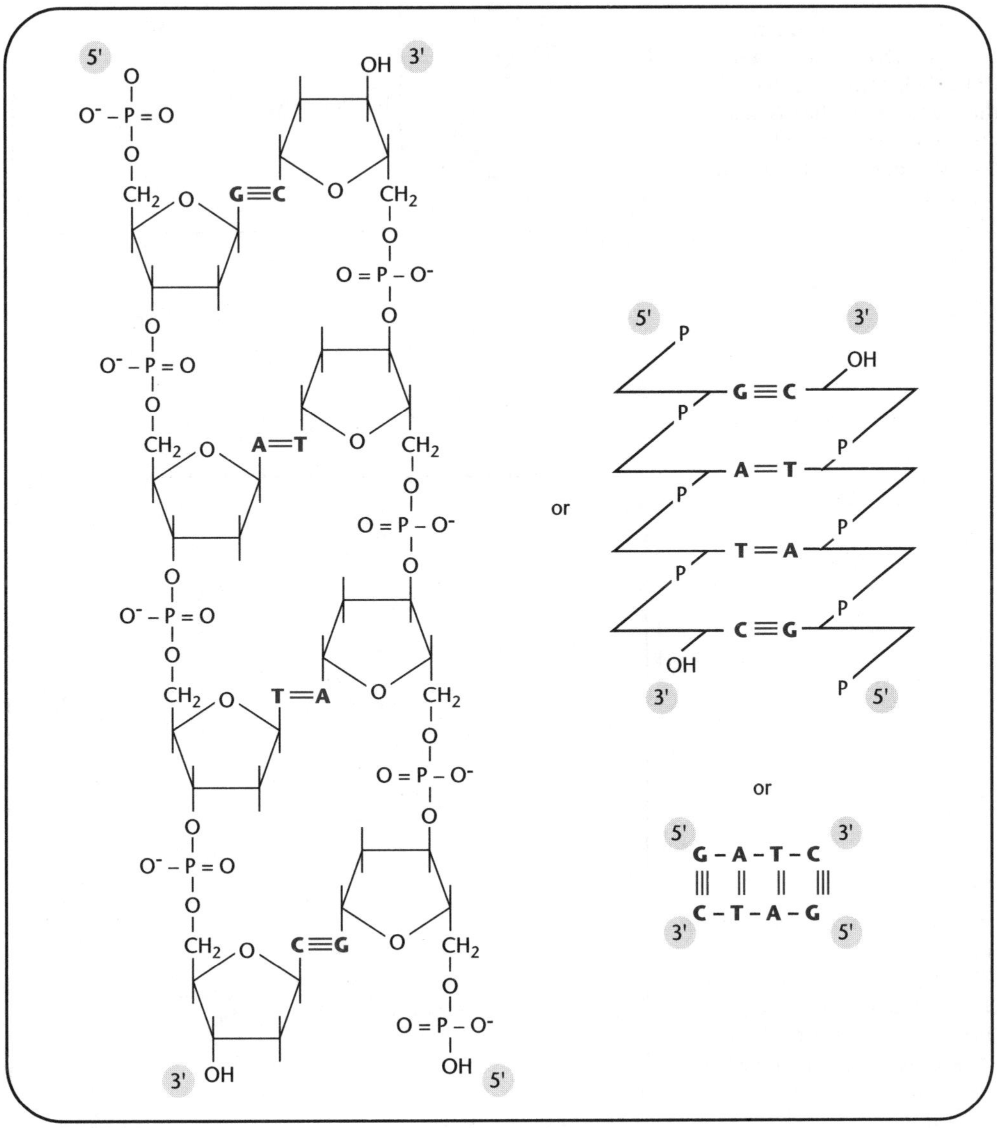

Fig. 2.6 The antiparallel orientation of a double-stranded DNA molecule. Three schematics illustrate the antiparallel orientation.

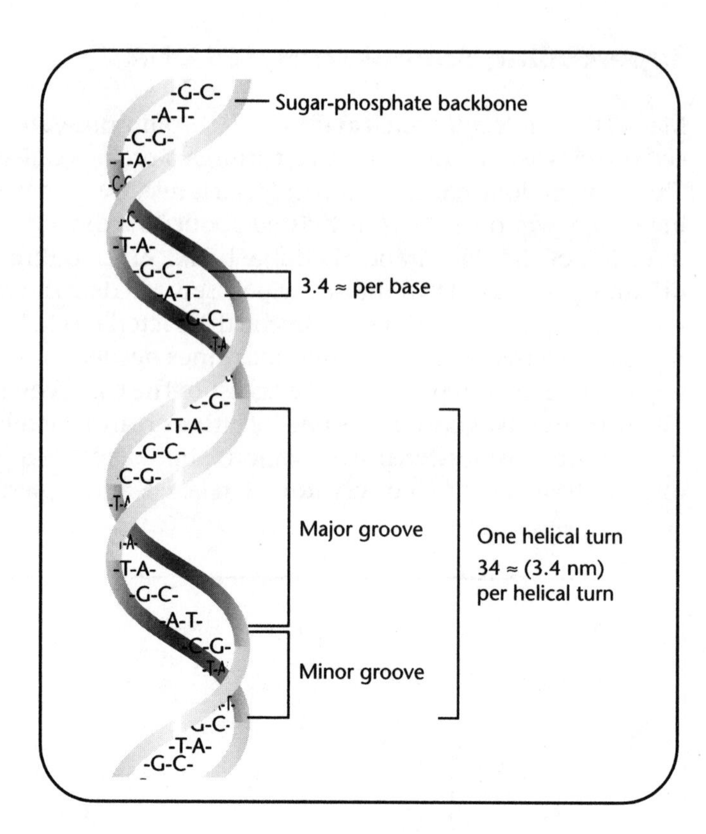

Fig. 2.7 Major and minor grooves in DNA. Watson and Crick's helical model depicting the double helix. The major and minor grooves are noted, as is the size of one helical turn and the amount of space between bases.

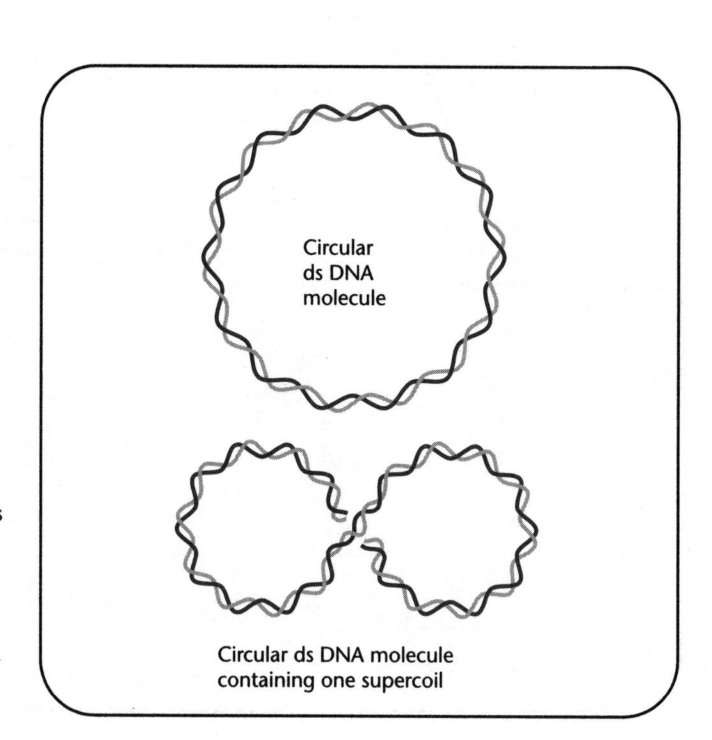

Fig. 2.8 Supercoiling introduces another layer of structure into a DNA molecule. It compacts the DNA for packing into the bacterial cell and brings into close proximity parts of the DNA molecule that are widely separated.

unknotted DNA, and catenated and decatenated circular DNA (Fig. 2.9a). Two classes of topoisomerases have been defined. This definition is based on how many DNA strands are cut and how many DNA strands are passed through the cut. Topoisomerase I cuts one strand of a double-stranded DNA molecule. The uncut strand is then passed through the cut strand before resealing the cut strand. Topoisomerase II cuts both strands of a double-stranded DNA molecule. Uncut sections from both strands pass through the cut section before the cuts are resealed (Fig. 2.9b). Topoisomerase activity differs from nuclease activity in the way it breaks and reseals phosphodiester bonds. Topoisomerases break the phosphodiester backbone and in the process become covalently attached to the DNA ends. The DNA is resealed using the energy stored in the topoisomerase—DNA covalent complex. Nucleases hydrolyze the phosphodiester bonds and release nucleotides or short segments of DNA. They do not become attached to the DNA in the process.

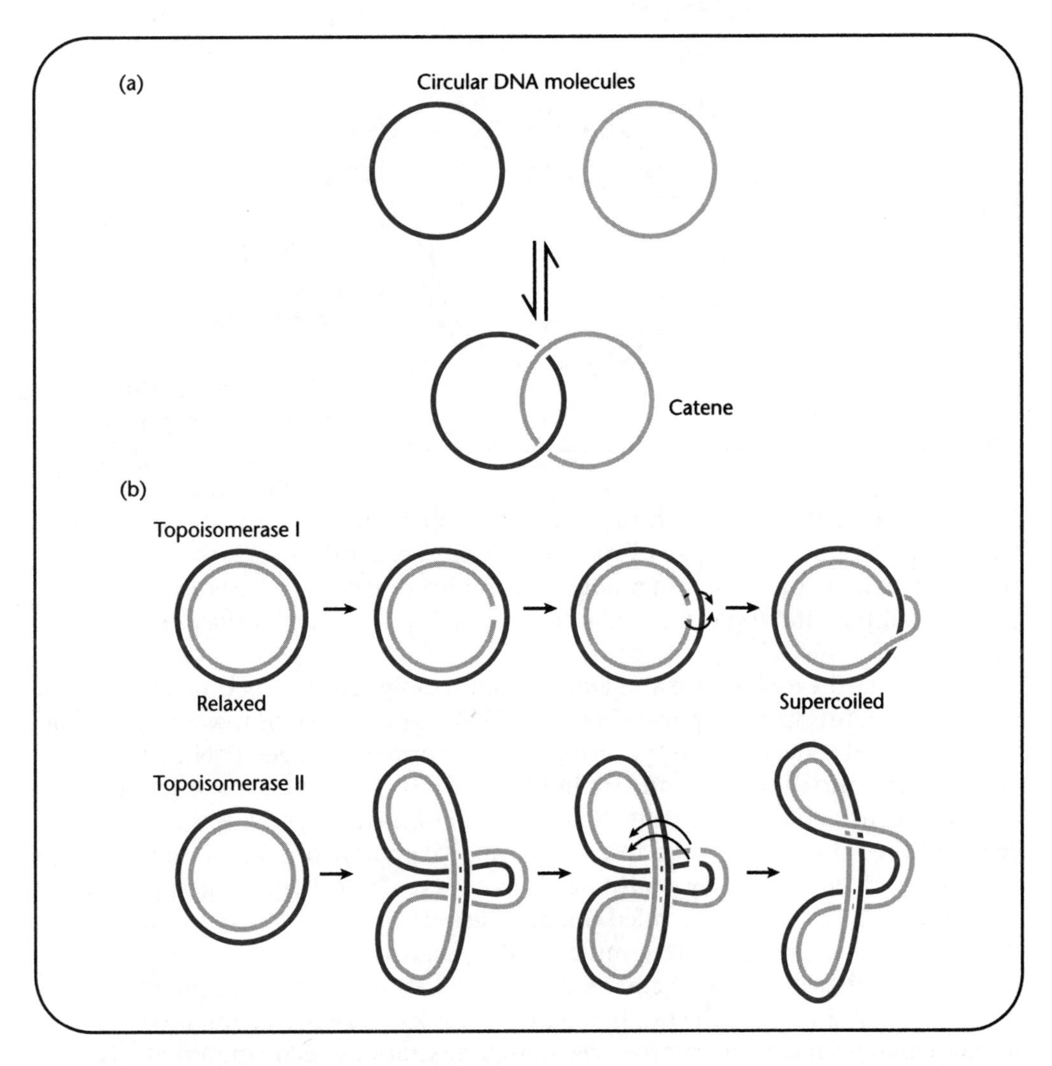

Fig. 2.9 The effects of topoisomerases. (a) Topoisomerases catalyze the interconversion of circular DNA molecules. These circular molecules can be relaxed or supercoiled, knotted or unknotted, catenated or decatenated by the action of topoisomerases. (b) Topoisomerases fall into two classes that are defined by how many DNA strands are cut and how many DNA strands are passed through the cut. Type I cut one DNA strand and Type II cut both strands.

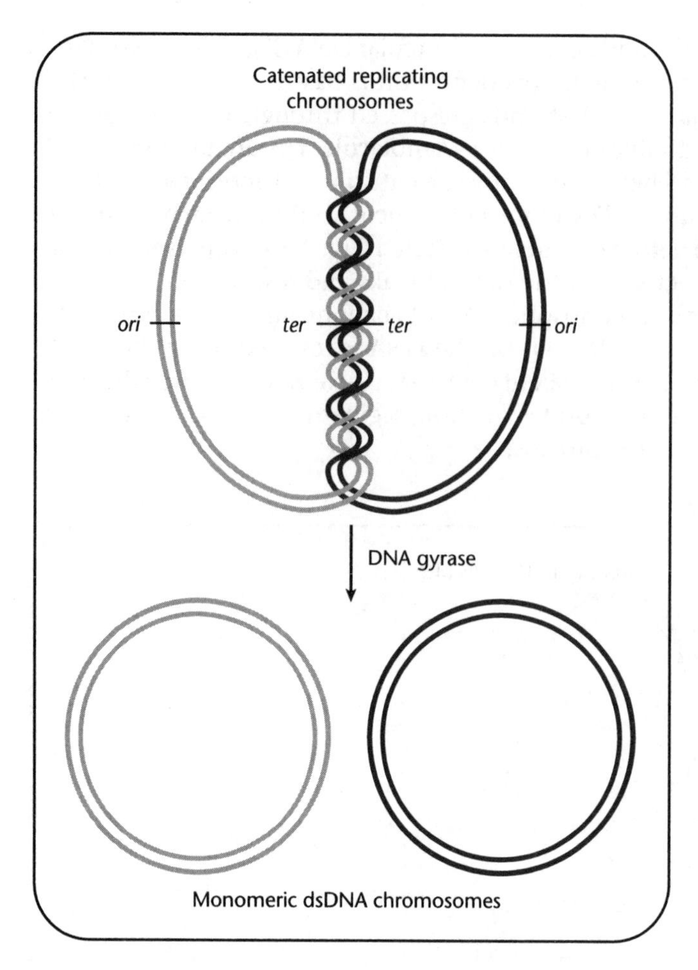

Fig. 2.10 DNA gyrase can untangle chromosomes both during DNA replication and after DNA replication has finished.

DNA gyrase is a type II topoisomerase. Its topoisomerase II activity is used to introduce negative supercoils into the DNA after replication and also to terminate DNA replication by decatenating or separating the two interconnected daughter chromosomes (Fig. 2.10). Nalidixic acid and novobiocin, two antibiotics used to treat bacterial infections, function by inhibiting the topoisomerase activity of DNA gyrase.

Replication of the *Escherichia coli* chromosome

In 1953, James Watson and Francis Crick presented a molecular model of the double-helix structure for DNA. Their structure for DNA hinted that the replication process was dependent on base complementarity and was **semiconservative**. Semiconservative means that the two daughter chromosomes produced by replication each consist of a strand from the original parental molecule hydrogen bonded to a newly synthesized daughter strand (Fig. 2.11a).

In 1958, Matthew Meselson and Franklin Stahl designed an experiment to test the idea of semiconservative replication (Fig. 2.11b). Their experiment was to grow bacteria for several generations in the ordinary, light form of nitrogen (^{14}N). The ^{14}N was incorporated into the purines and pyrimidines of DNA as well as into other nitrogen-containing molecules in the cells. A heavy form of nitrogen, ^{15}N, was added to the cells and the cells were allowed to grow for approximately one generation. The ^{15}N was incorporated into newly synthesized DNA, as well as other nitrogen-containing molecules. This technique is called **density labeling**. The DNA was isolated from the cells and separated using a technique that separates DNA molecules based on their density. The labeled products generated after replication revealed a semiconservative process: one strand from the original parental molecule, which contained ^{14}N, was found paired with a newly synthesized daughter strand, which contained ^{15}N. This experimental approach, along with others, provided the physical data to support a model for semiconservative replication.

In 1963, John Cairns provided the first visual evidence that the *E. coli* chromosome was circular and not linear through the use of autoradiographs. Cairns concluded

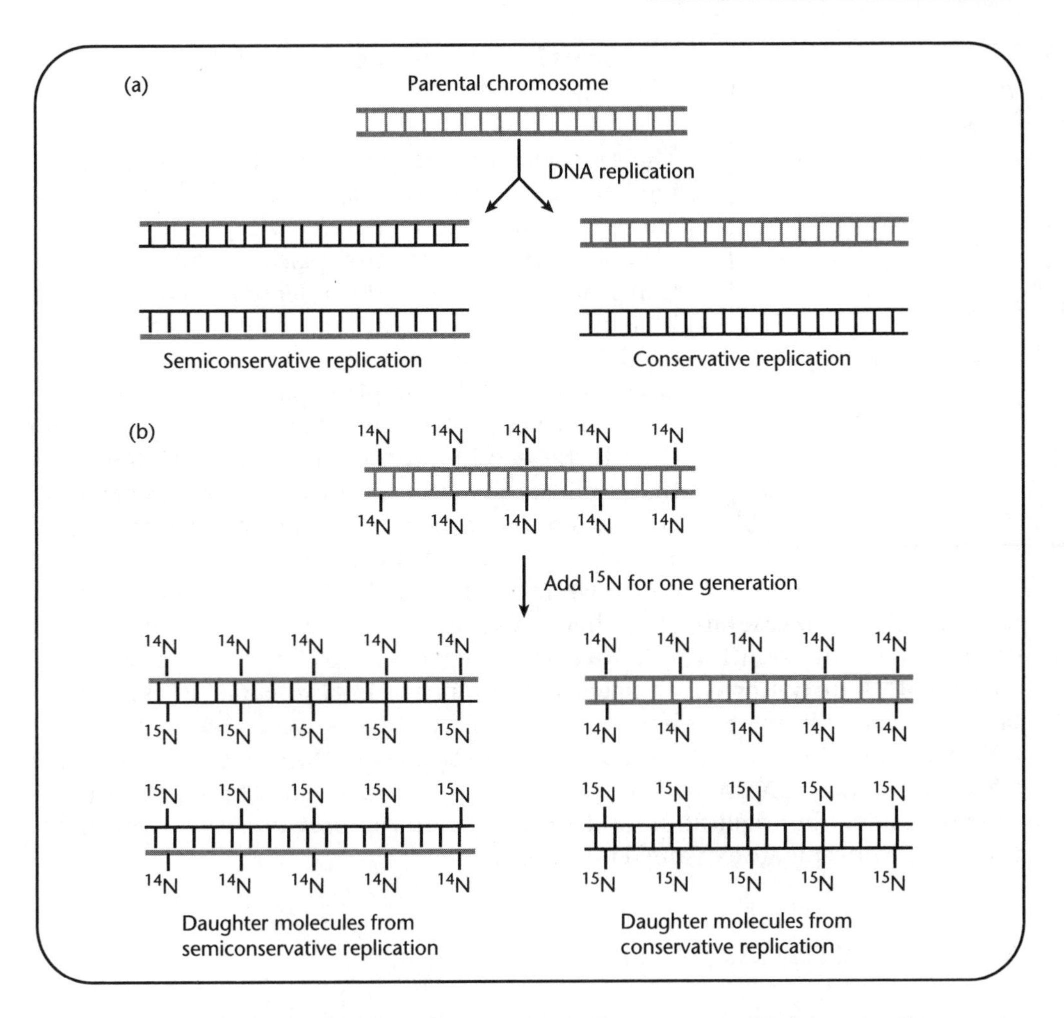

Fig. 2.11 (a) DNA replication could be either semiconservative or conservative. In conservative replication, one daughter would receive both strands that were present in the parent cell and the other daughter would receive two new strands. In semiconservative replication, both daughters receive one newly synthesized strand and one original template strand. (b) In the Meselson and Stahl experiment, density labeling was used to demonstrate that DNA replication occurs by a semiconservative mechanism.

from the visual data that the *E. coli* chromosome was double-stranded, circular, and that semiconservative replication began at a fixed point named the **origin for DNA replication** or ***ori***. While this experimental approach provided the first visual image for the structure of the *E. coli* chromosome, it failed to reveal that replication in *E. coli* is bidirectional. Subsequent experiments, provided evidence that replication of the *E. coli* chromosome did not proceed in one direction, but rather, proceeded in two directions (**bidirectional replication**) from a fixed place on the chromosome called *oriC*. Subsequently, it was found that *oriC* and termination sites for replication (*ter*) are located at opposite points on the circular DNA chromosome (Fig. 2.12). The *E. coli* chromosome can be replicated in 40 minutes under optimun growth conditions. The two replication forks must move at the rapid rate of ~970 nucleotides per second per replication fork to completely replicate a chromosome of 4,639,221 base pairs in 40 minutes.

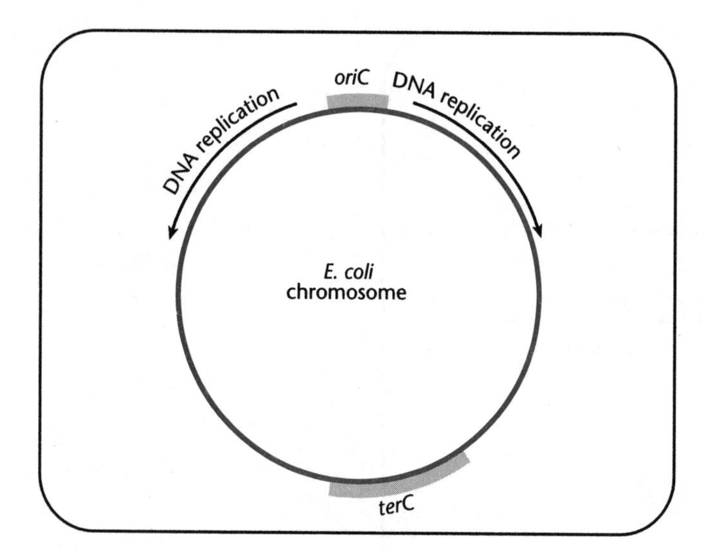

Fig. 2.12 The circular *E. coli* chromosome is replicated from a fixed point called *oriC*. Replication proceeds bidirectionally to the *ter* sites where replication is terminated. DNA gyrase and topoisomerase decatenate the daughter chromosomes so that they can be separated before cell division.

Constraints that influence DNA replication

There are several constraints that influence how DNA is replicated. The first constraint is how the nucleotides are selected for addition to the growing DNA chain. The complementarity of the base pairing defines the selection process by dictating the order of nucleotides in the DNA molecule to be synthesized. The newly synthesized daughter strand must contain the complementary base sequence of the template (parental) strand so that it may ultimately hydrogen bond with the template strand. The second constraint is a result of the enzymes that replicate the DNA. If DNA polymerases can only add nucleotides to an existing free hydroxyl group (the 3′ OH group), then an oligonucleotide with a free 3′ OH end must already exist for DNA Pol III to begin polymerization. If this is the case (and it is), then DNA polymerases need help to initiate replication. As we will soon learn, **DnaG** (DNA dependent RNA primase) or a similar enzyme, **RNA polymerase,** initiates replication. The final constraint is how orientation affects replication. Orientation is dictated by the antiparallel nature of the DNA molecule. Because of this orientation and because of the way DNA polymerase adds nucleotides, replication of one strand in a 5′ to 3′ direction can proceed in a continuous fashion, but replication of the other strand can only proceed discontinuously. These concepts are explained in detail below.

The replication machinery

Remembering complex genetic mechanisms is often easier when all of the components have been identified and their functions defined. For some genetic mechanisms we do not have this luxury because the functions of key components elude our understanding. However, in others, such as *E. coli* chromosomal replication, we do know quite a bit about the required machinery. Table 2.2 describes the DNA polymerases involved in replicating the *E. coli* chromosome and Table 2.3 describes the components of the *E. coli* DNA replication machine.

DNA polymerases

The most significant replication enzyme is DNA polymerase III (**DNA dependent DNA polymerase III** or DNA Pol III, Fig. 2.13). DNA Pol III is composed of many subunits and can also associate with other proteins called DNA polymerase accessory proteins. The α subunit of DNA Pol III, encoded by the *dnaE* (or *polC*) gene, provides the polymerizing function. The θ subunit, encoded by *holE* and the τ subunit encoded by *dnaX*, dimerize the subunits of DNA Pol III. The β subunit, encoded by *dnaN*, provides a sliding function which allows DNA Pol III to move along the template strand, while the γ, δ, δ', χ, ψ subunits, encoded by the *dnaX, holA, holB, holC, holD* genes, respectively, provide a clamping function that prevents DNA Pol III from falling off the template strand while it is polymerizing the new strand. An additional accessory activity, provided to DNA Pol III by the *dnaQ* gene product (ε subunit) is re-

Table 2.2 Fast facts for *E. coli* DNA polymerases.

1 All DNA polymerases use a complementary DNA template to polymerize deoxyribonucleotides on to the free 3′ OH end of an existing oligonucleotide. Growth of oligonucleotide is said to occur in a 5′ to 3′ direction.
2 All DNA polymerases have a 3′ to 5′ exonuclease activity which is referred to as an editing or proofreading activity. This activity can remove nucleotides from a free 3′ OH end on either strand of a duplex DNA molecule.
3 DNA Pol I and DNA Pol III have a 5′ to 3′ exonuclease activity. For DNA Pol I, this activity will digest either strand of duplex DNA from a free 5′ end. For DNA Pol III, this activity will only work on single-stranded DNA.

DNA polymerase	No. molecules/cell	V_{max} nucleotides/sec
DNA Pol III	10–20	250–1000
DNA Pol II	100	2–5
DNA Pol I	300–400	16–20

ferred to as an editing or proofreading function. The editing function is a **3′ to 5′ exonuclease** activity, which allows DNA Pol III to remove incorrectly paired nucleotides from the free 3′ OH end of DNA. DNA Pol III, in addition to its 3′ to 5′ exonuclease activity, has a 5′ to 3′ exonuclease activity, but this activity will only remove nucleotides from single-stranded DNA. Both the polymerizing and nuclease activities of DNA polymerase I, II, and III exert their effect on the phosphodiester bonds of the DNA molecule.

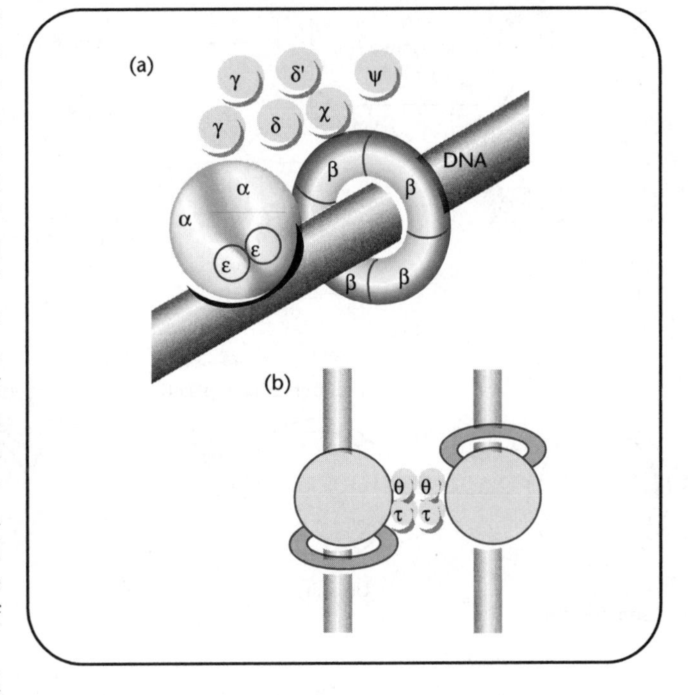

Fig. 2.13 The subunit composition of DNA Pol III. (a) DNA Pol III is composed of α_2, ε_2, β_4, γ_2, δ, δ', χ, and ψ subunits. Their functions are described in the text and in Table 2.3. (b) Two proteins, θ and τ, hold two DNA Pol III complexes together on each replicating strand.

Like DNA Pol III, DNA Pol I and DNA Pol II polymerize nucleotides to a free 3′ OH of an existing oligonucleotide (5′ to 3′ polymerase activity). They both have a 3′ to 5′ exonuclease activity (editing). DNA Pol I was discovered first. In 1969 John Cairns isolated an *E. coli* mutant defective in DNA Pol I activity. DNA Pol I (encoded by *polA*) appears to be primarily involved in editing activities, including removal of mismatch base pairs, removal of RNA oligonucleotides (primers), and gap fill-in reactions where an improper nucleotide had been incorporated into the newly elongating DNA strand. DNA Pol I, in addition to its other enzymatic activities, has a 5′ to 3′ exonuclease activity (repair activity) that can digest double-stranded DNA from an exposed 5′ end. In the DNA Pol I mutant described by Cairns, DNA Pol II and

DNA Pol III activities were detectable. DNA Pol II (encoded by *polB*) is involved in DNA repair.

DnaG primase

A short stretch of nucleotides or **oligonucleotide** with a free 3′ hydroxyl end has to be present for DNA Pol III to carry out its polymerizing reaction (Fig. 2.14). This means that DNA Pol III can only extend the length of an existing chain of nucleotides, it cannot join together the first two nucleotides of this chain. To accomplish this very first step, an enzyme that can prime the initiation of DNA replication is required. DNA dependent RNA primase (**DnaG** primase), encoded by the *dnaG* gene, primes the initiation of DNA replication. DNAG primase polymerizes an RNA primer of approxi-

FYI 2.2

Cairns' proof that *E. coli's* chromosome is circular

Actively growing *E. coli* was transferred to a medium that supported their growth yet contained ^{3}H-radioactively labeled thymidine. These cells were incubated for 1–2 cell doublings, approximately 60 minutes. Their chromosomal DNA was gently isolated. A prediction was made that any newly synthesized DNA should contain ^{3}H-radioactively labeled thymidine. The isolated DNA was spread over a thin membrane, which was placed onto a microscope slide. A photographic emulsion was added to the microscope slide to record emitted β-particles from the decaying incorporated ^{3}H radioactivity. After exposing the microscope slides for 1–2 months, the photographic emulsion was developed. The autoradiograph revealed a ^{3}H decay pattern of circular DNA molecules in a replicative state. This experiment revealed that replication of *E. coli's* double-stranded circular chromosome begins at a unique site called *oriC*. Each parental strand is replicated semiconservatively, so that the replication products each consist of one parental strand hydrogen bonded to one newly synthesized daughter strand.

FYI 2.3

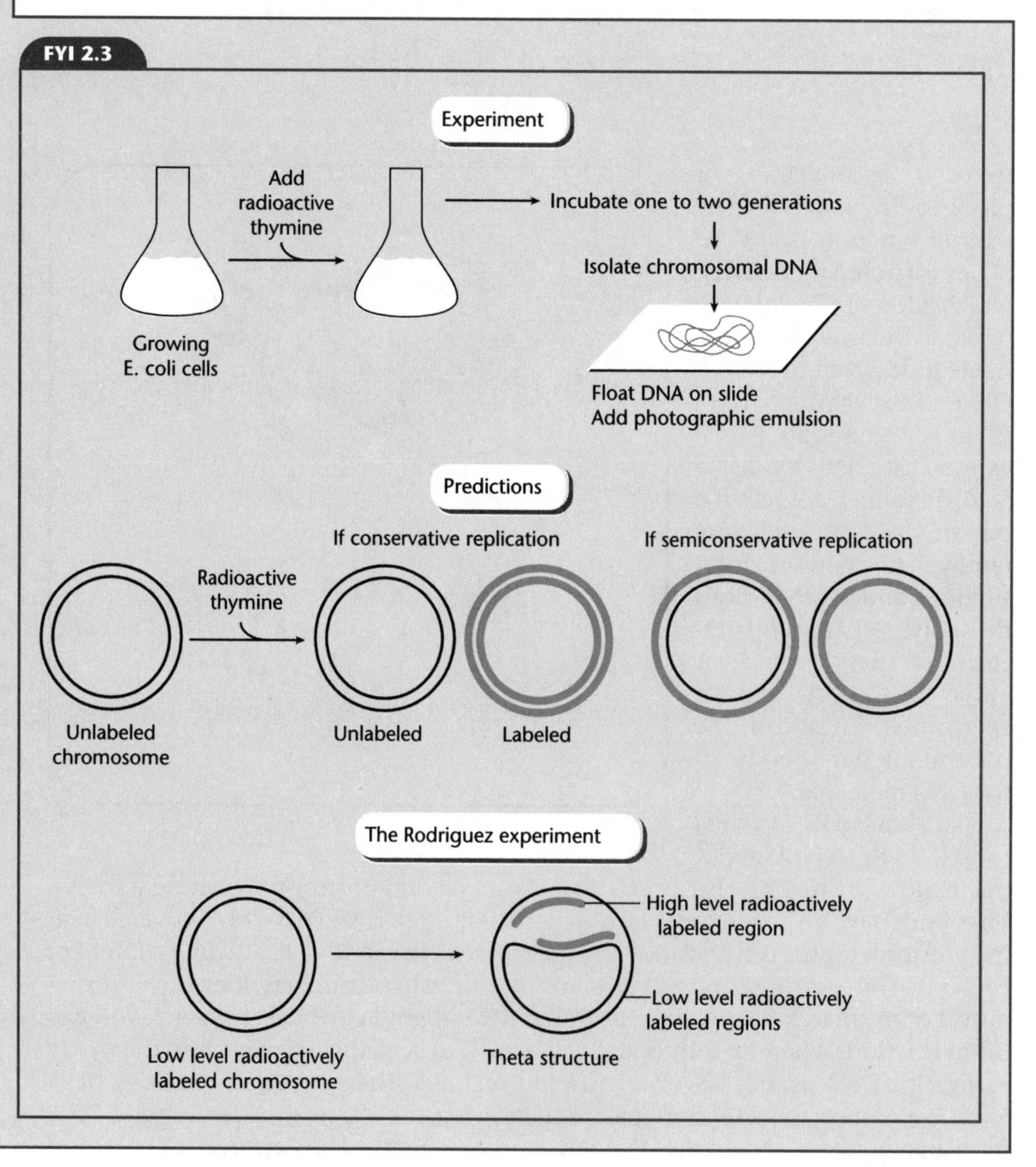

Table 2.3 Components of the *E. coli* DNA replication machine.

Enzyme/subunit	Mr	Gene name(s)	Function
DNA Pol III holoenzyme	740,000	–	–
α subunit	130,000	*polC, dnaE*	5′ to 3′ polymerization (catalytic subunit)
ε subunit	27,000	*dnaQ, mutD*	3′ to 5′ exonuclease (editing, proofreading)
θ subunit	10,000	*holE*	Component of core enzyme (αεθ)
τ subunit	71,000	*dnaZX*	Core dimerization
γ subunit	52,000	*dnaZX*	Processivity/clamp loading
δ subunit	32,000	*holA*	Processivity/clamp loading
δ′ subunit	–	*holB*	Processivity/clamp loading
χ subunit	14,000	*holC*	Processivity/clamp loading
ψ subunit	12,000	*holD*	Processivity/clamp loading
β subunit	40,600	*dnaN*	Initiator complex
DNA Pol II			
α subunit	120,000	*polB*	5′ to 3′ polymerization (catalytic subunit)
DNA Pol I			
α subunit	102,000	*polA*	5′ to 3′ polymerization (catalytic subunit); primer removal, editing, gap filling when associated with appropriate accessory proteins
DnaA	52,500	*dnaA*	Involved in primosome formation (initiator protein that recognizes *oriC*)
DnaB	52,300	*dnaB*	DNA helicase: opens up dsDNA to form a fork (priming)
DnaC	29,000	*dnaC*	Brings DnaB to the primosome (at *oriC*)
SSB	18,800	*ssb*	Single-stranded binding protein; binds and protects ssDNA
Primase	65,600	*dnaG*	Polymerizes ribonucleotides using ssDNA template, i.e. primes initiation of DNA replication
RNA polymerase	460,000	*rpo genes*	Major transcription enzyme, however, can prime DNA replication in place of primase
DNA ligase	74,000	*ligA*	Seals DNA nicks by forming phosphodiester bonds between two adjacent nucleotides
DNA gyrase (Topo II)	374,000		Introduce negative supercoils into closed circular DNA and relaxes (decatenates) supercoiled DNA
α subunit	97,000	*gyrA*	Nick closing
β subunit	89,800	*gyrB*	ATPase
DnaT	19,300	*dnaT*	DNA replication termination

mately 5–11 ribonucleotides in length from a single-stranded DNA template. RNA polymerase can also carry out this reaction. The short RNA oligonucleotide has a free 3′ OH that DNA Pol III can then use.

FYI 2.3

Rodriguez' proof that replication of *E. coli* is bi-directional

Ray Rodriguez provided evidence that replication of the *E. coli* chromosome did not occur in one direction, but rather, proceeded in two directions from a fixed *ori*. He modified Cairns' experimental approach by labeling synchronously growing *E. coli* cells in a medium consisting of a low activity ^{3}H-radioactive thymidine label, and before the replication cycle terminated, he switched the cells to a medium consisting of a high activity ^{3}H-radioactive thymidine label. He processed the DNA of these cells and viewed them by autoradiography in a manner similar to that used by Cairns. The autoradiograph revealed a theta structure with two replication forks originating out of the same *ori*. This experimental approach also revealed that the *ori* and termination sites for replication were located at opposite points on the circular DNA chromosome. Complete replication of *E. coli*'s circular DNA chromosome with nonadjacent start and stop sites could only occur with two replication forks proceeding in a bidirectional fashion. Thus, replication of the *E. coli* chromosome is bidirectional and replication ends when the two replication forks meet at a site opposite the *oriC* site.

DnaA, DnaB, and DnaC

Replication of the *E. coli* chromosome requires several proteins before DnaG can act. At least three accessory proteins, **DnaA**, **DnaB**, and **DnaC** comprise an initiator complex. Based on the most current model of the initiator complex in *E. coli* (Fig. 2.15), DnaA recognizes a specific base sequence (5′ TTATCCACA 3′) called a **DnaA box**. Several DnaA boxes are located in the *oriC* region of the *E. coli* chromosome. Ten

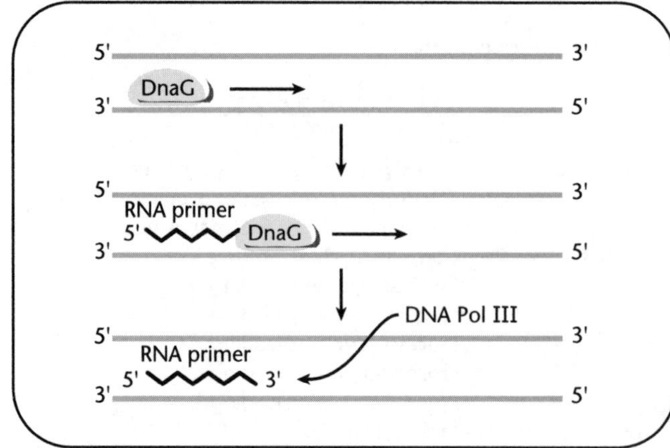

Fig. 2.14 DnaG can synthesize a small segment of RNA using DNA as a template. It makes a short segment of RNA with a 3′ OH that DNA Pol III can elongate.

to 12 molecules of DnaA bind to the DnaA boxes in *oriC*. By binding to *oriC*, the DnaA proteins disrupt hydrogen bonds between complementary base pairs in an area of *oriC* adjacent to the DnaA boxes that is A–T rich and opens the double helix (Fig. 2.15a,b).

Once the helix is open, DnaB protein, a **helicase**, binds to *oriC* with the help of the DnaC protein (Fig. 2.15c). DnaB further disrupts hydrogen bonds between base pairs and unwinds the DNA helix. Unwinding the two DNA strands allows DnaG primase to anchor to the template and subsequently synthesize a short RNA primer for use by DNA Pol III (Fig 2.15d).

Once DnaB helicase begins unwinding the double-stranded DNA, the single-stranded DNA that is generated must be protected. SSB (single-stranded binding) protein, encoded by the *ssb* gene, binds to the single-stranded DNA (ssDNA) and protects it from nuclease activity (Fig. 2.15d). SSB is used in many processes when ssDNA must be protected (see Chapter 5).

Replication of both strands

Replication of one template strand in a 5′ to 3′ direction can proceed in a continuous fashion, but replication of the other template strand can only proceed discontinuously. The strand being synthesized continuously is referred to as the **leading strand** whereas the strand being synthesized discontinuously is referred to as the **lagging strand** (Fig. 2.16). To synthesize the leading strand, only one RNA primer has to be available for DNA Pol III to add nucleotides. In *E. coli*, since there are two replication forks, there are two leading strands being synthesized, and the two leading strands only have to be primed once.

The lagging strand is synthesized in a discontinuous fashion (Fig. 2.16). As the replication fork moves through the template, the template strand oriented 3′ to 5′ is used by DnaG primase over and over to synthesize RNA primers. Priming reactions that occur outside of *oriC* do not require the DnaA, DnaB, and DnaC activities of the initiator complex. Using the free 3′ hydroxyl of the RNA primer, DNA Pol III polymerizes nucleotides (approximately 1000) until it reaches the previously synthesized RNA primer. The short DNA segments comprising the lagging strand are called **Okazaki fragments**. Each Okazaki fragment consists of a short stretch of RNA (5–11 bases) followed by a longer stretch of DNA (1000 bases). DNA Pol I uses its 5′ to 3′ exonuclease activity to recognize the free 5′ end of the RNA primer and digest the ribonucleotides of this primer. DNA Pol I then uses its 5′ to 3′ polymerizing activity to add deoxyribonucleotides in place of the excised ribonucleotide primer.

DNA polymerases can not generate the final phosphodiester bond between the 3′ hydroxyl and the 5′ phosphate of two Okazaki fragments. Rather, **DNA ligase**, encoded by the *lig* gene, links together all of the Okazaki fragments after the RNA primer has been removed. Ligase generates a phosphodiester bond between a free 5′ phos-

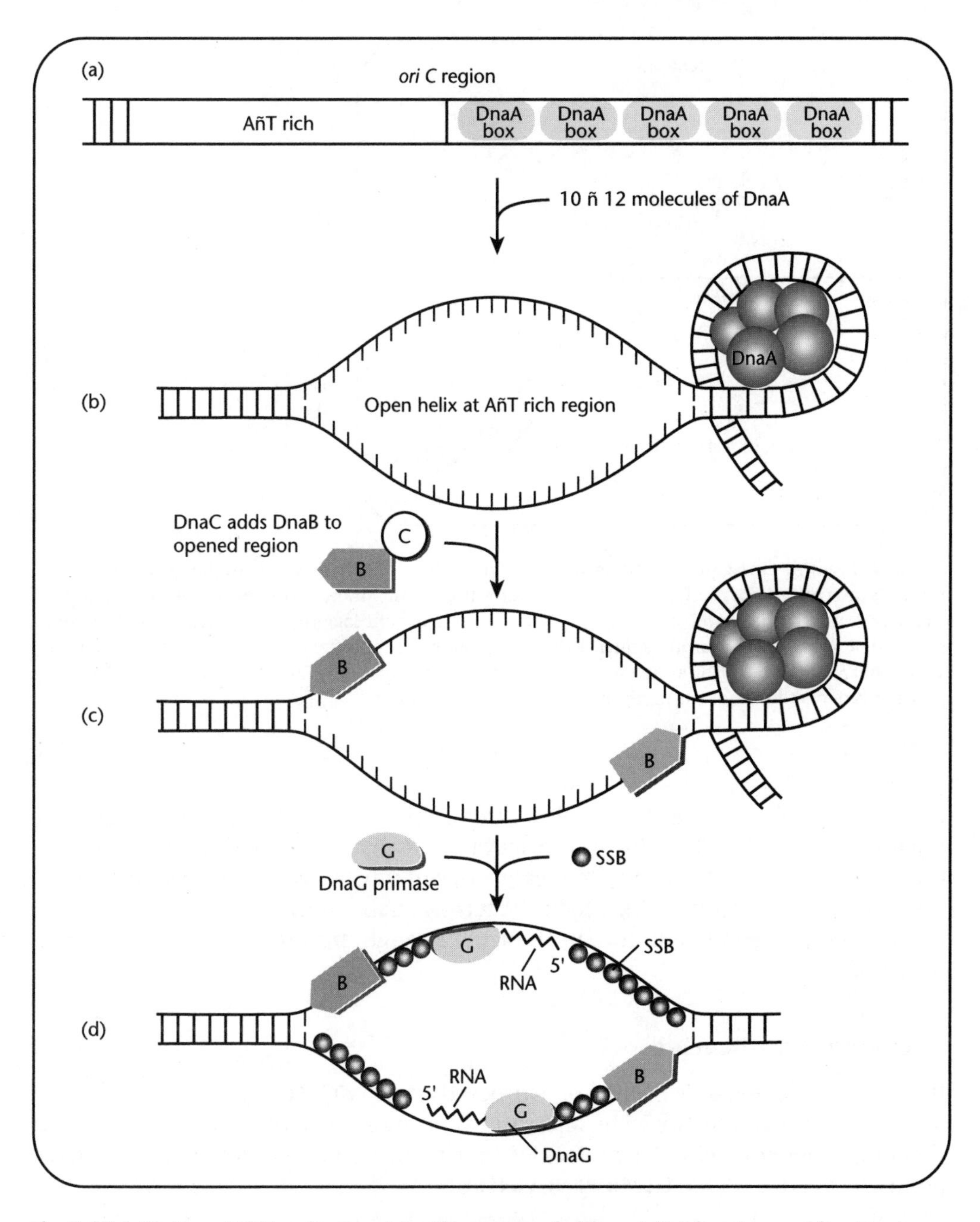

Fig. 2.15 Initiation of DNA replication. (a) *oriC* is composed of three A–T rich regions and four DnaA boxes. DnaA boxes consist of the sequence, 5′ TTATCCACA 3′. Ten to 12 molecules of DnaA protein bind the DnaA boxes to form a complex. (b) The bound DnaA protein forces the hydrogen bonds of the nearby A–T rich regions to dissociate, opening the double helix. (c) Once open, DnaC can add the DnaB helicase to further unravel this area. (d) Finally, DnaG primase initiates synthesis by polymerizing a short stretch of RNA.

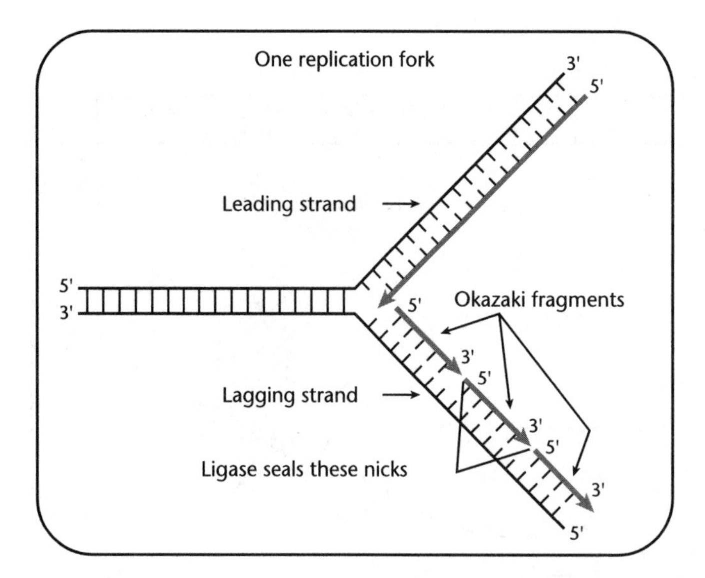

Fig. 2.16 Leading and lagging strand synthesis during replication. Synthesis of one daughter strand proceeds in a continuous fashion. This strand is called the leading strand. Synthesis of the other daughter strand proceeds in a discontinuous fashion. This strand is called the lagging strand. Ligase seals the single-stranded nicks between Okazaki fragments that have had the RNA primer removed to form a lagging strand that is contiguous. The antiparallel nature of the template strand and the way nucleotides are polymerized together require leading and lagging strand synthesis.

phate of one Okazaki fragment and the free 3′ hydroxyl of the proceeding Okazaki fragment (Fig. 2.16). These ligation reactions seal all of the single-stranded nicks in the sugar–phosphate backbone to create a newly synthesized lagging strand that is now contiguous. This building block effect is what allows the replicative machinery to move along a template strand synthesizing a daughter strand in a 5′ to 3′ direction (Fig. 2.17).

Theta mode replication

Replication of the *E. coli* chromosome proceeds from *oriC*. Two replication forks exit *oriC*, one in either direction. The replicating chromosome looks like the Greek letter theta (θ). Therefore, circular molecules that replicate bidirectionally from a fixed point are said to undergo **theta replication**.

Minimizing mistakes in DNA replication

Bacterial DNA polymerases are multifunctional when associated with accessory proteins. They can digest DNA in the 3′ to 5′ direction when the accessory protein, DnaQ (MutD), is sliding along the template strand with the polymerase. This 3′ to 5′ exonuclease activity, which is referred to as either **editing** or **proofreading**, is invaluable (Fig. 2.18). It guarantees that the genetic information is transmitted faithfully from generation to generation. Even with DNA Pol III's accuracy in selecting nucleotides during replication, it will make mistakes. The replication machine's error rate is approximately one out of every 10^{10} nucleotides incorporated. These mistakes, as we will learn in Chapter 4, are usually independent of DNA Pol III's polymerization

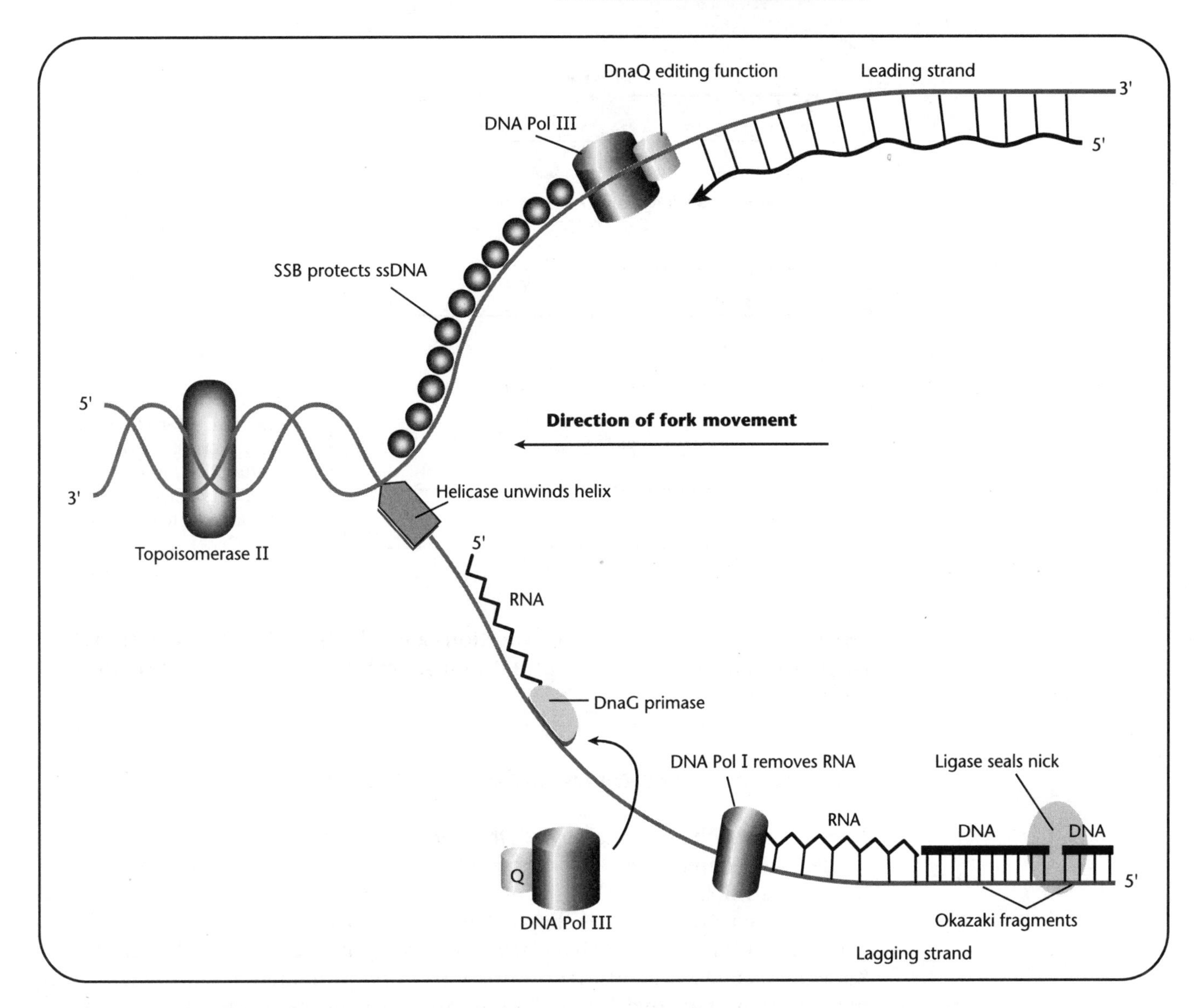

Fig. 2.17 Replication machinery at one replication fork. After initiation and followed by additional strand separation and unwinding, one DNA strand of the template molecule is replicated in a continuous fashion (top strand). The other DNA strand of the template molecule is replicated in a discontinuous fashion (bottom strand). The replication fork is illustrated as moving in a right to left direction. The leading strand is synthesized by the polymerization activity of one DNA Pol III, with DnaQ providing the editing function. The C-terminal domain of DnaQ appears to make contact with the polymerizing domain of DNA Pol III. The N-terminal domain of DnaQ contains the exonuclease activity (editing function). The lagging strand is synthesized as short DNA fragments (Okazaki fragments) which are ligated together by ligase after the RNA has been removed by DNA Pol I.

activity. Often mistakes are a function of changes in the hydrogen-bonding properties of the nucleotides comprising the template DNA. If a base in the template has a change in its hydrogen-bonding property DNA Pol III will place the wrong nucleotide across from the altered template base. This poor choice will create a **mismatched base pair**, which causes a distortion in the double helix. If an accessory protein such as DnaQ is sliding along the template strand with DNA Pol III, and a mismatch is

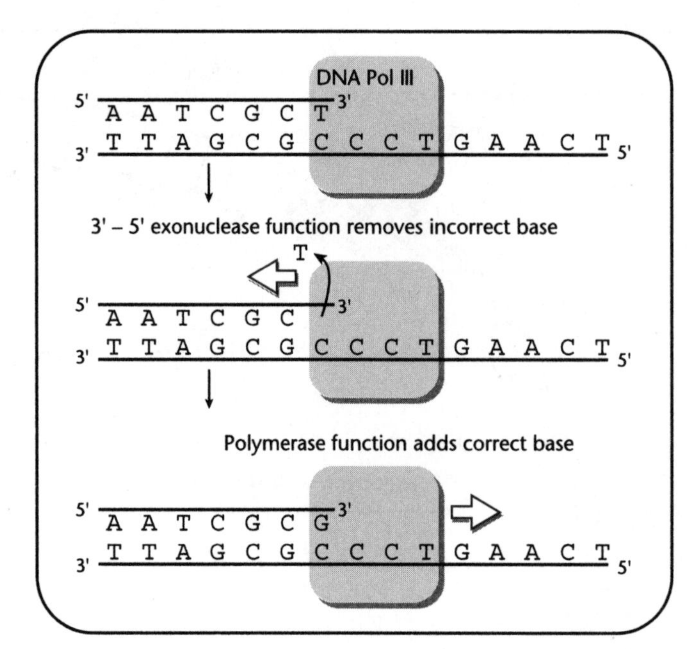

Fig. 2.18 DNA polymerases contain an editing function. The 3′ to 5′ exonuclease will remove incorrect bases and the 5′ to 3′ polymerization function will again attempt to insert the correct bases.

generated, DnaQ recognizes the distortion caused by the mismatch, and stops DNA Pol III from further replication until the incorrect base is excised and the correct one inserted.

The DNA replication machinery as molecular tools

Understanding the enzymology of DNA replication and other genetic mechanisms yielded the technical revolution referred to as biotechnology. Although molecular techniques will be comprehensively covered in Chapter 14, it is worth noting that what has been covered in this chapter has applications.

The activities of DNA polymerases and ligases are exploited in biotechnology. Knowing that ligase catalyzes the formation of a phosphodiester bond, and knowing the conditions that drive this catalysis, means we can ligate fragments of DNA in a test tube. Ligation is a necessary activity for cloning genes. Understanding the polymerization activity of DNA polymerase and the rules by which nucleotide addition occurs means we can synthesize DNA in a test tube. Knowing when editing activity needs to be functional and when it doesn't need to be functional means at times perfect copies of DNA can be synthesized and at other times imperfect copies of DNA can be synthesized. To the microbial molecular geneticist, imperfect DNA molecules are just as useful as perfect DNA molecules because they might contain mutations which could reveal something about the function of the corresponding encoded gene products.

Summary

DNA is composed of bases attached to a phosphodiester backbone. The four bases, A, C, G, and T, pair with each other in specific combinations, A with T and C with G. The chemical nature of the backbone results in the two strands of the double helix being antiparallel. The structure of the helix, as well as the enzymatic properties of the proteins involved, dictate how the DNA is replicated. It is not possible to replicate both strands in the same manner. One strand is replicated continuously (leading strand) and the other discontinuously (lagging strand). For most bacteria, replication starts at a specific place on the chromosome and proceeds in both directions from this origin. Replication is semiconservative, with each daughter inheriting one old strand and one newly synthesized strand. Many enzymes participate in the replication process, with the DNA polymerases providing both polymerizing and editing activities. DNA Pol III is the major replicating enzyme, while DNAPol I and DNAPol II provide activities needed in editing and repair.

Study questions

1 Draw two sets of complementary nucleotides bonded together. Label the important bonds. Which bonds are covalent and which are noncovalent?
2 Why do DNA molecules that are G–C rich not separate after heating as easily as those that are A–T rich?
3 What is the difference between a nucleotide and a nucleoside?
4 Where is a nucleotide added to a growing chain and why is it added there?
5 What bond(s) are affected by the exonuclease activity of DNA polymerases? How are these bonds affected? Describe other enzymatic activities that disrupt important bonds of the DNA molecule.
6 How do topoisomerases break and reseal the phosphodiester bond between two nucleotides? How are topoisomerases different from nucleases, such as the exonuclease activity of DNA Pol III?
7 What is semiconservative replication?
8 What is an *ori*, what takes place at an *ori*, and what would be found at an *ori*? What would happen to a DNA molecule without an *ori*?
9 What is the orientation of the two strands in a double-stranded DNA molecule and why are they oriented in this fashion?
10 What is an editing activity and why is it needed?

References

Baker, T. and Wickner, S. 1992. Genetics and enzymology of DNA replication in *Escherichia coli*. *Annual Review of Genetics*, **26**: 447.

Kornberg, A. and Baker, T. 1992. *DNA Replication*, 2nd edn. New York: W.H. Freeman and Co.

Marians, K. 1996. Replication fork propagation. In *Escherichia coli* and *Salmonella typhimurium: Cellular and Molecular Biology*, 2nd edn., eds. F.C. Neidhardt, R. Curtiss III, J.L. Ingraham, E.C.C. Lin, K.B. Low, B. Hagasanik, W.S. Rexnikoff, M. Riley, M. Schaechter, and H.E. Umbarger, pp. 749–63. Washington, DC: ASM Press.

Messer, W. and Weigel, C. 1996. Initiation of chromosome replication. In *Escherichia coli* and *Salmonella typhimurium: Cellular and Molecular Biology*, 2nd edn., eds. F.C. Neidhardt, R. Curtiss III, J.L. Ingraham, E.C.C. Lin, K.B. Low, B. Hagasanik, W.S. Rexnikoff, M. Riley, M. Schaechter, and H.E. Umbarger, pp. 1579–1601. Washington, DC: ASM Press.

Chapter 3

Mutations

A **mutation** is a physical change to one or more nucleotide pairs in the DNA of a cell. The change is inherited by every descendent of the mutant cell. The mutation can affect only one nucleotide pair or it can affect hundreds of kilobases. The affect of the mutation on the cell clearly depends on where in the DNA the mutation has occurred (Fig. 3.1). If a mutation affects a base pair that is located between two genes and does not have any regulatory or structural role, the mutation will be silent. If a mutation is within a gene, it can disrupt the function of the gene. Once a mutation has occurred, every time the mutated cell divides, the mutation is passed to the daughter cells. Thus, while mutations can affect cell growth, they also serve to mark a cell and its descendents. In this way, mutations can be used to delineate relationships among different populations of cells.

There are two basic types of mutations in a gene: loss of function and gain of function. As the name implies, a **loss of function mutation** results in the gene product not working. There are many alterations that can occur in the DNA that result in this type of mutation but the bottom line is that the gene product does not work as well as the wild- type gene product. If the gene product has only one function in the cell, then studying a loss of function mutation allows the geneticist to determine what the cell does without a specific gene. If a gene product has two or more functions in the cell, it is possible to separate them by mutating one of the functions and leaving the other intact.

Sometimes when a gene is taken away, nothing noticeable happens to the cell. It does not grow slower, change shape, or behave differently in any tests. This can pose a problem for determining its function. One way around such a dilemma is to use a **gain of function mutation**. In such a mutation, a characteristic of the gene is exaggerated. For example, if a gene product of interest binds to DNA, in a gain of function mutation this gene product might now bind DNA much more tightly. It is not possible to create a radically new function for a gene product by simple mutations. A protein that degrades the sugar lactose cannot be changed into a protein that replicates DNA. By studying what function is exaggerated by the mutant, the geneticist can infer a function that the wild-type gene product possesses.

When a mutation is first identified, it is not always obvious if it is a gain or loss of function. A simple test can help in this determination. Normally, *E. coli* contains a single copy of all of its genetic material or chromosome, a situation known as **haploidy**. Other organisms, such as bakers yeast, can contain two complete copies of their chromosomes or are **diploid**. While it is not possible to make an *E. coli* cell diploid for all of its genome, it is possible to introduce two copies of a part of the chromosome into the same cell. Such a cell is called a **merodiploid**, meaning partial diploid (see Chapter 9 for how to make a merodiploid). To determine if a mutation is a gain or loss of function, the mutant gene sequence is introduced into a cell so that a merodiploid is formed. The important characteristic of the merodiploid is that it contains the mutant gene and a corresponding wild-type copy of the same gene. The question is then, in a cell that contains both the mutant and the wild type, do you see the phenotypes of the mutant or the phenotypes of the wild-type gene? If the phenotypes of the mutant are seen, the mutant is said to be **dominant** to the wild-type gene and is a gain of function mutant. In the presence of the wild-type gene, the gained function is still visible. If the phenotypes of the wild-type gene are seen, the mutant is said to be **recessive** to wild type and is a loss of function mutation. In the presence of the wild-type gene, the lost function is not visible. By determining if a mutation is dominant or recessive, it is possible to know if it is a gain or loss of function and to more accurately interpret the mutation's affects on the cell.

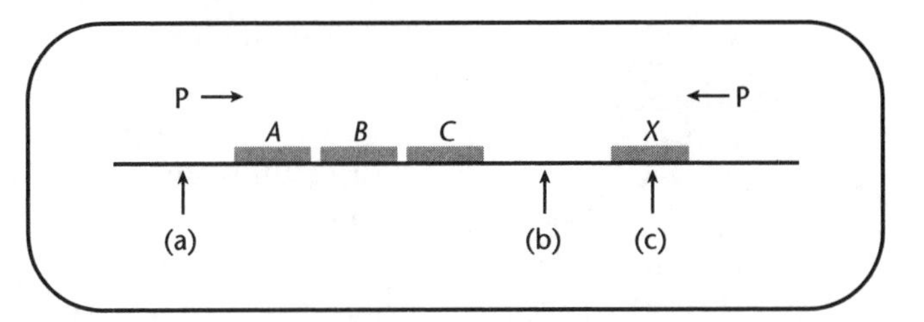

Fig. 3.1 The affect of a mutation depends on where in the DNA the mutation occurs. (a) A mutation in the promoter (P) for *genes A*, *B*, and *C* can affect the expression of these genes. (b) A mutation after *gene C* (but before *gene X*) will not affect gene expression and will most likely have no effect on the cell. (c) A mutation in *gene X* can affect the gene product of *gene X*.

Phenotype and genotype

Mutations can be described in two ways (Fig. 3.2). The first of these is the actual change in the bases of the DNA. The second is the affect that the change has on how the cell functions and grows. The change in the DNA is described by the **genotype** of a cell and the effect that the mutation has on cell function and growth is described by the **phenotypes** of the cell.

When working with a specific strain of bacteria it is very important to know if the strain is wild type or if it contains any known mutations. For example, if the goal is to study how cells metabolize the sugar arabinose, it is important that the starting strain does not contain any known mutations in arabinose metabolism. By examining the genotype of a cell, this can be determined. As a strain is studied, mutations that this strain is found to contain or any mutations that are made in it are listed in the strain's genotype.

Different experimental organisms have different conventions for writing a genotype. In *E. coli*, the genotype is indicated by three lower case, and italicized or underlined letters such as *lac*, *ara*, *mal* or lac, ara, mal. If more than one gene is involved in the same process, the genes are delineated by adding a capital letter, A, B, C, etc. Thus, *lacZ* encoded β-alactosidase, the enzyme that degrades lactose. *lacY* encodes a lactose-specific transport protein. Note that even if a genotype begins a sentence, the first letter is not capitalized.

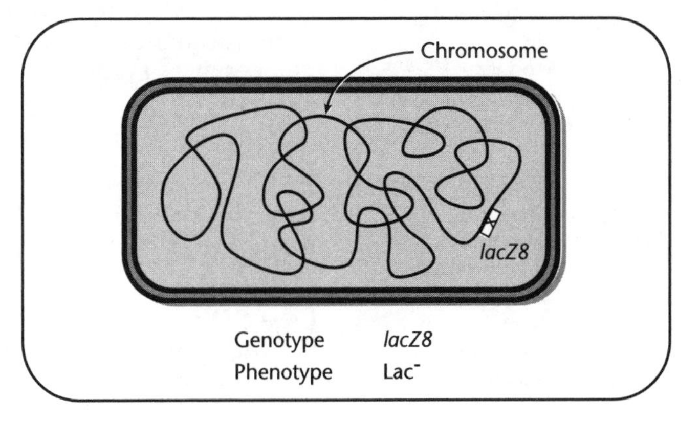

Fig. 3.2 Genotype and phenotype. The genotype refers to changes in the DNA base sequence of a cell. The phenotype refers to the physical characteristics of the cell that result from the change(s).

Following the three-letter code and the capital letter will be an allele number. The **allele number** is the name given to the specific mutation in that copy of the gene. Before the advent of DNA sequencing and site-directed mutagenesis, allele numbers were random numbers or sometimes gave information on how the mutation was induced. For example, lac_{U169} is a mutation in *lac* that was induced by ultraviolet light.

FYI 3.1

Symbols used in writing genotypes for *E. coli*

Δ = deletion, whatever follows the delta is missing from the cell

Δ*lac*$_{U118}$ deletion of *lac* allele # U118

Δ(*argF-lac*)$_{U169}$ deletion of the DNA from *argF* through *lac* allele # U169

:: = insertion, whatever follows the four dots is inserted into what preceeds them

lacZ$_{42}$::*tet* the *tet* gene has been inserted into the *lacZ* gene and the allele # for this insertion is 42 (the four dots are not italicized when writing this genotype)

lacZ$_{2235}$::Tn*10* the transposon Tn*10* is inserted into the *lacZ* gene (note in Tn*10*, only the *10* is italicized in the genotype)

Am, Oc, Op = amber, ochre or opal

These are specific types of mutations where a codon is changed so that the protein is truncated. Am = amber; Oc = ochre; Op = opal

lacZ$_{54(Am)}$ *lacZ* amber allele # 54

Ts, Cs = temperature sensitive or cold sensitive

These are specific characteristics of the mutation. Temperature sensitive means the mutation does not allow the protein to function at high temperatures and cold sensitive means that the mutation does not allow the protein to function at low temperatures.

lacZ$_{31(Ts)}$ *lacZ* temperature-sensitive mutation allele # 31

Now that it is expected that the DNA sequence of a mutated gene will be determined, allele numbers frequently give information on the exact change in the DNA or gene product. For example, a mutation labeled *lacZ*$_{A10V}$ indicates that the base pairs encoding amino acid 10 of the *lacZ* protein that used to encode alanine (A) now encode valine (V).

The phenotypes of a strain are directly dependent on the genotype of that strain. If a mutation in the DNA results in a physical change to the cell, this is described by the phenotype. For example, *lac*$_{U169}$ renders the cell unable to use lactose as a sole carbon source. The phenotype of a cell that carries *lac*$_{U169}$ is Lac$^-$. Many different mutations in the *lac* genes will result in a Lac$^-$ phenotype. Phenotypes are listed using the three-letter code. If the exact location of the mutation is known, the three-letter code plus the capital letter designation for the gene is used. Instead of using lower-case italicized letters, the first letter of the phenotype is capitalized and no part of the phenotype is italicized. The combination of the genotype and phenotype gives an accurate description of mutations in the DNA of a strain and the physical consequences of those mutations.

Classes of mutations

Mutations are classified as either having taken place at a single nucleotide or at multiple nucleotides. Mutations that affect a single nucleotide are called **microlesions**. Mutations that affect multiple nucleotides are known as **macrolesions**.

An example of a microlesion is a **point mutation**. Point mutations are a change in a single nucleotide base pair. An example of a point mutation is a **base substitution mutation**. If the point mutation results in a base substitution of a purine for a purine (A for a G or G for an A) or a pyrimidine for a pyrimidine (T for a C or C for a T), then it is known as a **transition** (Fig. 3.3a). If the point mutation results in a base substitution of a purine for a pyrimidine or a pyrimidine for a purine, then it is known as a **transversion** (Fig. 3.3a).

Another type of microlesion is a **frameshift mutation**. In a frameshift, the insertion or deletion of a single base pair within a gene will alter the reading frame of the gene and change the amino acid sequence that is translated from that gene (Fig. 3.3b). The translated polypeptide will contain the correct amino acid sequence up to the point of the frameshift mutation. After the frameshift mutation, the amino acid sequence will be altered because the codon reading frame has been altered. After the frameshift, translation will continue until a stop codon in the new frame is encountered. If a frameshift results from the insertion of a single base pair, it is known as a +1 frameshift. If a frameshift results from the deletion of a single base pair, then it is known as a −1 frameshift. Some frameshift mutations can also be classified as macrolesions. These would include insertion or deletion of many bases.

Other examples of macrolesions include deletions, duplications, insertions, and rearrangements such as inversions and translocations (Fig. 3.4). All involve major changes in the nucleotide sequence. When a macrolesion is described as a **deletion**, nucleotides are missing from a segment of DNA (Fig. 3.4a). The number of missing nucleotides can vary from a few to many kilobases. In a **duplication**, a number of nucleotides are directly repeated. For example, if a sequence contains AGATGGA, a duplication would be AGATGGA–AGATGGAAGATGGA (Fig. 3.4b). **Insertions** result from the addition of novel base pairs (Fig. 3.4c). In **rearrangements**, all of the base pairs are present but their order has been changed. In an **inversion**, a number of

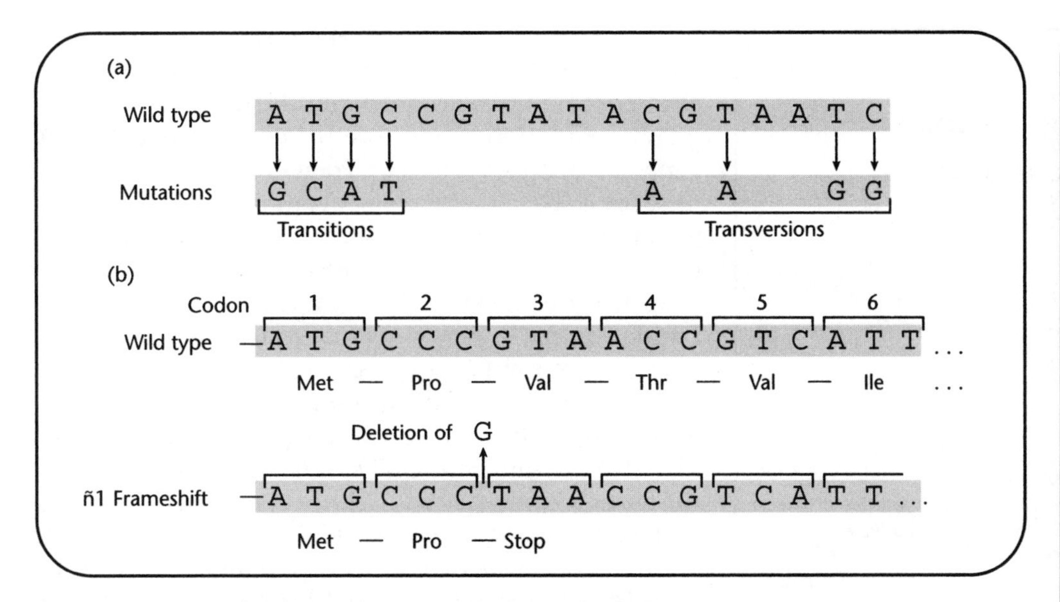

Fig. 3.3 Microlesions. (a) Transition mutations substitute a purine for a purine or a pyrimidine for a pyrimidine. Transversion mutations substitute a purine for a pyrimidine or a pyrimidine for a purine. (b) Frameshift mutations alter the reading frame and change the amino acid sequence of the encoded protein.

FYI 3.2

Writing phenotypes and genotypes in yeast

The genotypes of different experimental organisms are not always written in the same fashion. Scientists who work on a given organism decide the best way to express the genotype. For example, the bakers yeast, *Saccharomyces cerevisiae*, is normally diploid. The convention for writing *S. cerevisiae* genotypes lists both copies of gene and indicates which copy is dominant.

$LEU2^+$	*ade-6*	HIS^+
leu2-1	ADE^+	*his-45*

The genes listed above the line came from one parent and the genes listed below the line came from the other parent. The gene listed in capital letters is the dominant allele. For example, $LEU2^+$ is dominant to *leu2-1*. The number after the dash is the allele number.

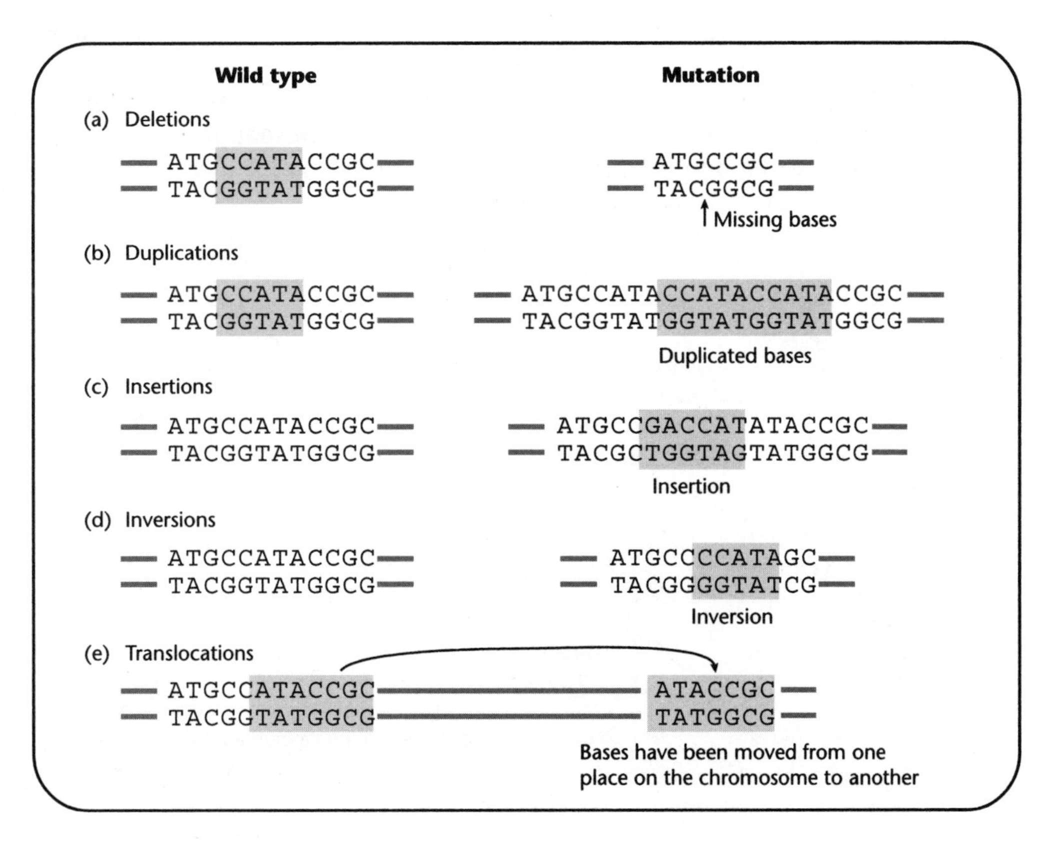

Fig. 3.4 Macrolesions include deletions, duplications, insertions, inversions, and translocations.

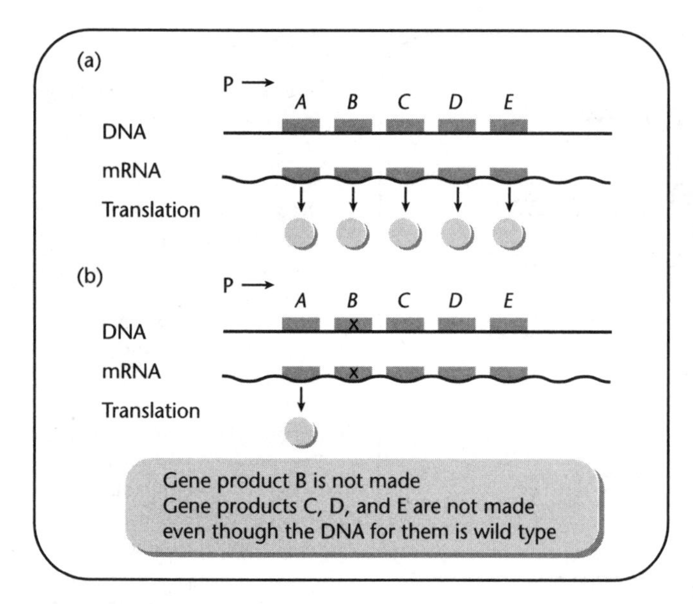

Fig. 3.5 Polar mutations affect downstream genes. (a) In wild type, there are no mutations in the DNA and all of the genes are translated into protein. (b) A mutation in *gene B* results in the failure to make gene product B. Because one promoter controls transcription of *genes A, B, C, D,* and *E,* a mutation in *gene B* blocks the transcription of *genes C, D,* and *E.* The gene products encoded by *C, D,* and *E* are not made even though the DNA sequence for *genes C, D,* and *E* is wild type.

base pairs have been flipped (Fig. 3.4d). In a **translocation**, the base pairs are moved to another place in the chromosome or to another molecule (Fig. 3.4e).

Some mutations are **polar** or have an effect on the genes downstream (Fig. 3.5). Polar mutations are unique to bacteria. Bacteria frequently contain operons or a single promoter followed by several genes. If a mutation occurs in an upstream gene, it can affect the expression of the downstream genes. For example, if the gene order is Promoter, *gene A, gene B, gene C, gene D,* an insertion in *gene A* can prevent the expression of *genes B, C,* and *D.*

Some mutations have an additional phenotype that is very useful; namely they cause the gene product to be sensitive to temperature (Fig. 3.6). The gene product or the encoded protein can be either **temperature sensitive** (Ts) or **cold sensitive** (Cs). If a gene product is temperature sensitive, it functions to some extent at the **permissive temperature** but is nonfunctional at the higher temperature. If a gene product is cold sensitive, it functions at the permissive temperature but is nonfunctional at a lower temperature. For *E. coli*, the permissive temperature is usually in the 25–30°C range, the Ts temperature is usually 37–43°C, and the Cs temperature is usually 15–25°C. The Ts and Cs phenotypes are usually the result of rare point mutations.

Point mutations and their consequences

Not all mutations have disastrous effects. In fact, many are quite innocuous. For example, a point mutation in a gene may not significantly alter its encoded protein and, thus, the cell would feel no effect. This type of point mutation is silent. Silent mutations can result from a base pair substitution without a corresponding amino acid change. The failure of a mutation in a gene to result in a change in amino acid sequence is because of the redundancy of the genetic code (Fig. 3.7). An example of a

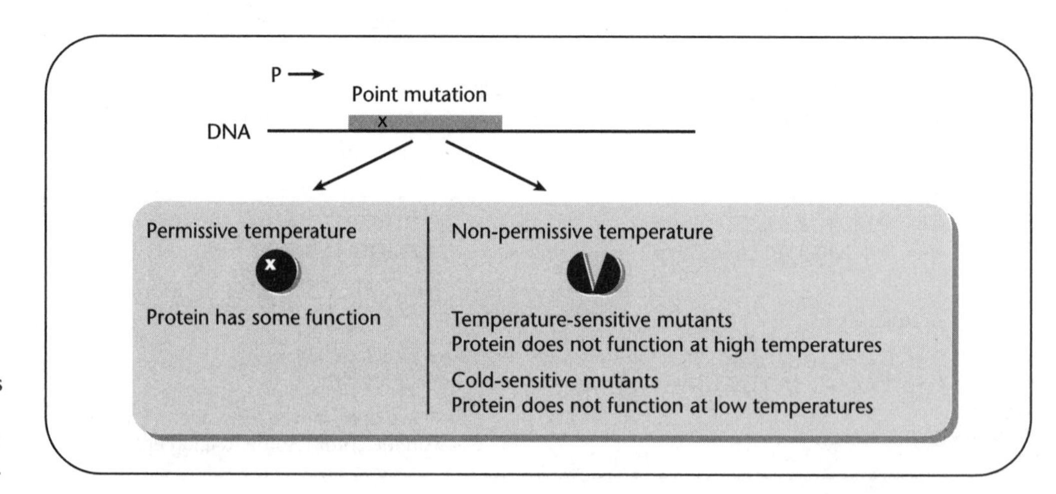

Fig. 3.6 Conditional mutations function at the permissive temperature but are defective at either high or low temperatures.

silent mutation or same sense mutation is described in Fig. 3.8a. The codon for leucine can be either UUA or UUG in the mRNA or TTA/TTG in the DNA. Thus, a base substitution in the nucleotide of the third position of TTA that resulted in a change from A to G (a transition) would be silent because this codon still encodes leucine.

Alternatively, a point mutation in a gene may significantly alter the amino acid sequence of a protein. One example is a missense mutation. A missense mutation occurs when there has been a base pair substitution with a corresponding amino acid change (Fig. 3.8b). The codon UGG in the mRNA or TTG in the DNA encodes tryptophan. A mutation in the first base of this codon, substituting a T for a G now results in a codon coding for glycine instead of tryptophan. Some missense mutations alter protein function severely whereas others do not. It depends on the location of the missense mutation and if the original and new amino acid residues have similar properties (see Fig. 1.2). If the missense mutation occurs in a domain of a protein responsible for catalytic function, and this base substitution resulted in the change of an acidic residue for a non-charged residue, then it is likely that the function of the protein will be altered. Neutral mutations result from an amino acid replacement that does not noticeably alter protein function. Sometimes a missense mutation gives rise to a protein that can still partially function. Mutations that result in partial function are called leaky mutations.

First letter	Second letter: U	C	A	G	Third letter
U	UUU Phe	UCU Ser	UAU Tyr	UGU Cys	U
	UUC Phe	UCC Ser	UAC Tyr	UGC Cys	C
	UUA Leu	UCA Ser	UAA Stop	UGA Stop	A
	UUG Leu	UCG Ser	UAG Stop	UGG Trp	G
C	CUU Leu	CCU Pro	CAU His	CGU Arg	U
	CUC Leu	CCC Pro	CAC His	CGC Arg	C
	CUA Leu	CCA Pro	CAA Gln	CGA Arg	A
	CUG Leu	CCG Pro	CAG Gln	CGG Arg	G
A	AUU Ile	ACU Thr	AAU Asn	AGU Ser	U
	AUC Ile	ACC Thr	AAC Asn	AGC Ser	C
	AUA Ile	ACA Thr	AAA Lys	AGA Arg	A
	AUG Met	ACG Thr	AAG Lys	AGG Arg	G
G	GUU Val	GCU Ala	GAU Asp	GGU Gly	U
	GUC Val	GCC Ala	GAC Asp	GGC Gly	C
	GUA Val	GCA Ala	GAA Glu	GGA Gly	A
	GUG Val	GCG Ala	GAG Glu	GGG Gly	G

Start codons
AUG Met (most frequently used)
GUG Val (occasionally used)

Stop codons
UAA Ochre
UAG Amber
UGA Opal

Fig. 3.7 The genetic code. The genetic code is redundant, meaning that for some amino acids, more than one combination of three bases can encode the same amino acid. For example, UUU and UUC both code for phenylalanine (Phe). The most frequently used codon for starting translation of mRNA into protein is AUG, encoding Met. An occasionally used start codon is GUG, encoding Val. The three translational stop codons are UAA (ochre), UAG (amber), and UGA (opal).

Nonsense mutations are base pair substitutions that change a codon into one of three possible chain termination triplets. The chain terminating codons are UAG (amber), UAA (ochre), or UGA (opal). If one of the chain termination codons appears as a result of a point mutation in the middle of a gene then the translation of this gene will terminate prematurely (Fig. 3.8c). When the translation machinery encounters the inappropriately placed nonsense codon, it will stop. The rest of the gene will not be translated. The effect on the function of the protein is determined by where the nonsense mutation occurred. If the nonsense mutation occurs early in the gene sequence, then the protein fragment will most likely not have any function. If the nonsense mutation occurs late in the gene, then the truncated protein may have no activity, some activity, or full activity. Frequently, the truncated protein will be recognized by the cell as abnormal and will be degraded.

FYI 3.3

Mutation vs. mutant

Mutation describes the physical change in the DNA and mutant describes a cell that has a physical change in its DNA. Thus, cells can be mutants but not mutations and DNA contains mutations but not mutants.

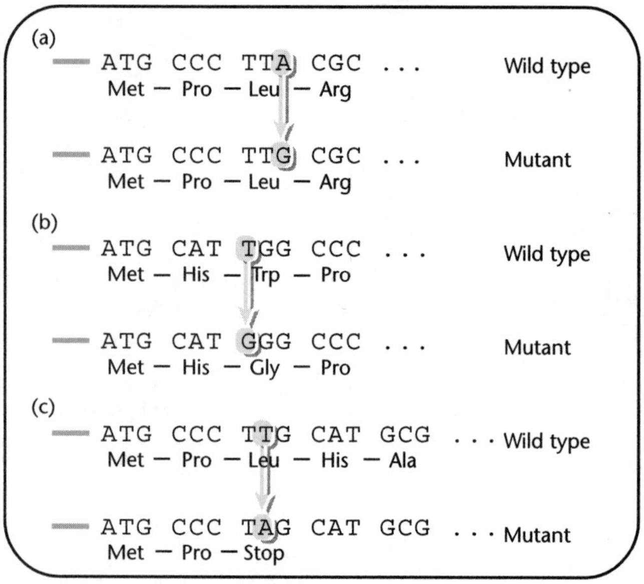

(a)
ATG CCC TTA CGC ... Wild type
Met — Pro — Leu — Arg

ATG CCC TTG CGC ... Mutant
Met — Pro — Leu — Arg

(b)
ATG CAT TGG CCC ... Wild type
Met — His — Trp — Pro

ATG CAT GGG CCC ... Mutant
Met — His — Gly — Pro

(c)
ATG CCC TTG CAT GCG ... Wild type
Met — Pro — Leu — His — Ala

ATG CCC TAG CAT GCG ... Mutant
Met — Pro — Stop

Fig. 3.8 (a) Silent mutations do not change the amino acids in the protein product. Example: both TTA and TTG code for leucine, thus a mutation that results in an A to G change has no effect. (b) Missense mutations alter the amino acid in the protein product. Example: a T to G change results in a change of the genetic code, from tryptophan to glycine. Function of a protein is more likely to be disrupted if the missense mutation results in amino acid changes of different classes (i.e. a basic amino acid for an acidic amino acid, etc.). (c) Nonsense mutations change a codon to a chain termination or stop codon.

Measuring mutations: rate and frequency

It is important to measure how often mutations occur. Two different measurements can be made. The most accurate measurement is the **mutation rate**. The mutation rate is the probability that a given gene will sustain a mutation in one generation. Mutation rates are expressed per cell per generation. For example, resistance to the antibiotic streptomycin occurs at approximately 5×10^{-9} per cell per generation.

Mutations can occur at any time in the growth of a population of cells. If the mutation occurs very early in the growth of the population, when the population of cells reaches stationary phase, many cells will be descendants of the original mutant. Because these cells are related, they are known as siblings. Having many siblings in one population is known as a **jackpot**. Mutation rates are measured using the fluctuation test so that jackpot events do not skew the numbers.

Mutation frequencies are a more inaccurate measure of how often a mutation occurs. For the frequency, the total number of cells in a population and the total number of mutants in the population are measured. The frequency is given by the number of mutants divided by the total number of cells. A jackpot event will dramatically affect the frequency. Measuring the frequency is very easy to do and so it is often used as a crude approximation. Mutation rates are the most accurate and reproducible measurement of how often a mutation occurs.

Spontaneous and induced mutations

Mutations can occur spontaneously with no externally applied agents. Mutations can also be induced by the application of chemical or physical agents to DNA or to intact cells. The chemical or physical agents are known as **mutagens**. Mutagens cause changes in the nucleotide sequence of the DNA molecule (Table 3.1).

Spontaneous mutations can arise as a result of DNA polymerase III incorporation errors during replication or repair, tautomeric shifts in the bases, low level radiation, spontaneous alterations in the bases (depurination or deamination), recombination or transposition. Most often spontaneous mutations result in a microlesion in the form of a mismatched base pair.

Errors during DNA replication: incorporation errors

Incorporation errors are the mispairing of bases during DNA replication or during DNA repair processes (see Chapter 4). For example, uracil can be misincorporated by DNA polymerase III during DNA replication resulting in U–A instead of T–A pairings. DNA polymerase III incorporates an incorrect base approximately one out of every 10^{10} bases it incorporates, which is an error rate of 10^{-4} base pairs/cell/generation. The editing function of DNA Pol III reduces mistakes so that the misincorporation rate drops to 10^{-6} to 10^{-7}/base pairs/cell/generation. The DNA repair

Table 3.1 Specificity of mutagens.

Base analogues	
5-bromouracil	AT–GC transitions
2-aminopurine	AT–GC transitions
Base modifiers	
Nitrous oxide	CG–AT and AT–GC transitions
Hydroxylamine	CG–AT transitions
Nitrosoguanidine	GC–TA transversions
Intercalating agents	
Acridine dyes	Frameshifts, small insertions, small deletions
ICR191	Frameshifts, small insertions, small deletions
Physical agents	
Ultraviolet light	All base substitutions, frameshifts, deletions, rearrangements
Ionizing radiation	Single- and double-stranded DNA breaks

mechanisms operate directly after replication and remove certain mismatches. These repair mechanisms further reduce the error rate to 10^{-9} to 10^{-10}/base pairs/cell/generation.

Mutations have been isolated in DNA polymerase which make it more efficient, however, these mutations significantly slow the speed of the enzyme. Conversely, mutations in DNA polymerase that increase its speed also increase the misincorporation rate. Wild-type DNA polymerase contains a balance of speed and accuracy.

Errors due to tautomerism

A **tautomer** is a structural isomer of each base. Tautomers form as a result of spontaneous, transient rearrangement of bonds through the shifting of hydrogen atoms in each of the bases. This rearrangement of hydrogen atoms alters the hydrogen bonding properties of the bases. If the rare tautomer is present as the template is being replicated, a wrong base may be incorporated by DNA polymerase III into the daughter strand. Adenine and cytosine usually exist in the amino (NH_2) form, but can exist in the tautomeric imino (NH) form (Fig. 3.9a). When adenine is in the tautomeric imino form, it can pair with cytosine. When cytosine is in the tautomeric imino form, it can pair with adenine.

Another tautomeric shift is observed when the C6 atom of guanine or the C4 atom of thymine shift in their bonding properties from a keto (C=O) to the rare enol (C–OH) form (Fig. 3.9b). A thymine in the enol form will pair with guanine. A guanine in the enol form will pair with thymine. A transition rather than a transversion is the most common base pair substitution as a result of tautomerism.

Spontaneous alteration by depurination

Depurination describes the removal of the purine from the 2-deoxyribose–phosphate backbone. In order to depurinate, the N, N-glycosidic bond holding the purine to the 2-deoxyribose has to be broken (Fig. 3.10). This can be accomplished naturally if environmental temperatures increase or the pH decreases. Hydrolysis of the N,N-glycosydic bond between a base and its 2-deoxyribose in a DNA molecule results in an

FYI 3.4

Hotspots

Mutations do not occur at the same rate for each base pair. Some bases are more likely to suffer a mutation than others. These regions of high mutability are called hotspots. It is not always clear why a hotspot is hypermutable. In some cases where microhomology is used to create insertions or deletions, small repeated DNA sequences may mark the hotspot.

FYI I3.5

Fluctuation tests

One use of fluctuation tests is to measure mutation rates. A culture of cells is diluted many times and used to inoculate many small cultures (up to 50–100 independent cultures). The cultures are incubated and the cells are allowed to divide many times. The number of bacteria mutated for a specific phenotype in each culture is determined. Some cultures will have no mutants, some will have one mutant, some two mutants, and so on up to those cultures which have a jackpot. The average number of mutations per cell division is calculated from the number of cultures with no mutants. From the Poisson distribution, the fraction of cultures with no mutations is:

$a = (-\ln p_0)/N$

a = mutation rate
p_0 = fraction of cultures in which no mutations occurred
N = number of cells in the culture

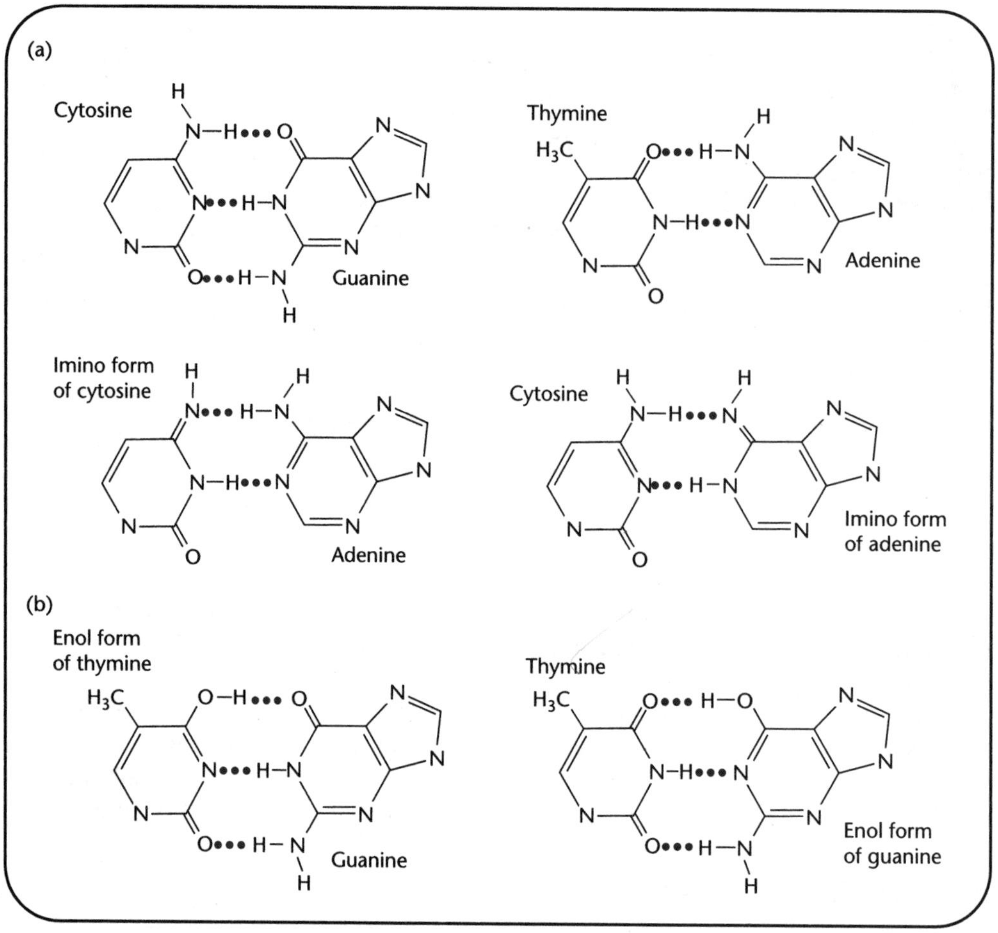

Fig. 3.9 Tautomeric shifts. Tautomeric shifts can change the hydrogen bonding properties of bases. (a) Adenine and cytosine are normally found in the amino form. Spontaneous rearrangement of bonds results in the imino form. Imino–cytosine base pairs with adenine and imino–adenine base pairs with cystosine. (b) Guanine and thymine are normally found in the keto form. Spontaneous rearrangement of bonds results in the enol form. The enol–guanine base pairs with thymine and the enol–thymine base pairs with guanine.

apurinic site or less frequently an **apyrimidinic site**. Spontaneous cleavage is usually rare and occurs at approximately one base/300/day/37°C. The rate of depurination can be increased by exposure to alkylating agents such as ethylmethanesulfonate (EMS), methylmethane sulfonate (MMS), mustard gases, and epoxides. These agents transfer an alkyl group to guanine, often weakening the N,N-glycosidic bond holding guanine to the 2-deoxyribose. Replication past an apurinic or apyrimidinic site can be mutagenic because DNA polymerase will not know what base is missing.

Spontaneous alteration by deamination

Cytosine, adenine, and guanine contain exocyclic amino groups (NH_2) that can be lost spontaneously or when exposed to base modifiers such as nitrous oxide. The loss of NH_2 groups is called **deamination** (Fig. 3.11). Deaminating C, A, or G changes their hydrogen bonding properties, creating a potential for mismatched base pairs.

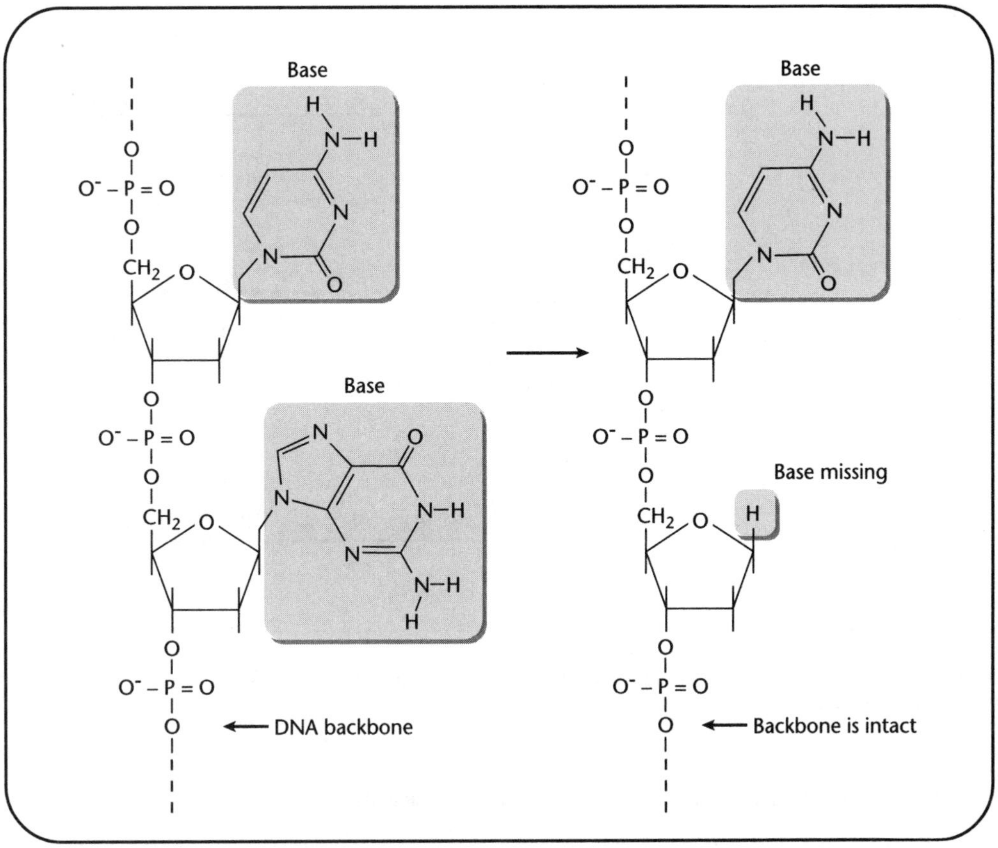

Fig. 3.10 Depurination is the spontaneous loss of a purine through the hydrolysis of the N,N-glycosidic bond between the base and the deoxyribose sugar. Depurination can occur naturally or upon exposure to alkylating agents.

Thymine is not affected because thymine does not contain an exocyclic amino group. Uracil is the product when cytosine is deaminated. Deamination of cytosine, in addition to DNA Pol III incorporation errors, explains the occurrence of uracil in a DNA molecule. Uracil pairs with adenine creating a C–G to A–T transition. Deamination of adenine to hypoxanthine results in an A–T to C–G transition because hypoxanthine pairs with cytosine instead of thymine. Deamination of guanine results in xanthine, which is unable to stably pair with either cytosine or thymine.

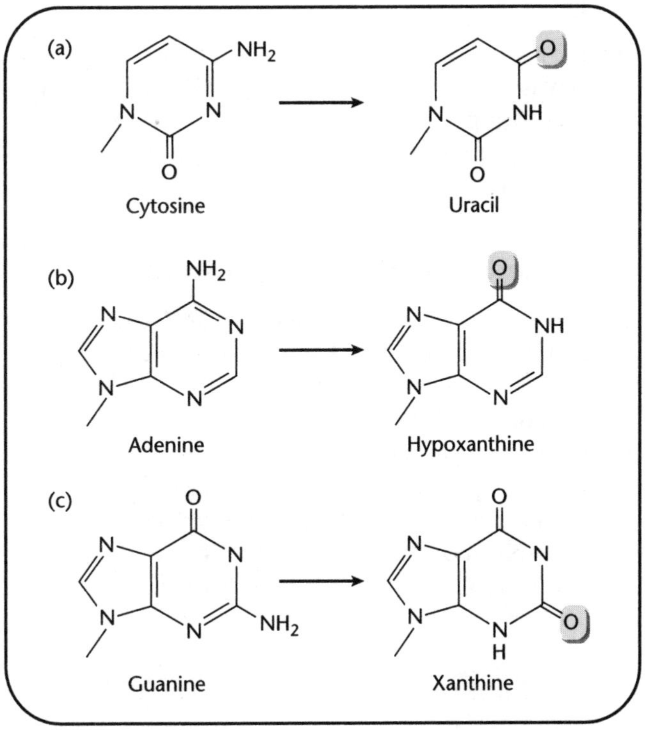

Fig. 3.11 Deamination. (a) Deamination of cytosine to uracil can occur either spontaneously or upon exposure to nitrous oxide (HNO_2). Uracil can base pair with adenine instead of guanine resulting in C–G to A–T transitions. (b) Deamination of adenine to hypoxanthine occurs upon exposure to HNO_2. Hypoxanthine base pairs with cytosine instead of thymine resulting in A–T to C–G transitions. (c) Deamination of guanine to xanthine changes the hydrogen bonding properties such that xanthine is unable to stably base pair with cytosine.

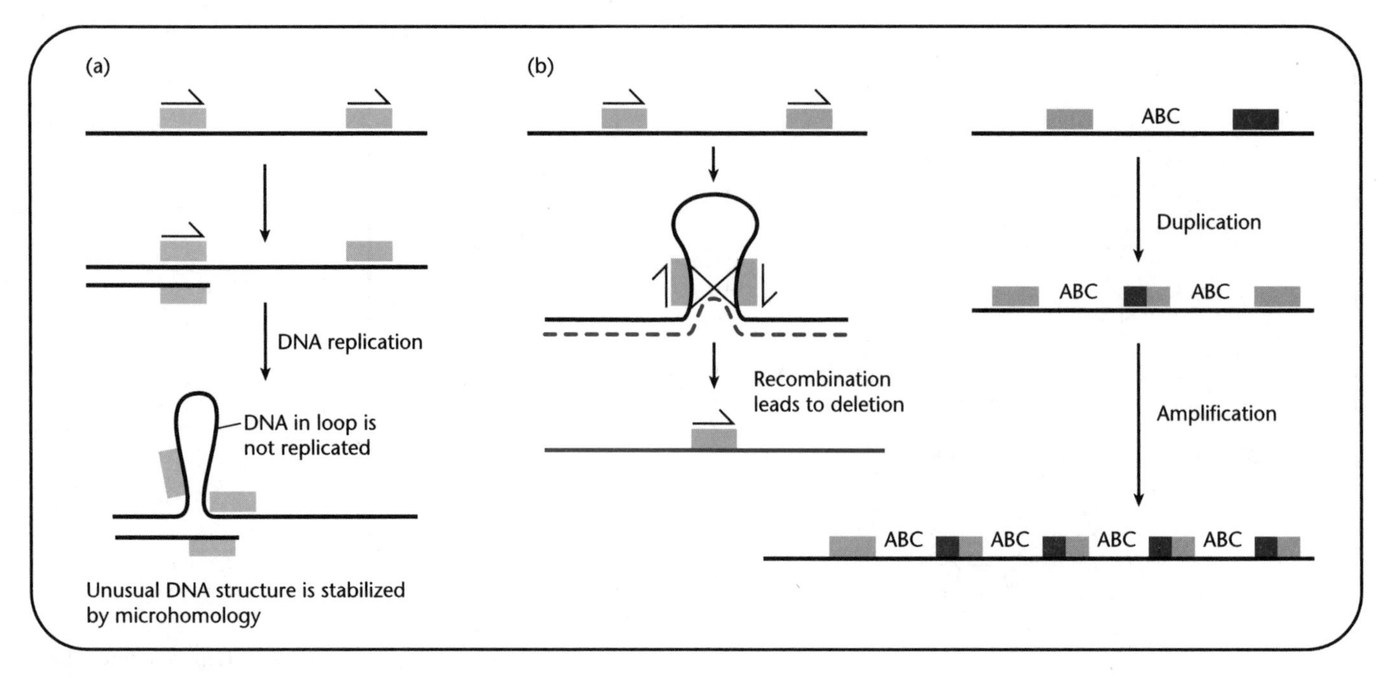

Fig. 3.12 Microhomology can lead to deletions, duplications, or amplifications. (a) Stabilization of slipped DNA strands by microhomology can lead to deletions. (b) Low-level recombination at microhomologies can result in rearrangements.

Alteration by spontaneous genetic rearrangement

Deletions, duplications, and amplifications can occur spontaneously in DNA. In many cases, small regions of homology are found at the endpoints of the rearrangements. The homology can be as little as 4 or 5 base pairs (bp) and is termed **microhomology**. Several mechanisms have been proposed to account for these rearrangements (Fig. 3.12). DNA strands can slip relative to each other. The microhomology is able to stabilize the slipped DNA. DNA replication through the slipped DNA would result in a deletion (Fig. 3.12a). Alternatively, low-level recombination at the microhomologies by known recombination systems can result in rearrangements (Fig. 3.12b).

Alterations caused by transposition

Transposons are mobile pieces of DNA that move themselves around the chromosome (Chapter 6). Every time a transposon moves it creates an insertion mutation (Fig. 3.13). If a transposon moves into a gene it will disrupt the gene and usually lead to a loss of gene function. If the gene is in an **operon**, frequently the transposon insertion will be **polar** on the downstream genes. This is because the transposon contains transcriptional termination signals that tell the transcribing RNA polymerase to stop, resulting in adjacent downstream genes not being expressed. When a transposon comes out of the chromosome to move to another place, several alterations can result. These alterations can include deletions, duplications, and rearrangements.

Induced mutations

Mutation rates can be increased by chemical or physical mutagens that act to induce

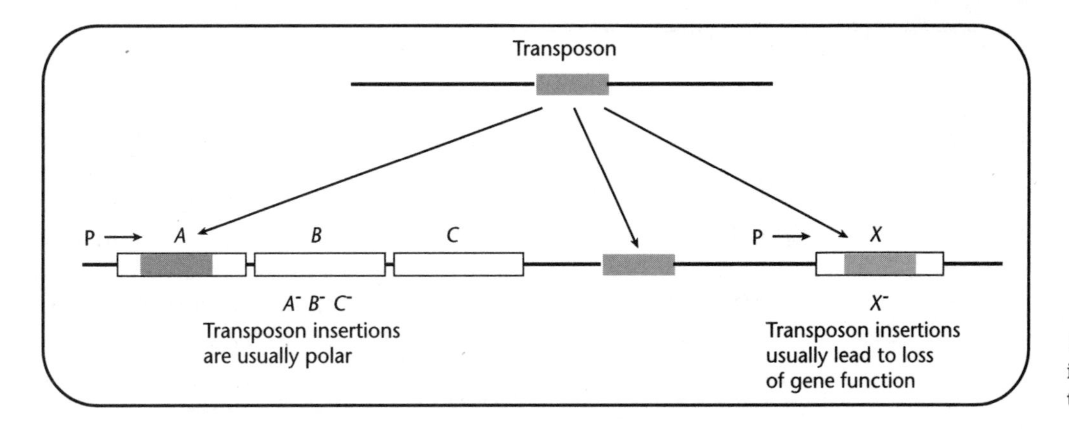

Fig. 3.13 Transposons create insertional mutations every time the transposon moves.

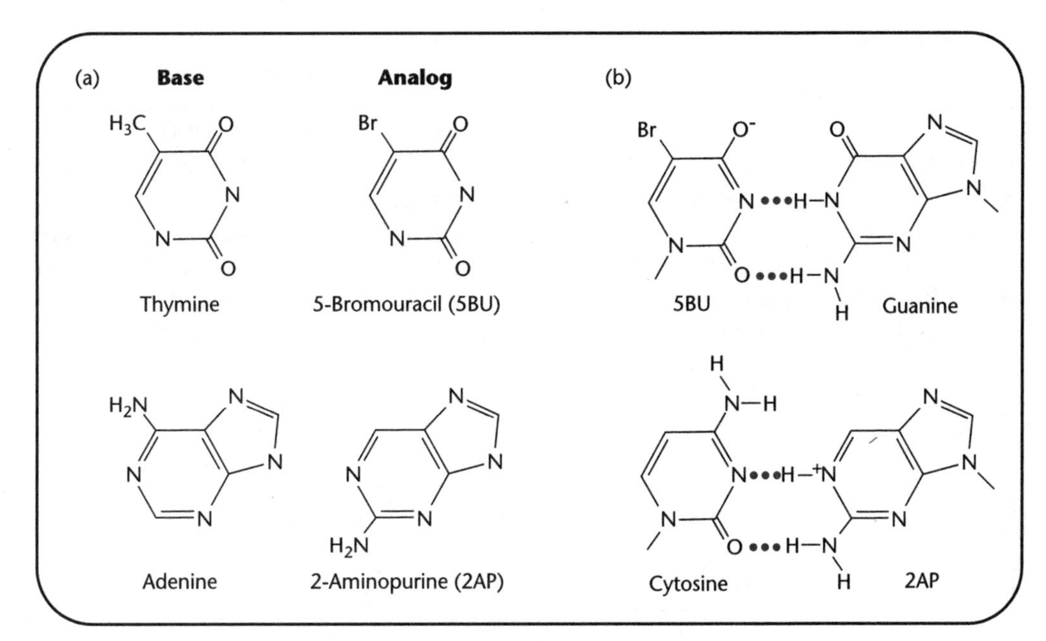

Fig. 3.14 Base analogs. (a) The analog for thymine is 5-bromouracil (5-BU) and the analog for adenine is 2-aminopurine (2-AP). (b) Base analogs mimic normal bases but can occasionally mispair. DNA Pol III incorporates 5-BU as if it were thymine and incorporates 2-AP as if it were adenine into a growing DNA strand. 5-BU normally pairs with adenine, but occasionally mispairs with guanine. 2-AP normally pairs with thymine but occasionally mispairs with cytosine.

nucleotide changes in DNA molecules. Mutagens can be divided into four categories: chemicals that mimic normal bases (base analogs); chemicals that react with bases (base modifiers); chemicals that bind DNA (intercalators); and agents that physically damage DNA (Table 3.1).

Chemicals that mimic normal DNA bases: base analogs

Base analogs such as 5-bromouracil (5-BU) and 2-aminopurine (2-AP) can be incorporated into a growing DNA strand by DNA polymerase III because these analogs are structurally related to bases (Fig. 3.14). 5-BU is incorporated in a similar manner as thymine yet occasionally pairs with guanine. 2-AP is incorporated in a similar manner as adenine yet occasionally pairs with cytosine. Both result in A–T to G–C transitions and occasionally G–C to T–A transversions.

Chemicals that react with DNA bases: base modifiers

Unlike base analogs which mimic bases, **base modifiers** react directly with the nucleotide bases, changing the bases' chemical structure. Nitrous oxide oxidatively

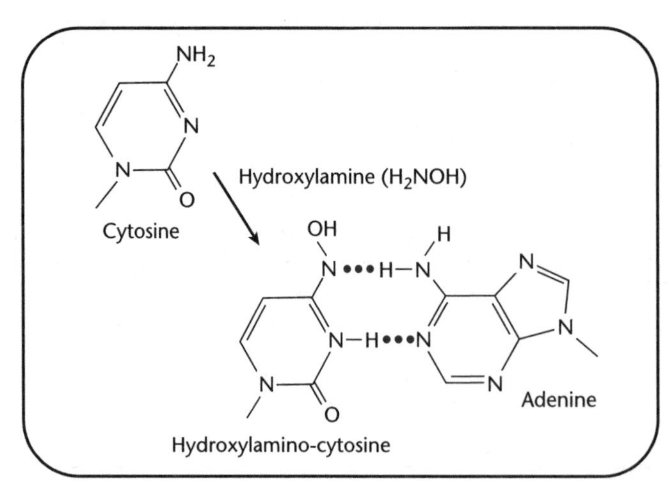

Fig. 3.15 Hydroxylamine, a base modifier, reacts primarily with cytosine. The HA-altered cytosine pairs with adenine instead of guanine.

deaminates cytosine, adenine, and guanine, which contain exocyclic amino groups (NH_2) (see above). Hydroxylamine (HA) reacts primarily with cytosine through a reduction reaction that adds a hydroxyl group (Fig. 3.15). The HA-altered cytosine pairs with A instead of G. Thus, HA induces only C–G to A–T transitions.

A common alkylating agent used in the laboratory to generate mutations is nitrosoguanidine (*N*-methyl-*N′*-nitro-*N*-nitrosoguanidine, NTG or MNNG). NTG primarily reacts with guanine, attaching an alkyl group (i.e. methyl or ethyl group, CH_3-, CH_3CH_2-) to the oxygen bonded to the sixth carbon of guanine (O6-methylguanine) (Fig. 3.16). This changes guanine's hydrogen bonding properties so that guanine pairs with thymine to give G–C to T–A transversions. NTG also adds alkyl groups to the oxygen bonded to the fourth carbon of thymine (O4-methylthymine), which changes its hydrogen bonding properties so that it pairs with guanine (Fig. 3.16). Other alkylating agents include nitrogen mustard gas, methyl-

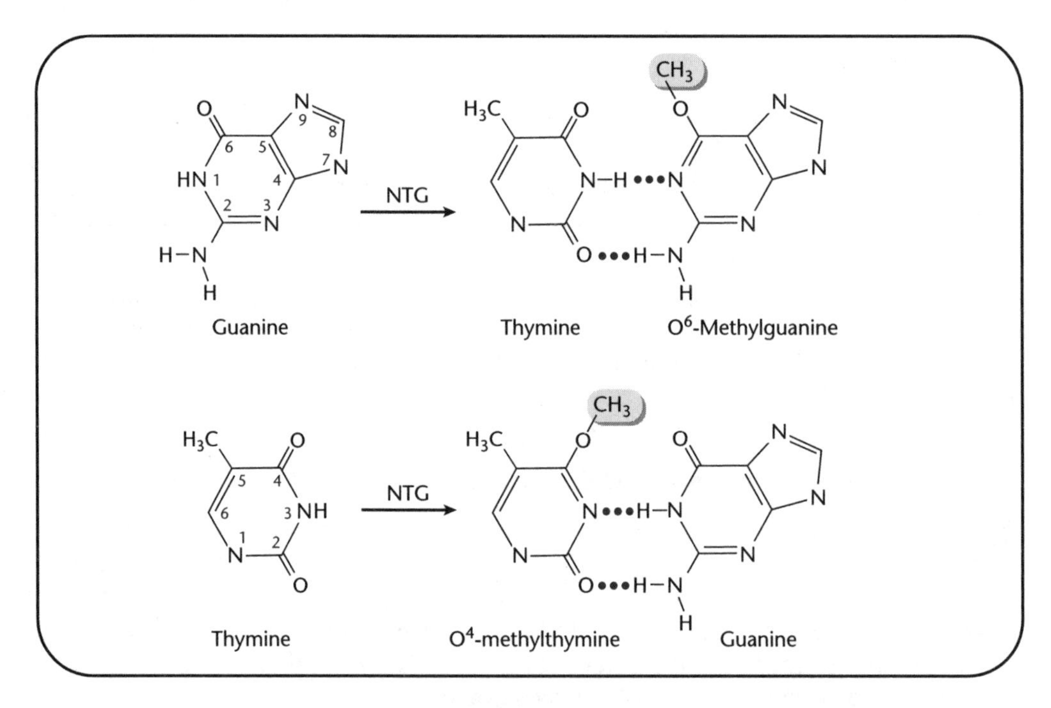

Fig. 3.16 Alkylation of guanine (primarily) and thymine (occasionally) by nitrosoguanidine (NTG) attaches an alkyl group (methyl, CH3-) to the oxygen bonded to the sixth carbon of guanine (O6-methylguanine) and to the fourth carbon of thymine (O4-methylthymine). As a result, the hydrogen bonding properties change. Methylated guanine base pairs with thymine instead of cytosine. Methylated thymine base pairs with guanine instead of adenine.

methane sulfonate (MMS), and ethyl methanesulfonate (EMS). EMS and MMS alkylate the vulnerable N7 of guanine (N7-methylguanine) and N3 of adenine (N3-methyladenine), causing distortions in the DNA molecule. This affects the hydrogen bonding properties of guanine and adenine. Alkylating agents can also add alkyl groups to the phosphates comprising the backbone of the DNA molecule.

Chemicals that bind DNA bases: intercalators

Flat aromatic molecules, such as acridine dyes and acridine-like derivatives (proflavin, ICR191, ethidium bromide) have the same approximate dimensions as a normal base. Because of the similar dimensions, these chemicals can slide between two adjacent base pairs in the double helix. This sliding action is called **intercalating**. An intercalated DNA molecule, when used as a template for replication, will synthesize daughter DNA molecules that have had either one or two base pairs added or deleted. This is because when DNA polymerase III encounters intercalated regions in the template DNA molecule, it does not know what to do. Should it or shouldn't it add a base across from the intercalated chemical? As a result of adding extra bases or deleting bases during this erroneous replication process, frameshift mutations are created.

Mutagens that physically damage the DNA: ultraviolet light and ionizing radiation

One group of lesions caused by ultraviolet (UV) light are pyrimidine dimers. This type of lesion results from the covalent linking of two adjacent pyrimidines on the same strand. This damage distorts the structure of the double helix, weakens or disrupts hydrogen bonding with bases on the complementary strand and impairs the movement of the replication machinery through the UV damaged bases. Two types of photoproducts form upon exposure to UV. The most common photoproduct is cyclobutane dipyrimidine (or pyrimidine dimer) (Fig. 3.17a). The most likely paired pyrimidines comprising cyclobutane dipyrimidines are thymine–thymine (T–T). Pairings exist in a ratio of: T–T (68%), C–T (13%), T–C (16%) and C–C (3%) in DNA irradiated with 254 nm UV light. The other photoproduct, 6-4 pyrimidine–pyrimidine lesions (or 6-4 photoproduct), causes major distortions in the double helical structure of DNA (Fig. 3.17b). The 6-4 photoproducts of C–C and T–C are most frequently observed, while T–T pairings are rarely observed and C–T pairings are never observed. UV also induces base substitutions, frameshifts, deletions, and rearrangements.

Ionizing radiation can cause DNA strand breaks, either in one strand (single-stranded breaks) or in both strands (double-stranded breaks) as well as interstrand covalent crosslinking. DNA strand breakage occurs in the phosphodiester backbone. Single-stranded breaks can be repaired using the other intact DNA strand. Double-stranded breaks cannot be repaired unless there is a second copy of that part of the chromosome in the cell, such as after replication of that part of the chromosome. Double-stranded breaks are usually lethal to the cell.

Interstrand covalent crosslinking means that the hydrogen bonds holding the two DNA strands together in the helix have been replaced with covalent bonds. Because interstrand covalent crosslinking prevents the two strands of the double helix from separating, the movement and action of replication and transcriptional machinery is blocked. Other mutagenic agents cause interstrand covalent crosslinking, including UV radiation, alkylating agents, nitrous acid, and some chemotherapeutic drugs such as cis-platinum.

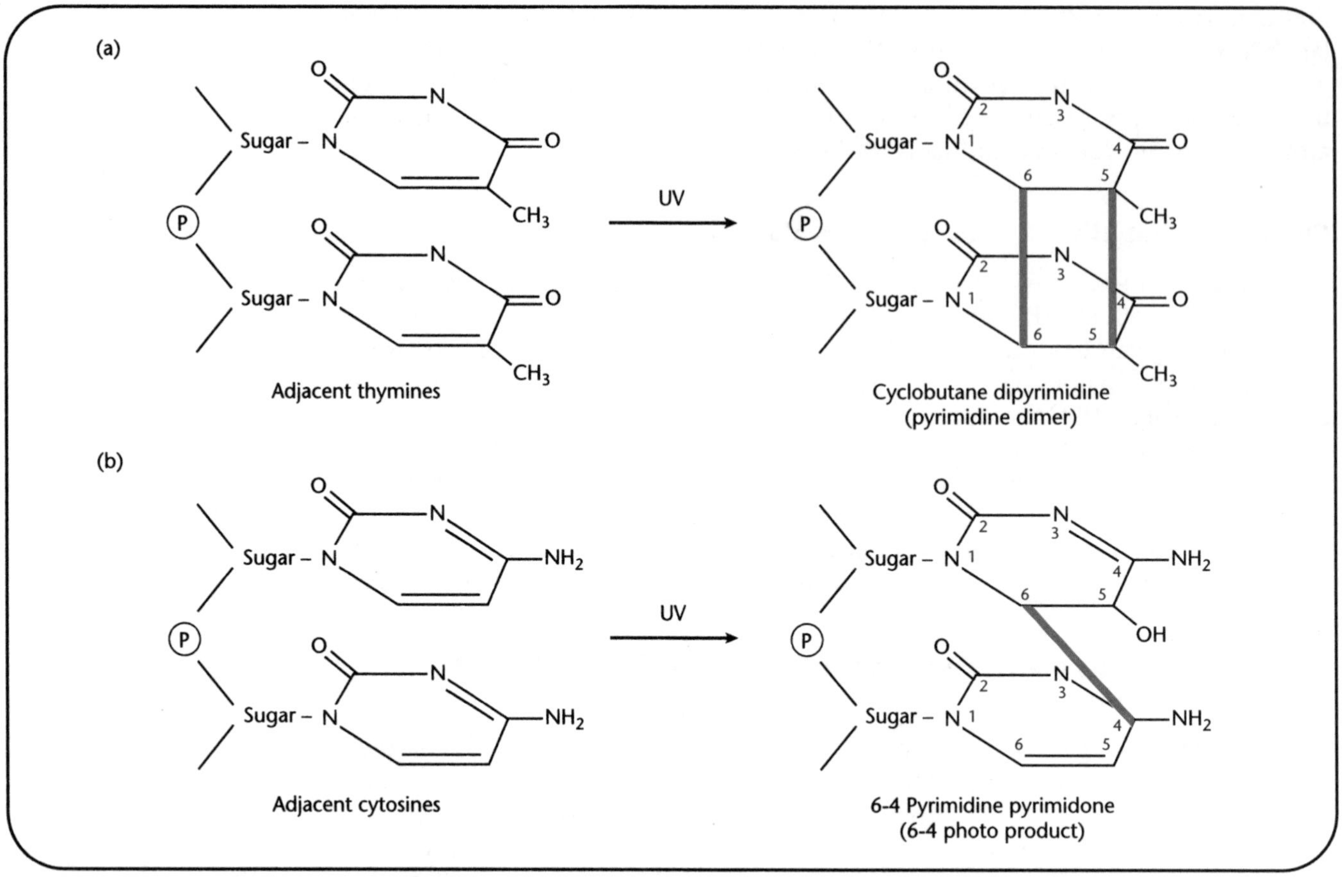

Fig. 3.17 Effect of ultraviolet light on adjacent pyrimidines. (a) Formation of cyclobutane dipyrimidine between two adjacent thymines. (b) Formation of a 6–4 photoproduct between two adjacent cytosines.

Mutator strains

As would be expected, mutations in certain genes will raise the general mutation rate of the cell. For example, a mutation in DNA polymerase III can lead to higher rates of misincorporation of the wrong base. The *mutD* gene encodes the epsilon subunit of DNA polymerase III (see Chapter 2). Certain mutations in *mutD* lead to an increase in all base substitutions and frameshift mutations in the cell. Additionally, cells mutant in mismatch repair systems also have an increase in the general mutation rate. For example, *mutH*, *mutL*, and *mutS* mutants lack the methyl-directed mismatch repair system. Mutations in any one of these genes leads to an increase in transitions and frameshifts. Cells containing mutations in *mutH*, *mutL*, *mutS*, or *mutD* are viable, even though they have high rates of mutagenesis.

Reverting mutations

Reversion of a mutation refers to changing the phenotype of the strain from mutant back to wild type (Fig. 3.18). Reversion requires a change in the DNA sequence to bring about the reversal of the phenotype. If the change in the DNA sequence is a direct reversal of the mutant base pair back to the wild-type base pair, it is known as a **true revertant**. If the revertant contains the original mutation and a second mutation that compensates, it is known as a **pseudorevertant**. The original mutation is known as the **primary mutation** and the second, compensatory mutation is called a **suppressor**. Pseudorevertants can be at another base pair in the same gene or in a completely different gene. For example, if the wild-type codon in a gene is CUU, which encodes leucine, a mutation can change the codon to UUU, which encodes phenylalanine. A pseudorevertant that changes the codon to UUA does not restore the original mutated base pair but because UUA also encodes leucine it will restore function to the gene product.

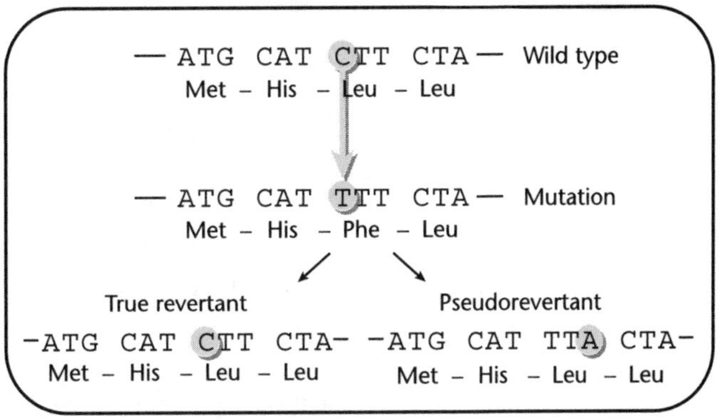

— ATG CAT CTT CTA — Wild type
Met – His – Leu – Leu

— ATG CAT TTT CTA — Mutation
Met – His – Phe – Leu

True revertant
–ATG CAT CTT CTA–
Met – His – Leu – Leu

Pseudorevertant
–ATG CAT TTA CTA–
Met – His – Leu – Leu

Fig. 3.18 Reversion of a mutation back to a wild-type phenotype can be the result of changing the mutant base back to the wild-type base (true reversion) or by changing a second base (pseudorevertant). In the pseudorevertant, the wild-type base pair sequence is not restored but the wild-type phenotype is restored.

Microlesions are easy to revert whereas macrolesions are usually more difficult. This is because it is much easier to restore the wild-type nucleotide sequence if only one base pair is involved. If base pairs are missing, true reversion is not possible. In the case of a deletion, if the phenotype is revertable, then it must be the result of pseudoreversion of the deletion. In this case, the pseudorevertant would have to be at another site in the chromosome.

Suppression

A **suppressor** is defined as a mutation that reverses the phenotype of another mutation but occurs at a location distinct from the first mutation. The reversal of the phenotype may be completely back to wild type or only some partial level of reversal can be seen. One test to distinguish **true revertants** of the primary mutation from suppressors is the genetic separation of the two mutations. Once separated, the suppressor mutation may have its own phenotypes or as is often the case, the suppressor may confer no detectable phenotypes.

Suppressor mutations are of two general types, depending on the location of the suppressor mutation (Fig. 3.19). **Intragenic suppressors**, which map within the same gene as the primary mutation, compensate for the original defect by internally altering the gene product and restoring function. **Extragenic suppressors**, which map outside the gene containing the primary mutation, can alleviate the primary defect in a number of different ways.

Informational suppressors alter the protein synthesis machinery to reverse the phenotypes of the primary mutation (Fig. 3.20). For example, certain tRNAs can be mutated so that a fraction of the time (usually no more than 1–2 percent of the time) they recognize stop codons and insert an amino acid. These suppressor tRNAs can suppress nonsense mutations, regardless of the gene that contains the nonsense mutation. Because suppressor tRNAs are inefficient, they do not lead to the general read through of all stop codons.

Bypass suppressors circumvent the primary mutation by activating a new pathway or altering another protein to perform the function of the mutant protein, thus alleviating the complete need for the mutant gene product. **Overproduction sup-**

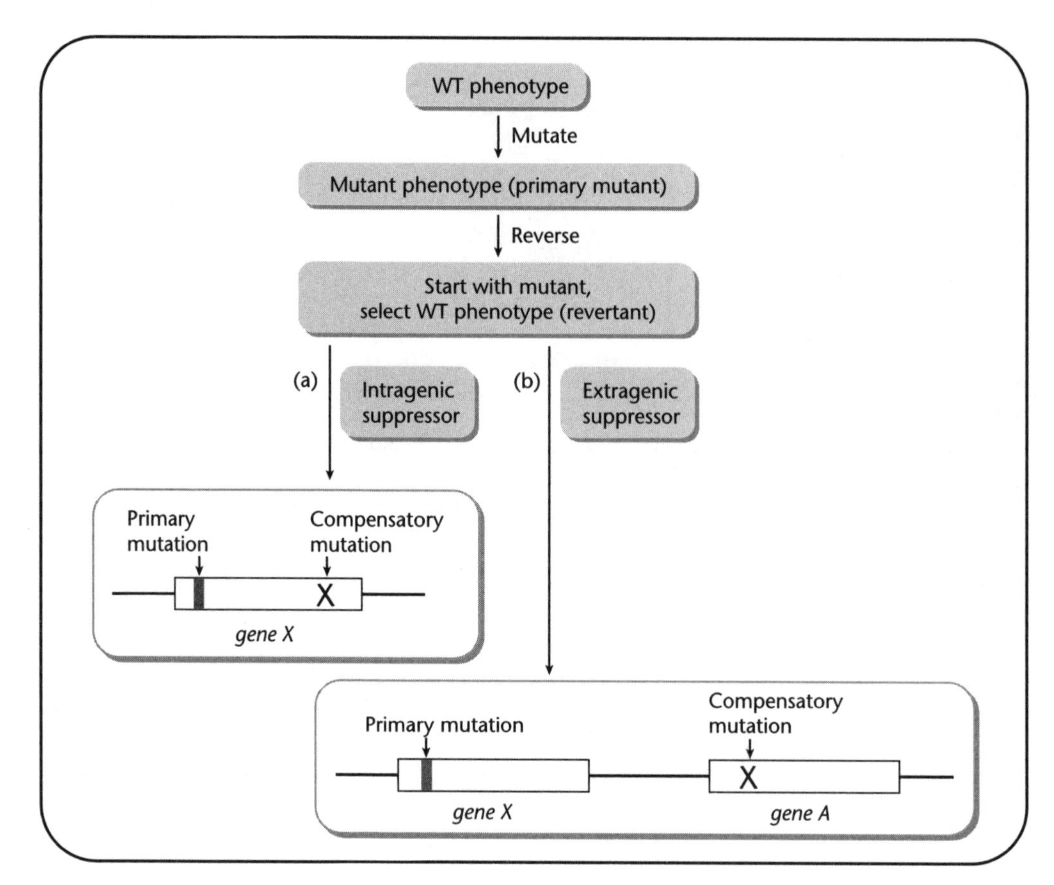

Fig. 3.19 Suppressor mutations are of two general types, depending on where the second mutation is located. (a) If the second mutation is in the same gene as the first, this is called an intragenic suppressor. (b) If the second mutation is outside the gene containing the first mutation, it is called an extragenic suppressor.

pressors increase expression of either the mutant gene product or the rate-limiting component of the affected pathway. **Environment suppressors** reverse the mutant phenotype by providing cytoplasmic conditions that positively affect the structure or function of the mutant gene product. For example, if the primary mutation makes the gene product more sensitive to degradation, removing the enzyme responsible for the degradation can suppress the primary mutation. **Interactive suppressors** restore function to the mutant gene product with a compensating change in a gene product that interacts (Fig. 3.21). For example, if proteins A and B interact, the primary mutation can be in A and prevent the interaction. The suppressor can be in protein B and restore the A–B interaction. Each type of suppressor provides valuable information about the primary mutation and the function of the gene products.

Ames test

Bruce Ames and his collaborators developed a test that allows the mutagenic potential of a particular chemical or physical agent to be determined. This test is very important as a first screen to determine if a particular agent can cause mutations. The Ames test eliminates the need to use animal models to screen for mutagenic potential. Rather, *Salmonella enterica* serovar *typhimurium* that is mutant for the ability to synthesis the amino acid histidine (His^-) is used in the Ames test (Fig. 3.22). Potential mutagens are then tested for their ability to reverse the His^- phenotype thus restoring the

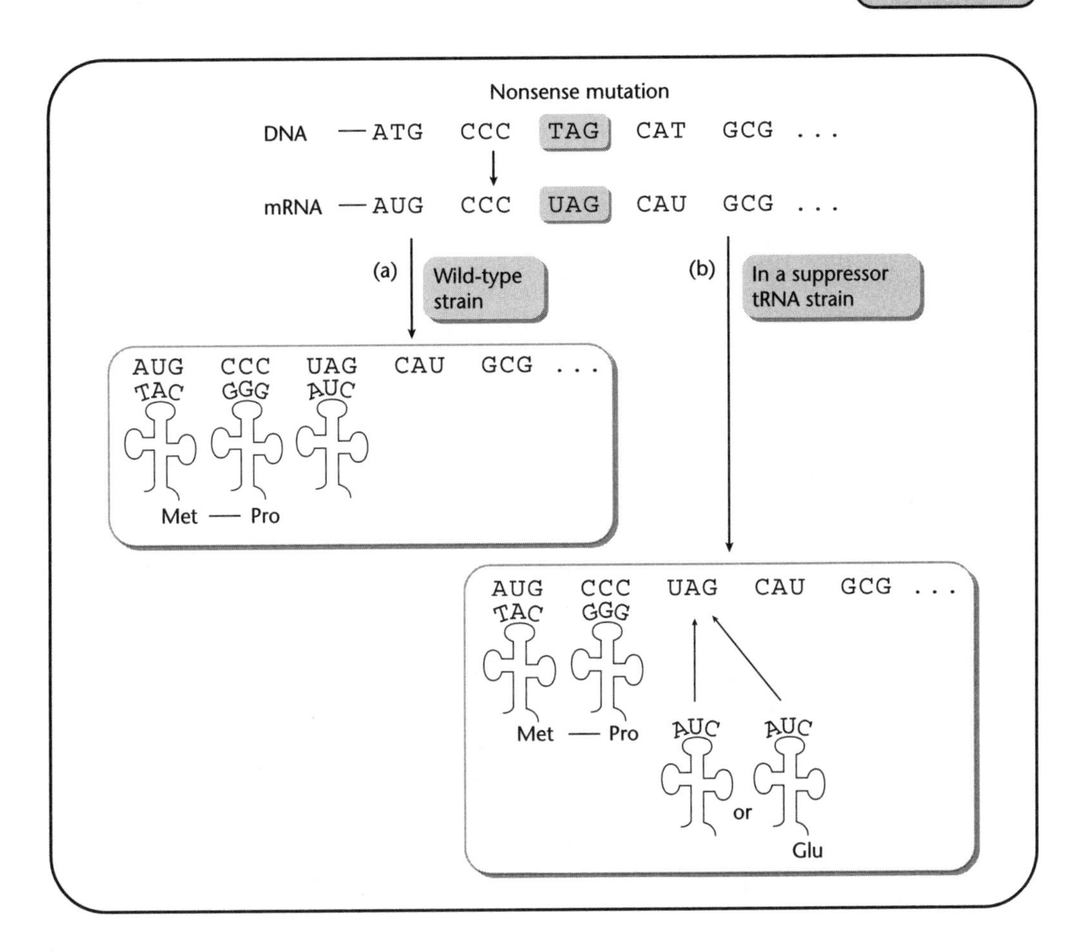

Fig. 3.20 Informational suppressors alter the protein synthesis machinery to reverse the phenotypes of the primary mutation. (a) In a wild-type strain, the tRNAs for stop codons are not charged with an amino acid. (b) In a tRNA suppressor strain, sometimes the tRNA for a stop codon is charged. If the suppressor tRNA that is charged with an amino acid is used, then translation will continue. If the uncharged suppressor tRNA is used, then translation will stop.

S. enterica serovar *typhimurium* to a His$^+$ phenotype. Sometimes, tryptophan auxotrophs of *E. coli* are used in this screen. Using different auxotrophic bacterial strains in the screen allows the demonstration that a certain mutagen causes a particular type of mutation. The auxotrophic strains used in an Ames test also lack a functional nucleotide excision repair mechanism, eliminating the strain's ability to repair any mutations that the mutagen may have caused. To allow for the possibility that some chemicals might not be mutagenic until they are broken down into their intermediate forms, the Ames test pretreats the potential mutagens with mammalian liver extracts to allow the mutagen to be metabolized. If a compound requires conversion before it exhibits mutagenic behavior, then the en-

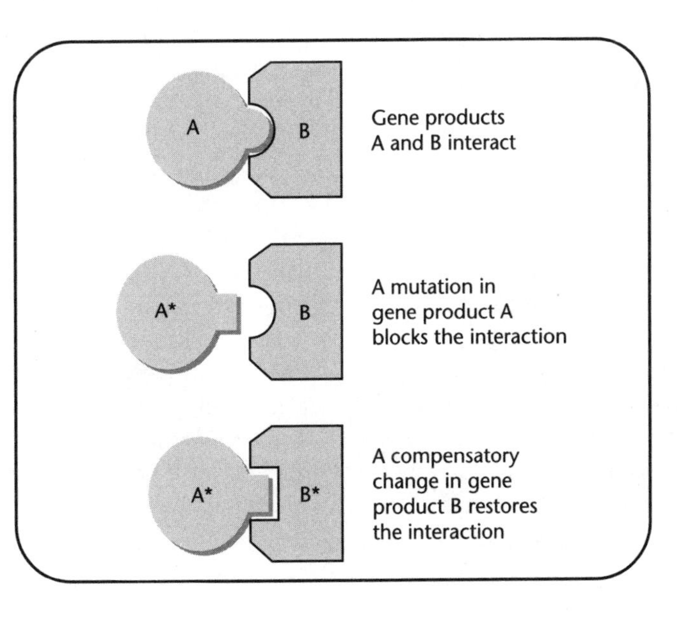

Fig. 3.21 Interactive suppressors restore function to the mutant gene product with a compensatory change in a gene product that the mutant gene product interacts with.

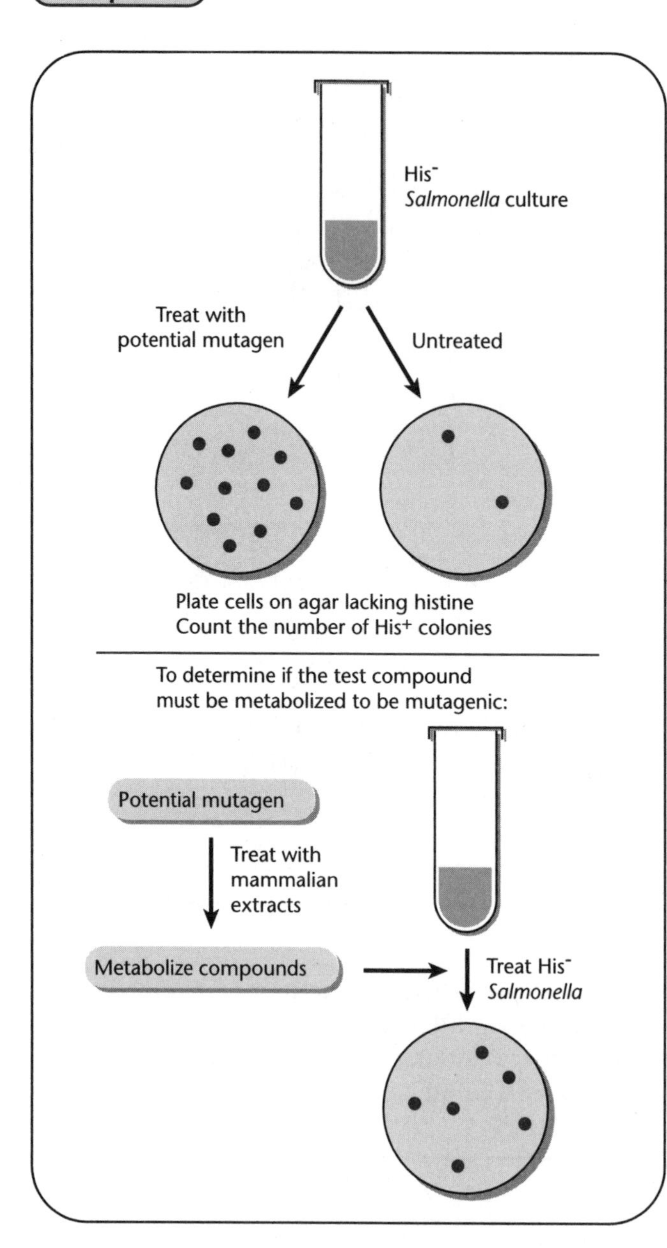

Fig. 3.22 In the Ames test, a *Salmonella* His^- mutant is reverted to His^+. If the number of His^+ revertants is greater after the His^- *Salmonella* mutant is treated with the test compound, then the test compound is mutagenic. To determine if the test compound must be metabolized to be mutagenic, the test compound is first treated with mammalian liver extracts before mixing with the *Salmonella* His mutant.

zymes in the liver extract will drive this conversion and reveal the mutagenic properties of the compound under study.

How have we exploited bacterial mutants?

Over the last century, bacterial mutants have been very valuable in delineating many different processes that make up a living cell. For example, the genes that encode many of the enzymes involved in DNA replication, DNA repair, transcription, translation, metabolism, and transport were first identified by mutations. The targets for many antibiotics were first defined by mutations. Determining how a gene is regulated has relied heavily on mutations. Without mutations, much of the information we have about how a cell produces energy, builds and tears down compounds, replicates and divides would not be known.

Summary

A mutation is a change in the base sequence of DNA. Mutations are described by how many base pairs are involved, what has happened to these base pair(s) and where in the DNA sequence the mutation has occurred. Some mutations have very dramatic consequences and others are silent. How mutations are induced and the rate at which they occur has been studied extensively. This information has been indispensable for studying bacteria and has also provided valuable insights into the physiology of all types of cells. For example, DNA repair enzymes have been extensively studied in *E. coli.* Many mutations that result in colon cancer are in DNA repair enzymes. Thus, the information and mutations available in *E. coli* were useful in interpreting the colon cancer mutations.

Study questions

1 What is the difference between a genotype and a phenotype?
2 How does a mutation in an upstream gene alter the expression of a downstream gene? What cell processes are disrupted? (DNA replication, transcription, translation?)
3 Can a missense mutation cause a temperature-sensitive phenotype? Why or why not?
4 Why is a mutation rate more accurate than a mutation frequency?
5 Draw each base and its corresponding tautomeric form. Show how the tautomer is not able to properly base pair with the corresponding complementary base.
6 Why are transitions more likely to occur than transversions? Give examples.
7 Why might uracil be found in a DNA molecule?
8 Which mutagen would potentially generate both microlesions and macrolesions? Why are microlesions revertible and macrolesions usually not revertible?
9 Design an experiment similar to the Ames test to explore the mutagenic potential of caffeine. What would be needed in your experiment to make the assay relative to humans?
10 What kind of suppressor can be used to indicate that two proteins interact? What kind of suppressor indicates two genes may have similar biological functions?

Further reading

Botstein, D. and Maurer, R. 1982. Genetic approaches to the analysis of microbial development. *Annual Review of Genetics*, **16**: 61.
Freese, E. 1959. The specific mutagenic effect of base analogues on phage T4. *Journal of Molecular Biology*, **1**: 87.
Gorini, L. and Beckwith, J. R. 1966. Suppression. *Annual Review of Microbiology*, **20**: 401–23.
Luria, S. and Delbruck, M. 1943. Mutations of bacteria from virus sensitivity to virus resistance. *Genetics*, **28**: 491.

Chapter 4

DNA repair

Repairing damaged DNA ensures an intact chromosome for DNA replication, mRNA transcription, and indirectly, protein translation. Because of its importance to the cell, specialized mechanisms to carry out DNA repair would be expected. These mechanisms must be able to remove errors, repair structural damage, and maintain the informational content of the parental DNA molecule. If DNA repair mechanisms were not available, then the ability of a cell to survive in most environments would be compromised. On the other hand, DNA repair mechanisms that are too efficient would impair a cell's ability to evolve. As we have learned in previous chapters, proofreading and editing functions associated with DNA polymerase contribute to fidelity during replication. But the damage that DNA can sustain goes beyond mistakes made by its replication machinery. What constitutes DNA damage and how damaged DNA is repaired will be explored in this chapter.

Lesions that constitute DNA damage

Simply defined, DNA is damaged when either the sequence of the bases or the phosphate backbone of the molecule is changed. Each change in the DNA molecule is described as a **lesion**. Lesions impact the damaged DNA's ability to convey its precise sequence of nucleotides to future daughter molecules. It has been conservatively estimated that *E. coli* acquires 3000 to 5000 lesions per chromosome per cell per generation. Most of these lesions are the result of oxidative damage, which is not too surprising since *E. coli* grows very well in the presence of oxygen and will be exposed to reactive oxygen species. The common types of DNA damage are summarized in Table 4.1 and Fig. 4.1 and are briefly described below. Common types of DNA damage include pyrimidine dimers, single- and double-stranded breaks, mismatch base pairs, covalent crosslinking of the double strands, and apurinic and apyrimidinic sites (see Chapter 3).

Ultraviolet (UV) light causes the formation of pyrimidine dimers by covalently linking two adjacent pyrimidines on the same strand of DNA. Mismatched base pairs arise when A fails to base pair with T, and G fails to base pair with C. Numerous mechanisms give rise to mismatched base pairs in a double-stranded DNA molecule. Covalently crosslinking the two DNA strands (inter-strand crosslinking) prevents the double helix from separating. This blocks the movement and action of replication and transcription machinery. Hydrolysis of the N, N-glycosidic bond between a base

Table 4.1 DNA damage and causative agents.

DNA damage	Causative agents
1. Pyrimidine dimers	Ultraviolet light
2. Single- and double-strand breaks	Ionizing radiation, endonucleases, peroxides, alkylating agents
3. Mismatched base pairs	Incorporation errors, tautomeric shifts, spontaneous deamination, alkylating agents, base analogs
4. Covalent crosslinks	Ultraviolet light, ionizing radiation, bifunctional alkylating agents, nitrous acid, cis-platinum
5. Apurinic and apyrmidinic sites	Spontaneous hydrolysis, alkylating agents

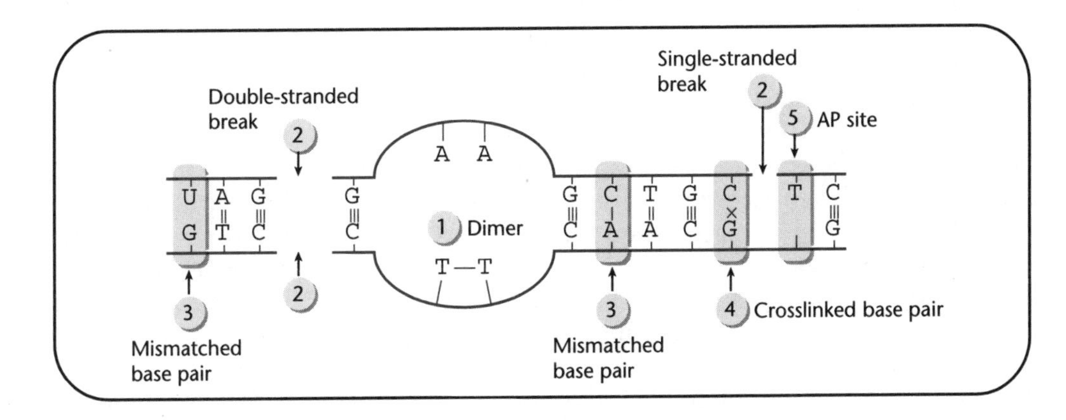

Fig. 4.1 The major types of lesions listed in Table 4.1 are illustrated. The numbers refer to those listed in the table.

and its deoxyribose sugar results in an apurinic or apyrimidinic site in a DNA molecule. The details of each of these lesions are described in detail in Chapter 3.

Single- and double-stranded breaks refer to broken phosphodiester bonds in either one (single) or both (double) of the DNA strands (Fig. 4.2). Double-stranded breaks can occur in either a blunt (generating **flushed ends**) or staggered (generating **sticky ends**) fashion. A double-stranded break that is **blunt** means that both DNA strands suffered breaks in a phosphodiester bond at the same point in the DNA molecule. The 3′ and 5′ termini are "flushed" with each other such that there are no single bases hanging at the ends of the broken DNA molecule. A double-stranded break that is **staggered** means that both DNA strands suffered breaks in a phosphodiester bond at different locations in the DNA molecule. The 3′ and 5′ termini are "sticky" rather than flush. Each

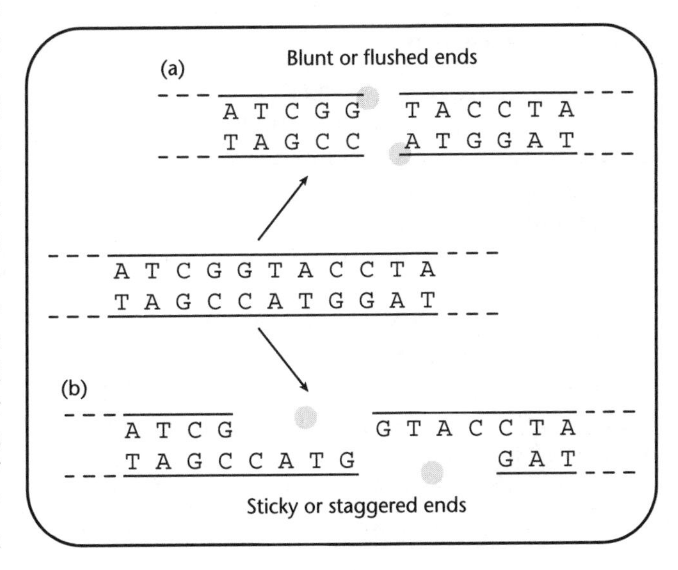

Fig. 4.2 The two types of double-stranded breaks. (a) The ends of the DNA are flush with each other and are called blunt or flushed ends. There is no single-stranded DNA on blunt ends. (b) The ends of the DNA contain short stretches of single-stranded DNA and are called sticky or staggered ends.

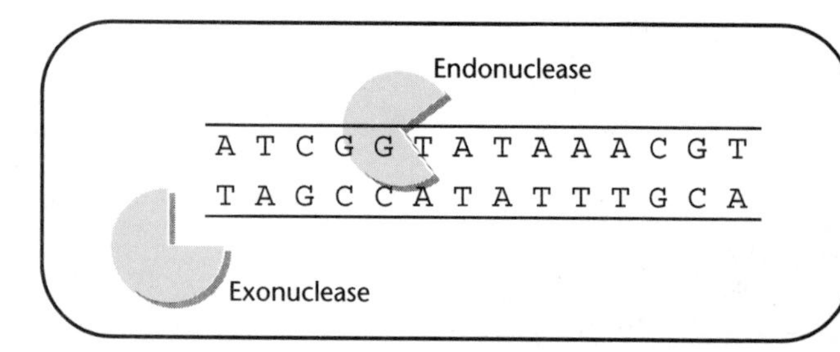

Fig. 4.3 Two types of nuclease can degrade DNA. Exonucleases need free DNA ends to degrade. Endonucleases cut DNA internally and do not need free ends.

sticky end contains a short stretch of single bases. Numerous physical and chemical agents break phosphodiester bonds, including ionizing radiation, peroxides, and alkylating agents. **Endonucleases** break phosphodiester bonds between bases located in the middle of a DNA molecule, whereas **exonucleases** break phosphodiester bonds only at the end of the dsDNA (Fig. 4.3).

Reverse, excise, or tolerate?

Many diverse DNA repair mechanisms have been identified. Table 4.2 summarizes this information. For the most part they can be categorized as those which reverse the damage or those which excise the damage. DNA repair mechanisms that reverse the damage are the least error prone because no new DNA synthesis is involved. In contrast, DNA repair mechanisms that excise damage are prone to error because upon excision of the lesion, the missing nucleotides are replaced by the 5′ to 3′ polymerizing activity of DNA polymerase. If the intact complementary strand still has the lesion or if DNA Pol I makes an incorporation error, then the repaired strand still has mistakes.

Several types of repair mechanisms can dispose of a similar lesion. For example, the pyrimidine dimer can be repaired by the reversal mechanism, photoreactivation, or by excision mechanisms such as UvrABC directed nucleotide excision repair or dimer-specific glycosylases. These mechanisms function usually before DNA replication takes place.

In the event that the damaged chromosome is being replicated, another group of mechanisms can tolerate the lesions so that the replication machinery can proceed through the damaged area. Tolerating mechanisms are often referred to as repair mechanisms, but the lesions are not actually repaired. Tolerating mechanisms include post replication repair (PPR, recombinational repair) and transdimer synthesis. They are described below.

FYI 4.1

DNA repair defects: impact from bacteria to man

The model bacterium, *E. coli*, has provided many insights into DNA repair/tolerating mechanisms. *E. coli* contains numerous DNA repair mechanisms to dispose of a variety of lesions. Understanding the way these mechanisms function, the type of lesions they dispose of, and the machinery involved, is providing new and exciting insights in certain human diseases that appear to be linked to DNA repair defects. A few of the diseases are highlighted in FYI sections to show how applicable an understanding of *E. coli* molecular genetics has been in our goal to conquer certain human diseases.

Mechanisms that reverse DNA damage

Photoreactivation

In 1949 Albert Kelner reported that when the bacterium *Streptomyces griseus* was irradiated with a large dose of ultraviolet light, its chance for survival increased several hundred fold if it was also exposed to visible light. This phenomenon is called **photoreactivation** (light repair), and its identification represented the first report of a DNA repair mechanism.

Photoreactivation is a three-step process that repairs cyclobutane dipyrimidine lesions (Fig. 4.4). First, the enzyme, **photolyase** scans DNA molecules. When photolyase identifies a cyclobutane dipyrimidine lesion it binds to it (Fig. 4.4a). Second, the photolyase-dimer complex absorbs a quantum of light in the wavelength of 350 to 500 nm (Fig. 4.4b). Energy generated from the absorption of light results in the uncoupling of the dimer. Finally, the bound photolyase is released from the DNA molecule (Fig. 4.4c).

Photoreactivation is essentially error free because nucleotide excision is not involved. DNA synthesis and an intact complementary DNA strand are not required.

Table 4.2 Summary of *E. coli* repair mechanisms.

Repair mechanism	Tactic	Damage recognized and repaired	Enzymes/genes involved
Photoreactivation	Reverse	Pyrimidine dimers: specifically cyclobutane dipyrimidines	Photolyase *phrA*, *phrB* genes
Methyltransferase	Reverse	Alkylated bases, specifically: O6-methylguanine and O4-methylthymine	Ada methyltransferase *ada* gene
UvrABC directed nucleotide excision repair (UvrABC NER)	Excise	Pyrimidine dimers cross-linked strands	UvrABC endonuclease *uvrA*, *uvrB*, *uvrC* genes
MutHLS methyl directed mismatch repair	Excise	Mismatch base pairs and GATC or CTAG are recognized and the nonmethylated daughter strand is cut	MutHLS complex (recognition and endonuclease functions) *mutH*, *mutL*, *mutS* genes
VSP repair	Excise	T-G mismatched base pairs in a specific sequence	Vsr endonuclease *vsr* gene
Glycosylases for:			
(1) uracil	Excise	Uracil in DNA	Uracil-N-glycosylase/*ung* AP endonuclease recognizes AP site generated by Ung
(2) deamination	Excise	Deaminated bases	Specific DNA glycosylase coupled with an AP endonuclease
(3) alkylation	Excise	Alkylated bases: specifically 3-methyladenine and 3-methylguanine	AlkA glycosylase/*alkA* coupled with an AP endonuclease
(4) oxidation	Excise	Oxidized bases: specifically 7, 8-dihydro-8-oxoguanine (also called: 8-oxoG, GO)	MutM/*mutM* is both a glycosylase and an AP endonuclease
(5) dimerization	Excise	Pyrimidine dimers	Dimer-specific glycosylase coupled with an AP endonuclease
Tolerating mechanisms			
(1) transdimer synthesis	Tolerate	Pyrimidine dimers	UmuC/*umuC* and UmuD/*umuD*
(2) PRR (post replication repair or recombinational repair)	Tolerate	Pyrimidine dimers	RecA/*recA*

Originally this repair mechanism was thought to be nonfunctional in the dark. However, recent evidence suggests that photolyase may enhance other repair mechanisms that operate in the dark. The photolyase-dimer complex may be recognized by excision repair mechanisms specific for dimers. Photoreactivation has been identified in many other bacteria and in plants and some animals. To date, photoreactivation has not been identified in humans.

O6-methylguanine or O4-methylthymine methyltransferase

The action of specific **methyltransferases** represents a second example of a rever-

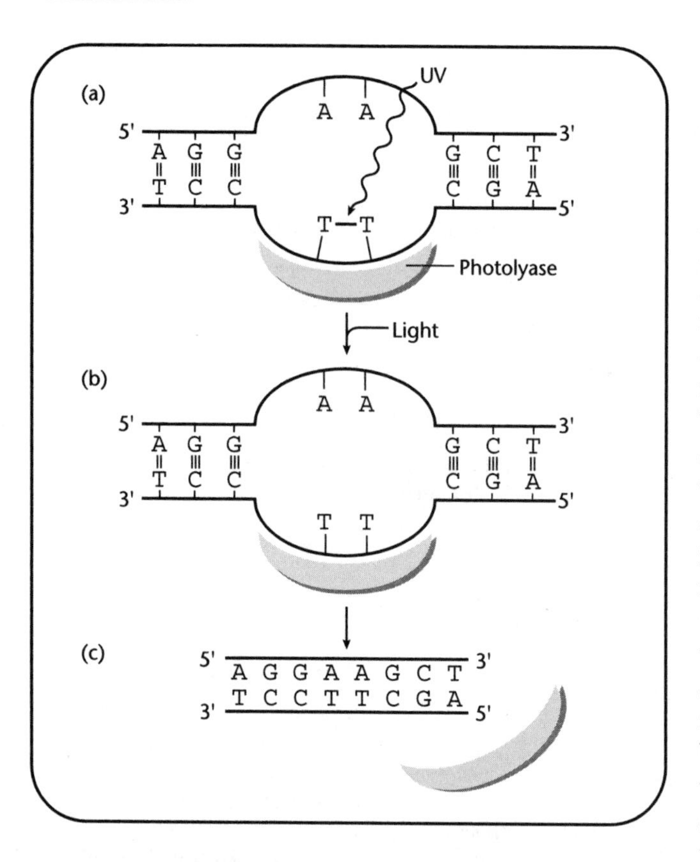

Fig. 4.4 Photoreactivation. This reversal mechanism, which only repairs cyclobutane dipyrimidine lesions, is a three-step process involving enzymatic uncoupling. (a) Photolyase-$FADH_2$ scans DNA molecule for cyclobutane dipyrimidine lesions. Upon finding the lesion, photolyase-$FADH_2$ binds to the lesion. (b) The photolyase-lesion complex absorbs a quantum of light (350–500 nm) resulting in the uncoupling of the pyrimidine dimer. (c) Photolyase is released from the DNA molecule, and resumes scanning for lesions. This repair process is error free because DNA synthesis is not required.

sal mechanism. This mechanism tackles lesions arising from alkylation. When DNA is alkylated by agents such as NTG, the most vulnerable bases are guanine and thymine. Alkylation of guanine and thymine give rise to O6-methylguanine and O6-methylthymine, respectively (see Fig. 3.16). The *ada* gene product, a methyltransferase, can transfer methyl or other alkyl groups from O6-methylguanine or O4-methylthymine to itself (Fig. 4.5). Upon removal of the alkyl group, guanine and thymine resume their normal hydrogen-bonding behavior. The *ada* methyltransferase also removes alkyl groups from the phosphate backbone of DNA. Once a methyltransferase molecule accepts an alkyl group, it is inactivated and subsequently degraded. Methyltransferases are not true enzymes because they only act as acceptors and do not catalyze a reaction.

Mechanisms that excise DNA damage

UvrABC directed nucleotide excision repair

Nucleotide excision repair (NER) was first reported by Robert Setlow in 1964. NER was referred to as "dark repair" because light was not needed for irradiated cells to repair their DNA nor was light needed for the cells to resume replication of their DNA after irradiation. NER is efficient, ubiquitous to most organisms, and apparently nonspecific in that it will repair several types of lesions. The NER system recognizes major distortions in the DNA molecule that have been created by UV or crosslinking induced lesions. It will also recognize major distortions created by some bifunctional alkylating agents. This system will not recognize minor distortions created by lesions such as mismatched base pairs, the methylated bases O6-methylguanine and O4-methylthymine, the oxidated base 8-oxoGuanine, or base analogs.

Pyrimidine dimer repair in *E. coli* by UvrABC directed NER is a three-step process (Fig. 4.6; for pyrimidine dimers see Fig. 3.17). There are two models for how this repair takes place. In the first model, the UvrABC endonuclease recognizes the distortion caused by the pyrimidine dimer and makes an incision 5′ to the dimer on the same strand as the dimer. This incision cuts the phosphodiester backbone. Steps 2 and 3 occur simultaneously. The 5′ to 3′ exonuclease activity of DNA polymerase I (DNA Pol

I) removes 7–20 nucleotides from the cut strand, including the dimer. DNA Pol I uses its 5′ to 3′ polymerizing activity to replace the missing nucleotides, using the uncut, undamaged complementary strand as a template. Thus, as DNA Pol I is removing nucleotides, it is polymerizing nucleotides. Finally, ligase seals the remaining single-stranded break.

In the alternative model (Fig. 4.6), upon recognition of the dimer distortion, UvrABC endonuclease makes two incisions on the same strand. These incisions are 5′ and 3′ to the dimer. The incision releases the damaged tract without the use of DNA Pol I's 5′ to 3′ exonuclease activity. DNA Pol I replaces the missing tract of nucleotides with its 5′ to 3′ polymerizing activity and ligase seals the single-stranded break. The difference between the two models is: (1) the number and location of the incisions; and (2) the need for exonuclease activity.

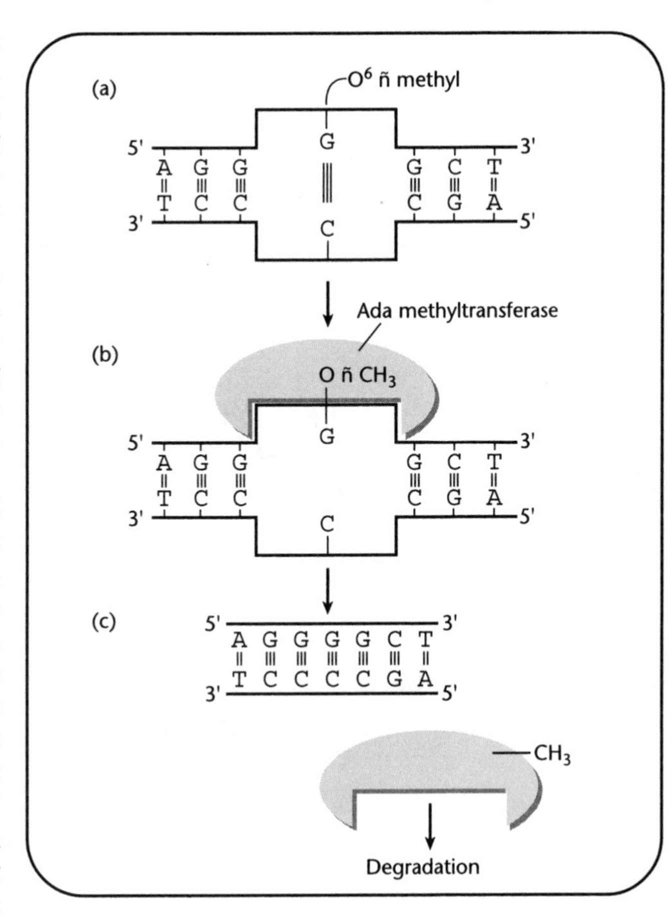

Fig. 4.5 O6-methylguanine methyltransferase. This reversal mechanism is a three-step process that is error free. (a) NTG alkylates guanine at the oxygen bonded at C6. (b) Ada methyltransferase scans DNA molecule for O6-methylguanine and then binds to this lesion. (c) The alkyl group is transferred from the guanine to the Ada methyltransferase. The Ada methyltransferase that accepts the alkyl group is degraded.

E. coli's UvrABC endonuclease is encoded by the *uvrA*, *uvrB*, and *uvrC* genes (*uvr* for ultraviolet repair). These genes were identified by mutation. *uvr* mutants display the phenotypes of increased sensitivity to UV irradiation and increased frequency of UV induced mutations. In *E. coli*, this repair mechanism has also been referred to as **short patch repair**.

Two subunits of UvrA interact with one subunit of UvrB to give rise to a protein complex that can bind any sequence of DNA, damaged or undamaged. The $UvrA_2B_1$ protein complex travels along a DNA molecule in search of major distortions such as those created by pyrimidine dimers or crosslinked bases (Fig. 4.6). Once it finds a recognizable distortion, UvrB binds to the distortion, and the two subunits of UvrA are replaced by a subunit of UvrC. Binding of UvrC to UvrB appears to induce UvrB to catalyze the first incision 3′ to the damage (4 nucleotides beyond the 3′ end of the dimer). Two possibilities arise, DNA Pol I may remove nucleotides including the dimer lesion or UvrC may catalyze a second incision 5′ to the damage (8 nucleotides beyond the 5′ end of the dimer). The helicase activity of a fourth gene product, *urvD*, assists in the removal of the tract of nucleotides containing the dimer lesion by breaking hydrogen bonds between base pairs. DNA Pol I uses its 5′ to 3′ polymerizing activity to add back the missing nucleotides while ligase catalyzes phosphodiester bond formation to seal the backbone.

FYI 4.2

Xeroderma pigmentosum

The first report of an inherited human disease described in terms of a DNA repair defect came in 1968 by James Cleaver. When cultured fibroblasts from the skin cells of a human with the disease **xeroderma pigmentosum** (XP) were exposed to UV irradiation, nucleotide excision repair (NER) in these cells was either nonfunctional or not present. Phenotypically, XP cultured cells showed an increased sensitivity to a wide variety of DNA damaging agents, including UV radiation. Clinically, XP patients exhibit early onset of extreme sensitivity to sunlight, especially in areas most likely to be exposed to natural sources of UV (skin, eyes, and tongue). Additionally, they display characteristic pigmentation abnormalities, a high frequency of skin cancers, and sometimes exhibit neurological abnormalities. The incident of XP is higher in Japan (1 in 40,000) and Egypt but much lower in Europe and the United States (1 in 250,000).

XP is described as a potentially multigene disease. To date, the human *XPA, XPB, XPC, XPG* genes, and quite possibly, the *XPF* gene have been cloned. Characterization of these genes reveals that the corresponding gene products could function as components of a nucleotide excision repair system. Defects in any one of these genes would explain the XP phenotypes.

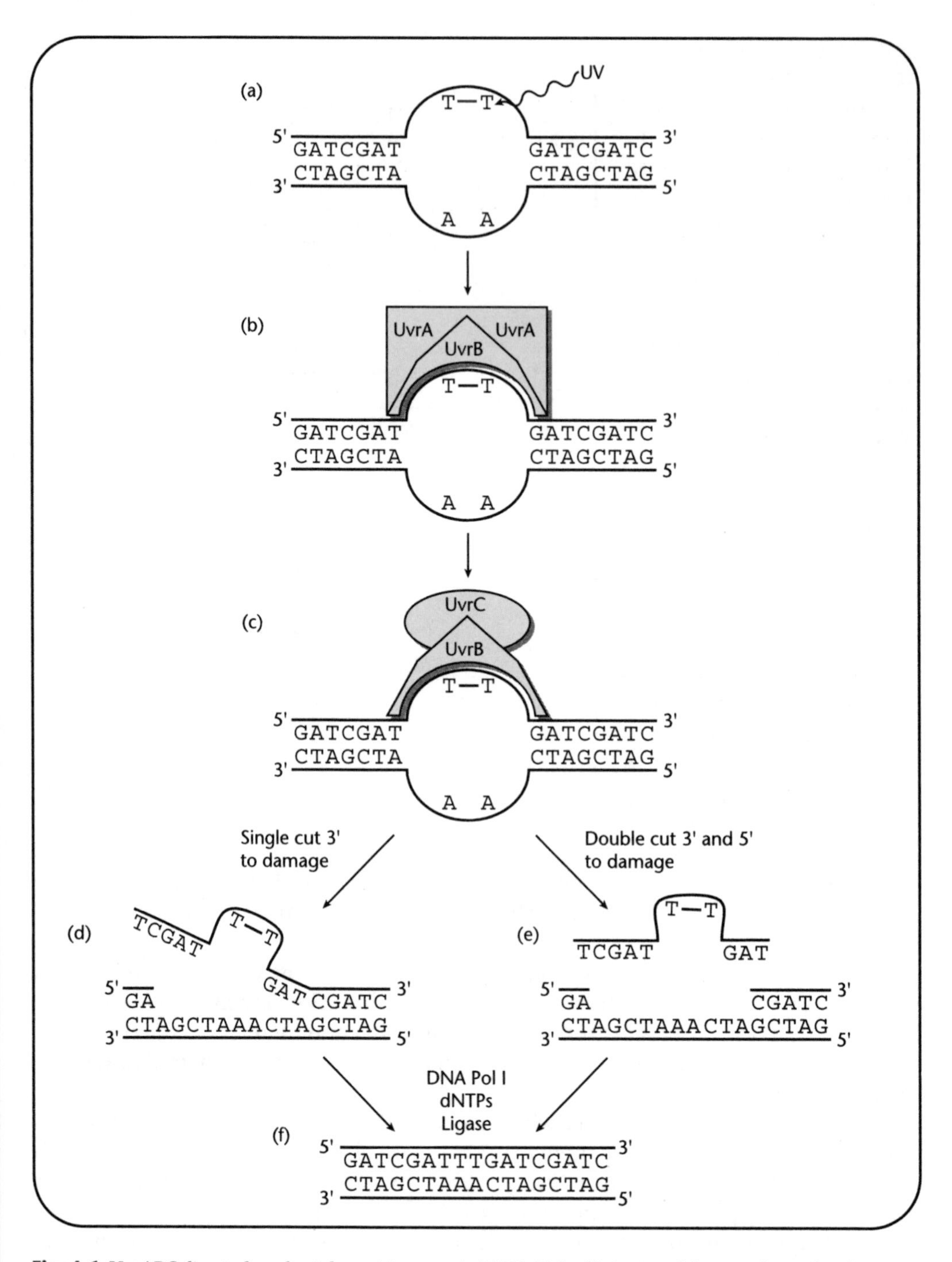

Fig. 4.6 UvrABC directed nucleotide excision repair (NER). This efficient excision repair mechanism recognizes and repairs several types of lesions that cause major distortions in the DNA molecule. Illustrated are two models for UvrABC NER. They differ by the number and location of the UvrABC incisions and the need for exonuclease activity. (a) UV damages the DNA. (b) $UvrA_2B_1$ complex scans the DNA molecule for lesions. (c) UvrB binds the lesion, while $UvrA_2$ is displaced by UvrC. Binding of UvrC to UvrB induces UvrB to cut 3′ to the damage, and possibly make a second cut 5′ to the damage. (d) If there is only one incision, then DNA Pol I 5′ to 3′ exonuclease activity removes nucleotides, including the dimer, from the cut strand. (e) If there are two incisions, then the cut fragment containing the dimer dissociates. (f) DNA Pol I 5′ to 3′ polymerizing activity fills in the gap and ligase catalyzes formation of the final phosphodiester bond to give an intact strand.

MutHLS methyl directed mismatch repair

The proofreading or editing activity of DNA Pol I and DNA Pol III does not always correct mismatched base pairs. An efficient mechanism to detect and eliminate mismatched base pairs relies on the hemimethylated state of newly replicated DNA. This mechanism is **methyl directed mismatch repair**, and involves the activities of the MutH: MutL: MutS protein complex (MutHLS), encoded by the genes, *mutH*, *mutL*, and *mutS* in *E. coli*.

MutHLS methyl directed mismatch repair is a major excision repair mechanism in *E. coli*. This repair mechanism can distinguish between parental template DNA and newly synthesized daughter DNA because the parental template DNA is methylated while the newly synthesized daughter DNA is not. MutHLS methyl directed mismatch repair recognizes lesions that cause only minor distortions in the DNA molecule, including those caused by mismatched base pairs, frameshifts, incorporated base analogs, intercalated dyes, and some types of alkyl damage.

MutHLS methyl directed mismatch repair is a four-step process (Fig. 4.7). First, MutS scans a hemimethylated DNA molecule in search of a minor distortion, such as a mismatched base pair and binds to it (Fig. 4.7a, b). Only one strand in a DNA molecule, the parent strand, is methylated immediately after replication. MutS takes advantage of asymmetry in methylation. Second, MutS interacts with two MutH proteins and one MutL protein, giving rise to a $MutS_1MutH_2MutL_1$ complex bound to DNA (Fig. 4.7c). The MutHLS complex spools DNA through it in both directions in search of the specific hemimethylated sequence, 5′GATC3′ where adenine on the parental strand is methylated. Adenine is methlyated at the C6 position by the enzyme, deoxyadenosine methylase (Dam methylase). Third, upon encountering a hemimethylated GATC sequence, one of the MutH proteins catalyzes a cut (breaks a phosphodiester bond) at this sequence in the non-methylated daughter strand (Fig. 4.7d). Finally, the cut daughter strand is degraded by a 5′ to 3′ exonuclease activity, removing nucleotides spanning the mismatched base pair (Fig. 4.7e). DNA Pol III, and not DNA Pol I, resynthesizes the stretch of nucleotides, including the mismatch, using the methylated parent strand as a template. Ligase seals the phosphodiester backbone.

MutHLS methyl directed mismatch repair always assumes that the sequence of the parental template strand is correct, and it is the newly synthesized daughter strand that is to be repaired. This repair mechanism identifies the parent strand through methylation signals that the daughter strand has yet to receive. The assumption that the parent is always right can be extended to other examples in biology!

Very short patch repair

Very short patch (VSP) **repair** is illustrated in Fig. 4.8. This excision mechanism is specialized, recognizing T-G mismatched base pairs in the specific sequence, 5′C^mCWGG3′/3′$GGWC^mC$5′ (where m = methylation and W = either A/T or T/A). How do T-G mismatches arise in this sequence? The methyltransferase, DNA cytosine methylase (DCM, encoded by the *dcm* gene) methylates the second cytosine in this sequence at the C5 position on the pyrimidine ring. T-G mismatches arise upon deamination of 5-methylcytosine to thymine, resulting in a thymine at a position where cytosine should be. If the DNA molecule containing this sequence is not being replicated, then this thymine will now be paired with guanine.

VSP is a three-step process which eliminates T-G mismatches in 5′C^mCWGG3′/

FYI 4.3

Trichothiodystrophy (TTD): another potential nucleotide excision repair defect

In approximately half of the trichothiodystrophy (TTD) cases, there is an associated photosensitivity, suggesting defects in DNA repair. Defects in nucleotide excision repair of UV lesions in cultured cell lines from TTD photosensitive patients have been demonstrated. TTD is clinically defined by the appearance of sulfur-deficient brittle hair with a "tiger tail" appearance under the microscope. Afflicted individuals may display fish-like scales on the skin (ichthyosis), and mental and physical retardation. Some TTD patients have protruding ears and a receding chin. In contrast to XP patients, TTD patients who exhibit the photosensitivity phenotype and increased mutation frequency do not necessarily have an increased risk for skin cancer nor do they display characteristic pigmentation defects.

FYI 4.4

DNA repair defects linked to cancer

As discussed previously, MutHLS methyl directed mismatch repair in *E. coli* repairs lesions that cause minor distortions in the DNA molecule, including those caused by mismatched base pairs. Human homologs of MutL and MutS have been identified, characterized, and evidence has been generated that these homologs are required for mismatch repair in human cells. Defects in three genes coding for MutL homologs and one gene coding for a MutS homolog have been associated with hereditary nonpolyposis colorectal cancer (HNPCC). HNPCC is a very common genetic disease, affecting 1 in 200 humans. In addition to developing tumors in the colon-rectal area, afflicted patients may also develop tumors of the endometrium, ovary, and other organs. HNPCC accounts for approximately 4 to 13% of all reported cases of colorectal cancer. That HNPCC is associated with defects in mismatch repair is very intriguing and truly represents how important our investigations of DNA repair systems in *E. coli* have been.

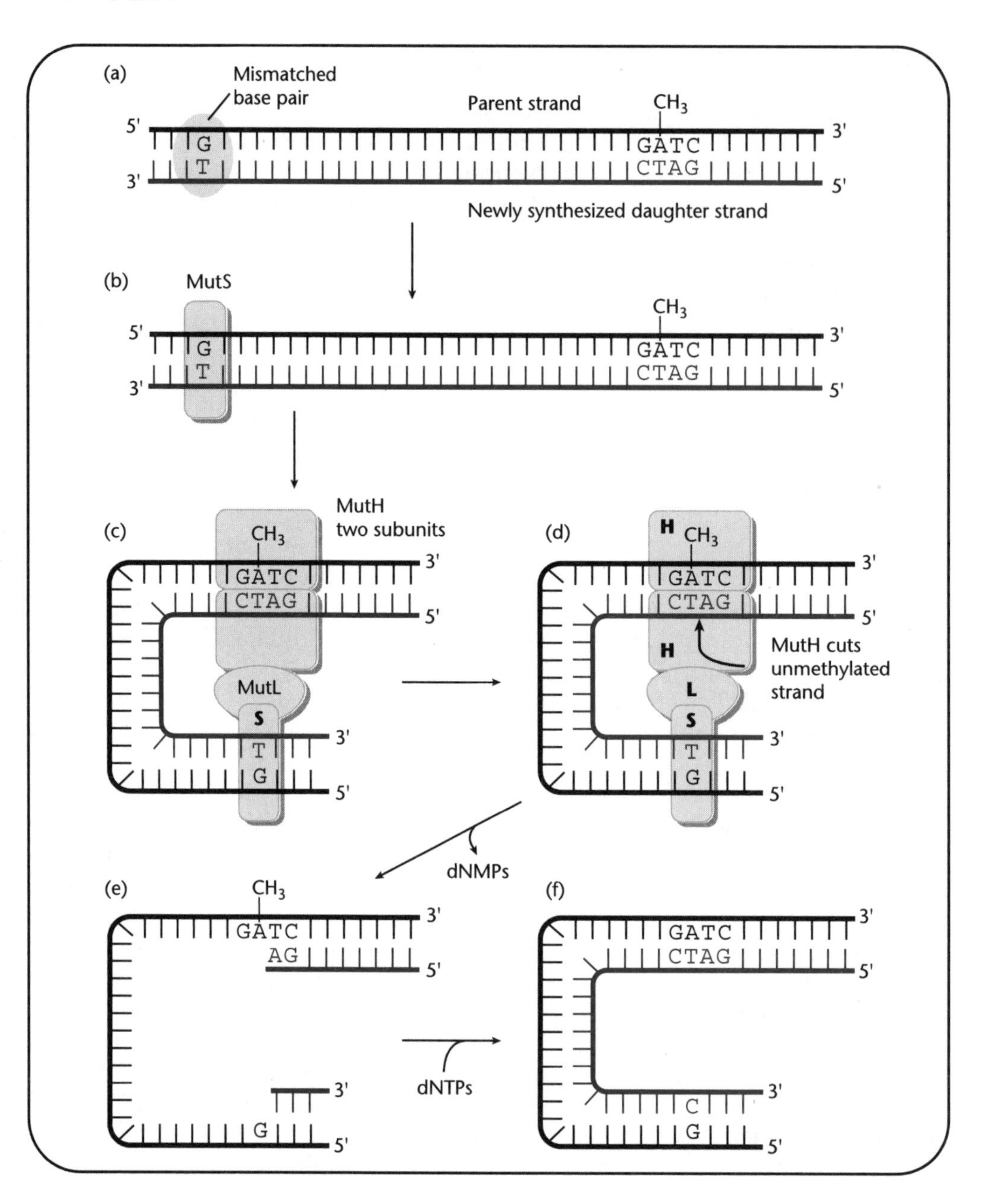

Fig. 4.7 MutHLS directed mismatch repair. This major repair mechanism is a four-step process that is only effective immediately after a DNA molecule has been replicated. (a) Newly replicated DNA consists of a parent strand, which is methylated and a newly synthesized daughter strand that is non-methylated. A DNA molecule in this state of methylation is referred to as being hemimethylated. (b) MutHLS identifies the DNA strand that has the wrong base in a mismatch based on the methylation pattern. MutS scans for mismatched base pairs, finds the mismatch, and then binds to it. (c) Three MutH proteins and one MutL protein bind to the bound MutS protein. The bound $MutS_1$-$MutH_2$-$MutL_1$ complex spools DNA through, stopping at a GATC sequence where the A is methylated. (d) MutH cuts the non-methylated daughter strand at GATC. (e) Exonuclease activity removes nucleotides from the non-methylated, cut daughter strand. The nucleotides from the GATC through the mismatch are removed. (f) DNA Pol III polymerizes nucleotides, filling in the gap created by the removal of the nucleotides, while ligase catalyzes the final phosphodiester bond to give an intact strand.

3′GGWC^{m}C5′. First, the Vsr endonuclease, encoded by the *vsr* gene, identifies and binds to a T-G mismatch in the 5′C^{m}CWGG3′/3′GGWC^{m}C5′ sequence. Second, bound Vsr catalyzes a break in the phosphodiester bond next to the T comprising the mismatched base pair. Finally, exonuclease activity removes the T and then DNA Pol I adds back the correct nucleotide, a cytosine.

T-G mismatches in the sequence, 5′C^{m}CWGG3′/3′GGWC^{m}C5′ could be repaired by MutHLS methyl directed mismatch repair. However, the creation of T-G mismatches by deamination of 5-methylcytosine to thymine can be independent of replication, meaning that the DNA molecule would probably not be in a hemimethylated state. How would MutHLS know which strand to target for repair without the appropriate hemimethylated signals? If the wrong strand were chosen, then a G–C to T-A transition would arise. Rather, the Vsr endonuclease targets the thymine that arose due to deamination, and in combination with exonuclease and polymerase activities, replaces the T with a C. To date, VSP has only been identified in *E. coli*. However, 5-methylcytosine is found in the DNA of other organisms, leading to the hypothesis that VSP repair may be ubiquitous.

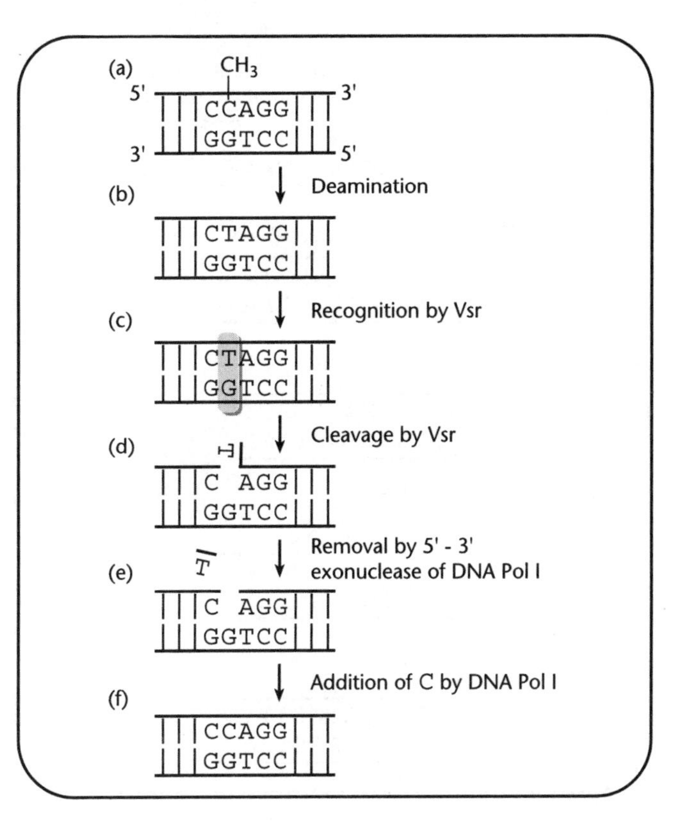

Fig. 4.8 Very short patch repair. (a) The second C in the sequence CCAGG can be methylated by DCM methylase. (b) Deamination of this modified C leads to thymine. (c) Vsp endonuclease recognizes the mismatch in CTAGG and binds to it. (d) Vsp endonuclease cleaves the DNA. (e) The 5′ to 3′ exonuclease activity of DNA Pol I removes the T and (f) replaces it with a C.

Glycosylases

E. coli contains many types of N-glycosylases (N-glycosidases), a class of repair enzymes that excise bases by hydrolyzing the N, N-glycosidic linkage between a base and its deoxyribose sugar leaving an AP site (AP = apurinic or apyrimidinic, depending on which base is removed, see Fig. 3.10). Five examples of base excision repair by glycosylases are given below.

Uracil-N-glycosylase coupled with AP excision repair

The best-characterized glycosylase is uracil-N-glycosylase, encoded by the *ung* gene. Uracil is normally found in RNA molecules, where it replaces thymine and base pairs with adenine. As mentioned in Chapter 3, uracil can be found in DNA molecules due to incorporation errors or due to deamination of cytosine to uracil. In the former case, an A-U base pair arises. In the later case, a U-G mismatch arises, leading to a C-G to T-A transition. Neither pairing creates a distortion recognizable by DNA Pol III or I proofreading activities or by MutHLS methyl directed mismatch repair.

Removal of uracil from DNA is a three-step process (Fig. 4.9). First, uracil-N-glycosylase recognizes uracil in single-stranded and double-stranded DNA, cleaving the N, N-glycosydic bond between the uracil and its deoxyribose sugar. Free uracil is released, creating an AP site. Second, the AP site is recognized by an **AP endonuclease** that catalyzes a break in the phosphodiester bond located on the 5′ side of the missing base. Finally, DNA Pol I uses its 5′ to 3′ exonuclease and 5′ to 3′ polymerase activities to excise a tract of DNA spanning the AP site and replace the nucleotides in this region using the uncut strand as template. Ligase seals the phosphodiester backbone.

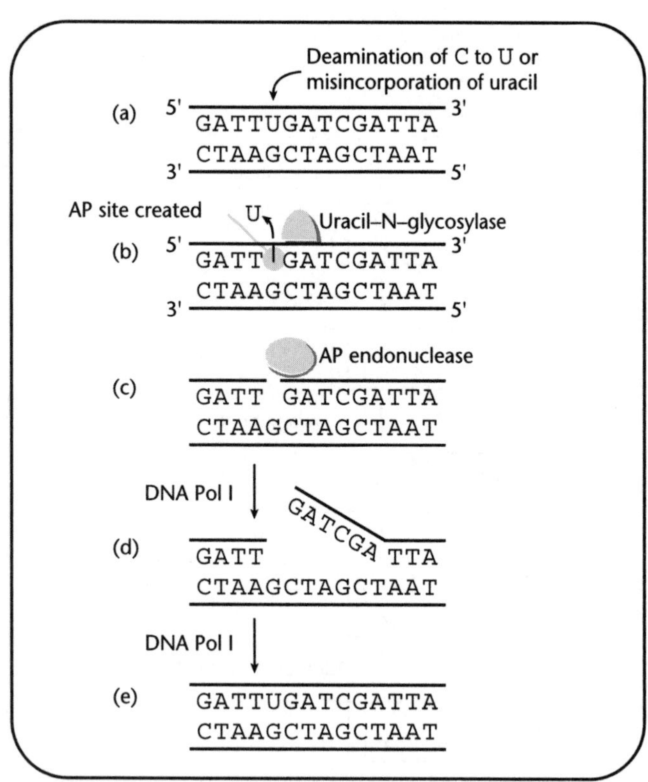

Fig. 4.9 Uracil-N-glycosylase (Ung) excision repair coupled with AP incision repair. (a) This three-step process recognizes uracil in both double- and single-stranded DNA molecules. (b) Uracil-N-glycosylase recognizes uracil in DNA. It cleaves the N, N-glycosydic bond between uracil and its deoxyribose sugar, creating an AP site. (c) AP endonuclease recognizes this site and cuts the phosphodiester bond on the 5′ side of the AP site. (d) DNA Pol I 5′ to 3′ exonuclease activity removes nucleotides spanning the AP site. (e) DNA Pol I polymerase adds nucleotides, filling in the gap created by the removal of the bases, while ligase catalyzes the final phosphodiester bond to give an intact strand. Additional glycosylases exist to recognize other lesions including deaminated, alkylated and oxidized bases and pyrimidine dimers. All glycosylase repair mechanisms are coupled with AP incision repair.

Deaminated bases removed by DNA glycosylase

For every specific type of deaminated base, a unique DNA glycosylase exists to remove it. The process used to remove deaminated bases is similar to that described above. The specific DNA glycosylase recognizes its specific deaminated base and cleaves the N, N-glycosidic linkage between that base and its deoxyribose sugar creating an AP site (see Fig. 3.10). An AP endonuclease catalyzes the cleavage of the phosphodiester bond 5′ to the AP site. DNA Pol I removes and resynthesizes the tract of nucleotides spanning the AP site. Ligase seals the phosphodiester backbone.

Alkylated bases removed by DNA glycosylase

As is the case with deaminated bases, specific N-glycosylases exist to remove alkylated bases. The process used to remove alkylated bases by N-glycosylases is identical to that described above for removing uracil and deaminated bases. In *E. coli*, a glycosylase encoded by the *alkA* gene removes 3-methyladenine and 3-methylguanine from DNA molecules.

MutM/MutY: oxidative damage

Both MutM and MutY are N-glycosylases. MutM targets the oxidized base, 7, 8-dihydro-8-oxoguanine (8-oxoG, GO) which arises as a result of exposure to reactive oxygen. Reactive oxygen includes superoxide radicals, hydrogen peroxide, and hydroxyl radicals, all of which have more electrons than molecular oxygen (O_2). 8-oxoG base pairs with adenine, creating G-C to T-A tranversions. MutM recognizes 8-oxoG in DNA, and acts as a glycosylase to hydrolyze the glycosidic linkage between this oxidized base and its deoxyribose. It then acts as the AP endonuclease to cleave the phosphodiester bond 5′ to the AP site. MutM is both an 8-oxoG glycosylase and an AP endonuclease. Following the activity of MutM, the nucleotides spanning the 8-oxoG lesions are removed by exonuclease activity and replaced by DNA Pol I. In contrast, the MutY glycosylase targets the adenine that base pairs with 8-oxoG. It targets any A-G mismatches, independent of oxidative damage.

Both *mutM* and *mutY* are part of a set of genes that comprise the **oxidative stress response regulon**. Another gene of this regulon, *mutT*, encodes a gene product that also deals with 8-oxoG lesions, but in a much different manner than MutM glycosylase. MutT is a phosphatase that specifically degrades an oxidized guanine precursor, 8-oxoGTP into 8-oxoGMP so that it cannot be incorporated during DNA replication.

N-glycosylases specific for pyrimidine dimers

Repair of pyrimidine dimers can also occur by cleavage of the glycosidic bonds holding the pyrimidine dimers to the deoxyribose. Specific N-glycosylases recognize pyrimidine dimers and, in combination with an AP endonuclease, excise the dimer in the three-step process described above.

Mechanisms that tolerate DNA damage

What happens if lesions in the DNA are not repaired before the replication machinery attempts to polymerize across a template containing lesions? Lesions involving crosslinked or chemically modified bases might prove daunting to the replication machinery, blocking its attempt to move through the lesion-containing area. The replication machinery can proceed through or circumvent certain types of lesions. These are the tolerating mechanisms. They do not actually repair the lesion, but provide the means by which the damaged DNA can be replicated. In both examples, we will look at pyrimidine dimers.

Transdimer synthesis

As DNA Pol III is polymerizing across the pyrimidine dimer lesion in the template strand, it may insert any nucleotide opposite the dimer. Replication is never stalled; no gaps form in the newly synthesized daughter strand, yet there is a high frequency of misincorporation which leads to mismatched base pairs. Mismatch base pairs are found in the daughter strand in the area corresponding to the dimer. We call this process **transdimer synthesis** or **SOS mutagenic repair**. This process is part of the SOS gene regulon (see below and Chapter 12).

What has happened to the proofreading functions of the DNA polymerases to allow insertion of any nucleotide opposite a dimer? Two gene products, *umuC* and *umuD* (*umu* = UV induced mutagenesis), of the SOS regulon appear to modify the proofreading activity of DNA Pol III. DNA Pol III is then able to tolerate the distortion created by the dimer and polymerize across this lesion.

A model describing how UmuC and UmuD modify DNA Pol III proofreading activity is shown in Fig. 4.10. During replication of a UV damaged template, DNA Pol III encounters the dimer lesion. Any nucleotide that DNA Pol III attempts to pair with the dimer would not be able to pair, thus alerting the editing function of DNA Pol III. As mentioned in Chapter 2, the β subunit of DNA polymerase is a component of the editing (proofreading) function; two β subunits form a circular clamp that anchors DNA polymerase to the DNA molecule. The β clamp will not allow DNA Pol III to polymerize the next nucleotide if there are distortions caused by previously inserted nucleotides. Cleaved UmuD (UmuD') and UmuC may substitute for the β subunits, forming the clamp for DNA Pol III. This substitution may cause a relaxation in the clamping function. The relaxed UmuD'—UmuC clamp allows DNA Pol III to polymerize across templates containing lesions such as dimers. The relaxation of the editing function would also explain the high rate of mutagenesis observed in UV damaged cells.

Obviously, this mechanism is a last ditch effort to save a cell whose chromosome has been intensely damaged. It does not repair the lesion and it is highly error prone. Newly synthesized daughter strands are generated containing misincorporation errors.

Post replication/recombinational repair (PRR)

In many ways the activities, incision, digestion, polymerization, and ligation, which constitute many DNA repair mechanisms also are part of another mechanism for tolerating lesions, homologous recombination. Homologous recombination involves processes that break nearly identical DNA molecules and then rejoin the DNA mole-

FYI 4.5

Cockayne's Syndrome (CS)

E.A. Cockayne originally described this rare syndrome in the mid-1930s. CS patients are arrested in both growth and development sometime between the infant and toddler stages. Other obvious symptoms include photosensitivity, deafness, optic atrophy, sunken eyes, large ears and nose, long arms and legs yet a short body, premature aging, mental deficiency, and neurological abnormalities. The average life span of CS patients is 12 years. The photosensitivity of CS patients suggest a defect in the repair of UV induced lesions. Studies with cultured cells from CS patients revealed defects in the repair of cyclobutane pyrimidine dimers but not in the repair of other types of photoproducts.

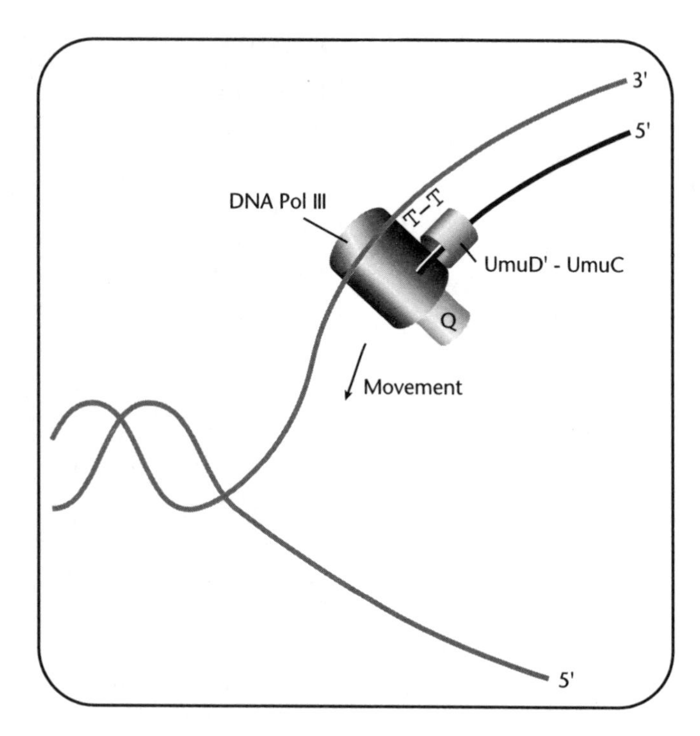

Fig. 4.10 Transdimer synthesis as a tolerating mechanism. Only one of the parental strands is illustrated as being replicated. UmuC and UmuD′ modify DNA Pol III editing/proofreading activity (Q = DnaQ) so that any nucleotide can be polymerized across from the dimer lesion. This mechanism does not repair the lesion in the parental strand, it only ensures that the newly synthesized daughter strand does not contain any gaps. Insertion of any nucleotide as the daughter strand is being synthesized will potentially create mutations, hence the name SOS mutagenic repair.

cules in different combinations. The enzymatic overlap between some DNA tolerating mechanisms and homologous recombination was observed with the identification of *E. coli* mutants defective in both repair of UV induced damage and in recombination reactions. The overlapping enzymatic activity that is defective in these mutants is a component of the recombinational machinery called RecA.

Although homologous recombination will be covered in detail in Chapter 5, a bit of enzymatic background will be provided so as to understand the **post replication repair (PRR or recombinational repair)** tolerating mechanism. RecA is a multifunctional enzyme. It binds single-stranded DNA (ssDNA). Using ATP hydrolysis for energy, RecA promotes ssDNA base pairing with homologous (similar yet not identical) sequences from a dsDNA molecule. RecA unwinds dsDNA molecules and it promotes hydrogen bonding between complementary base pairs. Another enzyme involved in homologous recombination reactions is RecBCD, which is also named Exonuclease V. RecBCD is a multifunctional enzyme. RecBCD is an ATP dependent exonuclease for dsDNA, an ATP dependent helicase for dsDNA, both an exonuclease and endonuclease for ssDNA, and a DNA dependent ATPase.

PRR begins with post dimer initiation (Fig. 4.11). Post dimer initiation occurs when the replicating DNA Pol III attempts to replicate through the lesion. DNA Pol III temporarily stalls upon encountering the lesion. It then commences polymerization, located some distance beyond the dimer (up to 800 basepairs away). A gap is left in the newly synthesized daughter strand; the gap is located opposite the dimer in the template strand. Both strands of the newly replicated DNA molecule now contain damage (the parent strand has the dimer, the complementary daughter strand has the gap). The gapped DNA has to be filled in or it will be degraded.

To fill the gap, a recombinational event occurs. As this recombinational event will occur post replication, we refer to this as post replication repair (even though the dimer is not repaired) or as recombinational repair (which is probably more appropriate, because at least the gap will be repaired).

PRR is a six-step process (Fig. 4.11). First, as mentioned, a gap forms (from post dimer initiation) in the newly synthesized daughter strand. Second, RecA binds to the single-stranded region of the gap. Third, the RecA coated gapped strand assimilates (aligns and then hydrogen bonds) with the complementary region on the undamaged parent strand that did not serve as template for the gapped strand. Fourth, a recombination associated endonuclease activity cuts the segment of the undamaged parent strand that is now hydrogen bonded with the single strand of the gapped region. Fifth, ligase seals the phosphate backbone creating a continuous strand. Finally, there still remains a gap in the undamaged parent strand that did not serve as template for the original gapped strand. Since its newly synthesized complementary strand does not contain any gaps or lesions, it can serve as template for the activities of repair synthesis (exonuclease, polymerase, and ligase activity), filling in the gap to give a contiguous strand.

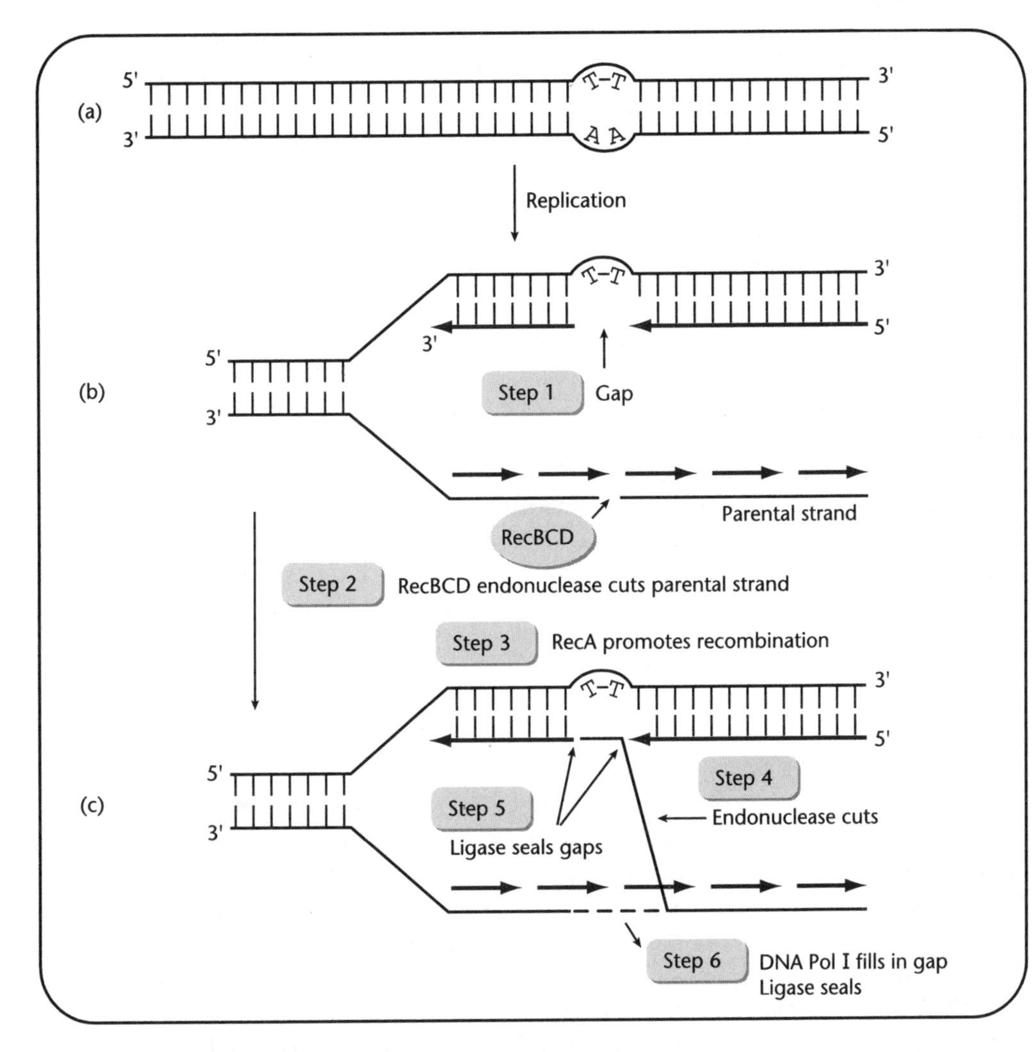

Fig. 4.11 Post replication/recombinational repair/PRR. This tolerating mechanism has also been linked to the post dimer initiation mechanism for jump-starting DNA Pol III after it skips over the dimer lesion in the parental strand during replication. PRR is a six-step process that borrows the enzymatic machinery of recombination to fill in the gaps left by DNA Pol III. (a) The T-T dimer forms. (b) DNA Pol III skips over the dimer and leaves a gap (step 1). (c) RecBCD cuts the parental strand (step 2). RecA promotes recombination of the single strand with the damaged strand (step 3). Endonuclease frees the parental strand with another cut (step 4). Ligase seals the gaps in the damaged dsDNA (step 5) and DNA Pol I fills in the gap in the parental strand (step 6).

FYI 4.6

Fanconi's Anemia

Patients with Fanconi's Anemia (FA) exhibit growth retardation, learning disabilities, hyperpigmentation, skeletal, urogenital, gastrointestinal and cardiac abnormalities, hearing loss, and ear abnormalities. Additionally, they are at increased risk for leukemia and solid tumors. Clinically, FA patients are defined by pancytopenia (depressed levels for all cellular components of the blood), short stature, hyperpigmentation, and radial aplasia. Defects in the repair of inter-strand DNA crosslinks were first presented as a plausible hypothesis for this disease. A model suggesting a defect in an error-prone pathway of DNA repair that involves replicative bypass (i.e. tolerating) of lesions has been advanced to explain the molecular basis of FA.

Introduction to the SOS regulon

Several of the repair/tolerating mechanisms discussed include gene products that are part of the SOS regulon. Simply defined a regulon is a group of genes located in different places on the chromosome that are coordinately regulated. The SOS regulon will be discussed in much more detail in Chapter 12, but a brief introduction at this point will serve to pull together some of the information we have learned about DNA repair/tolerating mechanisms.

Obviously the question arises, are DNA repair/tolerating enzymes always present in the cell at sufficient levels to combat intensive DNA damage? The answer to this question is no. Expression of some of the genes coding for DNA repair/tolerating enzymes has to be induced. Jean Weigle first demonstrated the inducibility of a DNA repair mechanism in response to UV damage. By inducibility it is implied that the genes encoding the components of this response are activated under some conditions.

In 1974, Miroslav Radman expanded our understanding of the inducibility of certain DNA repair/tolerating systems with two significant observations. First, he discovered that the inducibility of error-prone post replication repair (PRR) was dose dependent. High doses of UV irradiation induced this repair whereas low doses (500 nJ/mm^2 or less) did not. Second, he discovered that error-prone PRR was dependent on both RecA and a second gene product, LexA. Cells that are *lexA*$^-$ can undergo homologous recombination because they still express functional RecA.

Today it is known that the genes comprising the SOS regulon are controlled by the repressing activity of LexA. In response to high levels of DNA damage, LexA repressor is removed from the promoters of the SOS genes, thus allowing the transcriptional apparatus to activate high level expression of the SOS genes. As we will learn in Chapter 12, the function of the SOS gene products is to repair a genome that has been heavily damaged.

Summary

In the course of cell growth and division, DNA is damaged. Fortunately, DNA can be repaired. Common types of DNA damage include pyrimidine dimers, single- and double-stranded breaks, mismatched base pairs, covalent crosslinking of the double strands, and apurinic and apyrimidinic sites. In *E. coli*, three strategies are used to deal with damaged DNA: (1) reverse the damage; (2) excise the damage; or (3) tolerate the damage. For each strategy there are several types of mechanisms, depending on the damage. For example, photoreactivation reverses pyrimidine dimers. O6-methylguanine or O4-methylthymine methyltransferases remove methyl groups from guanine and thymine, respectively. UvrABC directed nucleotide excision repair (NER) excises pyrimidine dimers. MutHLS methyl directed mismatch repair excises mismatched base pairs, incorporated base analogs that are incorrectly base paired, intercalated dyes, and some types of alkyl damage. Tolerating mechanisms allow DNA replication to continue even if the parental template strand contains lesions. Such mechanisms include transdimer synthesis and post replication repair. Repair mechanisms have been found in all living organisms. Defects in repair mechanisms are linked to a number of human diseases.

Study questions

1 Describe how phosphodiester bonds can be broken. What is the difference between blunt and staggered cuts?
2 Give three examples as to how mismatched base pairs arise in a DNA molecule. Which repair mechanism would dispose of the mismatched base pair described in each example?
3 Draw a pyrimidine dimer. Show where photolyase would impact the dimer. Show where UvrB would impact the dimer. Contrast and compare the impact by these two repair mechanisms.
4 If an *E. coli* strain was a *phrA phrB* double mutant, would you predict that this mutant could repair pyrimidine dimers? Why or why not?
5 Contrast and compare a single incision to a double incision event by UvrB. What activity is needed in one event but not the other event, and why is this activity needed?
6 How does MutHLS determine which DNA strand to remove bases from? What are the roles of MutS, MutL, MutH, and MutHLS?
7 Which DNA repair mechanisms are error free and why are they error free?
8 Which DNA repair mechanisms are error prone and why are they error prone?
9 Why do tolerating mechanisms not result in the repair of a DNA molecule? Contrast and compare two tolerating mechanisms.

Further reading

Echols, H. and Goddman, M. 1991. Fidelity mechanisms in DNA replication. *Annual Review of Biochemistry*, **60**: 477.
Friedberg, E.C., Walker, G.C., and Siede, W. 1995. *DNA Repair and Mutagenesis*. Washington, DC: ASM Press.
Modrich, P. 1991. Mechanisms and biological effects of mismatch repair. *Annual Review of Genetics*, **25**: 229.
Rupp, W.D. 1996. DNA repair mechanisms. In *Escherichia coli* and *Salmonella typhimurium: Cellular and Molecular Biology*, 2nd edn., eds. F.C. Neidhardt, R. Curtiss III, J.L. Ingraham, E.C.C. Lin, K.B. Low, B. Hagasanik, W.S. Rexnikoff, M. Riley, M. Schaechter, and H.E. Umbarger, pp. 2277–94. Washington, DC: ASM Press.

Chapter 5

Recombination

Within living cells, the exchange of DNA sequences and genetic information can occur through an intricately regulated series of enzymatic reactions involving pairing of DNA molecules and phosphodiester bond breakage and rejoining. This type of sequence rearranging is known as **genetic recombination**. Genetic recombination is a fundamental process in all cells and is responsible for rearranging sequences between different pieces of DNA, shaping the genome by altering the sequences that are present, limiting the divergence of repeated DNA sequences, pairing chromosomes before cell division, and promoting DNA repair. Depending on the requirements for the DNA molecules and the proteins involved, three types of recombination have been described: homologous recombination, site-specific recombination, and illegitimate recombination.

Homologous recombination

Homologous recombination is the exchange of DNA sequences between DNA molecules that contain identical or nearly identical sequences along their length. The common stretch of bases that will be recombined is known as the **homology** between the sequences and can be as few as ~50–100 bp or as much as a whole chromosome. The greater the region of homology, the higher the frequency of recombination between the two sequences. Homologous recombination between sequences with as little as 25 bp of homology has been documented but it happens only rarely.

At an absolute minimum, the requirements for homologous recombination are: (1) two DNA sequences with similar or almost identical base pair sequences (homologous sequences); (2) the ability to form stable hydrogen bonds between the bases on one strand of one DNA sequence and the bases on the complementary strand on the other DNA sequence; and (3) the proteins needed to carry out recombination. These proteins include those that encourage the two DNA sequences to stay in close proximity to one another, enzymes that break phosphodiester bonds (endo- or exonucleases) and enzymes that rejoin phosphodiester bonds (ligase).

One way to identify components of the recombination machinery is to isolate mutations that impair the cell's ability to carry out recombinational reactions (Table 5.1). A mutation in *E. coli* was isolated in which the phenotypes were recombination deficiencies (Rec^-) and sensitivity to ultraviolet light (UV^S). In addition to abolishing homologous recombination, the mutation prevented *E. coli* from activating the

FYI 5.1

RecA

RecA has homologs in virtually every known species. Defining these homologs and their functions has indicated that homologous recombination has several very interesting roles in eukaryotes. While most bacterial cells have one *recA* gene in their chromosomes, many eukaryotes have multiple *recA* homologs. These homologs do not always have overlapping functions, rather each seems to be used for a specific process. The homologs are expressed differently in different tissues and defects in the RecA homologs have been shown to affect the cell division cycle in mammals and pattern formation and development in flies.

Table 5.1 Enzymes involved in homologous recombination.

Name (gene/protein)	Function
recA/RecA	Binds single-stranded DNA DNA-dependent ATPase Promotes base pairing (assimilation) Control of SOS system
recBC/RecBC	ssDNA and dsDNA exonuclease (part of ExoV) chi-specific endonuclease DNA helicase ATPase Initiator of recombination by unwinding a dsDNA molecule and creating a region of single strandedness in an area containing a chi site
recD/RecD	Exonuclease, part of ExoV
recE/RecE	5′ to 3′ dsDNA exonuclease, ExoVIII
recF/RecF	Binds ssDNA and ATP; can substitute for RecBCD
recG/RecG	Binds to Holliday junctions ATPase Involved in branch migration and resolution of Holliday junctions
recJ/RecJ	5′ to 3′ ssDNA exonuclease; can substitute for RecBCD
recN/RecN	Binds ATP; can substitute for RecBCD
recO/RecO	Binds ssDNA and promotes renaturation; can substitute for RecBCD
recQ/RecQ	dsDNA helicase can substitute for RecBCD
recR/RecR	Binds dsDNA and ATP; can substitute for RecBCD
recT/RecT	ssDNA binding and renaturation
rusA/RusA	Endonuclease; cleaves Holliday junctions
ruvA/RuvA	Binds to Holliday junctions; involved in branch migration
ruvB/RuvB	Binds to RuvA ATPase Involved in branch migration
ruvC/RuvC	Endonuclease specific for Holliday junction; involved in resolution

cellular response to DNA damage (the SOS response: see Chapter 12) and undergoing recombinational DNA repair (Chapter 4) in response to DNA damage by ultraviolet light. This mutation is in the *recA* gene whose product is very important for recombination reactions. For homologies between two sequences that are over 1000 bp, RecA is required for recombination. For homologies between 25 bp and 1000 bp, many of the recombinational events are RecA dependent but some are RecA independent.

Cells that are *recA*$^-$ are reduced in recombination 1000 fold. RecA binds to single-stranded DNA in the presence of ATP. One RecA protein molecule binds per five bases of DNA. This complex is called a DNA-protein filament. RecA is also a DNA-dependent ATPase, able to hydrolyze ATP in the presence of DNA. A RecA–ssDNA filament can bind to and unwind (by breaking hydrogen bonds) another double-stranded DNA, promoting base pairing of the nucleotides in the ssDNA with nucleotides in the complementary strand of the homologous dsDNA. This process is referred to as **assimilation** (Fig. 5.1). The ssDNA–RecA–dsDNA complex further stimulates unwinding of the dsDNA so that more nucleotides from the ssDNA can engage in pairing with bases in the complementary strand of the homologous dsDNA. Single-stranded DNA binding protein (SSB) assists RecA by protecting ssDNA from degradation.

Obviously, RecA does not comprise all of the machinery needed to carry out homologous recombination. RecA needs single-stranded DNA to initiate pairing

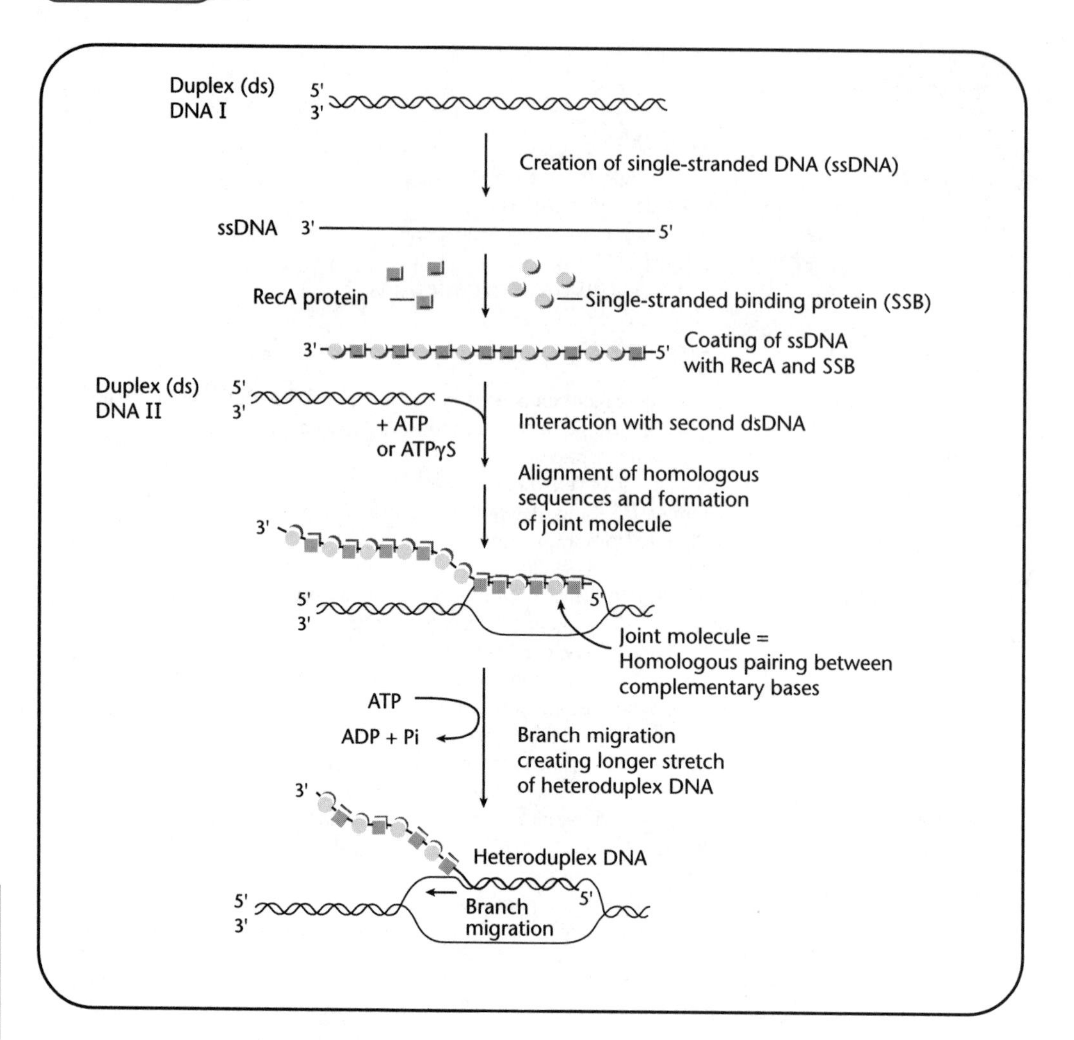

Fig. 5.1 RecA mediated assimilation between two double-stranded DNA molecules that contain homologous base sequences. Single-stranded binding protein (SSB) and energy (ATP) are required. Creation of regions of single strandedness is not mediated by RecA, and must rely on a different type of enzymatic process. Once regions of single strandedness are available, then RecA and SSB coat the single-stranded region in the DNA molecule. There is assimilation of this coated region with the homologous base sequence of a second DNA molecule. Once appropriate alignment has taken place, then a joint molecule is formed between the two homologous DNA molecules.

reactions. But what creates single-stranded DNA in the cell? Sources of ssDNA are nonenzymatic processes such as physical agents that break phosphodiester bonds. But this does not account for the majority of the ssDNA in the cell. Rather, there are specific enzymes in the cell responsible for creating ssDNA. Mutations isolated in *E. coli* that decreased recombination but do not affect DNA repair map to the genes *recB*, *recC*, and *recD*. The products of these three genes form an enzyme complex, named RecBCD. RecBCD is also called Exonuclease V. Mutations in any one of these three genes reduces recombination approximately 100 fold.

RecBCD has several enzymatic functions. It is a nuclease for dsDNA, ssDNA, and specific dsDNA sequences. It separates the strands of dsDNA, which is known as **helicase** activity and it also hydrolyzes ATP, which is known as **ATPase** activity.

FYI 5.2

Chi sequences

Chi sequences were first discovered in recombination studies using bacteriophage λ crosses. λ does not contain any naturally occurring chi sites but a single base pair mutation can create one. The mutation that creates the chi site was observed as a phage mutation that dramatically altered recombination frequencies in specific crosses. *E. coli* contains an unusually large number of chi sites in its genome. There is approximately one chi site every 5 kb, a greater frequency than would occur in random DNA sequences. The chi sites show a bias in how they are oriented relative to *oriC*. Nine chi sites are oriented so that the replication forks will encounter the 3′ end of the sequence first to every one chi site encountered from the 5′ direction.

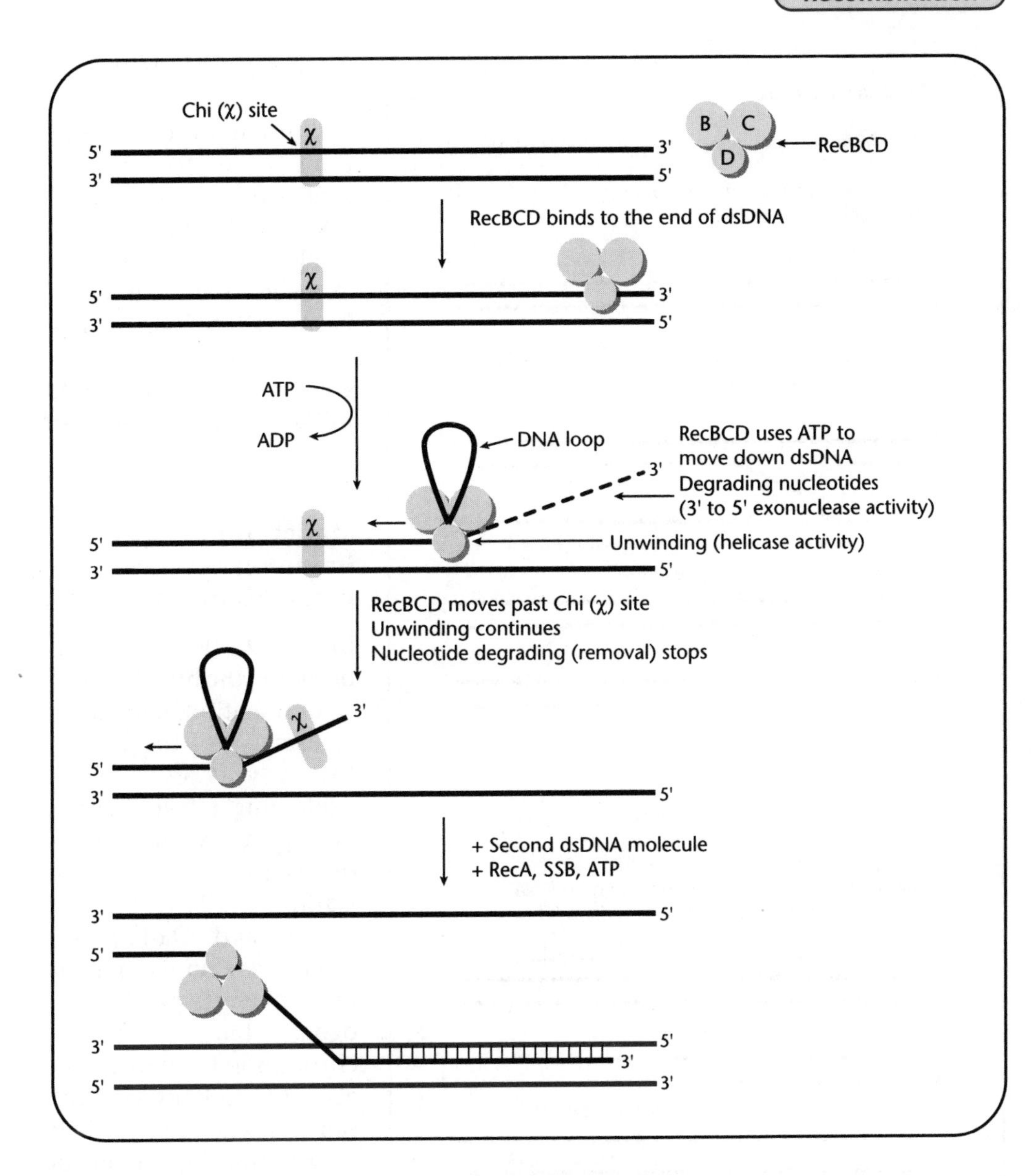

Fig. 5.2 The activity of RecBCD creates a region of single strandedness in a DNA molecule. RecBCD can initiate a homologous recombination reaction by creating a region of single strandedness on a double-stranded DNA molecule that contains a chi (χ) site (5′GCTGGTGG3′). Once a region of single strandedness is available, then RecA, along with SSB and ATP can mediate the assimilation process by promoting homologous base pairing between complementary bases on the two DNA molecules.

RecBCD's interesting array of activities readily contributes to its ability to perform the first function needed in homologous recombination, the creation of ssDNA (Fig. 5.2).

RecBCD binds to one end of a linear dsDNA molecule and initiates unwinding of the dsDNA using its helicase activity to break the hydrogen bonds holding the complementary base pairs together. Bound RecBCD uses ATP to move down the dsDNA molecule, unwinding and degrading nucleotides at the 3′ end on one of the displaced strands. Degradation requires the 3′ to 5′ exonuclease activity of RecBCD. Because RecBCD is unwinding DNA faster than it is degrading one strand, a RecBCD–DNA loop structure forms. The RecBCD keeps moving until it passes a **chi site**, which is a specific sequence of base pairs (5′GCTGGTGG3′). Once RecBCD passes chi, its 3′ to 5′ exonuclease activity is inhibited while its helicase activity is still functional. The inhibition of the 3′ to 5′ exonuclease activity of RecBCD means the 3′ end of the

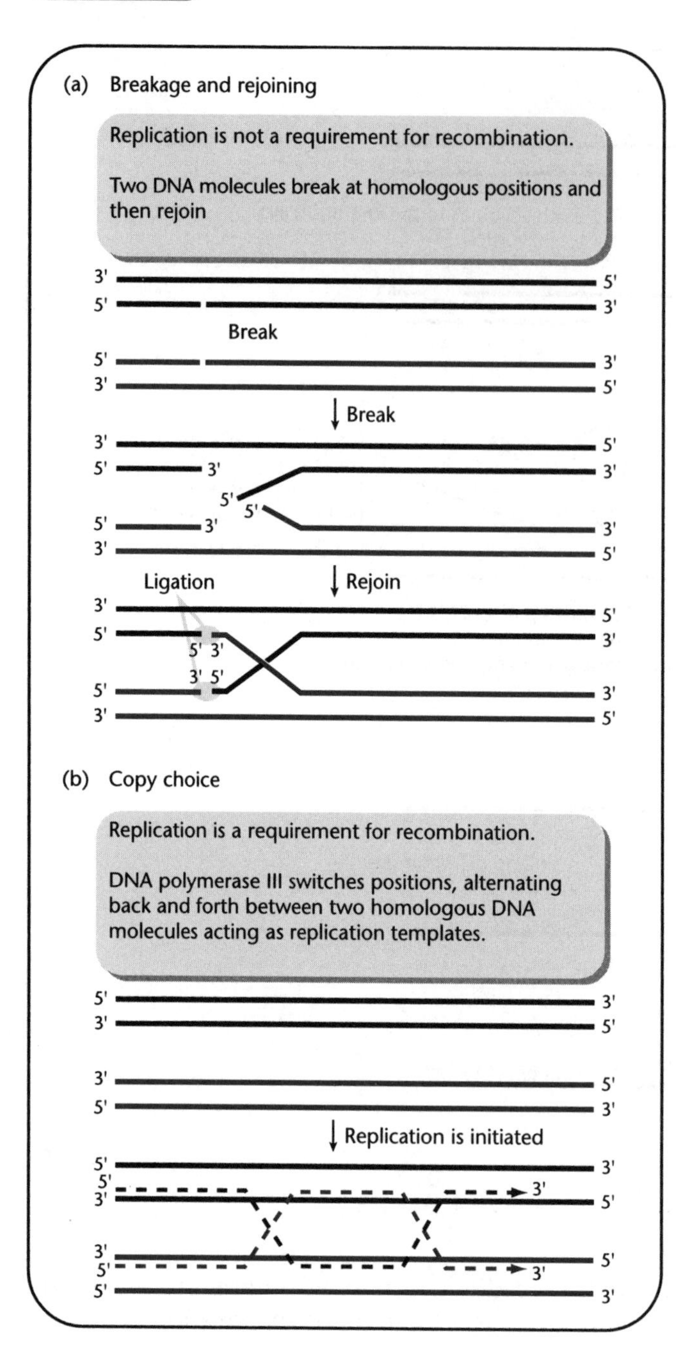

Fig. 5.3 Early models for homologous recombination. (a) Breakage and Rejoining Model: replication is not a requirement for recombination. Two DNA molecules break at homologous positions and then rejoin. (b) Copy Choice Model: replication is a requirement for recombination. DNA polymerase III switches positions, alternating back and forth between two homologous DNA molecules acting as replication templates.

displaced DNA strand will no longer be degraded. The end result of these reactions is the production of single-stranded regions of DNA. RecA binds to these single-stranded regions. Once RecA is bound, the RecA–ssDNA filament can invade another homologous dsDNA molecule to initiate the assimilation step of recombination.

Models for homologous recombination

In 1909, F.A. Janssens proposed the first theory of recombination, named the chiasmatype theory. In the early 1930s a model explaining recombination during meiosis in eukaryotes was proposed by Cyril Darlington. Both Janssens' theory and Darlington's model were similar in their requirements: two homologous, but non-related chromosomes needed to each break at exactly the same point along their length and then rejoin in a new combination (Fig. 5.3a). This was referred to as the Breakage and Rejoining Model. Although this model seemed reasonable at the time, there were several problems with it including the lack of the types of enzymes needed to carry it out and the breakage of two different molecules at exactly the same base pair.

The Copy Choice Model was first proposed by John Belling in 1931 and later revived by Alfred Sturtevant in 1949 and Joshua Lederberg in 1955 in an attempt to explain recombination between bacteriophage DNA and bacterial DNA. The Copy Choice Model proposed that DNA recombination was the result of DNA replication during which more than one parental chromosome acted as a template for the newly synthesized daughter chromosome (Fig. 5.3b). This model envisioned at least two homologous parental chromosomes being used alternatively as templates during replication to create a single daughter chromosome. Because the DNA replication machinery was bouncing back and forth between the two parental templates, the newly synthesized daughter strands would contain regions of new combinations of DNA sequence.

As with the Breakage and Rejoining Model, the Copy Choice Model had its critics. For the Copy Choice Model to be feasible, DNA replication had to be conservative. The model proposes that after replication is complete, the original parental strands that acted as templates would pair while the newly synthesized daughter strands would pair. In fact, this is not possible. DNA replication is semiconservative, with a newly synthesized daughter strand pairing with a parental template strand (see Chapter 2).

In 1961 Meselson and Weigle reported important results describing the recombination of mutations in genes that are distant from each other and physically located on two different bacteriophage λ DNA molecules. The λ DNAs were each labeled with a different radioactive tag and genetically marked by different mutations. The two chemically and genetically marked λ phage were mixed, used to infect *E. coli*, and allowed to recombine. The products from this experiment proved that DNA molecules were breaking and rejoining during recombination. A few years later, studies in eukaryotes led Harold Whitehouse and Robin Holliday to independently propose the first molecular models for recombination. The Whitehouse and Holliday models clearly described recombination as an enzymatic process in which phosphodiester bonds are broken and rejoined, creating short segments of heteroduplex DNA (see below).

The Holliday or double-strand invasion model of recombination

The Holliday junction is named for Robin Holliday who, in 1964, proposed a model for homologous recombination that became widely accepted. What was remarkable about Holliday's model was that at the time he proposed it, the enzyme functions (RecA, RecBCD, etc.) he described had not been fully characterized. The Holliday model for recombination is illustrated in Fig. 5.4. The basic steps for this model are breakage, disassociation and assimilation, ligation, branch migration, isomerization, resolution, and ligation.

In breakage, a single-stranded break is made at the exact same place on strands of like polarity in the two homologous DNA sequences. In disassociation and assimilation, the nicked strand from each DNA sequence disassociates from the non-nicked strand, invades (crosses over to) the other non-nicked strand of the opposite DNA sequence and base pairs (hydrogen bonds) with its complementary sequence to create a short segment of **heteroduplex DNA**. In ligation, phosphodiester bonds are generated, creating a cruciform-like structure called the Holliday junction. The crossed-over strands hold together the two double-stranded DNA sequences. In branch migration, once the two DNA sequences are joined by the crossed-over strands, the crossed connection can diffuse by a zipper-like action (breaking and reforming of hydrogen bonds) in which bases on the two DNA molecules exchange places thus creating a longer stretch of heteroduplex DNA. Isomerization is the process by which Holliday junctions rearrange without breaking the hydrogen bonds between bases. During resolution, the joined DNA sequences are resolved into two separate dsDNAs by a nuclease that breaks phosphodiester bonds. Ligation occurs to seal any remaining nicks so that the two separate dsDNAs do not have any gaps in their phosphodiester backbone.

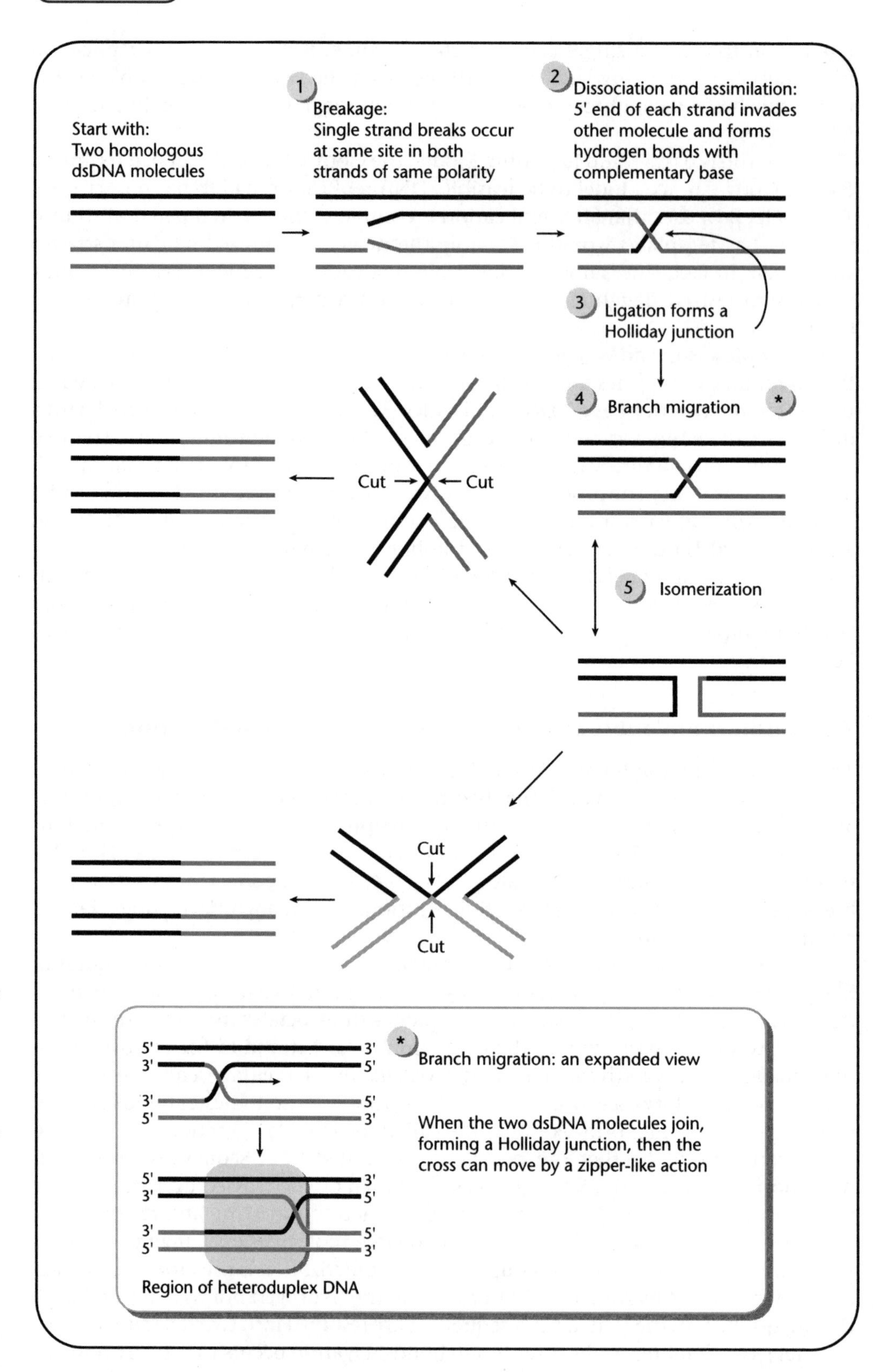

Fig. 5.4 The Holliday model for recombination. The Holliday model requires that a single-strand break occur at the same site in both strands of same polarity. After the strands break, there is dissociation followed by assimilation. Ligation occurs to form a joint molecule called a Holliday junction. Branch migration can now take place, leading to the formation of heteroduplex DNA. Branch migration is when the junction of the two joint DNA molecules moves in a zipper-like fashion.

An alternative to the Holliday model: the single-strand invasion model of Meselson and Radding

As is often the case with models, alternative models form when discrepancies arise. The Holliday model was no exception. The Holliday model proposed that both double-stranded DNA molecules would suffer single-stranded breaks in the same place and the nicked strand from each DNA molecule would invade the non-nicked strand of the other DNA molecule. This process predicted that a similar length of heteroduplex DNA would be found on both resolved DNA molecules. What is the likelihood that both homologous dsDNA molecules would suffer single-stranded breaks in the same place? Can an equivalent length of heteroduplex DNA always be found in all resolved DNA molecules? The answer to both these questions was no, indicating that alternatives to these steps were warranted.

In 1975 Meselson and Radding presented a model that retained several features of the Holliday model but addressed the issues of the number of breaks and the lengths of heteroduplexed DNA. Briefly, their model proposed the following steps (Fig. 5.5).

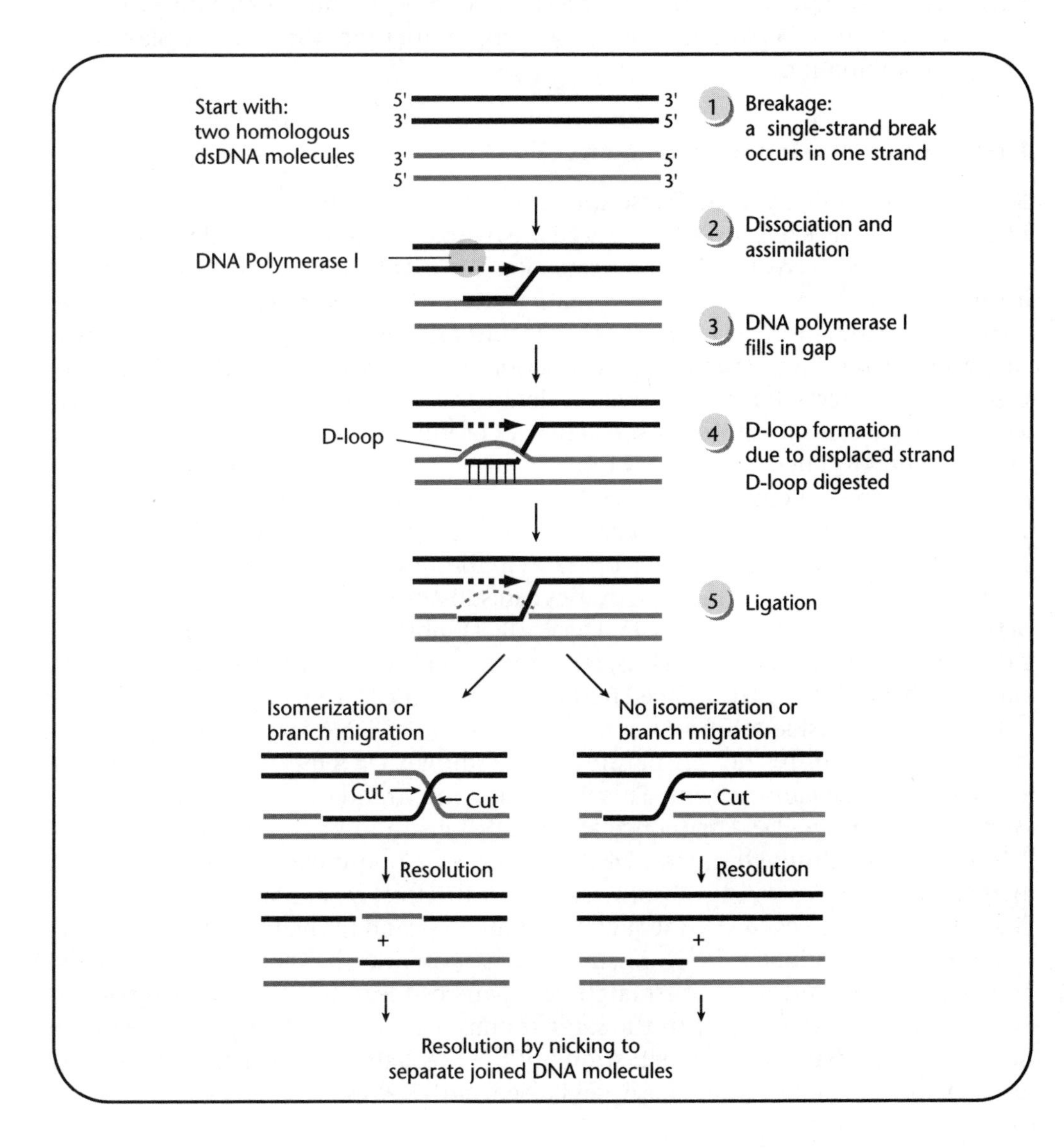

Fig. 5.5 The Meselson and Radding model for recombination. This model does not have the requirement that both DNA strands must initially be broken. Breakage only needs to take place in one DNA strand. This model explains the activity of DNA synthesis during recombination.

First, a single-stranded break is made in only one of the DNA sequences involved in the recombination reaction. Second, the nicked strand dissociates and invades another double-stranded DNA sequence, scanning the sequence until the nicked strand finds complementary bases. The single-stranded DNA assimilates (through hydrogen bonding), displacing the other strand and forming a D-loop structure. Third, the DNA that gave up a portion of its sequence to the other DNA sequence will have the gap filled in by the activity of DNA polymerase I. Fourth, the displaced strand is digested by nuclease activity. Fifth, the phosphodiester backbone is ligated back together. If the joined DNA sequences resolve without isomerization and or branch migration, then only one of the DNA sequences will have heteroduplex DNA. However, just like in the Holliday model, there is potential for the two joined DNA sequences to isomerize and/or branch migrate before being resolved into two separate DNA sequences.

The Meselson and Radding model can be described as either a single-strand invasion model or a break-copy model because of the single break that initiates recombination and the copying or gap filling that is needed to replace nucleotides lost from the dsDNA sequence that suffered the single break. Additional recombination models have arisen which retain features of the Holliday model. These alternatives address situations unique to a particular recombinational situation, for example plasmid or phage recombination.

Further enzymatic considerations

For either the Holliday or the Meselson and Radding recombination model, we can identify where the RecA and RecBCD activities would be needed. RecBCD initiates the nicking process to create ssDNA whereas RecA is involved in the dissociation and assimilation step. What are the enzymatic activities needed to carry out the other steps of recombination? The products of the *ruvA* and *ruvB* genes play a role in branch migration by binding to the Holliday junction. In the presence of ATP and the helicase activity of RecG, RuvA and RuvB accelerate the rate of the zipper-like action. The RecG helicase can substitute for the activities of RuvAB in branch migration in the event the *ruvAB* genes are mutant or missing. The *ruvC* gene product, which is a nuclease, resolves Holliday junctions by breaking phosphodiester bonds and releasing the joined DNA sequences from each other. DNA ligase is involved in the final step. Presumably, the isomerization step operates in the absence of enzymatic activities.

Although no other enzymatic activities can substitute for RecA, several enzymatic activities can substitute for RecBCD. The RecBCD substituting activities include RecF, RecJ, RecN, RecO, RecQ, RecR. Thus, recombination cannot take place in a *recA*$^-$ cell, but can still take place, at a reduced frequency, in a *recBCD*$^-$ cell.

One further consideration relates to the heteroduplex DNA formed during homologous recombination. The very nature of heteroduplex DNA means that there will be mismatched base pairs present. This is because one strand of the heteroduplex has originated from one DNA sequence and the other strand has come from the other DNA sequence. Although the two DNA sequences are homologous, it does not mean that their sequences will be absolutely identical. Thus, mismatched base pairs are likely to arise when each DNA sequence donates a strand in the formation of the recombinant heteroduplex DNA. Mismatch base pairs can be repaired, as discussed in Chapter 4. Repairing the mismatch base pairs can lead to **gene conversion** or both molecules ending up with the same mutant base pair at the same place in the sequence. This depends on which strand of the mismatched base pair is used to fix the other strand. If the mutant base pair is used, both DNA sequences end up carrying the mutant gene. Likewise, if the wild-type sequence is used, both sequences will carry the wild-type gene. In the absence of gene conversion, homologous recombination is **reciprocal**. If the mutant base pair is moved from DNA sequence 1 to DNA sequence 2 then the wild-type base pair is moved from DNA sequence 2 to DNA sequence 1. Genetic information is exchanged but not lost.

Site-specific recombination

In site-specific recombination, a specific type of protein called a **recombinase** acts on both participating DNA sequences at a very specific sequence of nucleotides. The sequence is often short, it must be present on both participating DNA segments, and the sequence is different for different site-specific recombinases. Recombinase-driven cleavage of the phosphodiester bonds is not by hydrolysis like a nuclease, but rather by transesterification, like topoisomerase. It uses this same toposiomerase activity to reseal the phosphodiester bonds. Often the gene encoding the recombinase is located close to the nucleotides comprising the specific site recognized by the recombinase. Site-specific recombination can take place between two separate DNA molecules as long as they both have the same specific site (**intermolecular**; Fig. 5.6a). It can also take place within the same DNA molecule (**intramolecular**; Fig. 5.6b) as long as the specific site is present at least twice on the molecule. Extensive regions of nucleotide homology and the Rec enzymes do not participate in site-specific recombination.

Site-specific recombination is described as conservative for a number of reasons. First, no nucleotides are added or lost as a result of a site-specific recombinational event. Second, DNA replication is not needed, and thus site-specific recombination typifies a breakage and rejoining mechanism rather than a copy-choice or break-copy mechanism. Third, the thermodynamics of a site-specific recombinational event are very energy efficient. Unlike homologous recombination, ATP is not needed for site-specific recombination.

A typical site-specific recombinational event

The specific sequence of nucleotides comprising the recombinase recognition site usually consists of 15 to 30 nucleotide pairs, in which there are at least two inverted repeat sequences acting as binding sites for at least two recombinase molecules. The first step in the reaction is the binding of the recombinase to its recognition sites. The next step, synapsis, is when the specific sites bound by recombinase come into close physical contact with each other and form a synaptic

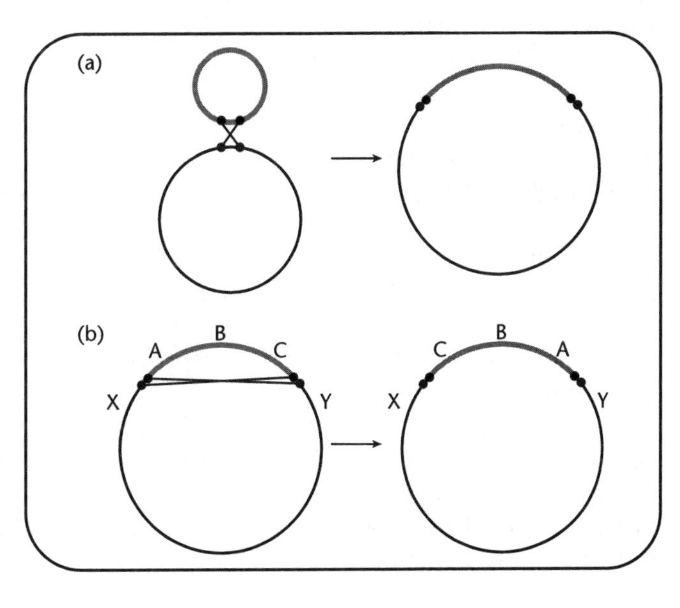

Fig. 5.6 Site-specific recombination can take place between three different combinations of molecule(s). (a) Intermolecular site-specific recombination occurs between two separate DNA molecules (example: λ integration). (b) Intramolecular site-specific recombination occurs within the same DNA molecule (example: inversion or λ excision).

structure consisting of recombinase–DNA complexes. The bound recombinase molecules catalyze the cleavage of four phosphodiester bonds such that both strands in the two partici-pating DNA segments are broken. Finally, the broken phosphodiester bonds are resealed by the bound recombinase. In this manner, the realignment of the broken ends occurs with one end of one DNA segment ligated to the other end of the other DNA segment.

Bacteriophage λ: a model for site-specific recombination

Bacteriophage λ is a virus that infects *E. coli*. As described in Chapter 7, it is capable of two developmental pathways, infecting cells and making more phage or infecting cells, recombining its genome into the bacterial chromosome and silencing the phage genome. After adsorption to an *E. coli* cell, λ injects its genome, which is a double-stranded linear DNA molecule. Once the λ genome is inside the cell, it is converted from a double-stranded linear DNA molecule to a double-stranded circular DNA molecule by a pairing reaction between the complementary single-stranded nucleotide sequence (***cos*** sites) found at each end of λ's linear chromosome (see Fig. 7.2). This circularized λ genome can integrate into the bacterial chromosome by a site-specific recombination event that requires the λ encoded recombinase protein Int (short for Integrase) and the recombination site on the phage (*attP*) and bacterial (*attB*) genomes.

Int recombinase recognizes the site-specific recombination locus, the *attP* site (*attP* = phage attachment site) found in λ and the *attB* site (*attB* = bacterial attachment site) found in *E. coli*. The *attB* site is located between the *E. coli* genes involved in galactose utilization (*gal*) and biotin synthesis (*bio*). The *attP* and *attB* sites are also referred to as *POP′* and *BOB′*, respectively, and contain the same 15 bp core sequence, 5′GCTTTTT TATACTAA3′ (the *O* in *BOB′* and *POP′*) flanked by two dissimilar sequences (Fig. 5.7). Int recombinase uses its topoisomerase activity to mediate cutting within a 7 bp sequence, 5′TTTATAC3′, of the core sequence (Fig. 5.7). The *attB* site of the *E. coli* chromosome contains two binding sites for Int recombinase whereas the *attP* site of the λ chromosome contains five binding sites for Int recombinase. An accessory protein needed for λ's site-specific recombination mechanism is provided by *E. coli*. The *E. coli* encoded protein is named IHF for integration host factor. IHF functions to bend the sequences flanking the *attP* core sequence into closer proximity to the core nucleotide sequence. Three IHF binding sites are located in the *attP* site but no IHF binding sites are found in the *attB* site.

Once the *attP*, bound by Int and IHF, and the *attB*, bound by Int, are in close proximity, Int catalyzes four phosphodiester bond breaks in a staggered fashion in the core 5′TTTATAC3′ sequence of both *attP* and *attB*. The right end of the *attP* core is joined to the left end of the *attB* core and vice versa, with Int catalyzing the formation of the necessary phosphodiester bonds. What were once two distinct DNA molecules have been joined together to form one larger DNA molecule. The site-specific recombination is reciprocal, preserves all preexisting DNA, and no mismatched base pairs are formed that need to be repaired. The entire process proceeds without ATP hydrolysis.

Just as the λ genome can be integrated into the *E. coli* genome through an intermolecular site-specific exchange between the *attP* and *attB* sites, this same λ genome can be removed or excised with a second site-specific exchange at these same sites. This process is called excision, and it requires the activities of Int, IHF, and a second protein, Xis (an excisionase) (Fig. 5.7).

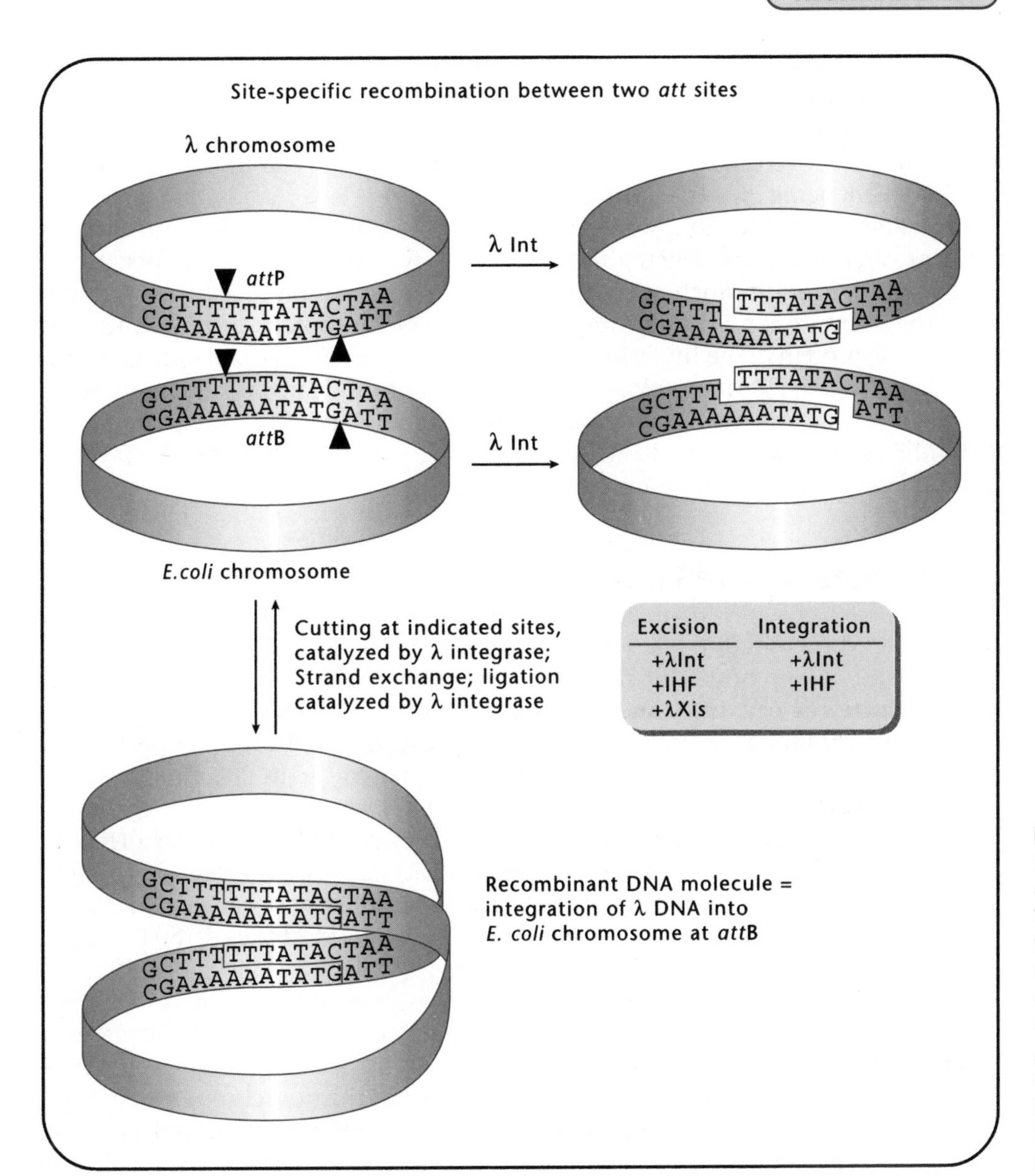

Fig. 5.7 A description of the molecular events for integration and excision of bacteriophage λ. To integrate the λ genome into the *E. coli* chromosome, two specific sites (*attP*, *attB*) and the activities of λ Int, and *E. coli* IHF are required. To excise the λ genome from the *E. coli* chromosome, two specific sites (*attR*, *attL*) and the activities of λ Int, *E. coli* IHF, and λ Xis are required.

Other microbial examples of site-specific recombination

Other bacteriophages, such as HK22, P1 and P2, have site-specific recombinational mechanisms similar to λ. Transpositional events mediated by the transposons Tn*3*, Tn*1000*, and Tn*21* use an intramolecular site-specific recombinational event after replicating to separate the replicated transposons during transposition (see Chapter 6). Multicopy plasmids with a ColEl origin of replication use an *E. coli* encoded site-specific recombinational system (*xer*) to resolve dimers or multimers of the plasmid as it replicates and to enhance the efficiency of segregation of the ColE1 plasmids into daughter cells.

Another type of intramolecular site-specific recombination is called inversion. Inversion involves recombination between specific core sites that lie in inverted orientation (Fig. 5.6b). Organisms use inversion as a means to alternate between two different states of gene expression. For example, inversion can be used to change the orientation of a promoter. In one orientation the promoter is active and drives the expression of the gene it controls. In the other orientation the promoter is not active, effectively silencing the gene it controls.

Inversion can be used to change the sequence of a gene. Inversion contributes to antigen variation by influencing the type of cell surface proteins that appear on a pathogenic bacterium, such as trypanosomes. Phase variation in *Salmonella* results from the alternate expression (through the *hin* inversion system) of genes encoding its flagellar proteins. The inversion system, *vin*, of *Citrobacter freundii* mediates the expression of two alternate colony morphology forms. The inversion *fim* system of *E. coli* controls expression of alternate forms of pili, which act as virulence factors to bind *E. coli* to eukaryotic cells. The bacteriophages, Mu and P1, control expression of different sets of tail fiber proteins by inversion.

Illegitimate recombination

Sometimes recombination takes place in the absence of RecA, large regions of DNA sequence homology, or specific sites. These recombinational events are referred to as **illegitimate recombination**, and include spontaneous DNA arrangements such as deletions, duplications, or the formation of specialized transducing phages. While little is known about the mechanisms and proteins involved in illegitimate recombination, several interesting observations have been made.

Illegitimate recombination events are frequently associated with very small repeated sequences. These **short sequence repeats** (SSR) can be as small as 3 or 4 bp. The frequency of illegitimate recombination decreases as the SSR are located further apart. The frequency increases exponentially as the SSR get longer. At SSR of 8 bp in length, illegitimate recombination can be quite frequent.

Illegitimate recombination can occur in a RecA-independent fashion. However, some illegitimate recombination events are stimulated or enhanced by the presence of RecA. The other recombination functions can influence illegitimate recombination. In assays that look at induction of a phage out of the host chromosome, RecE, RecT, RecJ, RecO, and RecR stimulated illegitimate recombination and RecQ inhibited it. Topoisomerases have been implicated in illegitimate recombination in both bacteria and yeast.

Summary

Why would living organisms need genetic recombination mechanisms? Considering that *recA* gene homologs have been identified in all living organisms examined, a need for recombination in species survival and evolution might be suggested. Rearranging genetic material permits species to adapt to changing environments or respond to environments that no longer provide optimal conditions. Genetic recombination creates new combinations of genes enlarging the genetic repertoire of a given species (genetic variation). Genetic recombination is involved in DNA repair processes thereby allowing cells to retrieve DNA sequences lost through DNA damage. Genetic recombination can control gene expression by affecting the location and orientation of genes in their chromosome.

Homologous recombination is the major mechanism that bacteria use to rearrange their genetic material. Homologous recombination requires homologous DNA sequences, the enzymatic machinery of the Rec system and energy in the form of ATP hydrolysis. In contrast, site-specific recombination is very specialized, involving short sites of specific nucleotide base pairs recognized by a recombinase that functions like a topoisomerase rather than a nuclease. Site-specific recombination is energy efficient and does not require ATP hydrolysis.

Depending on the DNA sequences that are being rearranged, a significant portion of the recombination events is the result of illegitimate recombination. While the mechanisms and enzymes involved are still being defined, some of the Rec functions and short sequence repeats have been implicated in the process.

Study questions

1 What are the similarities and differences between the Holliday model and Meselson and Radding model for homologous recombination?
2 Why are *recA⁻ E. coli* more deficient in homologous recombination than *recB⁻ E. coli*?
3 What does branch migration accomplish?
4 Compare and contrast the enzymatic machinery needed to carry out the processes of homologous recombination and site-specific recombination.
5 If an *E. coli* strain mutant for the gene encoding IHF is infected by λ, what would be the outcome of this infection? Why?
6 What function(s) does RecA carry out?
7 What is the function(s) of RecBCD in recombination?
8 What sizes of substrates require RecA for homologous recombination?

Further reading

Clark, A.J. and Sandler, S.J. 1994. Homologous genetic recombination: the pieces begin to fall into place. *Critical Reviews in Microbiology*, **20**: 125–42.

Lloyd, R.G. and Low, K.B. 1996. Homologous recombination. In *Escherichia coli* and *Salmonella typhimurium: Cellular and Molecular Biology*, 2nd edn., eds. F.C. Neidhardt, R. Curtiss III, J.L. Ingraham, E.C.C. Lin, K.B. Low, B. Hagasanik, W.S. Rexnikoff, M. Riley, M. Schaechter, and H.E. Umbarger, pp. 2236–55. Washington, DC: ASM Press.

Nash, H.A. 1996. Site-specific recombination: integration, excision, resolution, and inversion of defined DNA segments. In *Escherichia coli* and *Salmonella typhimurium: Cellular and Molecular Biology*, 2nd edn., eds. F.C. Neidhardt, R. Curtiss III, J.L. Ingraham, E.C.C. Lin, K.B. Low, B. Hagasanik, W.S. Rexnikoff, M. Riley, M. Schaechter, and H.E. Umbarger, pp. 2363–76. Washington, DC: ASM Press.

Ptashne, M. 1992. *A Genetic Switch: Phage λ and Higher Organisms*, 2nd edn., Cambridge: Cell Press, Blackwell Scientific Publications.

Smith, G.R. 1988. Homologous recombination in procaryotes. *Microbiological Reviews*, **52**: 1–28.

Chapter 6

Transposition

A **transposon**, or **transposable element**, is a segment of DNA that is capable of moving itself from one location in a DNA molecule to another. Transposons have discreet ends—the same piece of DNA, down to the single base pair, is always moved as a unit. Transposons can move from the DNA molecules they reside on to virtually any other DNA molecule. One restriction that is placed on transposon movement is that transposons generally cannot transpose into themselves. In this chapter, we will examine the different kinds of transposons, the mechanisms they use to move from once place to another, and the consequences and uses for that movement.

Transposons were discovered by Barbara McClintock in the 1940s. She carefully documented the inheritance of kernel color in maize and proposed that a mobile piece of DNA would best explain her data. At the time Dr. McClintock was conducting her experiments, DNA was thought to be a static repository for information that could not be rearranged. Not until several decades later, when transposons were discovered in bacteria, was the maize story revisited. Dr. McClintock had indeed discovered the first transposons. She was awarded the Nobel Prize in 1983 for her pioneering work.

The structure of transposons

Transposons exhibit varying degrees of complexity in their structures. The simplest are known as **insertion sequences (IS)** or insertion elements (Fig. 6.1). Insertion sequences are small and compact; some are as small as 750 base pairs and they range to ~2000 base pairs. They encode only proteins needed for transposition and contain the cis-acting DNA sequences, called **inverted repeats (IR)** at the ends of the element that are needed for transposition. The inverted repeats are usually ~15 to 25 base pairs in length. Well-studied examples of insertion sequences include IS*1*, IS*3*, IS*10*, IS*50*, and IS*903*.

Composite transposons contain two IS elements at either end of a centrally located piece of DNA (Fig. 6.2). The central piece of DNA can encode a variety of different proteins, such as **antibiotic resistance determinants** and **virulence factors**. The IS elements on the ends are found as either **direct repeats** or inverted repeats. Well-studied examples of composite transposons include Tn*5*, which contains IS*50* at the ends and Tn*10*, which contains IS*10* at the ends.

Non-composite transposons are complex transposons that are not composed

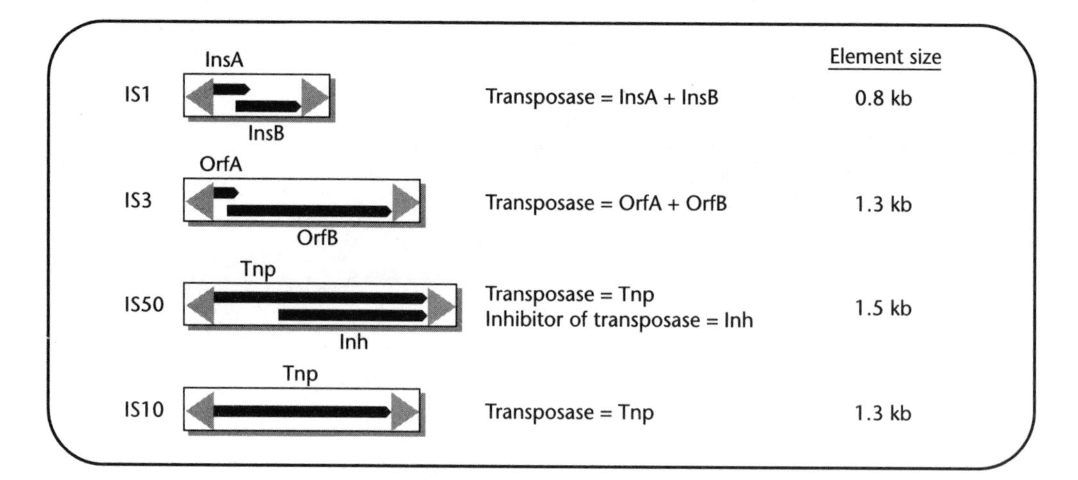

Fig. 6.1 The structure of some *E. coli* insertion sequences. The triangles indicate the cis-acting sequences at the ends of the elements. They are required for transposition. The solid lines indicate the proteins encoded by the element. Some transposases are a complex of two separate polypeptides (those in IS1, IS3) and some are a single polypeptide (those in IS50, IS10). IS50 encodes an inhibitor of transposition that is actually a mutant form of transposase. This inhibitor is missing the DNA binding activity of transposase.

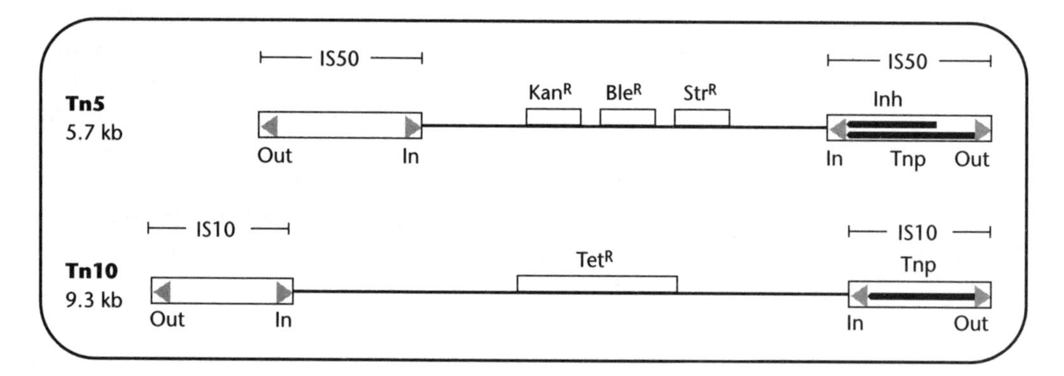

Fig. 6.2 Two examples of composite transposons. Tn*5* is bounded by copies of IS*50*. Only one of the IS*50* makes the transposase and inhibitor proteins. Tn*5* encodes kanamycin resistance (Kan^R), bleomycin resistance (Ble^R), and streptomycin resistance (Str^R). Tn*10* is bounded by IS*10* elements. Tn*10* encodes tetracycline resistance (Tet^R). Because the transposons are bounded by IS elements that can move independently, composite transposons have four reactive ends (two outside ends and two inside ends).

of smaller modules (Fig. 6.3). They encode transposition proteins but do not have IS elements at their ends. Non-composite transposons usually carry genes for antibiotic resistance determinants, virulence factors, and **catabolic enzymes**. These elements also use a more complex set of DNA breaking and joining reactions to move from place to place. Well-studied examples of non-composite transposons are Tn*3* and Tn*7*.

The most complex transposons known are actually bacteriophage, which use transposition as part of their lifestyle. The best-studied example is bacteriophage **Mu** (Fig. 6.4). Mu uses transposition to integrate into the bacterial chromosome and establish

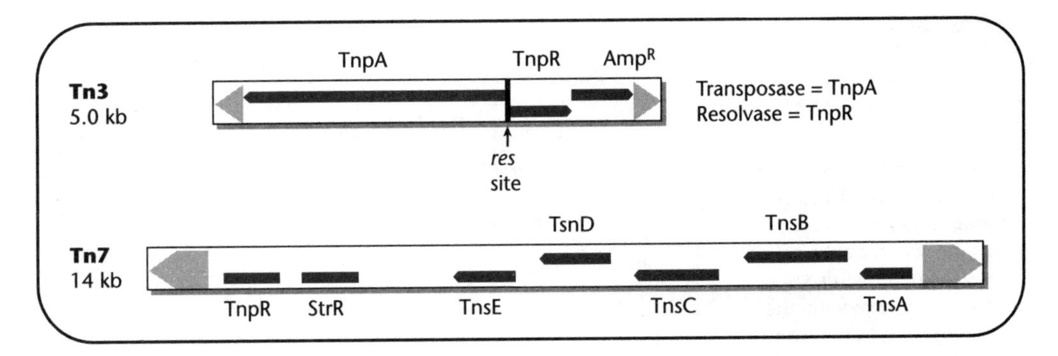

Fig. 6.3 Two examples of non-composite transposons. While non-composite transposons do not have IS elements at their ends, they do have cis-acting sequences required for transposition. In Tn*7*, these sequences are larger than in Tn*3* and other transposons. Tn*3* contains an internal site-specific recombination site (*res*) to resolve cointegrate structures. TnpR (resolvase) is the enzyme needed for cointegrate resolution. In Tn*7*, the five transposon proteins work in varying combinations to choose the target site and provide transposase activity. Amp^R is ampicillin resistance, Tnp^R is trimethoprim resistance and Str^R is streptomycin resistance.

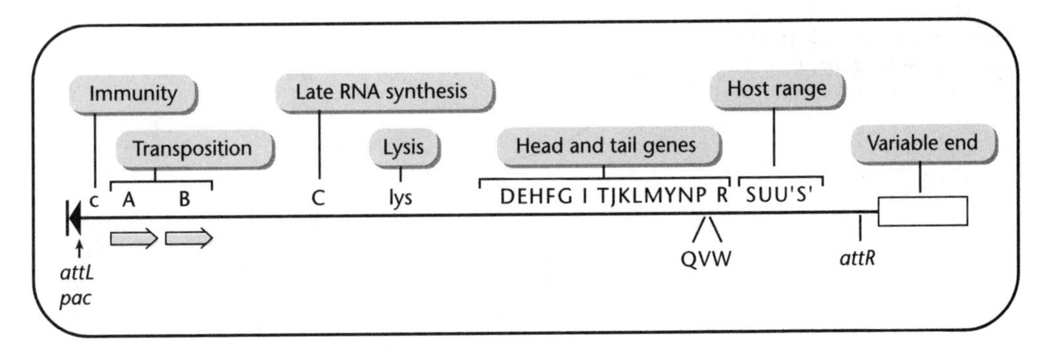

Fig. 6.4 The genetic map of bacteriophage Mu. The phage recognizes the *pac* site on the left end of the genome and begins packaging DNA. Mu packages DNA by a headful mechanism, thus the right end of the phage will have a different sequence of chromosomal DNA in each phage particle. The rest of the phage genome is arranged in modules with functions that are needed at the same time usually expressed together. A good example of this is the head and tail genes.

its DNA in a silenced state as part of the chromosome. In order to produce more phage particles, Mu transposes at a very high frequency to many different sites in the chromosome. The phage DNA is packaged into phage heads directly from these new chromosomal locations (see Chapter 7 for more details on phage development).

The frequency of transposition

The frequency of tranposition can vary widely, depending on the transposon. It is generally between 10^{-6} to 10^{-3} per transposon per generation. The frequency must be low enough that transposition does not significantly harm the bacterial chromosome but high enough so that the transposon can disseminate among the population.

FYI 6.1

Measuring transposition

Several different assays have been developed to measure transposition. Most take advantage of the antibiotic resistance gene carried by transposons to monitor the location of the transposon. In one of the most widely used assays, the transposon is carried on the chromosome and the cell contains an F plasmid. The cells are allowed to grow for a certain amount of time and then the F plasmid is mated into another strain that does not carry any transposons. Transposition of the element from the chromosome to the F plasmid is followed by screening of the exconjugants for the antibiotic marker from the element.

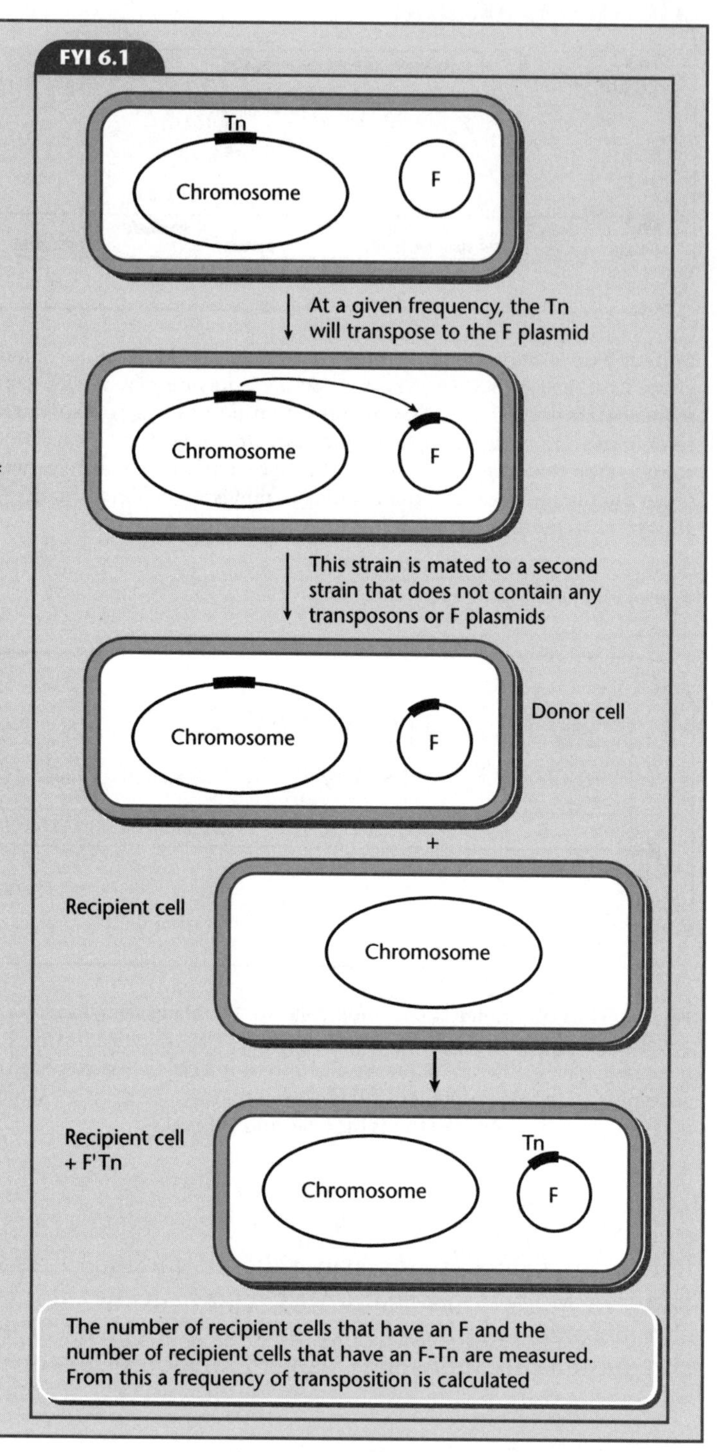

The two types of transposition reactions

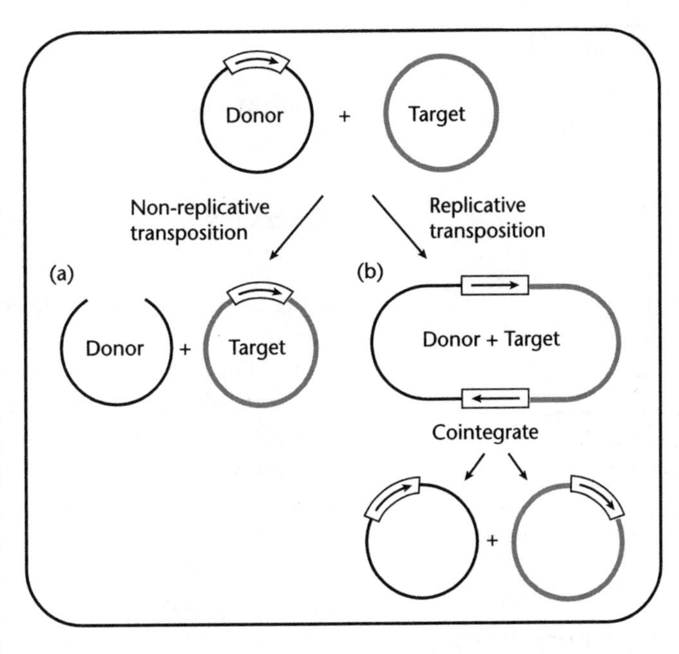

Fig. 6.5 The two types of transposition reactions. (a) Non-replicative transposition results in the simple insertion of the transposon into the target DNA. (b) Replicative transposition results in a joining of the target and donor DNA by two copies of the transposon. This structure is called a cointegrate. Cointegrates can be resolved by general recombination or by site-specific recombination depending on the transposon.

Transposons move from one piece of DNA to another using two distinct mechanisms. In one mechanism, **non-replicative** or "**cut-and-paste**" transposition, the transposon is "cut" out of the donor DNA and "pasted" into the target DNA (Fig. 6.5a). A limited amount of DNA replication is needed to repair the joints between the transposon and the target DNA. In the second, **replicative transposition**, DNA replication is much more extensive (Fig. 6.5b). During translocation of the transposon from one site to another, the entire transposon is replicated to generate an intermediate called a **cointegrate**. The cointegrate contains two copies of the transposon. Cointegrates can be subsequently processed by other types of recombination reactions to yield what appears to be a simple insertion of the transposon and a copy of the donor molecule (see below).

The transposition machinery

Transposition requires three DNA substrates, the two inverted repeats at the ends of the transposon and the **target DNA** (Fig. 6.6). The inverted repeats are required for two reasons. First, they are the sites of breaking and joining of the transposon DNA to the target DNA. Second, the inverted repeats bind transposition proteins to bring them into close proximity. One of the distinct features of transposons is their ability to insert into many different target sites. For most transposons, there is little or no apparent sequence specificity for the target. Other transposons preferentially insert at certain sequences, or **hotspots**.

The major enzyme required for transposition is **transposase**. Transposase mediates specific recognition of the transposon ends, as well as breaking and joining of the transposon to the target DNA. Transposase is encoded by the transposon. The amount of transposase protein made influences how

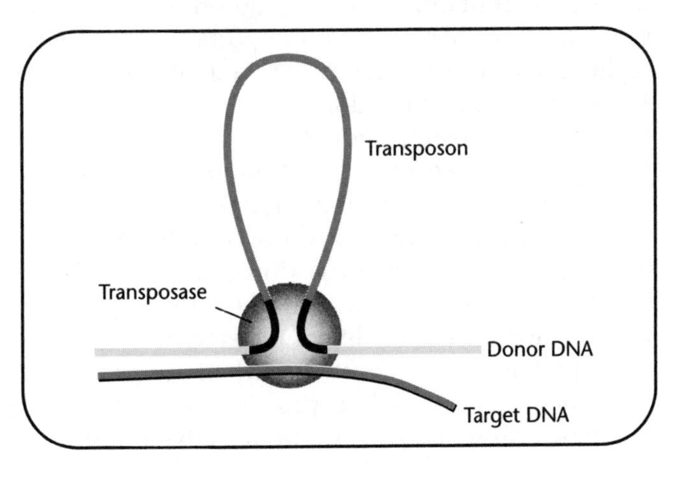

Fig. 6.6 Transposition requires three DNA substrates, both ends of the transposon and the target site. Transposase protein and the three DNA substrates must be brought together into a complex. This complex is necessary to ensure proper breakage and joining of the DNA.

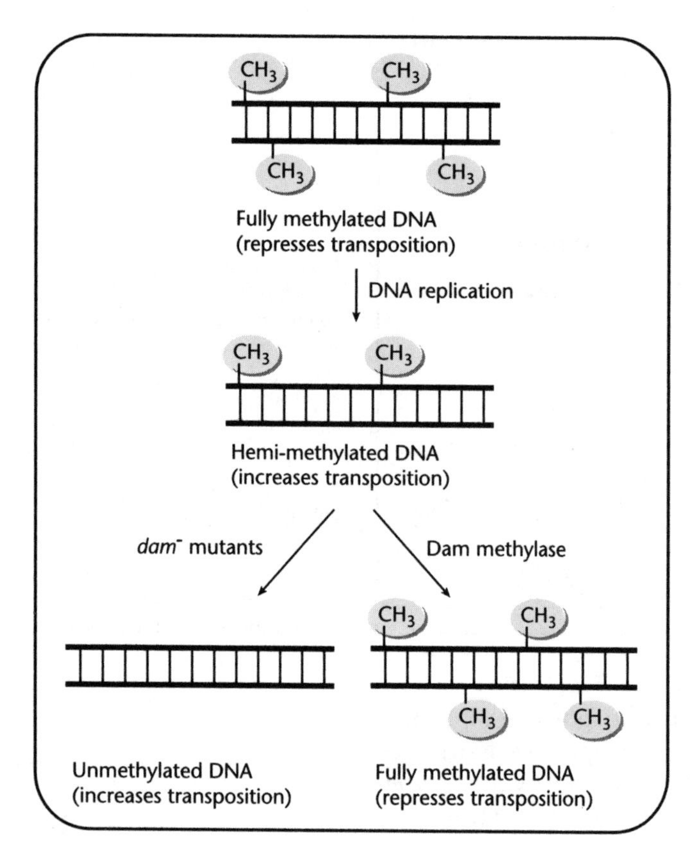

Fig. 6.7 The methylation state of the DNA affects the transposition frequency of some transposons.

many transposition reactions can take place. By regulating the expression of transposase, the frequency of transposition can be set at a level that is tolerated by the host cell.

Accessory proteins encoded by the transposon

Some transposons encode additional proteins that play a role in moving or regulating the movement of the element. Tn*7* has one protein that mediates recognition of the ends of the element and another protein that is responsible for breaking and joining the DNA molecules. Several other Tn*7* proteins control the end-binding activity and the breaking and joining activities of these proteins. IS*50* contains a negative regulator of transposition. This negative regulator is actually a truncated form of the transposase that lacks DNA binding activity. The truncated negative regulator interacts with wild-type transposase and prevents it from working properly. Mu contains a protein, MuB, which is a positive activator of transposition. MuB binds to the target site and plays a key role in juxtaposition of the transposon ends, transposase, and target DNA.

Accessory proteins encoded by the host

Host proteins assist in transposition at several different levels. They can regulate transposition, directly affect the breaking and joining reactions, or help clean up the DNA after transposition has taken place. The *E. coli* protein, Dam methylase is responsible for methylating newly replicated DNA (Fig. 6.7). Fully methylated DNA represses the transposition of several elements, including IS*10*, IS*50*, and IS*903*. In *dam* mutants, the DNA is unmethylated. In newly replicated DNA, the DNA is only methylated on one strand (hemi-methylation). The non-methylated DNA of *dam* mutants and newly replicated DNA are subject to an increased transposition frequency. This mechanism also allows the transposon to monitor the replication status of the DNA. If the DNA containing the transposon is newly replicated, there are two copies of the donor transposon. This is important for those transposons that move by non-replicative mechanisms and leave a double-stranded break in the DNA (see below). The second copy of the donor transposon can serve as a template for homologous replication to repair the double-stranded break.

E. coli contains several, small DNA-binding proteins that are involved in constructing and maintaining DNA–protein complexes. They frequently help bend the DNA and bring the components of the complex into the correct positions. These DNA-binding proteins, named IHF (integration host factor), HNS (histone-like protein), HU (histone-like protein), and FIS (regulator of transcription), aid in the transposition reactions of several different elements including Mu, IS*1*, IS*10*, and IS*50*. Not all of the proteins are necessary for each of the elements.

The best examples of host proteins required for transposition are DNA replication and repair proteins. With non-replicative transposons, small, single-stranded gaps are generated in the DNA at the target site. DNA repair proteins come in and synthesize the complementary strands at these gap sites (Fig. 6.8). With replicative transposons, DNA replication proteins replicate across the entire transposon to generate the cointegrate structure.

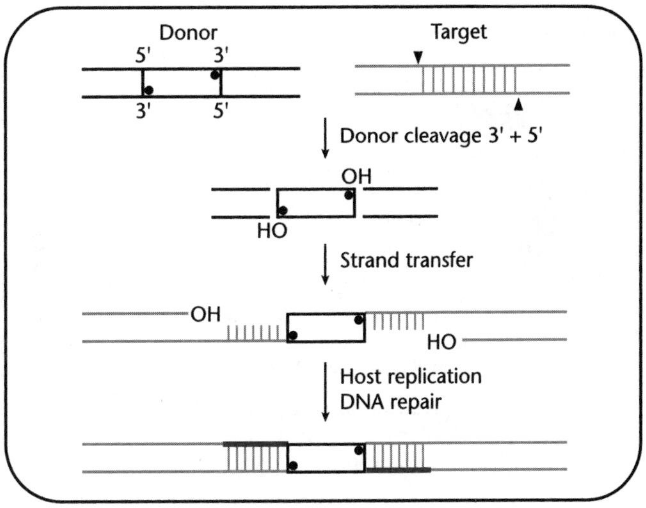

Fig. 6.8 Non-replicative transposition. The transposon ends are cleaved at the 3′ and 5′ ends. A strand transfer reaction takes place that joins the 3′ OH to the asymmetrically cleaved target site. Host DNA repair enzymes carry out limited DNA synthesis of the target site DNA. This produces the characteristic target site duplications.

Non-replicative transposition

Transposition initiates with transposase binding to the ends of the inverted repeats of the transposable element and to the target DNA (Fig. 6.8). For non-replicative insertions, the transposon ends are cleaved at the 3′ ends, leaving a 3′ OH, and at the 5′ ends. The target DNA is cleaved asymmetrically. The number of asymmetric bases in the cleavage is indicative of a given transposon. These bases are duplicated in the process of transposition. After both the transposon and the donor DNA have been cut, an exchange of DNA strands takes place. The 3′ OH groups exposed on the transposon are covalently attached to the 5′ ends of the donor DNA. Host DNA repair enzymes replicate the small patch of target DNA that is single stranded. This leads to the characteristic target site duplication found in most transposition events. In this manner, a simple excision and insertion of the transposon takes place. Tn*10* and Tn*7* are examples of transposons that move by non-replicative transposition.

Replicative transposition

Replicative transposition begins in the same manner as non-replicative transposition, with the binding of transposase to the element's ends and to the target DNA (Fig. 6.9). The 3′ ends of the transposon are cleaved, as is the target DNA. Unlike non-replicative transposition, the 5′ ends of the transposon are not cleaved. Strand transfer takes place and the 3′ OH groups of the trans-

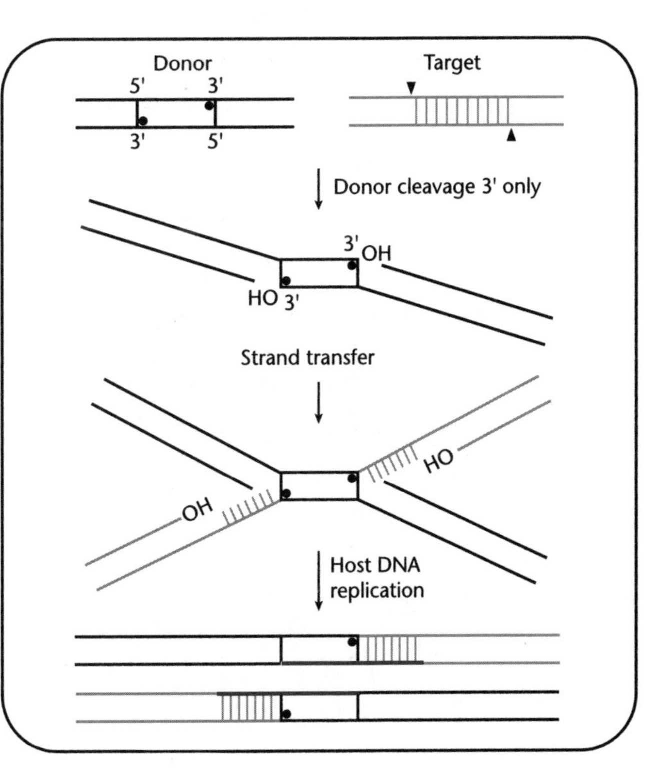

Fig. 6.9 Replicative transposition. The transposon ends are cleaved at the 3′ ends only. Strand transfer joins the 3′ OH of the transposon to the asymmetrically cleaved target site. Extended host DNA replication travels across the entire transposon as well as the target site.

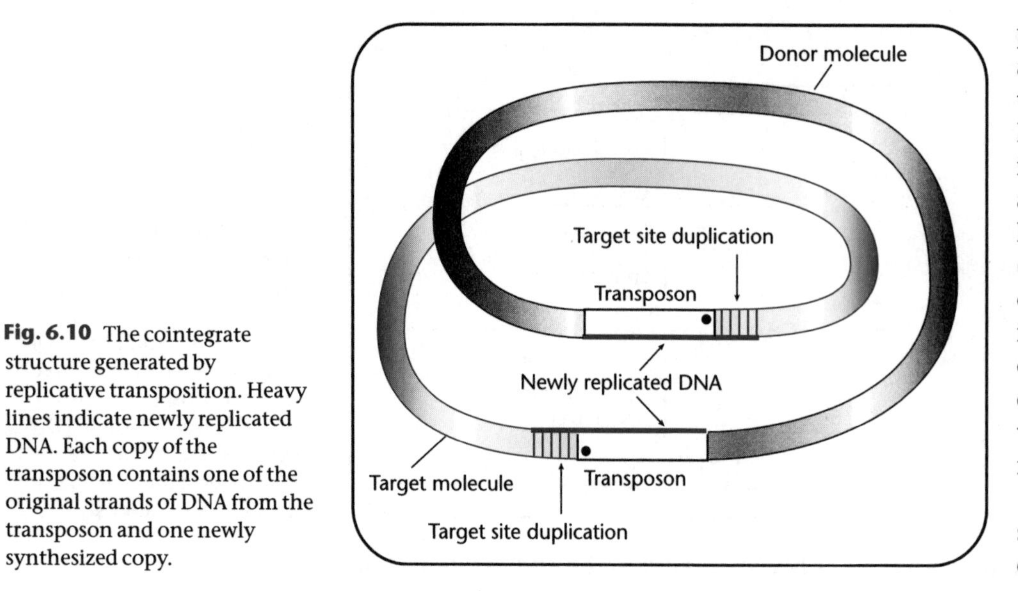

Fig. 6.10 The cointegrate structure generated by replicative transposition. Heavy lines indicate newly replicated DNA. Each copy of the transposon contains one of the original strands of DNA from the transposon and one newly synthesized copy.

poson are joined to the asymmetrically cleaved target DNA. Host-mediated DNA replication polymerizes across this structure and generates what is known as a cointegrate (Fig. 6.10). There are two copies of the transposon in the cointegrate. Both copies have one strand of DNA from the original transposon and one newly replicated strand.

A cointegrate can be resolved into two DNA molecules that contain a single copy of the transposon by one of two mechanisms. Some transposons, like Tn*3*, encode a specific protein called **resolvase** (Fig. 6.11a). Resolvase binds to a specific site internal to the transposon, called the ***res* site**, and promotes site-specific recombination between the two copies of the transposon. Other transposons rely on the host's homologous recombination enzymes to resolve cointegrates (Fig. 6.11b).

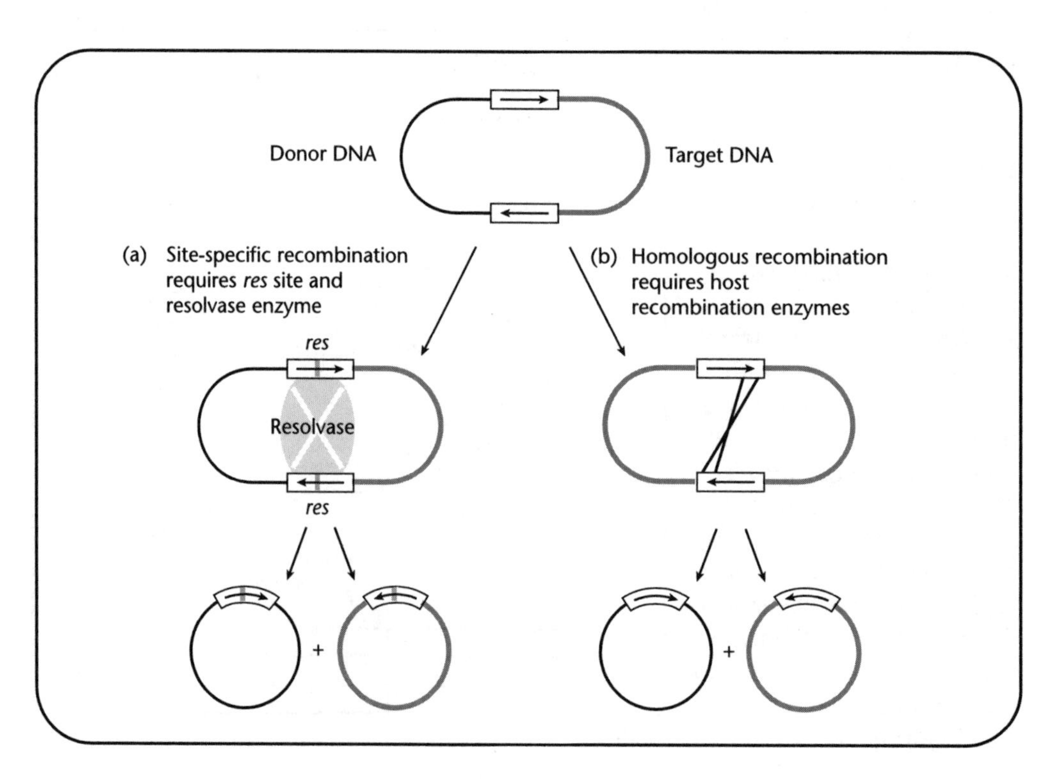

Fig. 6.11 Resolution of the cointegrate structure. (a) If the transposon carries a *res* site and encodes a resolvase enzyme, it will use a site-specific recombination mechanism to resolve (separate) the cointegrate. (b) If the transposon does not contain a site-specific recombination system, the cointegrate will resolve by homologous recombination. The homologous recombination event can take place anywhere along the transposon.

Does the formation of a cointegrate predict the transposition mechanism?

On the surface, it would seem that if a cointegrate structure is one of the products of the transposition event that this predicts the transposon must have moved by a replicative mechanism. In fact, a cointegrate can be formed by several different mechanisms (Fig. 6.12) and does not indicate the mechanism used. In order to determine the transposition mechanism, additional experiments must be carried out.

The fate of the donor site

When a transposon moves by a non-replicative mechanism, it is completely excised from one place (the donor site) and moved to a second place (the target site). In this reaction, the donor site is left with a double-stranded DNA break that has no homology on either side (Fig. 6.13a). What happens to the donor molecule? The donor molecule will be degraded if the double-stranded break is not fixed. If the donor molecule is the chromosome and the cell has only one copy of that chromosome, degradation is lethal to the cell. While degradation can occur, it cannot be the fate of all donor molecules or it is unlikely that transposons would have survived evolution. The broken donor DNA can be repaired if the cell carries a second copy of the donor DNA with the

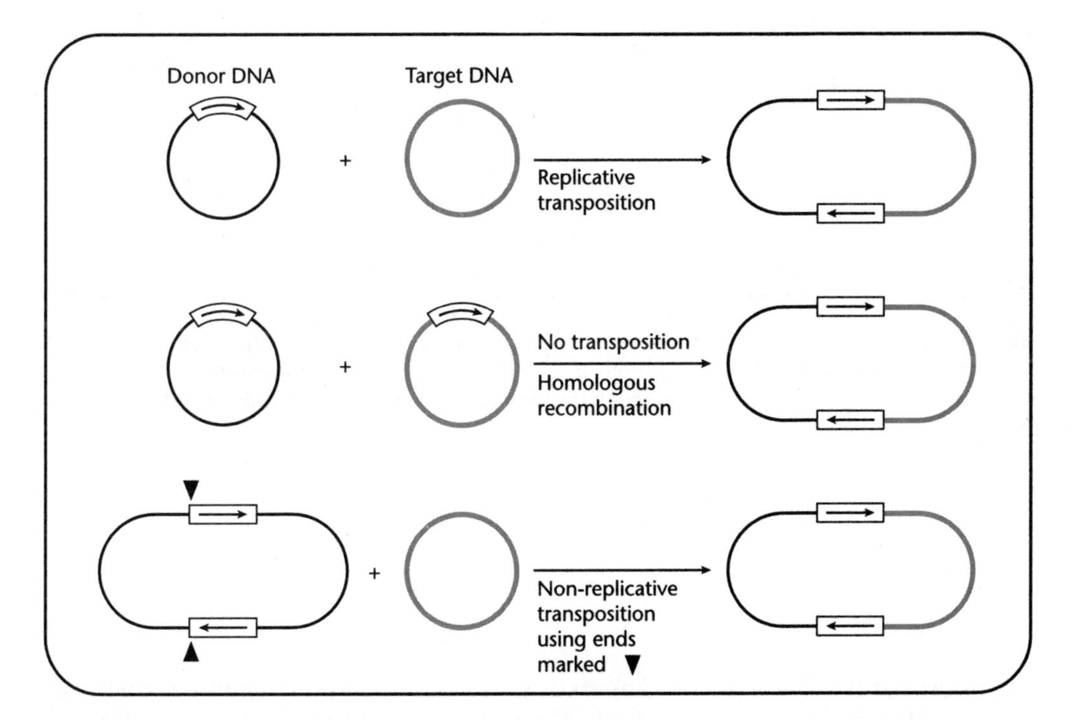

Fig. 6.12 Cointegrates can be formed by several different mechanisms. (a) Cointegrate formation by replicative transposition. (b) Cointegrate formation by homologous recombination between two molecules that carry transposons. No transposition takes place in this example. (c) Cointegrate formation by non-replicative transposition from a donor molecule that is itself a cointegrate. In this example the ends marked (▲) are involved in the transposition reaction.

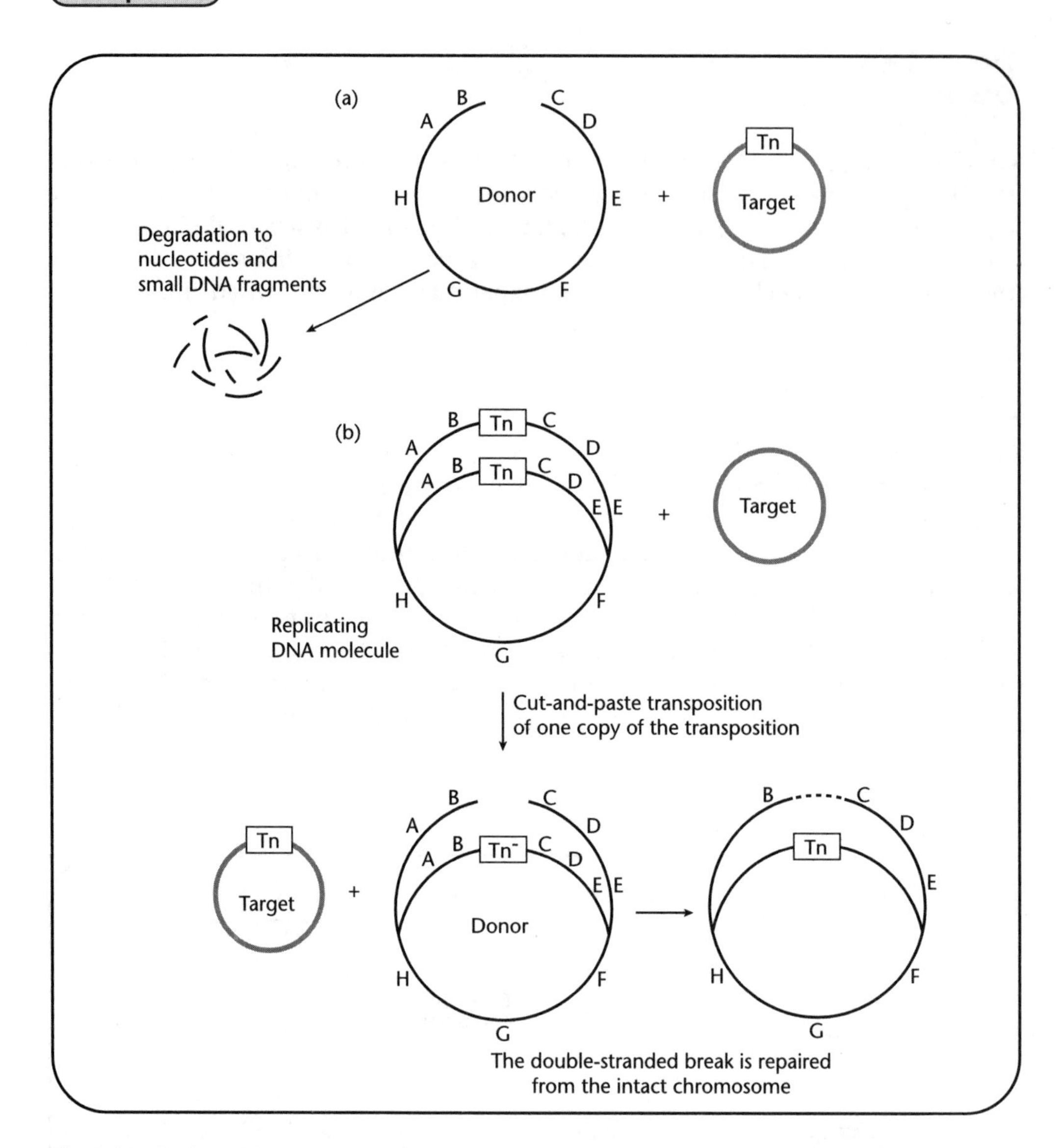

Fig. 6.13 The fate of the target DNA after a cut and paste transpositon event. (a) If the double-stranded break is not fixed the donor DNA molecule will be degraded. (b) If the donor molecule has undergone DNA replication of the region containing the transposon, the double-stranded break will be fixed by homologous replication.

transposon still in place (Fig. 6.13b). In this case, homologous recombination would use the intact copy of the donor site to repair the damaged donor site. When a transposon moves by a replicative mechanism, no gap is left in the donor DNA. Instead an intact copy of the transposon remains in the donor site. Once DNA replication has restored the DNA to a double-stranded duplex and the cointegrate is resolved, nothing more need occur.

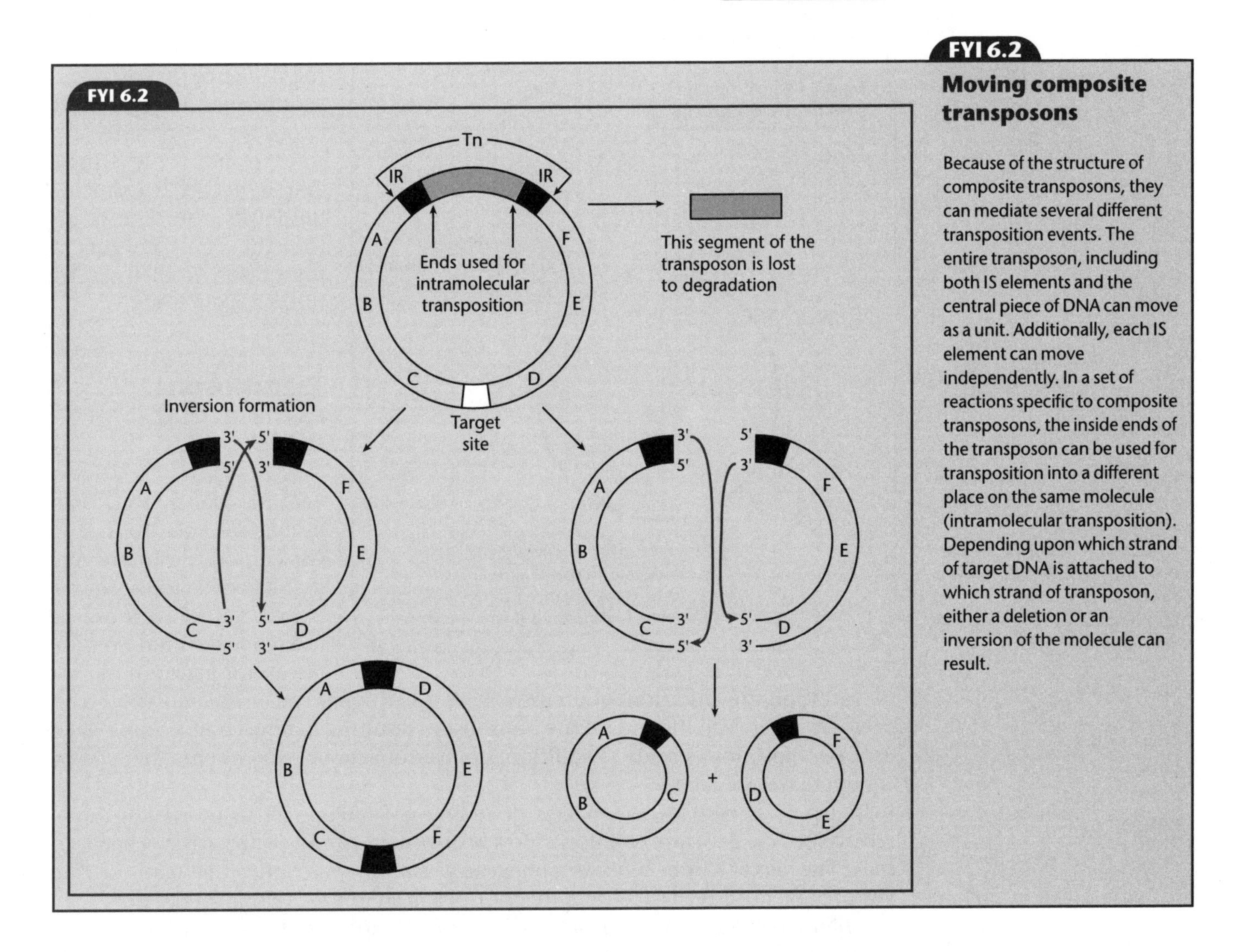

FYI 6.2
Moving composite transposons

Because of the structure of composite transposons, they can mediate several different transposition events. The entire transposon, including both IS elements and the central piece of DNA can move as a unit. Additionally, each IS element can move independently. In a set of reactions specific to composite transposons, the inside ends of the transposon can be used for transposition into a different place on the same molecule (intramolecular transposition). Depending upon which strand of target DNA is attached to which strand of transposon, either a deletion or an inversion of the molecule can result.

Target immunity

Some transposons, such as Tn*3*, Tn*7* and Mu, exhibit an unusual property when choosing target sites. If a target piece of DNA already contains a transposon, then a second copy of that same transposon cannot insert into that target DNA in the vicinity of the first element. This phenomenon is called transposition immunity or **target immunity** (Fig. 6.14). Target immunity can be effective over very long distances. For example, Tn*7* can provide immunity over distances of approximately 175 kb. Immunity is only provided to the DNA molecule that contains the transposon. If a cell carries a transposon on the chromosome and a plasmid with no transposon, only the chromosome will have transposition immunity. The transposon can still move to the plasmid.

Target immunity is specific to a particular element. Tn*3* provides immunity only to another Tn*3* and not to any other elements, even if they are closely related.

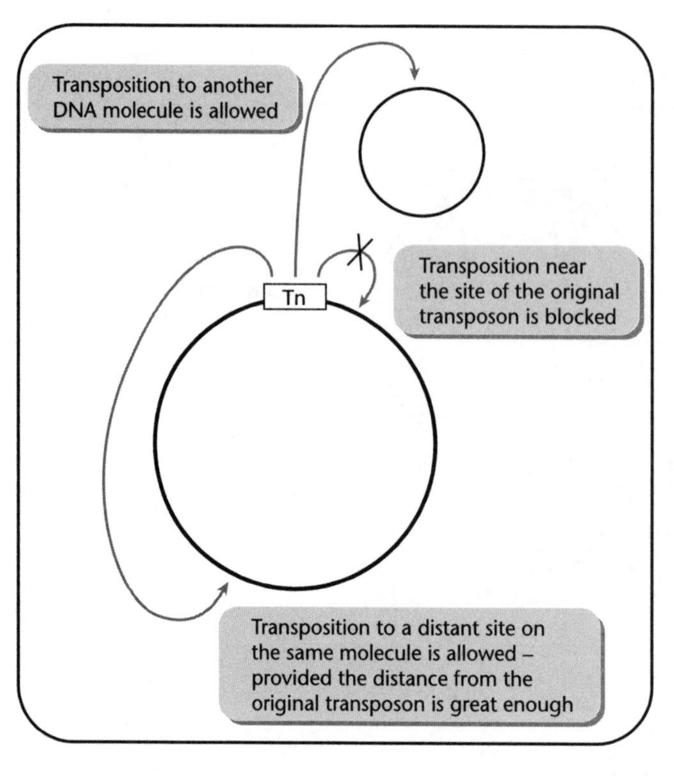

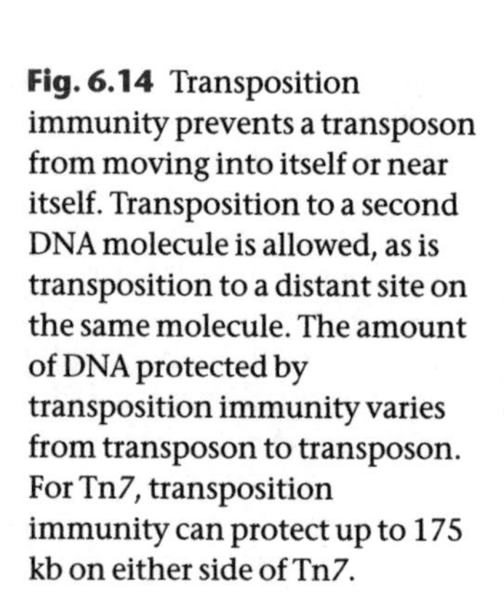

Fig. 6.14 Transposition immunity prevents a transposon from moving into itself or near itself. Transposition to a second DNA molecule is allowed, as is transposition to a distant site on the same molecule. The amount of DNA protected by transposition immunity varies from transposon to transposon. For Tn*7*, transposition immunity can protect up to 175 kb on either side of Tn*7*.

Immunity is provided by the sequence that resides at the transposon ends. While two ends are required for transposition, only a single end is needed to provide immunity. Immunity is thought to prevent a transposon from inserting into a copy of itself.

Transposons as molecular tools

Because transposons can insert randomly around the chromosome, they are very useful genetic tools. For example, transposons can be used to make mutations by disrupting gene function at the insertion site. Insertions of a transposon near any gene can be isolated. Genes can be inserted between the ends of the transposon, resulting in novel transposons. Each of these applications leads to additional powerful genetic experiments that can be conducted on bacteria.

Transposons, with few exceptions, do not move from one cell to another by themselves (see FYI6.3 for an exception). Most transposons need a vector to travel between cells. The vector can be a bacteriophage, a plasmid, or a conjugative plasmid. The vector moves from cell to cell and the transposon simply tags along for the ride.

When a transposon moves from one piece of DNA to another, it pays little attention to the function of the target DNA. If the target DNA is in the middle of a gene, then the insertion of the transposon usually disrupts the gene (Fig. 6.15). This usually results in a nonfunctional gene that is known as a null mutation. When isolating this type of null mutation, a transposon that carries an antibiotic resistance determinant is used. The advantage of this approach is that the mutation has two properties; it is null for the gene that has been disrupted and it is marked with an antibiotic resistant determinant. The antibiotic resistance determinant can be used to move the null mutation from cell to cell (see Chapter 8). The antibiotic resistance determinant can also be used to clone and identify the disrupted gene using standard molecular techniques (see Chapter 14).

While transposons can insert into a gene, they can also insert next to genes (Fig. 6.16). Isolating a transposon insertion next to a gene of interest is very useful for genetic manipulations, especially if that gene has phenotypes that are difficult to score. A transposon next to a gene can be used to move that gene from one cell to another using generalized transduction, conjugation, or other methods that transfer DNA from one cell to another. In this example, the antibiotic resistance determinant on the transposon is selected in the cross and subsequently the phenotype of the gene of interest is scored (Fig. 6.17). The transposon near a gene can be used in the localized

FYI 6.3

Conjugative transposons

Conjugative transposons were first identified in Gram-positive bacteria. They are unique because they have the ability to move from the chromosome of one cell to the chromosome of a different cell, without the need for a vector such as a plasmid. Conjugative transposons encode the functions needed to excise themselves from the donor molecule, transfer themselves to another cell, and transpose into the target site in the second cell. These elements do not duplicate any bases from the target site and can bring non-transposon sequences from the donor site along for the ride.

mutagenesis technique described in Chapter 8. The transposon is used as the selectable marker to move the mutagenized fragments of DNA. Once a transposon's insertion site on the chromosome has been characterized, it can be used in mapping experiments to help determine where uncharacterized genes are located on the chromosome (Chapter 8). Because of the usefulness of having a transposon insertion near the gene being studied, it is frequently one of the first techniques applied to a newly identified gene.

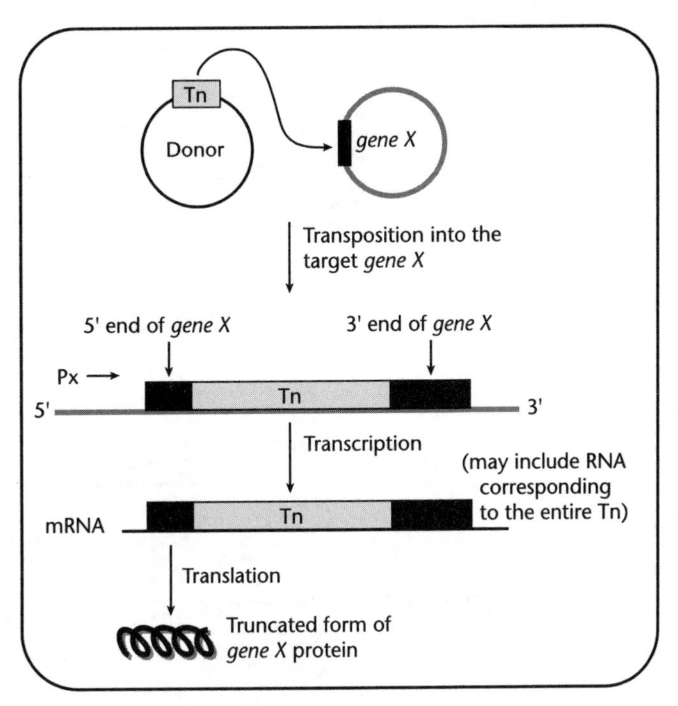

Fig. 6.15 Transposition into a gene usually results in a nonfunctional gene. The translation of *gene X* in this example into mRNA will begin in a normal manner from the correct Shine–Delgarno site and AUG start signals. It will be terminated at the first stop codon encountered in the transposon.

Several transposons, including Mu, Tn*5* and Tn*10*, have been developed into sophisticated genetic tools. Because a transposon moves any piece of DNA that is between its two ends, novel transposons with useful genetic features can be constructed. In the case of Tn*10*, normally a tetracycline resistance gene is encoded by the element. Using molecular biology, a variety of different genes have been inserted in place of tetracycline resistance. These include genes for resistance to kanamycin or chloramphenicol, a uracil biosynthesis gene from yeast, a suppressor tRNA gene, and the *lacZ* gene (Fig. 6.18). For the antibiotic resistance genes, the uracil biosynthesis gene and the suppressor tRNA, the complete gene with transcription and translation signals were placed between the two ends of the transposon. These genes will be expressed from any chromosomal location.

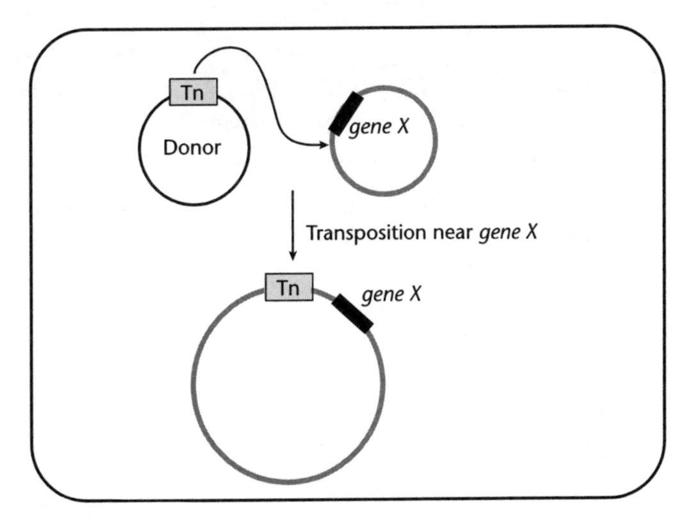

Fig. 6.16 Transposition near *gene X*.

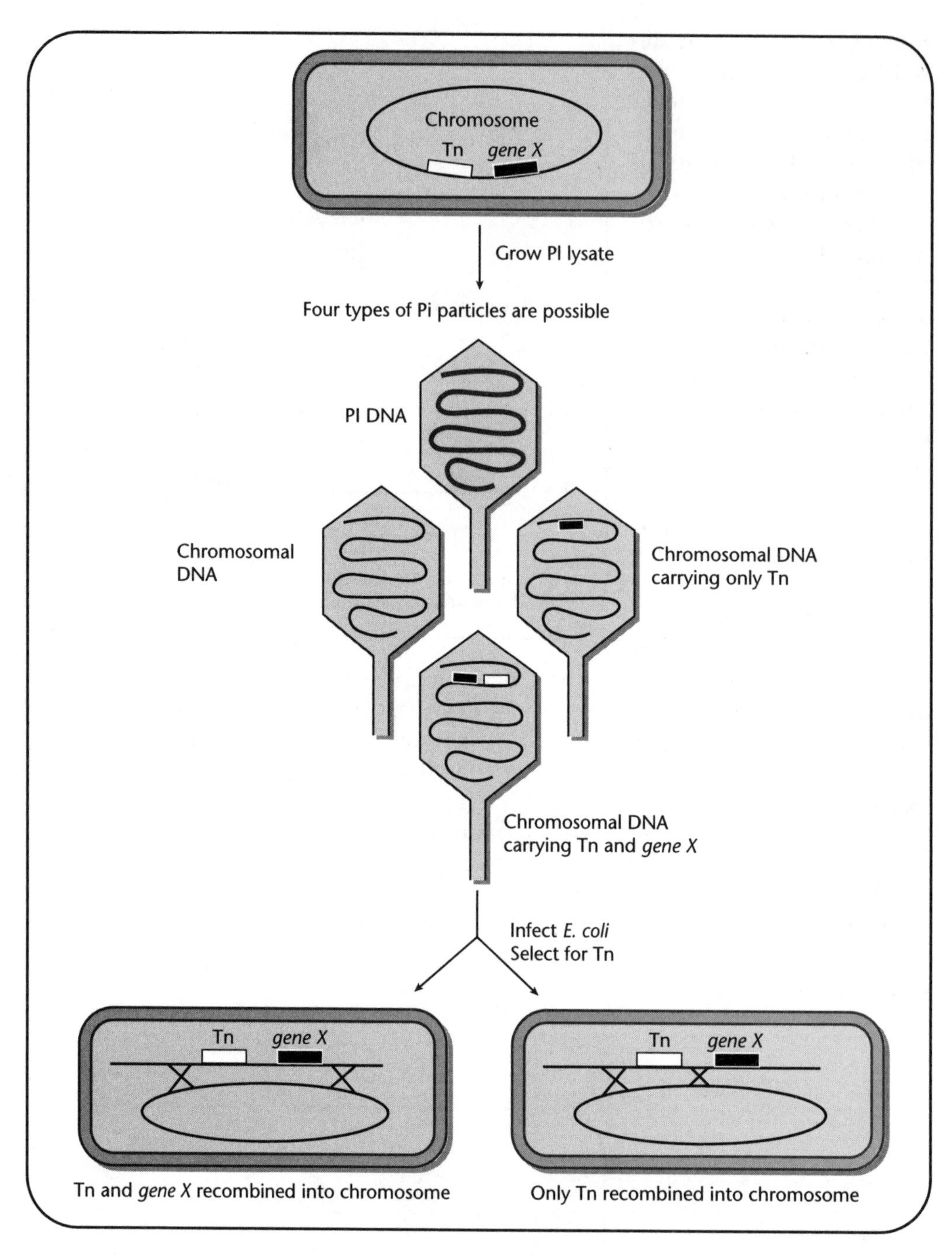

Fig. 6.17 Moving *gene X* by generalized transduction using bacteriophage P1. The antibiotic resistance determinant from the Tn is used as the selectable marker to track the movement of *gene X* (see Chapter 8 for details of transduction).

FYI 6.4

Eukaryotic transposons

Eukaryotic cells contain a wide variety of transposable elements. The majority of eukaryotic transposons move through cut-and-paste mechanisms. No eukaryotic elements that move by replicative transposition have been identified.

One class of elements that is novel to eukaryotes are the retroviruses and retrotransposons. These elements are first copied from the DNA version of the element into an RNA intermediate. The RNA intermediate is then reverse transcribed back into a DNA copy of the element and the DNA copy is inserted into the genome at a novel place.

A family of eukaryotic transposons known as mariner elements was originally found in the fly, *Drosophila mauritiana*. The mariner elements are small (~1.3 kb) and require no host proteins for their cut-and-paste transposition mechanism. What makes them most unusual is that they are found in a wide variety of organisms, including fungi, plants, ciliates, nematodes, arthropods, fish, frogs, and humans. In fact, they will transpose in organisms from all three kingdoms, Eukaryotes, Archeae, and Prokaryotes. This is in contrast to the *E. coli* transposons, which only transpose in a few closely related bacterial species. One member of the mariner family, called Himar1, has been used as a genetic tool to generate transposon insertion mutations in bacterial species where no other element will transpose. Himar1 also transposes in vitro and thus can be used for in vitro mutagenesis of DNA. The limited requirements for Himar1 transposition allow the genetic manipulations that rely on transposons to be unleashed on many fascinating but genetically intractable species.

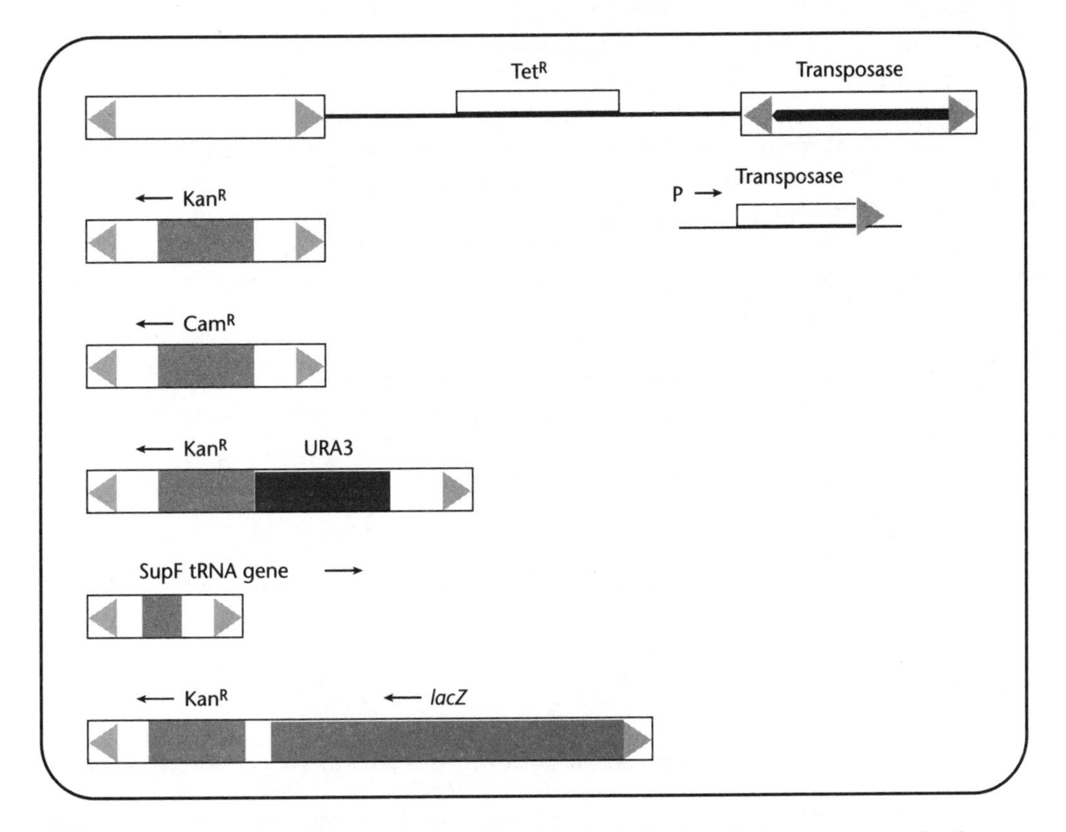

Fig. 6.18 Genetically engineered derivatives of Tn*10*. In the derivatives, the transposase gene has been deleted. Deletion of the transposase allows the elements to be smaller and more importantly, to be more stable. Once one of the Tn derivatives has transposed, it can be separated from the source of transposase. Without transposase, the derivative cannot transpose again. To provide a source of transposase, a plasmid with transposase under the control of an inducible promoter is used. This source of transposase can be used with all of the Tn*10* derivatives.

Summary

Transposons are pieces of DNA that can move themselves from one DNA site to another. The simplest transposons encode all of the information needed for this movement. The more complex transposons require host proteins. Transposons use one of two mechanisms for transposition: non-replicative or replicative. While both mechanisms have many features in common, there are a few steps that distinguish one mechanism from another. Because transposons make mutations when they move, the frequency of transposition must be low enough that the chromosome does not become disabled by mutations.

The fact that transposons can insert randomly into the bacterial chromosome has led to their exploitation as genetic tools. Virtually any piece of DNA can be inserted between the inverted repeats of a transposon and transposed to novel sites on the chromosome. Antibiotic resistance determinants, virulence factors, reporter gene constructs, and even binding sites for specific proteins have been moved around the chromosome in this way. Transposons have led to the development of sophisticated genetic techniques for many different bacterial species that have not been amenable to other genetic analyses.

Study questions

1 What type of transposons are composed of modular units? What types are not composed of modular units?
2 What factors govern how frequently a transposon moves?
3 What steps are different between cut-and-paste transposition and replicative transposition? What steps are the same?
4 How many ways can a cointegrate structure be generated?
5 Describe the ways that a cointegrate structure can be resolved?
6 When a transposon moves by cut-and-paste transposition, what can happen to the donor site?
7 What is target immunity and what is its purpose?
8 Design an experiment using a transposon conferring tetracycline resistance to select for an *E. coli* mutant incapable of using lactose as a carbon source (phenotypically Lac^-).
9 Describe the impact of the methylation state of DNA on the frequency of some transposition events.

Further reading

Bender, J. and Kleckner, N. 1986. Genetic evidence that Tn*10* transposes by a nonreplicative mechanism. *Cell*, **45**: 801–15.

Craig, N.L. 1996. Transposition. In *Escherichia coli* and *Salmonella typhimurium: Cellular and Molecular Biology*, 2nd edn., eds. F.C. Neidhardt, R. Curtiss III, J.L. Ingraham, E.C.C. Lin, K.B. Low, B. Hagasanik, W.S. Rexnikoff, M. Riley, M. Schaechter, and H.E. Umbarger, pp. 2339–62. Washington, DC: ASM Press.

Grindley, N.D.F. and Reed, R.R. 1985. Transpositional recombination in prokaryotes. *Annual Review of Biochemistry*, **54**: 863.

Shapiro, J.A. 1979. Molecular model for the transposition and replication of bacteriophage Mu and other transposable elements. *Proceedings of the National Academy of Science, USA*, **76**: 1933–7.

Chapter 7

Bacteriophage

Bacteriophage or **phage** for short are viruses that infect only bacteria. In contrast to cells that grow from an increase in the number of their components and reproduce by division, **viruses** are assembled from pre-made components. Viruses are nucleic acid molecules surrounded by a protective coating. They are not capable of generating energy and reproduce inside of cells. The nucleic acid inside the coating, called the **phage genome** in a bacteriophage, encodes most of the gene products needed for making more phage. The phage genome can be made of either double- or single-stranded DNA or RNA, depending on the bacteriophage in question. The genome can be circular or linear. The protective coating or **capsid** surrounding the phage genome is composed of phage-encoded proteins.

Many important discoveries have been made using phage as model systems. From the discovery that a nonsense codon stopped protein synthesis to the first developmental switch to be understood at the molecular level, phage have proven to be very useful. In this chapter, we will look at phage development using T4, λ (lambda), P1, and M13 as examples. Each of these phage infect *E. coli*. We will examine specific discoveries using these phage or specific properties of the phage that have made them particularly useful to biologists.

The structure of phage

All phage have a chromosome encased in a capsid that is composed of phage-encoded proteins. For many phage types, the capsid is attached to a tail structure that is also made from phage-encoded proteins. T4 and P1 contain a linear double-stranded DNA genome enclosed in a capsid and attached to a tail (Fig. 7.1a). The T4 genome is 172 kb, while P1 is a smaller phage with a genome of 90 kb. The T4 capsid is an elongated icosahedron. T4 has a very elaborate tail structure including a collar at the base of the head and a rigid tail core surrounded by a contractile sheath. The core and sheath are attached to a hexagonal base plate. Also attached to the tail plate are tail pins and six kinked tail fibers. P1 also has an icosahedral capsid, a tail with a contractile sheath, a base plate, and tail fibers. λ contains a linear double-stranded DNA genome of 48.5 kb, a capsid, and a tail (Fig. 7.1b). The finished capsid is again shaped like an icosahedron whereas the tail is a thin flexible tube that ends in a small conical part and a single tail fiber. M13 contains a circular single-stranded DNA genome of 6407 nucleotides surrounded by five phage-encoded proteins (Fig. 7.1c). The M13 chromosome is coated

FYI 7.1

Discovery of phage

Phage were first described in 1915 by Frederick Twort and 1917 by Felix d'Herelle. Both men discovered phage when the bacteria they were working with lysed. The agent responsible for the lysis was transferable from culture to culture, invisible by light microscopy, and would go through the smallest filter they had. d'Herelle coined the term "bacteriophage", signifying an entity that eats bacteria.

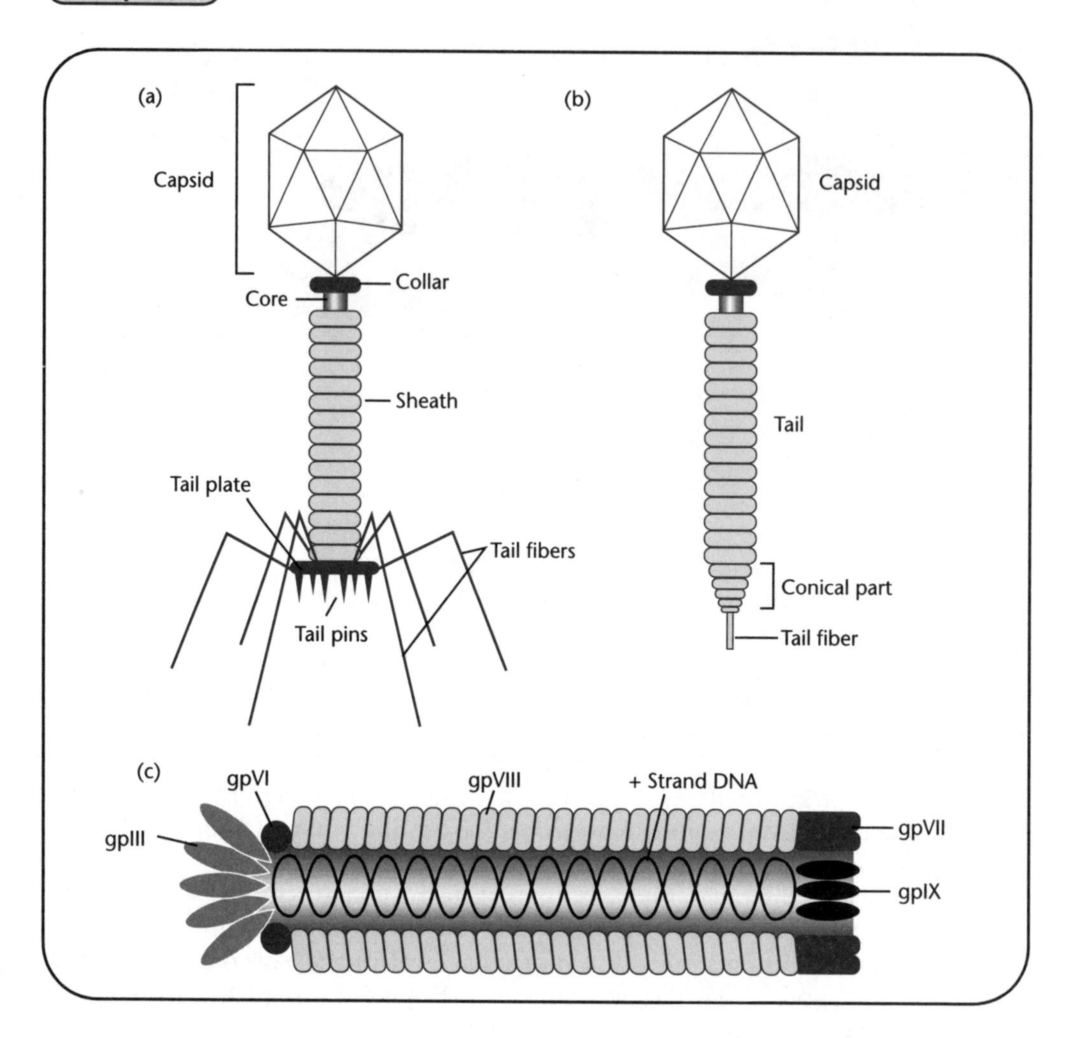

Fig. 7.1 The structures of (a) T4, (b) λ, and (c) M13.

by a single layer of ~2700 subunits of gene VIII encoded protein (gpVIII) giving it a filamentous appearance, the reason M13 is also known as a **filamentous phage**. At one end of the filament are bound the M13 proteins encoded by genes VII and IX (gpVII and gpIX) and at the other end are bound the M13-encoded gene III and VI proteins (gpIII and gpVI).

The lifecycle of a bacteriophage

All phage must carry out a specific set of reactions in order to make more of themselves. First, the phage must be able to recognize a bacterium that it can multiply in by binding to the bacterial cell surface. Next, the phage must inject its genome and the genome must be protected from the bacterial nucleases in the cytoplasm. The phage genome must be replicated, transcribed, and translated so that a large number of genomes, capsid proteins, and tail proteins, if present, are produced at the same or nearly the same time. Complete phage particles are then assembled and the phage must get back out of the bacterium. Different phage use different strategies to carry out each of these reactions.

Phage are very choosy as to what bacteria they infect. This is referred to as the **host range** of the phage. For example, λ only infects certain *E. coli*, whereas Spo1 phage infect only *Bacillus subtilis*. Several phage types may infect a single bacterial species. *E. coli* can be infected by λ, M13, P1, T4, and Mu phages, to name a few.

The number of phage that can be released from one bacterium after infection and growth by one phage is known as the **burst size**. Every phage has a characteristic burst size. Different phage also take different amounts of time to go through one growth cycle. We know when a phage has successfully reproduced when we are able to detect **plaques** or circular areas with little or no bacterial growth on an agar plate covered with a thin layer of bacteria.

Once bound to the cell, the phage must get its genome into the cytoplasm. The rate of phage DNA transport can be very rapid. It is different for different phages but can reach values as high as 3000 base pairs per second. In contrast, two other methods for getting DNA from the outside of the cell to the cytoplasm (conjugation, Chapter 10 and transformation, Chapter 11) transfer the DNA at a rate of approximately 100 bases per second. In many cases the details of how a phage genome gets into the cytoplasm are not known. From the information we do have, it is clear that not just one mechanism is used.

Lytic–Lysogenic options

The process of a phage infecting a bacterium and producing progeny is referred to as a **lytic** infection. Some phage, like T4, are only capable of lytic growth. Some phage are also capable of maintaining their chromosome in a stable, silent state within the bacteria. This is called **lysogeny**. Phage that are capable of both a lytic and lysogenic pathway are called **temperate phage**. P1 and λ are temperate phage. M13 is unusual because phage continually exit from a bacterium without killing it. For this reason, M13 is not considered to have a true lysogenic state and is not a temperate phage. When the bacterium contains a silent phage chromosome, it is referred to as a **lysogen**. The incorporated phage genome is referred to as a **prophage**.

The λ lifecycle

λ adsorption

Phage identify a host bacterium by binding or adsorbing to a specific structure on the surface of the cell. Many different cell surface structures can be used as binding sites. The basics of adsorption are that a specific structure on the surface of the phage interacts with a specific structure on the surface of the bacterium. λ binds to an outer membrane protein called LamB via a protein that resides at the tip of the λ tail called the J protein. LamB normally functions in the binding and uptake of the sugars maltose and maltodextrin.

λ DNA injection

Initially, λ binds to LamB and the binding is reversible. This step requires only the λ tail and the LamB protein. Next, the bound phage undergoes a change and the binding to LamB becomes irreversible. The nature of the change is unknown but it requires that a phage head be attached to the phage tail. Next the λ DNA is ejected from the phage and taken up by the bacterium. The DNA in the phage head is very tightly packed. If the condensed state of the phage DNA is stabilized, ejection of the DNA does not occur. The addition of small positively charged molecules such as putrescine to the phage counteracts the negatively charged DNA and stabilizes the DNA in the phage head. This implies that the tight packing of the DNA is used to help eject the DNA from the phage particle. When λ DNA is put into the capsid, one end known as the left end is inserted first. When the λ DNA comes out of the phage head, the right end exits first. Unlike phage T4 (see below), no change in the λ tail structure is seen

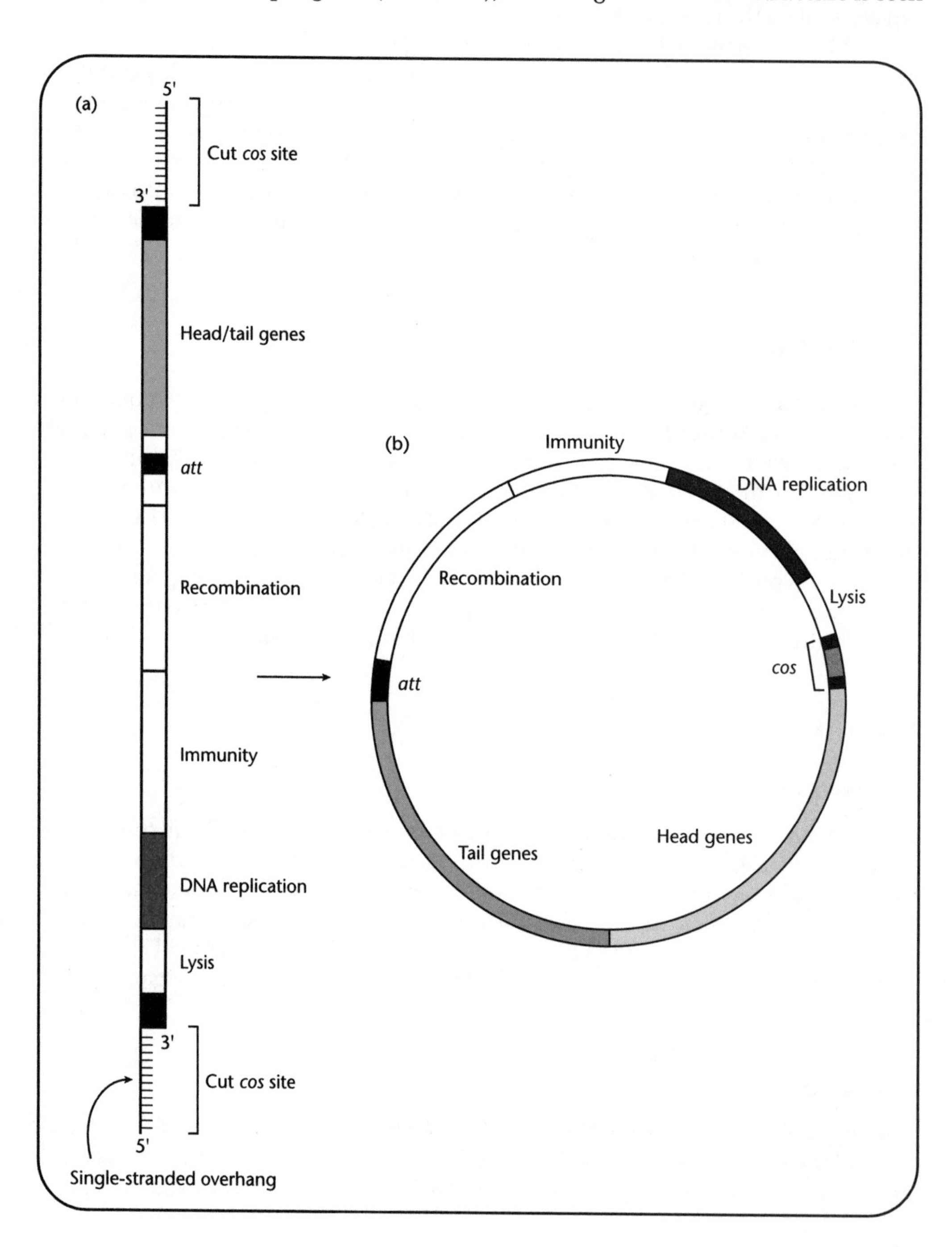

Fig. 7.2 The structure of the λ DNA in the phage capsid (a) and after circularization in the cytoplasm (b). The DNA circularizes via the *cos* site.

when the DNA is ejected. In addition to LamB, λ also uses an inner membrane protein called PstM to gain entry to the cytoplasm. How the λ DNA physically traverses the peptidoglycan and periplasm and gets through PtsM is not known.

Protecting the λ genome in the bacterial cytoplasm

What protection the phage genome needs in the cytoplasm depends on the physical state of the injected nucleic acid. λ contains a linear double-stranded DNA molecule in its capsid. In the bacterial cytoplasm, dsDNA molecules are subject to degradation by exonucleases that need a free end to digest the DNA. The first event that happens to newly injected λ DNA is that the DNA circularizes to prevent it from being degraded.

λ has a specific site on its DNA, termed the *cos* site, which it uses to circularize the DNA (Fig. 7.2). The *cos* site is a 22 bp sequence that is cut asymmetrically when the λ DNA is packaged (see below). The cut *cos* site has a 12 bp overhang. There is one cut *cos* site at the left end of the λ genome and another cut *cos* site at the right end of the λ genome (Fig. 7.2a). When the λ DNA is injected into the cytoplasm, the cut *cos* sites at either of the linear λ genome anneal (Fig. 7.2b). A host enzyme, DNA ligase, seals the nicks at either end of the *cos* site generating a covalently closed, circular λ genome. The host encoded enzyme, DNA gyrase, supercoils the λ molecule.

What happens to the λ genome after it is stabilized?

The λ genome contains six major promoters known as P_L for promoter leftward, P_R for promoter rightward, P_{RE} for promoter for repressor establishment, P_{RM} for promoter for repressor maintenance, P_I for promoter for integration, and $P_{R'}$ for secondary rightward promoter (Fig. 7.3). After the genome is circularized and supercoiled, transcription begins from P_L and P_R. A series of genes known as early genes are transcribed and translated. These gene products are the initial proteins needed for further phage development. *E. coli* RNA polymerase interacts with P_L to give rise to a short mRNA transcript that is translated into the N protein (Fig. 7.4a). *E. coli* RNA polymerase interacts with P_R to give rise to a short mRNA transcript that is translated into the Cro protein (Fig. 7.4a).

The N protein is able to extend transcription when RNA polymerase encounters a sequence in the DNA that tells it to stop. For this reason, N is called an anti-termination protein. N allows RNA polymerase to transcribe through the t_L and t_{R1} termination signals resulting in the synthesis of longer mRNA transcripts (Fig. 7.4b). The longer transcripts from P_R encode the O, P, and CII proteins, and a small amount of another anti-terminator, the Q protein. From P_L, CIII, the recombination proteins Gam and Red and a small amount of Xis and Int are made.

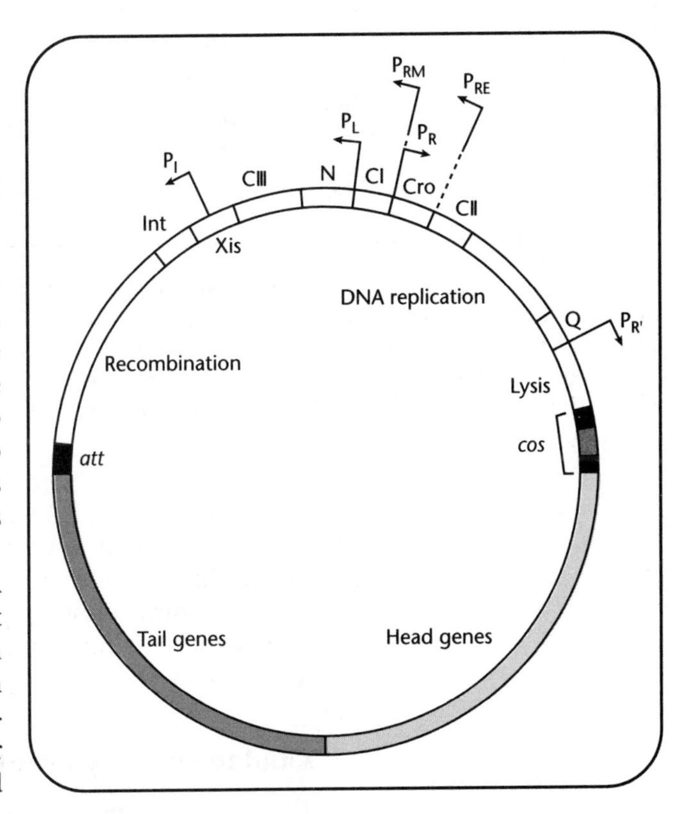

Fig. 7.3 The location of the six major promoters on the λ genome and the direction in which they specify mRNA production.

The N protein anti-terminates by binding to RNA polymerase after a specific base pair sequence, located upstream of the transcriptional termination site, has been

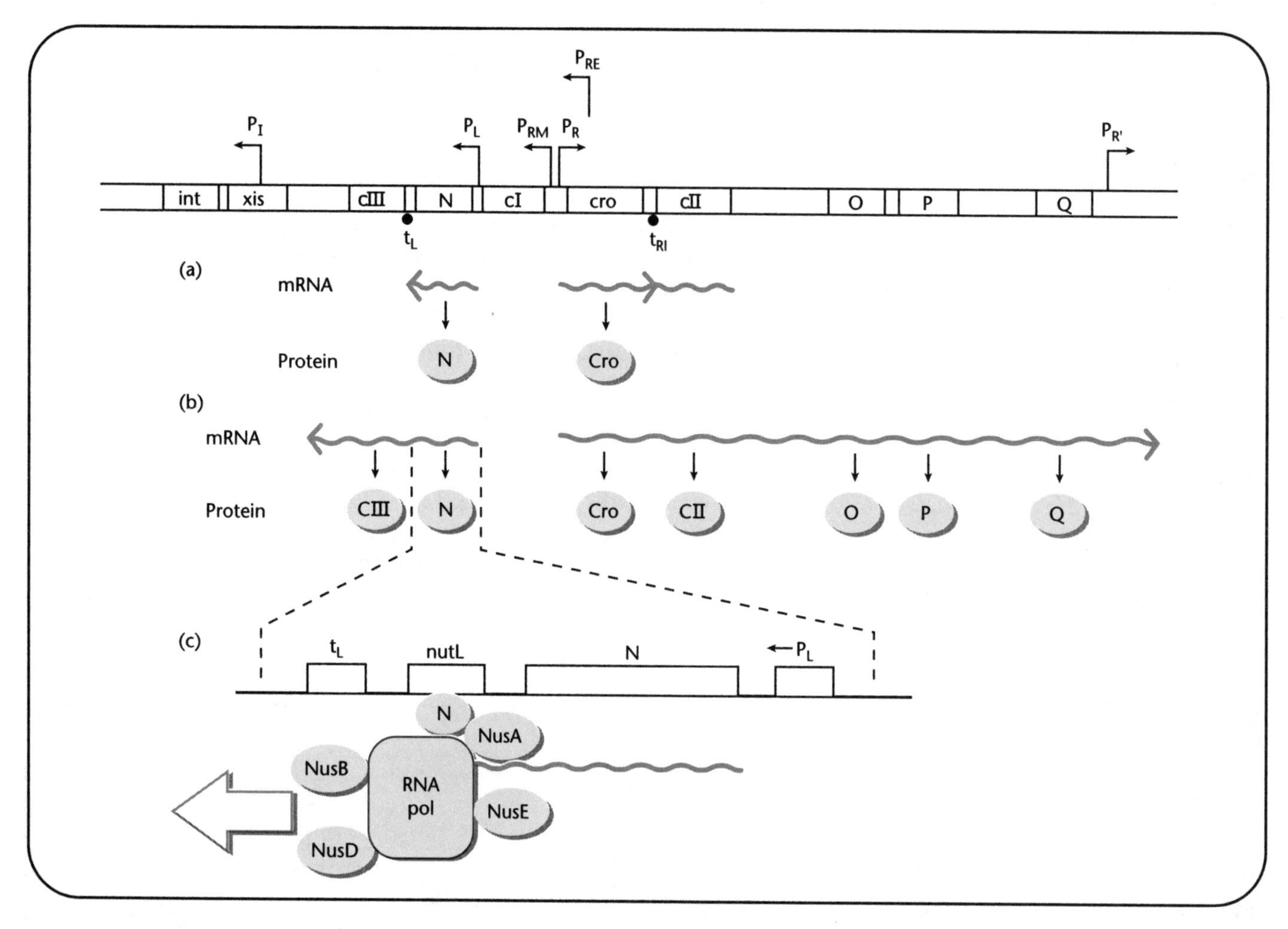

Fig. 7.4 The first transcription and translation events that take place on the λ genome after infection. (a) Transcription from P_L leads to the production of N protein. Transcription from P_R leads to Cro protein. (b) N is an anti-terminator that allows RNA polymerase to read through the t_L and t_{R1} terminators. From P_L, N and CIII proteins will be produced. From P_R, Cro, CII, O, P, and Q proteins will be produced. (c) N binds to the *nutL* site on the DNA. In conjunction with four bacterial proteins, NusA, NusB, NusD, and NusE, N allows RNA polymerase to read through the terminator t_L.

transcribed into mRNA (Fig. 7.4c). This sequence is called *nut* for N utilization. Other *E. coli* proteins contribute to anti-termination. These proteins have been named Nus, for N utilization substance.

At this point, all of the players needed to make the lytic–lysogenic decision have been made. CII and CIII are needed for lysogenic growth. Cro and Q are needed for lytic growth. The O and P proteins are used for replicating the λ DNA.

λ and the lytic–lysogenic decision

The decision between lytic or lysogenic growth for λ was the first developmental switch understood at the molecular level. At the most basic level, the decision depends on the amounts of two phage-encoded proteins called CI (pronounced C-one) and Cro, and their binding to their promoter control regions (Fig. 7.5). When CI is bound, the expression of the lytic genes is repressed and the phage follows the

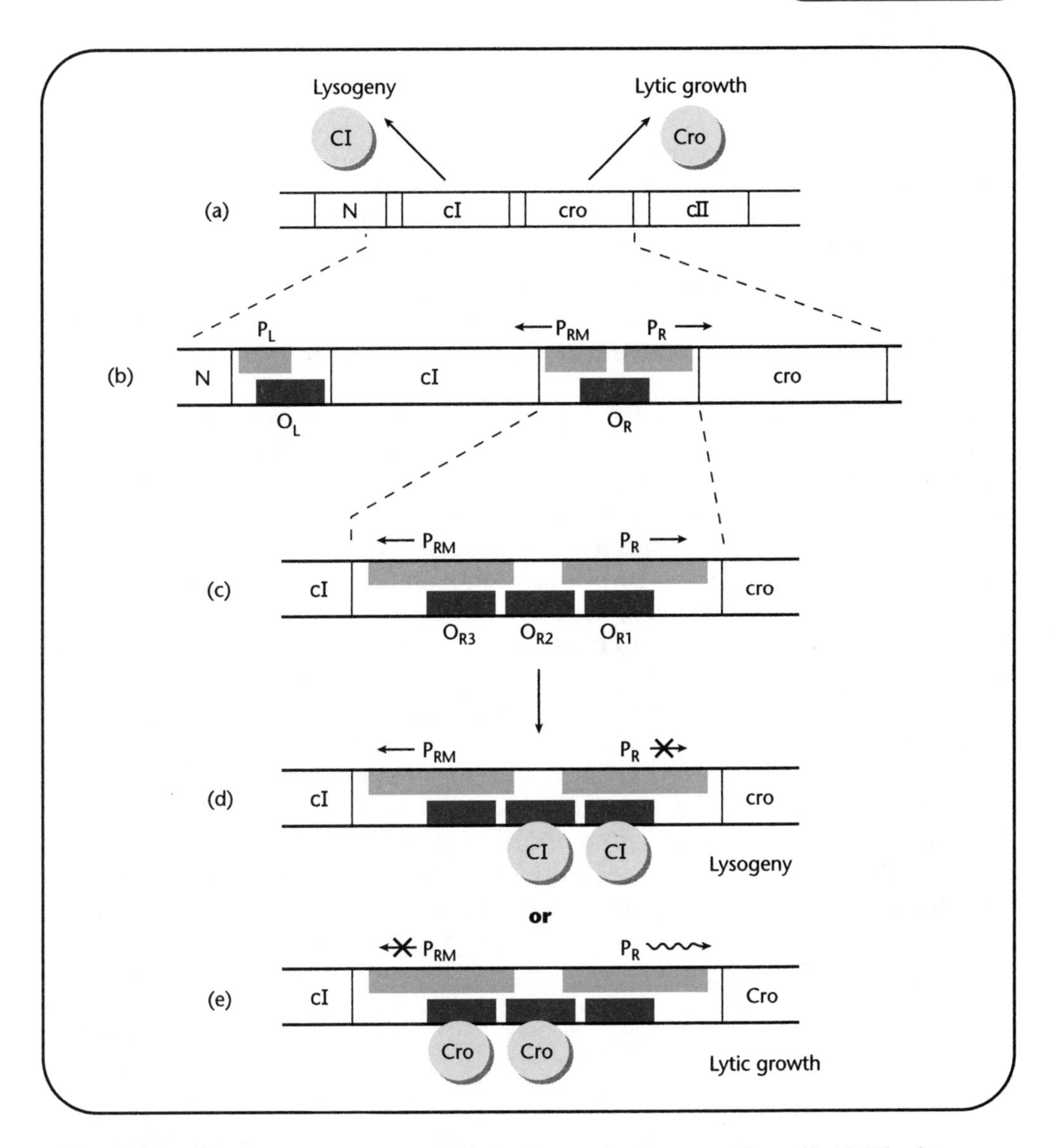

Fig. 7.5 CI and Cro are the proteins responsible for the two developmental fates of λ. (a) CI leads to lysogeny and Cro leads to lytic growth. (b) Both CI and Cro bind to two operator regions, O_R and O_L. O_R overlaps with both P_R and P_{RM}. O_L overlaps with P_L. (c) O_R is required for the switch between developmental pathways. It is composed of three 17 base pair sequences called O_{R1}, O_{R2}, and O_{R3}. They are similar in sequence but not identical. (d) CI binds to O_{R1} first then O_{R2}. It will bind to O_{R3} but only at very high concentrations. When CI binds to O_R, it represses transcription from P_R and activates it from P_{RM}. CI binding to O_R is actually required for P_{RM} to be activated. CI binding leads to lysogeny. (e) Cro also binds to O_{R1}, O_{R2}, and O_{R3} but in the opposite order from CI. Cro binds to O_{R3} first then O_{R2} and at high concentrations O_{R1}. Cro binding to O_{R3} inhibits P_{RM} and leads to lysogeny.

lysogenic pathway (Fig. 7.5a). For this reason, CI is also known as CI repressor or λ repressor. The expression and binding of Cro leads to lytic development.

Cro is made from P_R and CI is made from either P_{RE} or P_{RM}. Both Cro and CI bind to the same DNA sequences called operators (Fig. 7.5b). λ contains two operators that bind Cro and CI. One, called O_R, overlaps the P_{RM} and P_R promoters. The other, called

O_L, is behind the P_L promoter. O_R is a major player in the lytic–lysogenic decision, while O_L is not part of the decision.

O_R is composed of three 17 base pair sequences called O_{R1}, O_{R2}, and O_{R3} (Fig. 7.5c). CI repressor binds to O_{R1} 10 times better than it binds to O_{R2} or O_{R3}. At increasing concentrations of CI, it will bind to O_{R2} and eventually to O_{R3}. When CI is bound to O_R, it stimulates the P_{RM} promoter and the production of CI repressor and inhibits the P_R promoter and the production of Cro, leading to lysogeny (Fig. 7.5d). Cro also binds to O_{R1}, O_{R2}, and O_{R3} but in the reverse order from CI repressor. Cro binds to O_{R3} first, then O_{R2}, and finally at high concentrations to O_{R1}. When Cro is bound to O_R, it inhibits the P_{RM} promoter and the production of CI, leading to lytic growth (Fig. 7.5e). This is the basis for either lytic or lysogenic growth.

How does the phage switch between these two developmental pathways? The major protein involved in the switch is another phage-encoded protein called CII (pronounced C-two, Fig. 7.6). CII activates the P_{RE} and P_I promoters. This leads to the production of repressor and the Integrase protein, which is also needed for lysogeny (Fig. 7.6b). The gene for CII (*cII*) resides just to the right of the *cro* gene. When λ infects a cell, transcription automatically begins from P_L and P_R using host proteins. Transcription from P_R leads to production of both the Cro and CII proteins. If CII is active it will lead to production of CI and Integrase and lysogeny. If CII is inactive then Cro will repress P_{RM}, preventing expression of CI and leading to lytic growth.

The CII protein is inherently unstable. Several factors influence this feature of the protein. CII is degraded by the bacterial-encoded HflA protease. When cells are actively growing in nutrient-rich conditions, the amount of HflA in the cell is high, leading to degradation of CII and lytic growth. When cell are growing slowly, HflA levels are low, leading to stabilization of CII, production of CI, and lysogeny. In this manner, CII is used to monitor the health of the cell and impact the lytic–lysogenic decision accordingly. It is thought that λ wants to produce more phage when cells are healthy, nutrients are plentiful, and the prospect of completing phage development is good. Lysogeny is a better bet when cells are growing poorly. CII is also stabilized by a phage-encoded protein called CIII. CIII is produced from P_L by infecting phage.

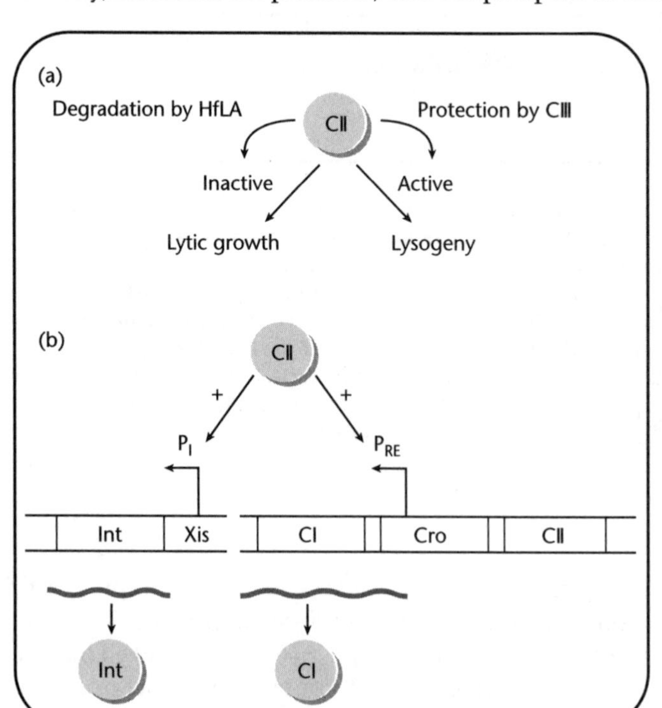

Fig. 7.6 The CII protein is the major player in the switch between lytic and lysogenic growth. CII is unstable and rapidly degraded by the host-encoded HflA protease. Inactive CII leads to lytic growth. CII can be protected by the phage-encoded CIII protein. Active CII leads to lysogenic growth.

The λ lysogenic pathway

If CII prevails, CI will be produced, initially from the P_{RE} promoter and eventually from the P_{RM} promoter. CI activates P_{RM} ensuring that a continuous supply of CI is made. CI also activates the P_I promoter, leading to the production of the Integrase protein. The recombination of λ DNA into the chromosome occurs at a specific site in the λ DNA called *attP* and at a specific site in the bacterial chromosome called *attB* (Fig. 7.7). The recombination of λ DNA into the chromosome requires Integrase and the host-encoded IHF protein (for *i*ntegration *h*ost *f*actor). Once in the chromosome, the phage DNA is bounded by hybrid *att* sites called *attL* and *attR*. The reverse of this reaction, recombination of λ DNA out of the chromosome requires Int, IHF, and a third protein Xis (for exc*is*ion and pronounced excise). Because the recombination always occurs at specific sites and requires very specific enzymes, it is known as a **site-specific recombination** event (Chapter 5). Once the λ DNA is recombined into the chromosome, it is replicated and stably inherited by daughter cells as part of the bacterial chromosome. The *attB* site on the chromosome lies between the *gal* and *bio* genes and does not disrupt either gene. When λ DNA has recombined into the bacterial chromosome it is quiescent, except for the continued production of CI from P_{RM}.

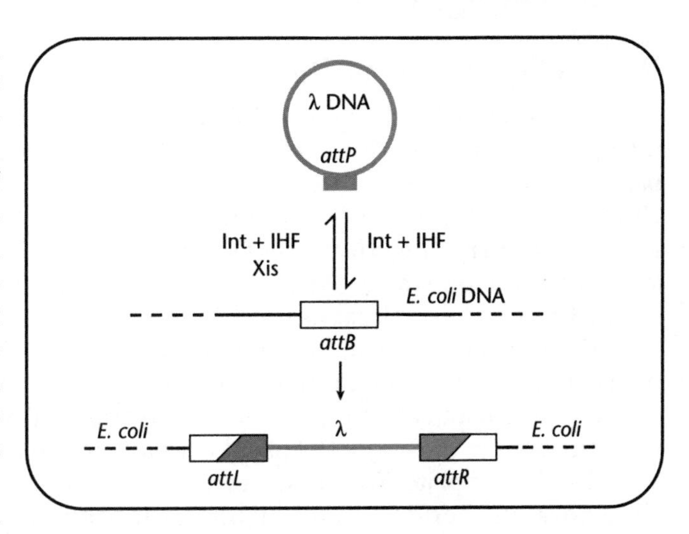

Fig. 7.7 λ recombines into the chromosome using a specific site on the phage called *attP* and a specific site on the bacterial chromosome called *attB*. When the λ DNA is in the chromosome, it is bounded by *attL* and *attR*, which are hybrid *attP/attB* sites.

What prevents the expression of the late genes coding for lytic function? The expression of late genes is prevented by the action of the λ repressor. λ repressor binding to the operator sequences O_R and O_L blocks transcription from P_L and P_R. Since P_R is blocked, the λ Q protein is not made and transcription of the late genes does not occur.

The λ lytic pathway

If enough of the Q protein accumulates in the cell, RNA polymerase will continue its transcription from a third promoter, $P_{R'}$, located in front of the Q gene (Fig. 7.8). This extends transcription into the late genes located downstream of Q. The late genes encode the proteins needed to complete the lytic infection including the head, tail, and lysis proteins.

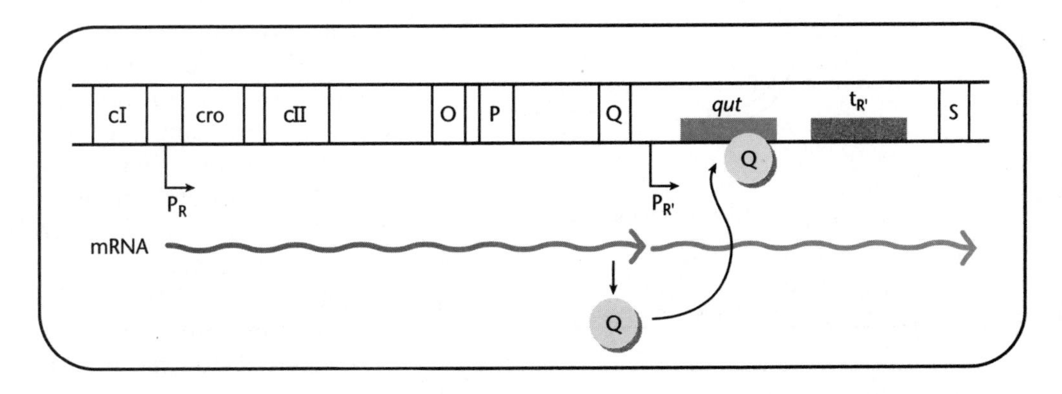

Fig. 7.8 The Q protein which is made from P_R when N is present is a second anti-termination protein. It acts on the *qut* site and allows transcription through $t_{R'}$. Q is necessary for synthesis of the head and tail genes.

DNA replication during the λ lytic pathway

After the infecting λ DNA has been converted to a double-stranded circular molecule, it replicates from a specific origin using both the phage-encoded O and P proteins and bacterial-encoded proteins. Replication proceeds bidirectionally, much like the *E. coli* chromosome. This form of replication produces molecules that look like the Greek letter theta and is called **theta replication** (Fig. 7.9a). Later in lytic development, λ switches to a second mode of replication called rolling circle replication.

Rolling circle replication of λ DNA commences when an endonuclease, encoded by λ *exo*, cuts one strand of the covalently closed circular double-stranded DNA molecule (Fig. 7.9b). The cut strand is called the plus strand. The 5′ end of the cut plus strand is peeled away from the intact minus strand. DNA polymerase adds deoxyribonucleotides to the free 3′ OH of the cut plus strand using the intact circular minus strand as the template. This produces new plus strands through a process of continually elongating the original plus strand. The new plus strands are used as a template to synthesize new minus strands. Rolling circle replication produces long DNA molecules containing multiple phage genomes called **concatomers**.

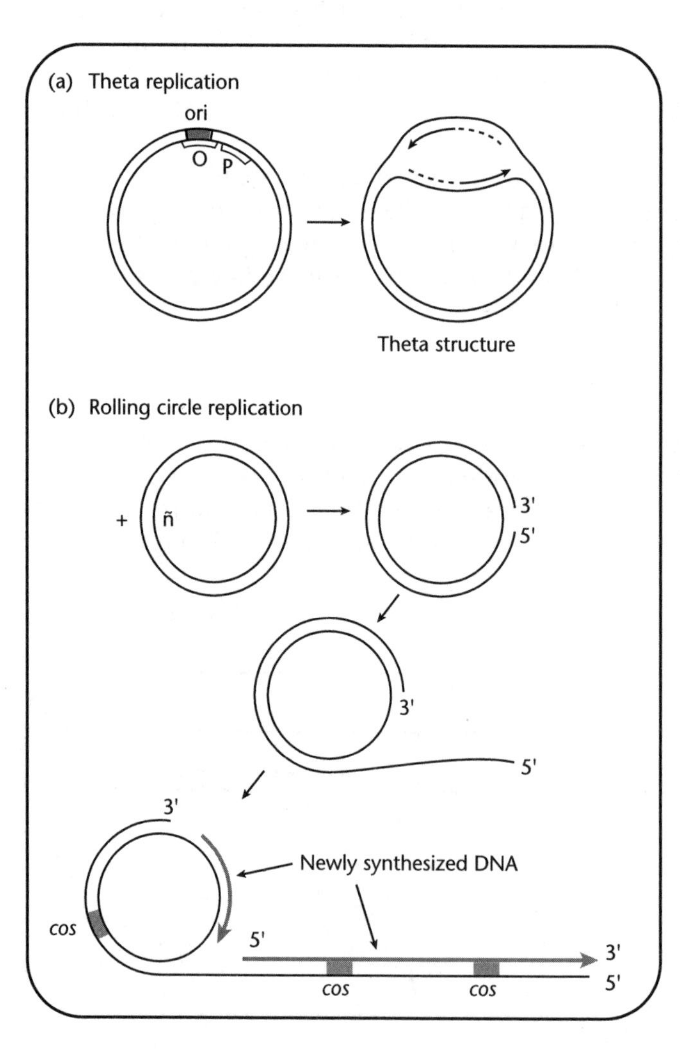

Fig. 7.9 λ has two modes of DNA replication: theta replication (a) and rolling circle replication (b). Theta replication occurs early in infection and rolling circle replication occurs late in infection. Rolling circle replication produces concatomers for packaging into phage heads.

Making λ phage

The structure of the finished capsid is determined by the physical characteristics of the structural proteins that they are made from and the phage and host proteins used for assembly. Assembly of the capsids requires at least 10 phage-encoded proteins and two host-encoded proteins. The final capsid is made up of eight proteins, E, D, B, W, FII, B*, X1, and X2. Initially, B, C, and Nu3 (all phage proteins) form a small, ill-defined initiator structure (Fig. 7.10a). This structure is a substrate for the host-encoded GroEL and GroES proteins. GroEL and GroES act on proteins or protein complexes and help remodel them. The major coat protein, E, is added to this structure to form an immature phage head (Fig. 7.10b). The immature phage head is converted to the mature

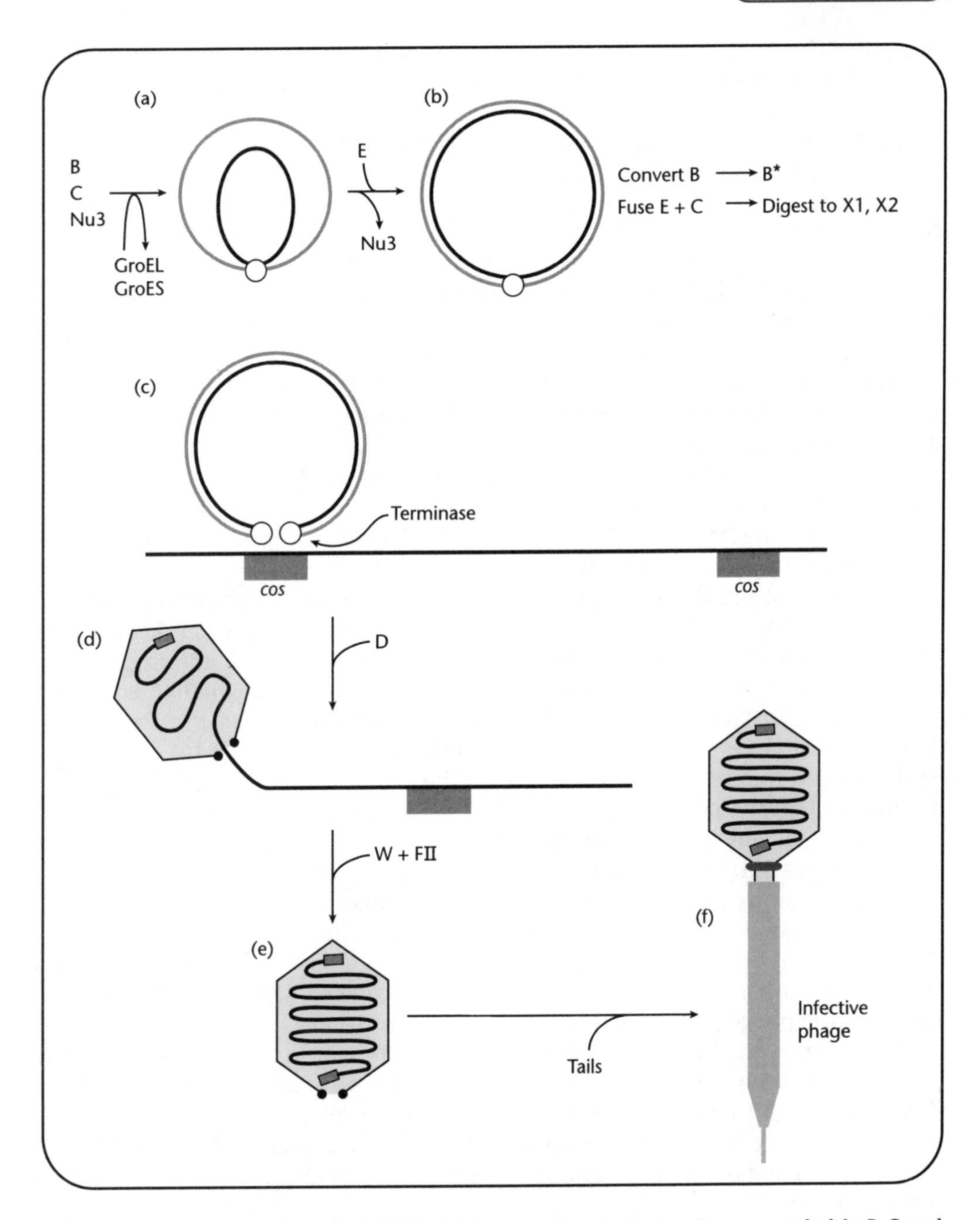

Fig. 7.10 The assembly pathway for λ. (a) The initiator structure for the head is composed of the B, C, and Nu3 proteins. (b) E, the major head protein, is added to this structure. Nu3 is degraded, B is cleaved to a smaller form (B*), and E and C are fused and cleaved at a new position to form X1 and X2. This forms the immature phage head. (c) The immature phage head is now ready for DNA from a concatomer. The D protein is added to the capsid at this point. (d) Packaging starts at a *cos* site and proceeds to the next *cos* site. (e) The DNA is inserted into the capsid and sealed inside by the W and FII proteins. (f) Tails are added to the full capsid to form a phage.

phage head by the degradation of Nu3, the cleavage of B to B*, and the fusion of C protein and some E protein followed by the cleavage of the fused protein into two new proteins, X1 and X2 (Fig. 7.10c).

The mature phage head is now ready for DNA. As the DNA is inserted into the phage head, it expands and the D protein is added to the surface of the capsid. λ DNA cannot be packaged from a monomer of λ DNA but only from concatomers usually produced by rolling circle replication (Fig. 7.10c). The DNA is cut at one *cos* site by a λ encoded enzyme and put into the phage head. Terminase binds to a *cos* site and to a phage head, cuts the *cos* site, and inserts that end of the λ DNA into the phage head (Fig. 7.10d). Terminase cuts the *cos* site asymmetrically, leaving the 12 base pair overhang. The terminase enzyme then tracks along the concatomer of λ DNA until it reaches a second *cos* site. As terminase tracks, the DNA is inserted into the phage head. When a second *cos* site is reached, terminase cuts the DNA and the last bit of DNA is inserted into the phage head (Fig. 7.10e).

This phage head with newly inserted DNA is unstable and not able to join to phage tails. The W and FII proteins are added to the base of the full head (Fig. 7.10e). This both stabilizes the DNA-containing head and builds the connector to which the tail will bind. Tails add spontaneously to this structure (Fig. 7.10f).

Tails are constructed from 12 gene products. Like the capsids, the tails are formed from an ill-defined initiator complex. This complex requires the J, I, L, K, H, G, and M phage proteins. They are added to the complex in the order listed beginning with the J protein. For this reason, it is thought that tails are built from the tip that recognizes the bacterium towards the end that binds to the phage head. Once the initiator structure is formed, the major tail protein, V, is added. The H protein is used as a measuring stick and determines how long the tail will be. Once the tails reaches the correct length, the U protein is added to prevent further growth and the H protein is cleaved. The Z protein is added last and is required to make an infectious phage. A tail without Z will bind to a full phage head but the resulting particle is not infectious.

The λ phage packaging system packages DNA molecules on the basis of the *cos* sites rather than on the basis of the length of the DNA molecule. Varying lengths of DNA molecules, within set limits, can be packaged as long as the molecule contains a *cos* site at both ends. If the distance between the two *cos* sites is less than ~37 kb, the resulting phage particle will be unstable. When the DNA is inside the capsid, it exerts pressure on the capsid. Likewise the capsid exerts an inward force on the DNA. If there is not enough DNA inside the capsid, it will implode from the inward force of the capsid. If the distance between the two *cos* sites is too far (~52 kb), then the capsid will be filled before the second *cos* is reached. The tail cannot be added because the DNA hanging out of the capsid is in the way and no infectious phage particle is produced.

Getting out of the cell—the λ S and R proteins

The λ R and S proteins are required for λ to release progeny phage into the environment. The R protein is an endolysin that degrades the peptidoglycan cell wall and allows the phage to escape from the cell. The S protein forms a hole in the inner membrane to allow the endolysin to gain access to the cell wall. After the hole is formed, approximately 100 intact λ phage particles are released into the environment. The entire lytic cycle lasts ~35 minutes.

Induction of λ by the SOS System

When a λ lysogen is treated with ultraviolet light (UV), ~35 minutes later the cells lyse and release phage. What does the UV do to the cell? UV damages the DNA and triggers a cellular response called the SOS response to deal with this damage (Fig. 7.11). The RecA protein, which is normally used for homologous recombination, is activated and becomes a special kind of protease. The activated RecA interacts with LexA, leading to cleavage of LexA. This leads to the activation of a number of genes whose products repair the DNA damage in the cell. λ has tapped into this system through the CI protein. CI repressor can interact with activated RecA, leading to the cleavage of CI. This leads to expression of the phage lytic genes and phage production. The rational for this response is that λ does not want to risk staying in a cell that has DNA damage and may not survive.

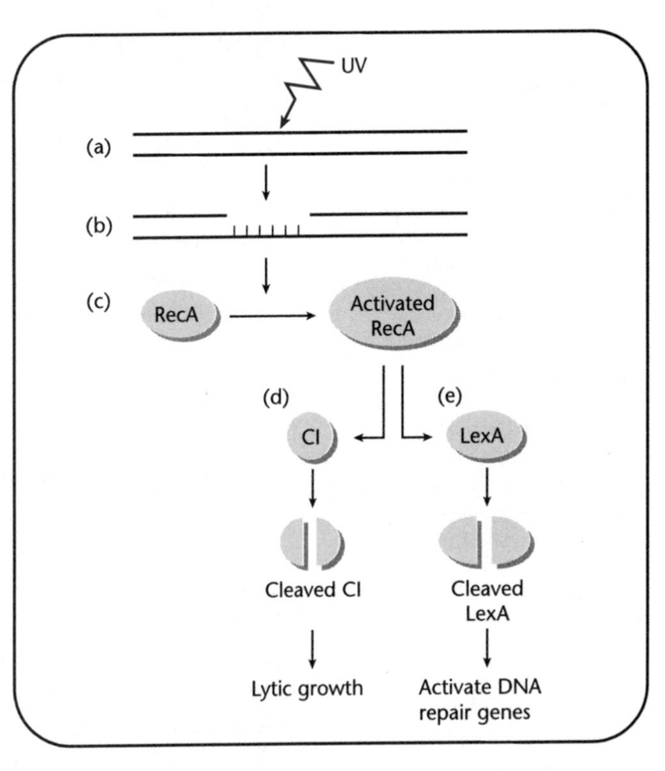

Fig. 7.11 The SOS response induces λ. UV treatment of cells (a) damages the DNA and leaves stretches of single-stranded DNA (b). The single-stranded DNA activates RecA (c). Activated RecA interacts with CI, leading to cleavage of CI and induction of the λ lysogen (d). Activated RecA also interacts with LexA and leads to LexA inactivation (e). LexA inactivation leads to expression of a number of genes, including some DNA repair enzymes.

Superinfection

If a cell is a λ lysogen, another λ phage that infects is not able to undergo lytic development and produce phage. The incoming phage can inject its DNA, however, the DNA is immediately shut down and no transcription or translation of the λ initiates. λ lysogens are immune to infection by another λ phage particle, which is called **superinfection**. Superinfection is blocked because the lysogen is continuously producing CI repressor. The lysogen actually produces more repressor than it needs to shut down one phage. This extra repressor binds to the superinfecting phage DNA at O_L and O_R and prevents transcription from P_L and P_R.

Restriction and modification of DNA

A simple experiment with λ leads to the discovery of how bacteria tell their own DNA from foreign DNA. λ is capable of making plaques on two different types of *E. coli*, *E. coli* K12 and *E. coli* C. If λ is grown on *E. coli* K12, it will form plaques on *E. coli* K12 or *E. coli* C with equal efficiency. If λ is grown on *E. coli* C, it will form plaques on *E. coli* C but if it is plated on *E. coli* K12, only a few phage will form plaques. The efficiency of forming plaques or **efficiency of plating** (EOP) is decreased by 10,000-fold. This is known as **restriction**. If the *E. coli* C grown phage that did plaque on *E. coli* K12 are replated on *E. coli* K12, the EOP is 1. This is known as **modification**. The few phage that survive the replating on *E. coli* K12 have been modified so that they can efficiently plate on *E. coli* K12.

While this originated as a curiosity of phage growth, it has proven to be essential for many molecular techniques. Further investigation showed that the protein responsible for restriction, a **restriction enzyme** or **restriction endonuclease**, actually recognizes a specific DNA sequence and cleaves the DNA on both strands. The cut or digested DNA is sensitive to nucleases that degrade DNA. The modification part of the system is a protein that specifically modifies the DNA sequence recognized by the restriction enzyme and prevents the DNA from being digested. *E. coli* K12 has a restric-

FYI 7.2

Pathogenicity in Vibrio cholera

Cholera is caused by the bacterium, *Vibrio cholera*. Many of the genes that make this bacteria pathogenic or disease causing are part of a prophage located in the *V. cholera* chromosome. This prophage bears striking resemblance to M13 and other filamentous phage. It is possible that the transmission of these pathogenic genes is as simple as the phage moving from one bacterial species to another sensitive bacterial species.

tion/modification system and *E. coli* C does not. This explains the original observation with λ growth. If a bacterium carries the restriction enzyme, it must also carry the modification enzyme so that the bacterial chromosome is not digested and degraded. The restriction/modification system allows a bacterium to tell DNA from its own species from foreign DNA. Many different bacteria contain restriction/modification systems that recognize different DNA sequences. The restriction enzymes are purified and used in vitro to cleave DNA at specific DNA sequences, depending on the recognition sequence of the enzyme in question. Restriction enzymes are used to cleave and clone DNA fragments as described in Chapter 14.

The lifecycle of M13

M13 adsorption and injection

M13 adsorbs to the tip of the F pilus, a hair-like structure on the surface of some bacteria. It can only infect bacteria that carry an F or F-like conjugative plasmid that encodes the proteins that make up the F pilus (see Chapter 10). For the filamentous phage, it is known that infection is initiated by the binding of gpIII to the tip of the F pilus. GpIII then interacts with the inner membrane protein TolA. Two additional facts about gpIII suggest a mechanism for phage DNA entry. GpIII contains amino acid sequences that are fusogenic or promote localized fusion of two membranes and gpIII is capable of forming pores in membranes that are large enough for DNA to go through. If each of these properties of gpIII are important for phage entry, then the phage could bind to the F pilus, promote fusion of the membranes, and use gpIII to form holes in the membrane to gain entry into the cytoplasm.

Protection of the M13 genome

The M13 DNA that ends up in the cytoplasm is a circular single-stranded DNA molecule. The strand present in phage particle is known as the plus or + strand. After entry into the cytoplasm, the + strand DNA is immediately coated with an *E. coli* single-stranded DNA binding protein known as SSB. The SSB coating protects the DNA from degradation.

M13 DNA replication

The M13 plus strand is converted to a double-stranded molecule immediately upon entry into *E. coli* (Fig. 7.12). Synthesis of the complementary strand is carried out entirely by *E. coli*'s DNA synthesis machinery. The complementary strand is called the minus or – strand. Only the minus strand is used as the template for mRNA synthesis and ultimately it is the template for the translation of the encoded M13 gene products. The SSB that coats the plus strand upon entry of the DNA into the *E. coli* cytoplasm fails to bind to ~60 nucleotides of the molecule (Fig 7.12c). These nucleotides form a hairpin loop that is protected from nuclease degradation. M13 gpIII from the phage is found associated with the hairpin loop. The hairpin loop is recognized by *E. coli* RNA polymerase as a DNA replication origin and is used to initiate transcription of a short RNA primer (Fig. 7.12d). The RNA primer is extended by *E. coli* DNA polymerase III to create the minus strand (Fig. 7.12e). The RNA primer is eventually removed by the exonuclease activities of *E. coli* DNA polymerase I. The gap is filled in by the 5′ to 3′ polymerizing activity of the same DNA polymerase. *E. coli* ligase forms the final phosphodiester bond resulting in a covalently closed double-stranded circular M13 chromosome. The double-stranded form of M13 chromosome is called the replicative form (RF) DNA.

The RF form is replicated by rolling circle replication similar to the mechanism used by the λ chromosome (see Fig. 7.9b). The M13 gene II encoded protein is an endonuclease that nicks the plus strand of the RF DNA at a specific place to initiate the replication process for M13 RF DNA. Approximately 100 copies of M13 RF DNA are made. While the M13 chromosome is being replicated, the genes encoding the coat proteins are being transcribed and translated. When M13 gpV protein accumulates to sufficient levels, a switch from synthesizing RF DNA to synthesizing the plus strand occurs. GpV blocks the synthesis of the minus strand, presumably by displacing SSB on the plus strand and preventing the plus strand from being used as a template. The plus strand is circularized.

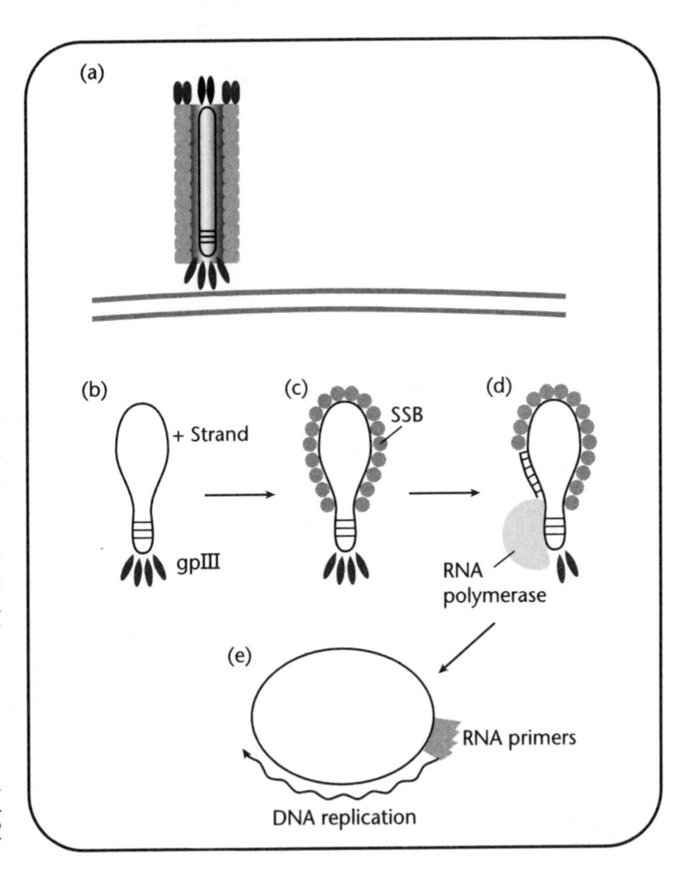

Fig. 7.12 The conversion of the M13 plus strand to a double-stranded DNA molecule. The plus strand enters the cell (a and b) with gpIII attached. It is immediately coated with host SSB (c). RNA polymerase synthesizes a short primer (d) and DNA polymerase synthesizes the minus strand.

M13 phage production and release from the cell

M13 phage particles are assembled and released from *E. coli* cells through a process that does not involve lysing *E. coli* or disrupting cell division (Fig. 7.13). The gpV coated plus strand makes contact with the bacterial inner membrane (Fig. 7.13a). This interaction requires a specific packaging sequence on the DNA and gpVII and gpIX. The protein-coated DNA traverses the membrane and gpV is replaced by gpVIII in the process (Fig. 7.13b). GpVIII is found in the membrane. When the last of the phage particle crosses the membrane, gpIII and gpVI are added. M13 phage are continually released from actively growing infected *E. coli*.

The lifecycle of P1

Adsorption, injection, and protection of the genome

P1 adsorbs to the terminal glucose on the lipopolysaccharide present on the outer surface of the outer membrane. The P1 tail can contract, suggesting that P1 might inject its DNA into the cell like T4 (see below). Once inside the cell, P1 DNA circularizes by homologous recombination. Circularization can occur by recombination because when the phage DNA is packaged, 107% to 112% of the phage genome is incorporated into a capsid. This ensures that every phage DNA molecule has between 7 and 12% homology at its ends; a property called **terminal redundancy** (Fig. 7.14). The terminal redundancy is used to circularize the genome.

P1 DNA replication and phage assembly

Like λ, early P1 replication takes place by the theta mode of replication. Later in infection, P1 switches to rolling circle replication, again like λ (see Fig. 7.9). At

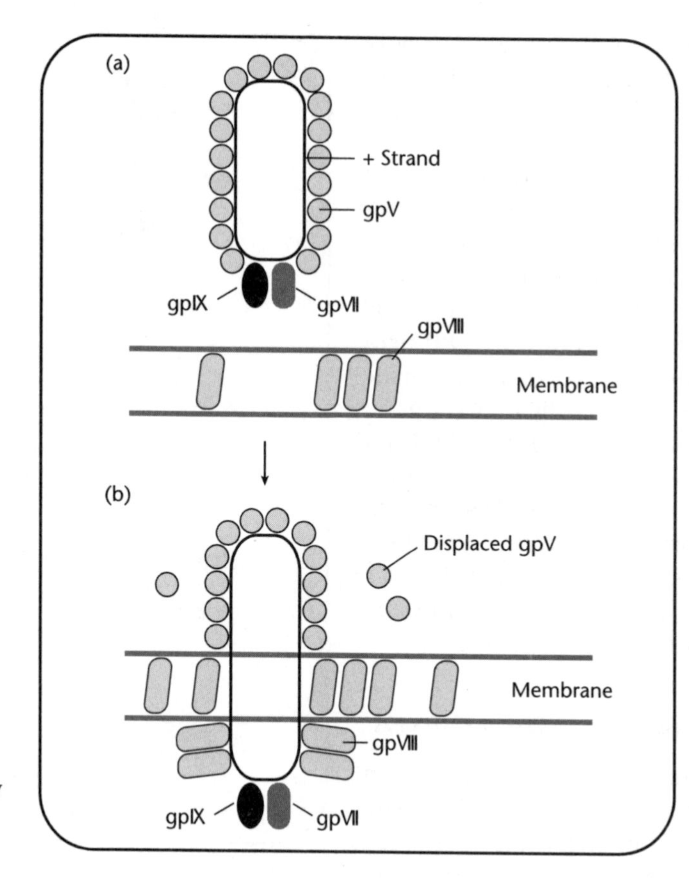

Fig. 7.13 M13 is released from the cell without lysing the bacterium. (a) The plus strand, coated with gpV interacts with the membrane through gpVII and gpIX. (b) As the DNA traverses the membrane, the gpV is replaced by gpVIII.

approximately 45 minutes after infection, the cells are filled with concatomers of phage DNA, assembled phage heads, and assembled phage tails. Now assembly of the complete phage must take place. A protein made from the phage genome recognizes a site on the concatomers of phage DNA called the ***pac* site** (Fig. 7.15). The protein cuts the DNA, making a double-stranded end. This end is inserted into a phage head. The DNA continues to be pushed inside the head until the head is full, a process termed **headfull packaging**. Once the first phage head is filled, another empty phage starts packaging. Experiments indicate that up to five headfulls of DNA can be packaged sequentially from a single *pac* site at 100% efficiency. An additional five headfulls of DNA can be packaged although the efficiency gradually decreases over these last five headfulls to only about 5%. While each phage head contains the same genes, the gene order changes. This is known as **circular permutation** of the genome (Fig. 7.14). After the head is full of DNA, a double-stranded cut is made and a tail is attached. This part of phage development is very much an assembly line. P1 is thought to encode an endolysin and holin to use in lysing the cell, similar to those described for λ.

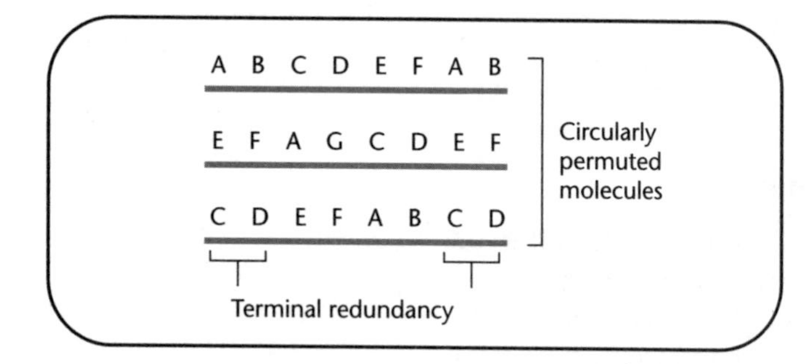

Fig. 7.14 P1 genomes are both circularly permuted and terminally redundant. Terminal redundancy means that the same sequences are present on both ends of one DNA molecule. Circular permutation means that the order of the genes on each DNA molecule is different but every DNA molecule contains the same genes.

The location of the P1 prophage in a lysogen

Prophages can be physically located in one of two places in a lysogen. In the case of λ, the phage genome is recombined into the bacterial chromosome. P1 is maintained in the cytoplasm as a stably inherited extrachromosomal piece of DNA or **plasmid** (see Chapter 9). P1 contains an origin for DNA replication and once the phage genome is converted to circular, double-stranded DNA, it can be established as a plasmid.

P1 transducing particles

One unusual aspect of P1 development is the formation of **transducing particles** or phage particles that contain chromosomal DNA instead of phage DNA. The *E. coli*

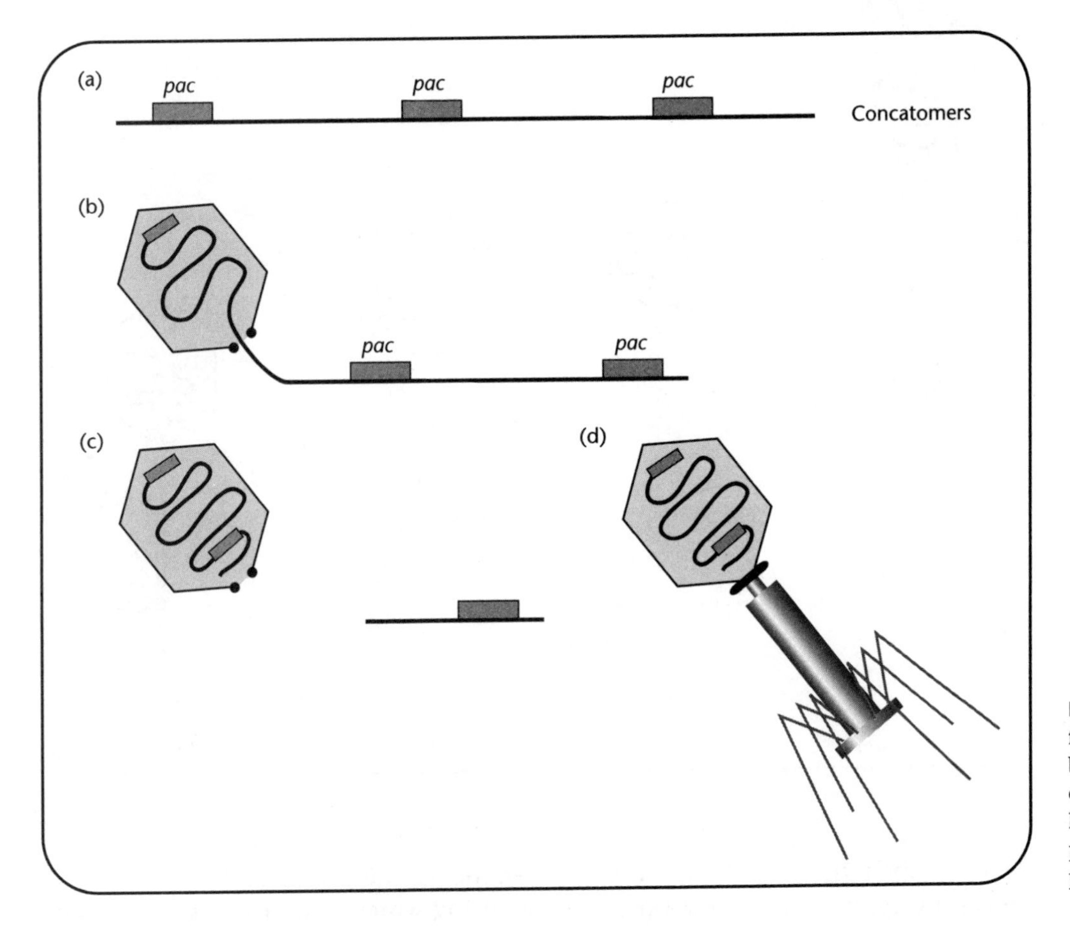

Fig. 7.15 P1 packages DNA from a *pac* site (a) and packages between 7 and 12% more than one P1 genome, until the phage head is full (b and c). Once the phage head is full, a preassembled head is added (d).

chromosome contains many **pseudopac sites** or sites that can be used to initiate packaging of host chromosomal DNA into maturing phage. These pseudopac sites are used much less frequently than the phage *pac* sites but they are used. The resulting phage carry random pieces of the chromosome in place of phage genomes. The ability to package any piece of chromosomal DNA instead of phage DNA makes P1 a **generalized transducing phage**. Transducing particles are used to move pieces of host chromosomal DNA from one strain to another for the purposes described in Chapter 8.

The lifecycle of T4

T4 adsorption and injection

For T4, the phage binds to the lipopolysaccharide. The tips of the tail fibers contact the cell first (Fig. 7.16). Once the phage has bound to the cell, the base plate rearranges creating a hole in the base plate. The outer sheath contracts and the internal tube goes through the outer membrane, peptidoglycan, and periplasm and comes close to the cytoplasmic membrane. The DNA is injected and crosses the cytoplasmic membrane in about 30 seconds. Not all phage that have the structure of T4 inject their DNA this way. Some phage such as T7, have tails that cannot contract. The T7 genome is only 40 kb but takes 9 to 12 minutes to cross into the cytoplasm. For T7, a small portion of

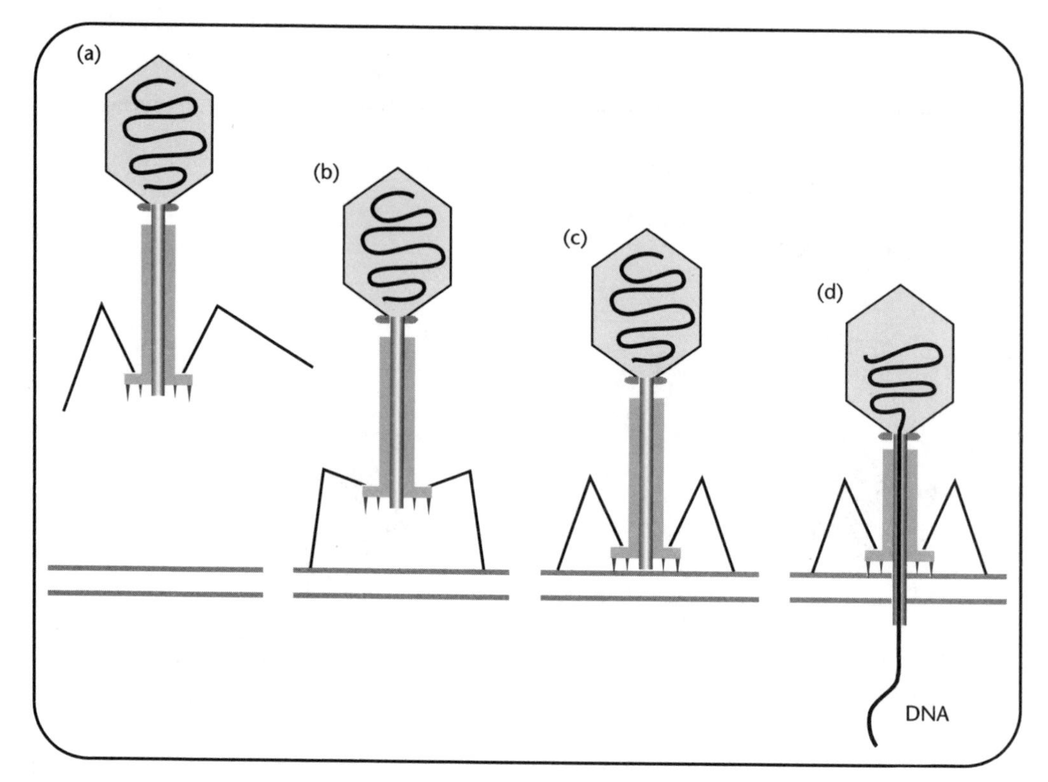

Fig. 7.16 Injection of T4 DNA into the cell. (a) T4 "looks" for a susceptible bacterium with its tail fibers. (b) The tail fibers recognize the membrane first. (c) The tail spikes interact with the membrane. (d) The tail sheath contacts, driving the internal tail tube through the outer membrane, peptidoglycan, and to the inner membrane where the DNA is released.

genome (about 8%) crosses both membranes, the peptidoglycan and periplasm, and enters the cytoplasm. After a 4-minute lag during which two proteins encoded by this piece of DNA are synthesized, the rest of the phage DNA enters the cytoplasm. Binding of these two phage proteins to the DNA is thought to pull the DNA into the cytoplasm.

Once T4 DNA is in the cell cytoplasm, it specifies a highly organized and coordinated program of gene expression. A group of genes with similar promoters, called the **early genes**, are transcribed and translated by host enzymes. One early gene encoded protein activates a second set of promoters for the **middle genes**. A different early gene encoded protein shuts off synthesis of the early genes. One product of middle transcription is required to activate the **late genes**. The early genes encode the proteins needed for DNA synthesis and late genes encode the proteins needed to build the capsid and tail structures. Many phage stage the expression of their genes in this temporal fashion to ensure proper construction of the phage particles.

The T4 genome does not contain cytosine residues. All of the cytosines are modified by methylation. Several of the early genes encode proteins that degrade the cytosine-containing host DNA. The phage DNA is protected from degradation. The T4 genome, like P1 is both circularly permuted and terminally redundant. T4 has about 3% terminal redundancy. Unlike P1, T4 does not appear to use this terminal redundancy to circularize upon infection. T4 begins replication immediately after the early gene products are made. T4 replicates as a linear molecule and uses replication and recombination to both replicate the entire genome and to make concatomers to package into phage heads (Fig. 7.17). Like other phage, capsids, tails, and concatomers of phage DNA are premade and assembled into infectious phage particles late in phage development.

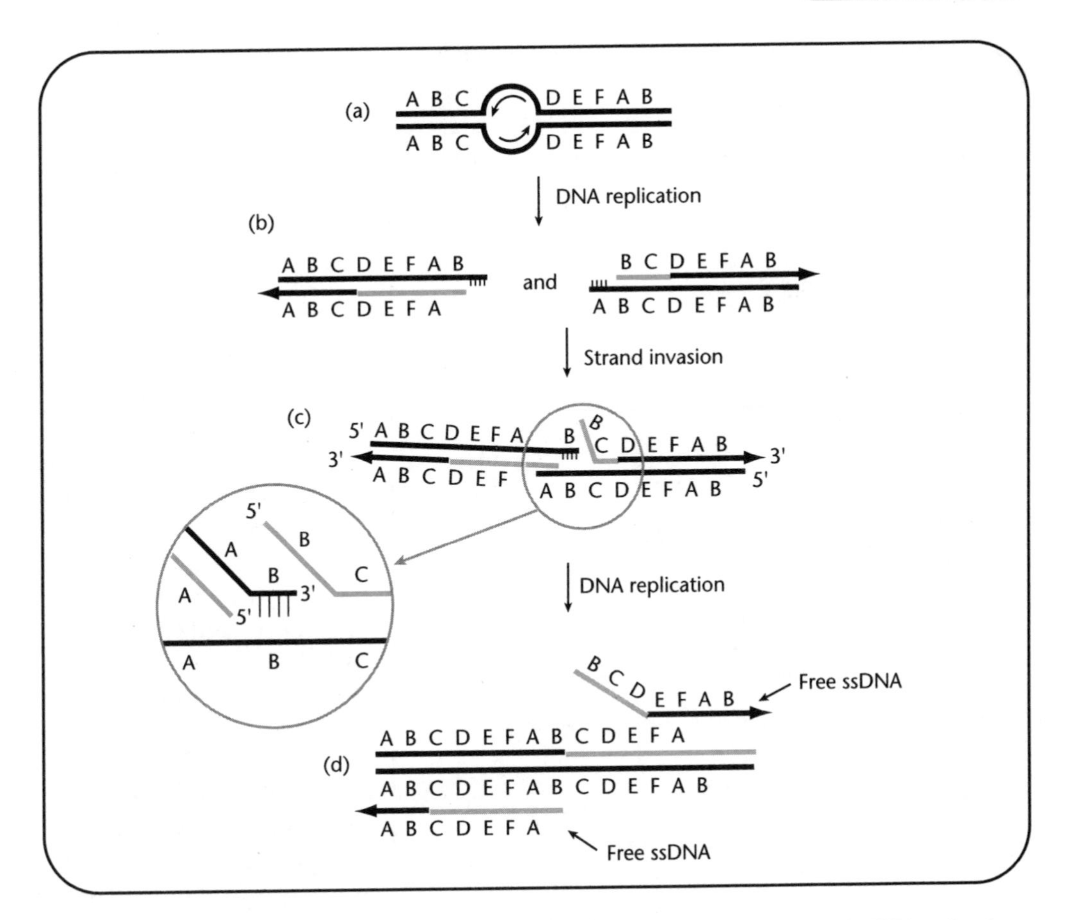

Fig. 7.17 T4 replicates its DNA using both replication and recombination. (a) Linear T4 DNA molecules are injected into the cytoplasm of the host. (b) DNA replication begins at an origin and proceeds bidirectionally to the ends. However, because of DNA polymerase's requirement for a primer, a piece of the DNA at one end of the molecule cannot be replicated and remains single stranded. (c) This piece of single-stranded DNA can invade duplexed DNA at any place where it has homology, like the initial reaction in recombination. (d) DNA replication of this molecule will lead to concatomers of the phage genome. Depending on where strand invasion takes place, branched molecules can also be formed. T4 packages its DNA out of the concatomers. The displaced single strands are free to strand invade the concatomer structures.

T4 rII mutations and the nature of the genetic code

The study of two genes in T4 has contributed significantly to our understanding of the genetic code and the nature of the gene. In the late 1950s and 1960s the understanding of the nature of the gene was in its infancy. The prevailing thought was that the gene was the smallest genetic unit and it was inherited as a unit. The chemical nature of DNA had just been described by Watson and Crick. The relationship between DNA and the gene was not understood. Seymour Benzer used the T4 rII locus and genetic logic to describe several key features of the gene.

rII encodes two proteins, A and B. Several thousand point mutations and deletions were isolated in these two genes. These mutations were put to good use. A phage carrying one mutation and a phage carrying a second mutation were mixed together and

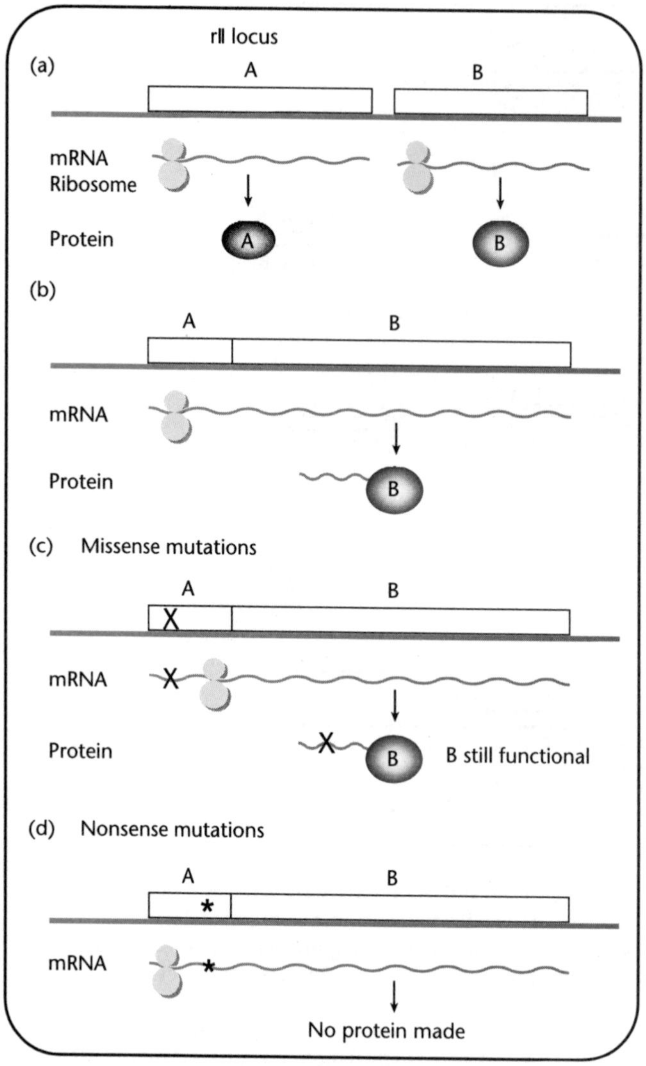

Fig. 7.18 The T4 rII locus was used to conduct studies on the nature of the gene. (a) rII is composed of two genes A and B that are normally made into separate proteins. (b) A specific deletion was described from the mapping studies. This deletion (called r1589) fused the A and B genes, leaving A nonfunctional but B functional. (c) Some mutations in A, called missense mutations did not interfere with B function. (d) Other mutations in A did interfere with B function. These mutations were said to be "nonsense" and interfered with the production of B.

grown on an appropriate host to determine if any of the offspring had recombined back to wild type. Many of these phage crosses were carried out and used to construct a map of where the mutations resided in the rII locus.

Several conclusions were drawn from these studies. Deletion mutations were defined as mutations that could not generate wild-type recombinants when crossed with more than one of the other mutations. Deletion mutations were missing a part of the gene. In the thousands of crosses that were carried out, the frequency of obtaining wild-type recombinants was predictable based on the positions of the starting mutations. This led to the conclusion that DNA was a linear molecule across the length of the gene and not a branched molecule. If DNA was branched, then the recombination frequencies should be very different when one mutation resided on one side of a branch and the other mutation resided on the opposite side of the branch. Using the recombination frequencies, it was determined that recombination can take place within a gene and not just outside of it as had been thought. This changed the definition of a gene from the unit of heredity that mutated to altered states and recombined with other genes. It is now recognized that the gene is a functional unit that must be intact in the DNA to lead to a specific characteristic or phenotype. Each of these studies shed more light on the behavior of the gene and the nature of mutations.

One of the deletions fused the A gene to the B gene such that the A gene was not functional but the B gene was (Fig. 7.18). Some of the point mutations in A could be crossed onto the same phage that carried the deletion. If a point mutation in A did not affect the activity of B in the fused genes then the point mutation was interpreted to be a missense mutation. Missense mutations could cause a change in the genetic code without affecting the production of the protein product. Other point mutations in A did affect the activity of B. These were interpreted to be nonsense mutations or changes that stopped the production of the B protein. These studies led directly to the modern concepts of a gene and how it functions.

FYI 7.3

The RNA phage MS2

MS2 is a typical phage containing an RNA genome. MS2 binds to the F pilus. It has a genome of 3569 nucleotides and encodes only four proteins: the coat protein, an RNA-directed RNA polymerase, a lysin and an adsorption protein. MS2 has an icosahedral capsid composed mainly of one type of protein with a few molecules of a minor protein in it.

Summary

Bacteriophage are a very diverse group of viruses. Their genomes can be made from either DNA or RNA. They can be linear or circular, single stranded or double stranded. Phage have evolved many different ways to carry out the limited number of steps in a phage infection. All phage must recognize the correct bacterium to infect, get their genome inside the cell, replicate the genome, transcribe and translate the genome, and assemble phage particles. The relative simplicity of phage have made them favorite model systems to study many biological processes. While it may appear that phage carry out some processes using baroque mechanisms, it usually turns out that other biological systems share these mechanisms. For example, the unusual mechanism used to replicate T4 DNA is also used to help maintain bacterial and eukaryotic chromosomes.

Study questions

1 What processes must be carried out by all phage to produce progeny?
2 What is the phenotype of a λ mutant containing a defective *cI* gene?
3 Which regulatory proteins and promoters are crucial in λ's decision-making process? Which regulatory proteins and promoters are crucial in λ's lytic pathway? Which regulatory proteins and promoters are crucial in λ's lysogenic pathway? Describe the roles for all identified participants.
4 A new phage from local sewage was recently isolated that infects laboratory strains of *E. coli*. How would you determine if this new phage is a temperate or lytic phage using simple genetic tests?
5 Contrast and compare the lytic pathway for λ and M13 phage. What do they do that is similar? What do they do that is different?
6 Contrast and compare rolling circle replication and theta mode replication. What components of the machinery are similar? What components of the machinery are different? When would one type of mechanism be preferable to the other type? Why?
7 How does T4 gets its DNA from the phage head into the cytoplasm?
8 How do restriction/modification systems function?
9 How do different phage protect their DNA in the cell cytoplasm?
10 Why is M13 not considered a temperate phage?

Further reading

Campbell, A.M. 1996. Bacteriophages. In *Escherichia coli* and *Salmonella typhimurium: Cellular and Molecular Biology*, 2nd edn., eds. F.C. Neidhardt, R. Curtiss III, J.L. Ingraham, E.C.C. Lin, K.B. Low, B. Hagasanik, W.S. Rexnikoff, M. Riley, M. Schaechter, and H.E. Umbarger, pp. 2325–38. Washington, DC: ASM Press.

Ptashne, M. 1993. *A Genetic Switch*. 2nd edn. Cambridge, MA: Blackwell Scientific.

Young, R., Wang, I.-N., and Roof, W. 2002. Phage will out: Strategies of host cell lysis. *Trends in Microbiology*, **8**: 120–8.

Zaman, G., Smetsers, A., Kaan, A., Schoenmaters, J., and Konings, R. 1991. Regulation of expression of the genome of bacteriophage M13. Gene V protein regulated translation of the mRNAs encoded by genes I, II, V and X. *Biochimica et Biophysica Acta*, **1089**: 183–92.

Chapter 8

Transduction

Transduction is the process of moving a piece of chromosomal DNA from one cell to another using a bacteriophage to carry the DNA. Transduction was described first by Zinder and Lederberg in 1952 as a way to move "heritable traits" (physical characteristics or phenotypes that could be inherited by daughter cells) from one strain of *Salmonella enterica* serovar *typhimurium* to another. They noted that the agent responsible for moving the chromosomal DNA could pass through a filter with a pore size smaller than any known bacterium. We now know that this agent is a **bacteriophage** or phage (see Chapter 7). Phage are viruses that infect and grow in bacterial cells. In most documented cases, phage grow on a single bacterial species or at most a few closely related species. The process of transduction allows a bacterium to acquire a segment of DNA approximately 50 to 100 kb in length. This can speed the pace of evolution by restoring a mutated section of the chromosome or allowing the acquisition of a block of new genes. Transduction is usually a byproduct of the phage life cycle without an obvious benefit for the phage.

Generalized transduction vs. specialized transduction

There are two types of transducing phage:
1 generalized;
2 specialized.
Generalized transducing phage are capable of moving any piece of chromosomal DNA from one cell to another (Fig. 8.1a). Specialized transducing phage move the same piece of DNA every time (Fig. 8.1b). Different specialized transducing phage can carry different segments of the chromosome but each phage and the progeny it produces carry the same chromosomal region. The difference between these two types of phage can best be illustrated by looking at a well-studied example of each.

P1 as a model for generalized transducing phage

One of the best-studied generalized transducing phage is P1. P1 infects *E. coli* and has a double-stranded DNA genome of approximately 90 kb. As described in Chapter 7, P1 packages DNA into a phage head from a specific site on the P1 DNA known as a *pac*

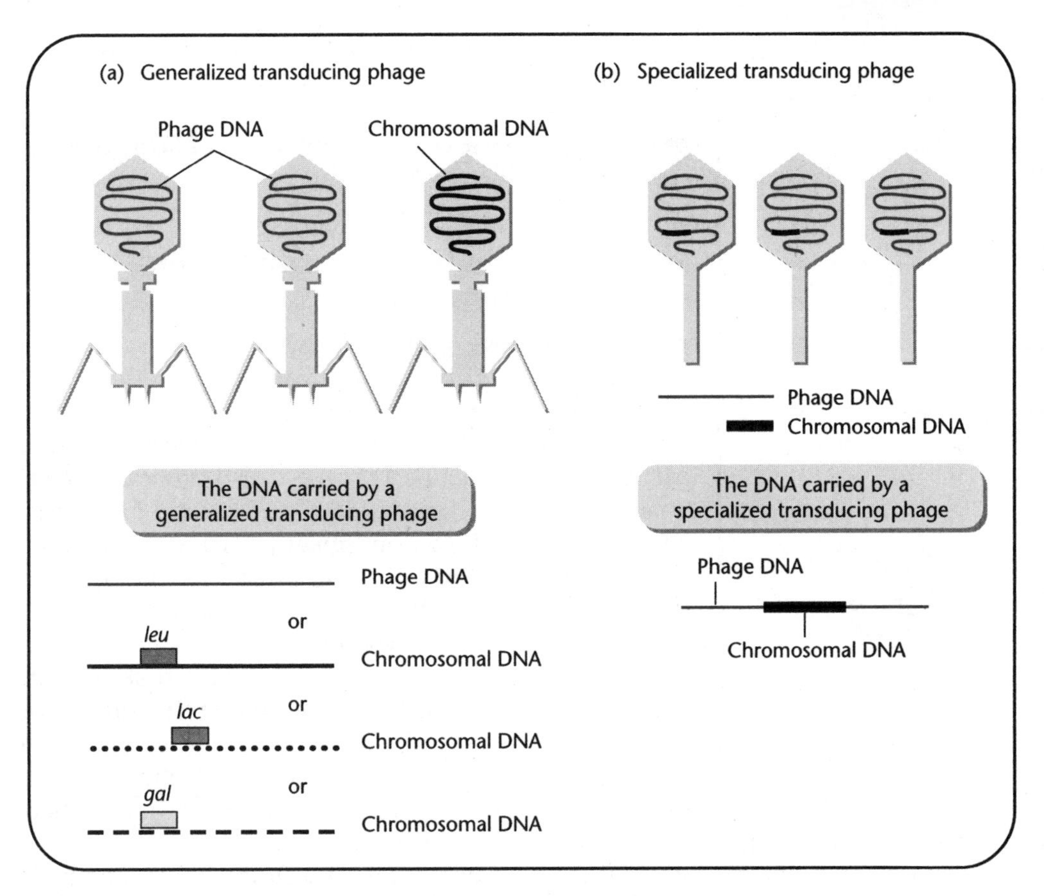

Fig. 8.1 The DNA carried by transducing particles. (a) Generalized transducing phage. (b) Specialized transducing phage.

site. Once the DNA is inserted into the phage head, a P1 tail is added to make a complete phage that is capable of infecting another *E. coli* cell.

Packaging the chromosome

If the phage DNA is put into the phage head from the *pac* site on the phage genome, how does the chromosomal DNA used for transduction ever find its way into a phage? While a specific sequence for the *pac* site has been determined, it is also known that the requirement is not absolute. Other related sequences can function as *pac* sites. The *E. coli* chromosome has many **pseudo-*pac* sites** that can be utilized by P1 (Fig. 8.2). The pseudo-*pac* sites are located randomly around the chromosome. P1 packages chromosomal DNA from pseudo-*pac* sites the same way it packages P1 DNA from *pac* sites located in concatomers of P1 DNA (see Chapter 7).

As expected, P1 packages from pseudo-*pac* sites at a reduced frequency. For every ~500 P1 phage genomes that are correctly packaged to form infectious phage particles, one phage containing chromosomal DNA or a **transducing particle** is made. While this may seem like a small number of transducing particles, consider these additional numbers. An average lysate of P1 contains 10^9 infectious P1 phage per milliliter and making a 5 ml lysate is trivial. This means that the lysate also contains 10^6 to 10^7 transducing particles per milliliter. A transducing particle contains ~90 kb

FYI 8.1

Isolating phage

The best place to find phage is, not surprisingly, where you find the bacteria. Most of the phages that infect *E. coli* have been isolated from sewers. One of the most famous *E. coli* phage, λ, came straight out of a Parisian sewer! The first step in identifying phage is to collect samples (soil, water, etc.) from the appropriate environment and expose them to bacteria that have been spread on the surface of an agar plate. If there are phage present, they will lyse the bacteria in a small circular area (usually a few millimeters). Once phage have been identified, they can be tested to determine if they are capable of generalized transduction. What tests would you use to determine if a phage can move pieces of chromosomal DNA?

Generalized transducing phage have been identified for many bacteria. In *Salmonella enterica* serovar *typhimurium*, the phage P22 is the transduction workhorse. It transduces fragments of chromosomal DNA several orders of magnitude better than P1. P22 is structurally related to λ, yet capable of generalized transduction. FP43 carries out generalized transduction in many *Streptomyces* species. It is a temperate phage that packages 56 kb of DNA per phage head. A partial list of bacteria for which there are characterized generalized transducing phage includes: the Gram-positive bacteria, *Bacillus subtilis*; the plant symbiont, *Rhizobium meloti*; the causative agent of tuberculosis and its relatives, the Mycobacteria; and two bacteria that undergo complex developmental pathways, *Caulobacter crescenti* and the Myxococcus.

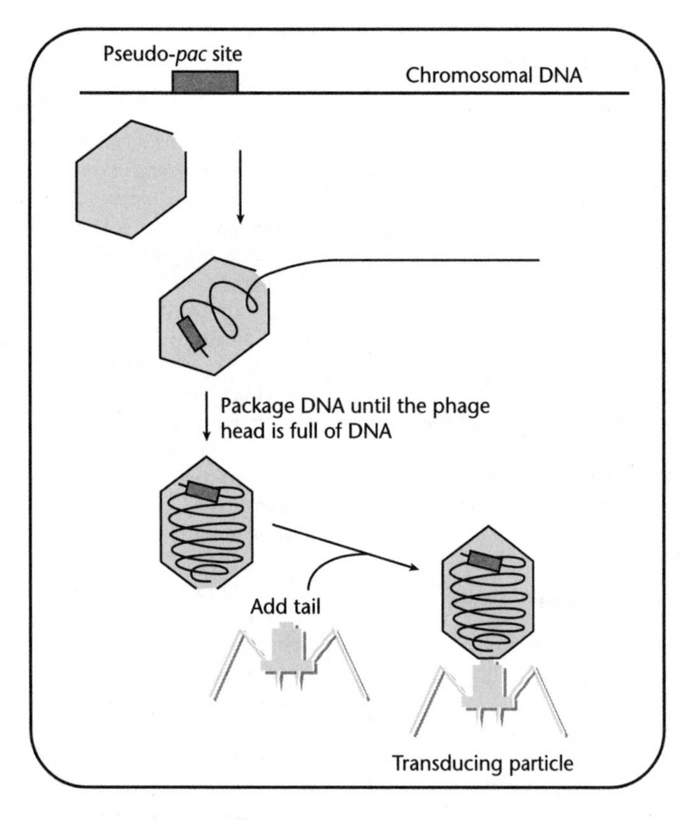

Fig. 8.2 The packaging of chromosomal DNA by P1. Pseudo-*pac* sites in the chromosome are used to package chromosomal DNA instead of P1 DNA.

and the *E. coli* chromosome is 4639 kb. It takes ~52 transducing phage to cover the entire chromosome. Thus, in every milliliter of P1 lysate, there are enough transducing particles so that the entire chromosome of *E. coli* is represented many times.

Moving pieces of the chromosome from one cell to another

As described above, growing a lysate of P1 on a particular bacterial strain results in the chromosome of that strain being packaged into P1 phage particles. These pieces of bacterial chromosome can then be moved from one bacterium to another by a process called **transduction**. Simply put, the phage lysate is mixed with the bacteria in equal numbers of infectious phage particles and bacterial cells or slightly more cells than phage. The P1 particles that contain the chromosomal DNA attach to the cells in the mixture and the chromosomal DNA is injected into the cell. At the same time, P1 particles that contain authentic P1 DNA infect cells. Because of the number of phage and bacteria used, on average, each cell is infected by one phage particle. This results in a cell being infected by either a particle containing phage DNA or a particle containing chromosomal DNA.

The fate of the chromosomal DNA is somewhat different from that of the phage genome. While the phage genome circularizes and is either established as a plasmid or used to make more phage, the chromosomal DNA must either be recombined into the chromosome or it is degraded. If the incoming chromosomal DNA recombines into the chromosome, it does so by a double recombination event only at a site of homology to the chromosome (Fig. 8.3). Because this recombination requires homology between the two pieces of chromosomal DNA, it also requires the enzymes of

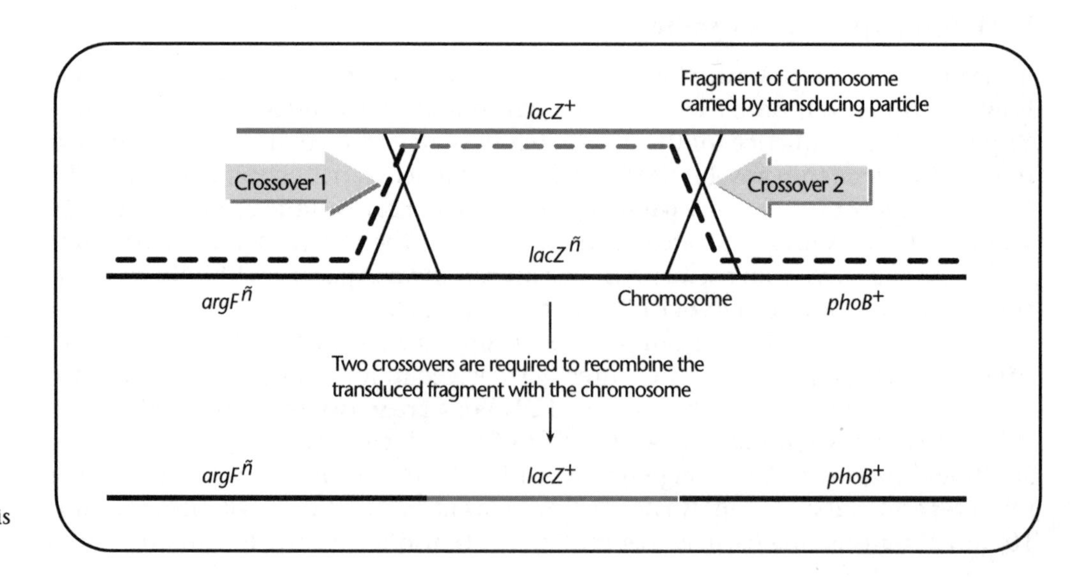

Fig. 8.3 A double recombination event recombines the transduced fragment of DNA into the chromosome of the host cell. The host DNA that is removed is degraded.

homologous recombination as described in Chapter 5. The double recombination event replaces a piece of the infected bacterium's chromosome with the piece of chromosomal DNA that was brought in by the transducing particle. The piece of infected bacterium's chromosome that is removed by the homologous recombination event is subsequently degraded.

If the chromosomal DNA brought in by the P1 does not recombine, eventually the DNA will be degraded by cellular nucleases. The length of time needed for degradation of each fragment can vary widely, with most of the DNA being degraded within hours, before many cell divisions take place. Occasionally, a piece of transduced chromosomal DNA will persist in a cell long enough so that the transduced cell grows and divides. Upon prolonged growth, these cells will eventually lose the transduced fragment of chromosomal DNA.

Identifying transduced bacteria: selection vs. screening

To identify those cells that have inherited a tranducing particle, several factors must be considered. Because the P1 lysate contains only 10^4 to 10^5 transducing particles per ml, simply plating the mixture of phage and cells on growth media and sorting through cells to find the ones that have received a transducing phage would be too time consuming. Examining each individual bacterium is known as a **screen** and the process is called screening. In order to find one cell that had been transduced, 10,000 to 100,000 cells would have to be screened. Screening is time consuming and tedious and is carried out only when an event occurs very frequently.

To identify transduced cells, they must be selected out of the mixture of cells and phage. In a **selection**, cells are grown on media that only allows those with a specific characteristic or phenotype to grow. This type of media is known as **selective media**. Many different media can be selective. Media that contains lactose as the sole carbon source requires that cells be Lac^+, media without any added amino acids require that cells be able to synthesize all of their own amino acids and media containing the antibiotic ampicillin require that cells be resistant to ampicillin in order to grow, to name just a few. All cells without the specific phenotype required for growth on the selective media either do not grow or are killed when they are plated on this media. If the cells without the specific phenotype fail to grow, the selection is known as a **nonlethal selection** and the selective condition is called **bacteriostatic**. If the cells without the specific phenotype are actually killed by the selective media, the selection is known as a **lethal selection** and the selective conditions are **bacteriocidal**. Growing cells on media containing lactose as a sole carbon source is an example of a nonlethal selection or a bacteriostatic condition. Growing cells on media containing the antibiotic ampicillin is an example of a lethal selection or a bacteriocidal condition. On a standard petri plate that is 15 cm in diameter and contains media solidified with the agar, up to 10^{10} bacteria can be plated. From this large number of cells, as few as one or two cells that possess the selected phenotype can be identified. These cells will be able to grow and divide and will form a visible group of cells or a **colony**.

Carrying out a transduction

Selecting transduced cells out of the mixture of P1 phage and bacteria requires that a specific fragment of interest is identified and that something is known about the genes contained in the fragment of interest. It is not possible to simply select for a bac-

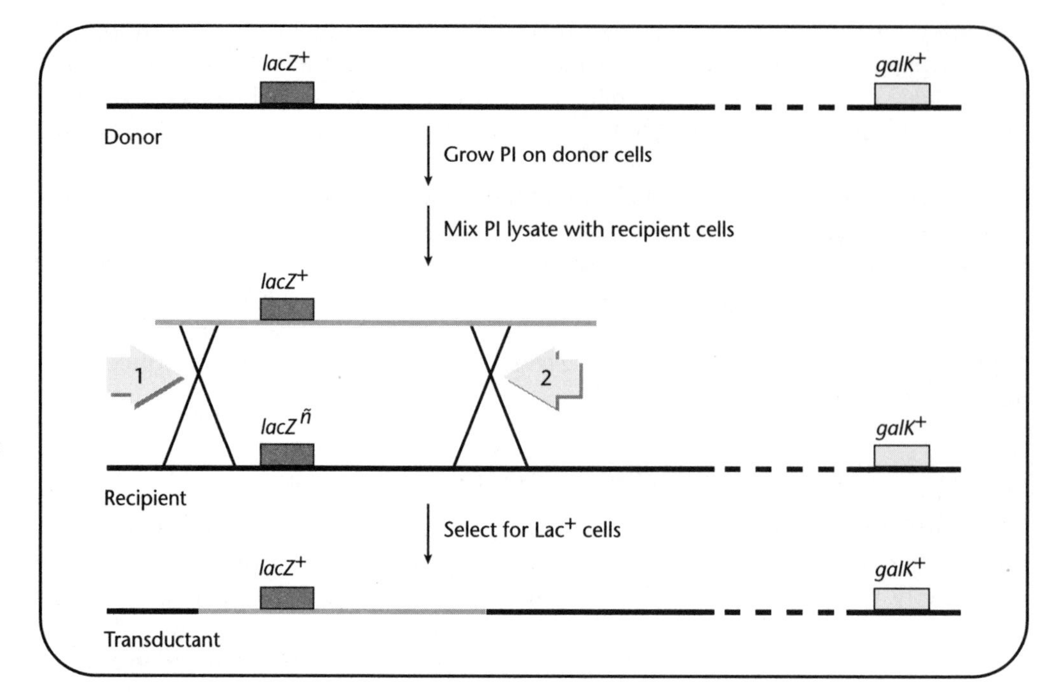

Fig. 8.4 Transducing Lac⁻ cells to Lac⁺ using P1. Only genes that are contained on the transduced fragment of DNA have the possibility of being altered by the transduction. In this example, the *galK* gene is too far away from the *lacZ* gene and cannot be altered by this transducing fragment.

terium that has been transduced by any fragment of chromosomal DNA. Once the specific region of the chromosome to be moved has been identified, the phenotypes of the genes in that region are examined to determine if any of them can be used in a selection. If there are no known genes on a given piece of chromosomal DNA or no identifiable phenotypes of the genes on that piece of chromosomal DNA, then that fragment DNA cannot be identified in a P1 transduction. Because P1 can package 90 kb of DNA and so many genes have been characterized in *E. coli*, there are very few regions of the chromosome that cannot be moved by P1. For regions without known genes, a selectable marker, such as the antibiotic resistances found in some transposable elements (see Chapter 6), can be isolated in the region and used to move the region by P1.

An example of a transduction is shown in Fig. 8.4. The *lacZ* gene encodes β-galactosidase, the enzyme needed to break down lactose for use as a carbon source. If the starting cells are unable to grow on lactose as a sole carbon source (Lac⁻) because of a mutation in the *lacZ* gene, the Lac⁻ defect can be "fixed" by transduction. P1 is grown on cells that are *lacZ*⁺. The resulting phage lysate is mixed with the Lac⁻ cells and the mixture spread on agar containing lactose as the sole carbon source (minimal lactose agar). The Lac⁺ transductants are purified on the selective agar plates (in this example minimal lactose agar) to remove any contaminating phage or cells. After they have been purified twice, the transductants are ready for use in other experiments. Any wild-type or mutant gene that has a selectable phenotype can be moved by P1 transduction.

Uses for transduction

Transduction is one way to move a defined piece of DNA from one cell to another. Chapters 10 and 11 describe two additional ways to move DNA, conjugation and transformation, respectively. The decision on which method to use hinges on what is available for the bacteria being studied. For *E. coli* and *Salmonella enterica* serovar *typhimurium*, transduction is a commonly used technique. Moving a defined piece of DNA from one cell to another has many uses. It is used to map the position of newly identified genes and to determine the order of the new gene with respect to known genes in a region of the chromosome. When new chromosomal mutations are isolated, they are mapped to determine where and how many different loci are involved in the phenotype. Strains with a specific complement of genes can be constructed at will. What follows are descriptions of some of the more common uses for transduction.

Two-factor crosses to determine gene linkage

P1 transduction can be used to determine if two genes are near each other on the chromosome (Fig. 8.5). The phenotype of one of the genes is selected for in a transduction. A large number of transductants that inherited the first gene can be screened for inheritance of the second gene. The percentage of transductants that have inherited the second, unselected, gene indicates the **linkage** or closeness of the two genes. If two genes are inherited together 98% of the time, then the genes must be physically located very close together on the chromosome. If the genes are inherited together only 10% of the time, then the genes are physically located much further apart. Thus, the frequency of **coinheritance** gives a relative measure of the distance between two genes. Measuring the frequency of coinheritance of two genes is known as a **two-factor cross**.

In this type of coinheritance experiment, several technical factors must be taken into consideration. First, in order for the coinheritance percentages to be accurate, 100 to 500 transductants must be screened for the presence or absence of the second gene. If 500 transductants are screened for the phenotype of the second gene and none containing the appropriate phenotype are found, then the two genes are not linked. Second, if the same transduction is repeated several times, the percentage of

FYI 8.2

Relating genetic and physical distances

Genetic mapping can determine how close or far apart two genes are located on the chromosome. However, the numbers obtained from crosses are expressed in percent linkages not the number of kilobases between the two genes. To convert from cotransduction frequencies to kilobases, the following formula is used:

$C = (1 - d/L)^3$

C = cotransduction frequency expressed as a decimal
d = distance between the selected marker and an unselected marker in minutes
L = size of the transducing fragment in minutes
If *gene A* and *gene B* are coinherited 50% of the time, then $C = 0.5$, $L = 2$ min for P1 (P1 carries ~90 kb or 2 minutes of DNA).
Solving for d
$d = 1.58$ minutes.
If there are 90 kb in 2 minutes then *gene A* and *gene B* are ~71 kb apart.

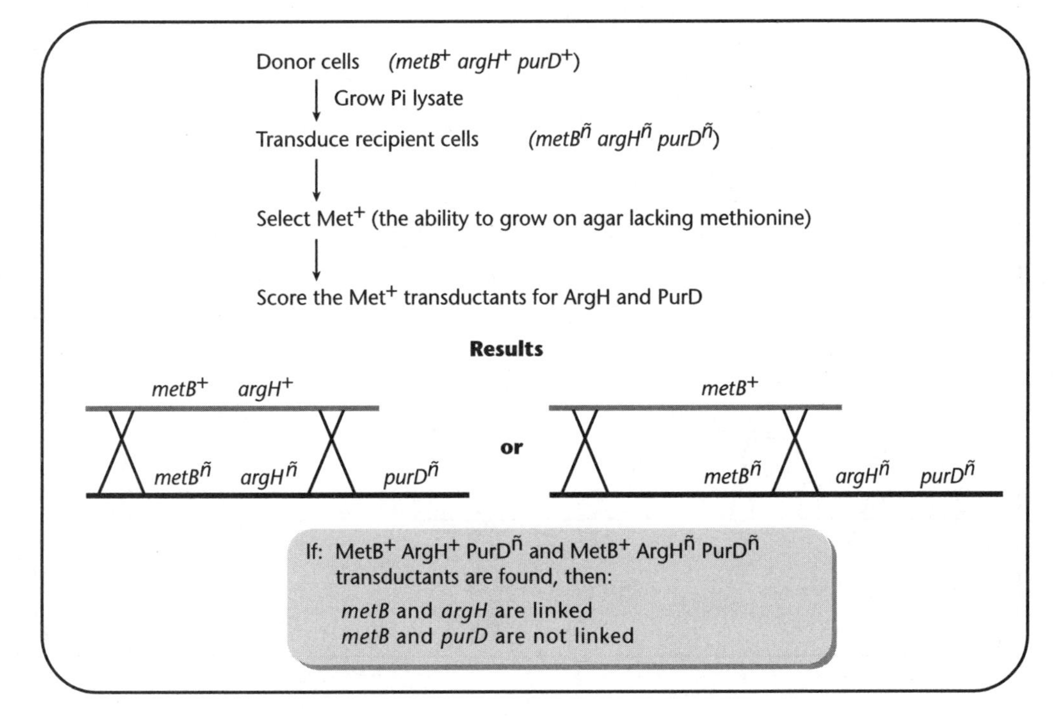

Fig. 8.5 Using transduction to determine if two genes are linked. In this example, the donor cells are phenotypically $MetB^+$ $ArgH^+$ and $PurD^+$ and that the recipient cells are $MetB^-$ $ArgH^-$ and $PurD^-$.

coinheritance can vary by ~10–15%. Thus, a frequency of 45% is not significantly different from a frequency of 55%. However, a frequency of 10% is distinguishable from a frequency of 45%.

Mapping the order of genes – three-factor crosses

Measuring coinheritance frequencies can be taken a step further by including three or even four genes in the cross. By determining the percentage of transductants that have inherited every combination of two of the three genes, the order of the three genes relative to one another can be determined. Monitoring three genes in a transduction is known as a **three-factor cross**.

An example of a three-factor cross is shown in Fig. 8.6. The phenotype associated with *gene A* is selected for in the initial plating of the transduction. Then, the number of transductants that have inherited *gene A* and *gene B* or *gene A* and *gene C* are measured by screening the transductants for the phenotypes associated with *gene B* and *gene C*. The transductants are classified by the phenotypes associated with all three genes:

- those that have inherited *gene A*, *gene B*, and *gene C*;
- those that have inherited *gene A* and *gene B* but kept *gene C* from the recipient strain;
- those that have inherited *gene A* and *gene C* but kept *gene B* from the recipient strain.

The number of transductants in each class is determined. The class that occurs most frequently and the class that occurs least frequently give the most information. *The frequency indicates the number of recombinational crossovers needed to generate that class. The most frequently obtained classes require the least number of crossovers to generate them and the least frequent require the most crossovers* (Fig. 8.7). Thus, the easiest way to determine the correct gene order is to draw out every possibility and count the number of crossovers that are required to give each class of transductants. In most cases, this approach

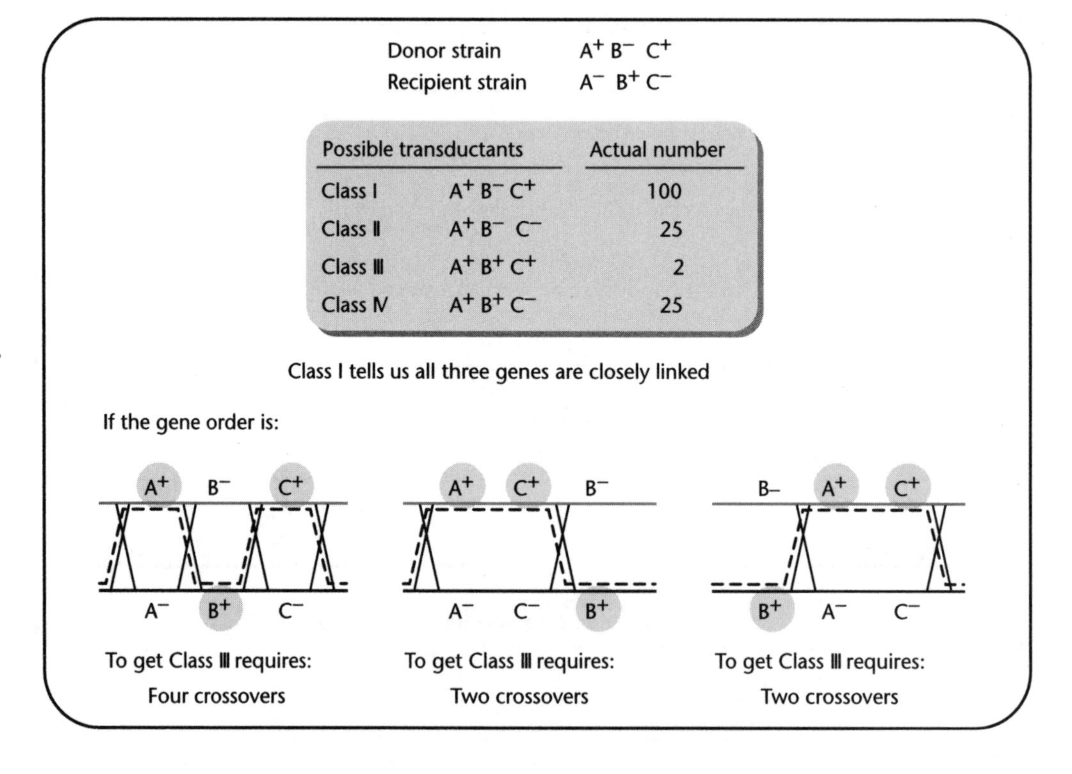

Fig. 8.6 A three-factor cross to determine the gene order of *gene A*, *gene B*, and *gene C*. When drawing crossover events, begin and end on the recipient DNA. A crossover to the donor DNA, must be followed by a crossback to the recipient to keep the chromosome intact. Because Class III is the least frequent, it requires the most crossovers to accomplish. This leads to a gene order of *geneA–geneB–geneC* because this order requires four crossovers.

will give the gene order of the three genes. Occasionally, the gene order cannot be unambiguously assigned. In these instances, a different three-factor cross must be designed.

Strain construction

One of the major uses for transduction is in constructing strains with a specific set of mutant or wild-type genes. In many cases, an experiment requires that the starting strain has a specific complement of genes. The speed and accuracy by which bacterial strains can be constructed is one of the major advantages of working with bacteria. If the exact strain needed is not available, it is usually possible to build the strain using one or several successive transductions. If the mutation to be moved cannot be selected for directly, then it can be moved using a linked marker that can be selected (Fig. 8.7).

Localized mutagenesis

Transduction is also used to mutagenize a small specific region of interest in the chromosome by a technique called **localized mutagenesis** (Fig. 8.8). This technique is very useful if mutations in a specific gene are needed and if that gene:

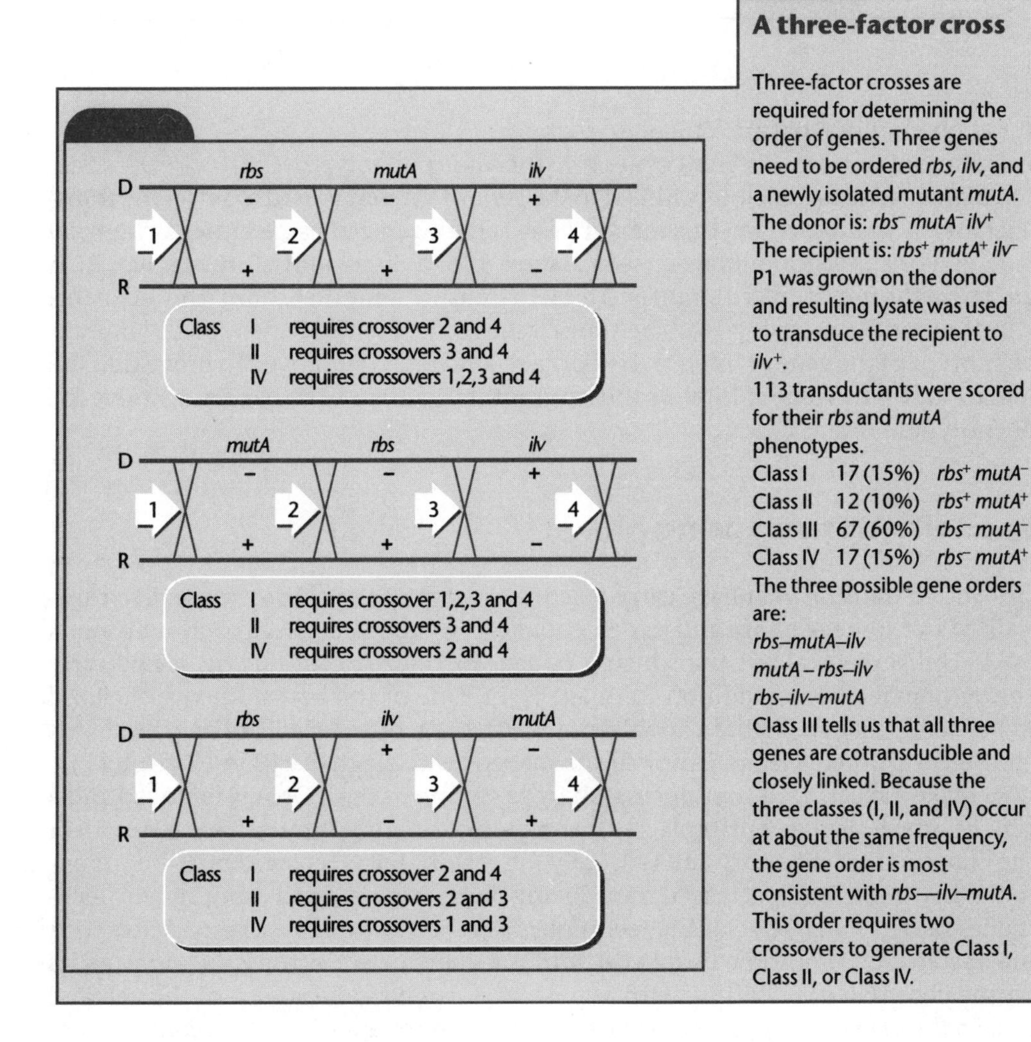

FYI 8.3

A three-factor cross

Three-factor crosses are required for determining the order of genes. Three genes need to be ordered *rbs*, *ilv*, and a newly isolated mutant *mutA*.
The donor is: rbs^- $mutA^-$ ilv^+
The recipient is: rbs^+ $mutA^+$ ilv^-
P1 was grown on the donor and resulting lysate was used to transduce the recipient to ilv^+.
113 transductants were scored for their *rbs* and *mutA* phenotypes.

Class I	17 (15%)	rbs^+ $mutA^-$
Class II	12 (10%)	rbs^+ $mutA^+$
Class III	67 (60%)	rbs^- $mutA^-$
Class IV	17 (15%)	rbs^- $mutA^+$

The three possible gene orders are:
rbs–mutA–ilv
mutA – rbs–ilv
rbs–ilv–mutA
Class III tells us that all three genes are cotransducible and closely linked. Because the three classes (I, II, and IV) occur at about the same frequency, the gene order is most consistent with *rbs—ilv–mutA*. This order requires two crossovers to generate Class I, Class II, or Class IV.

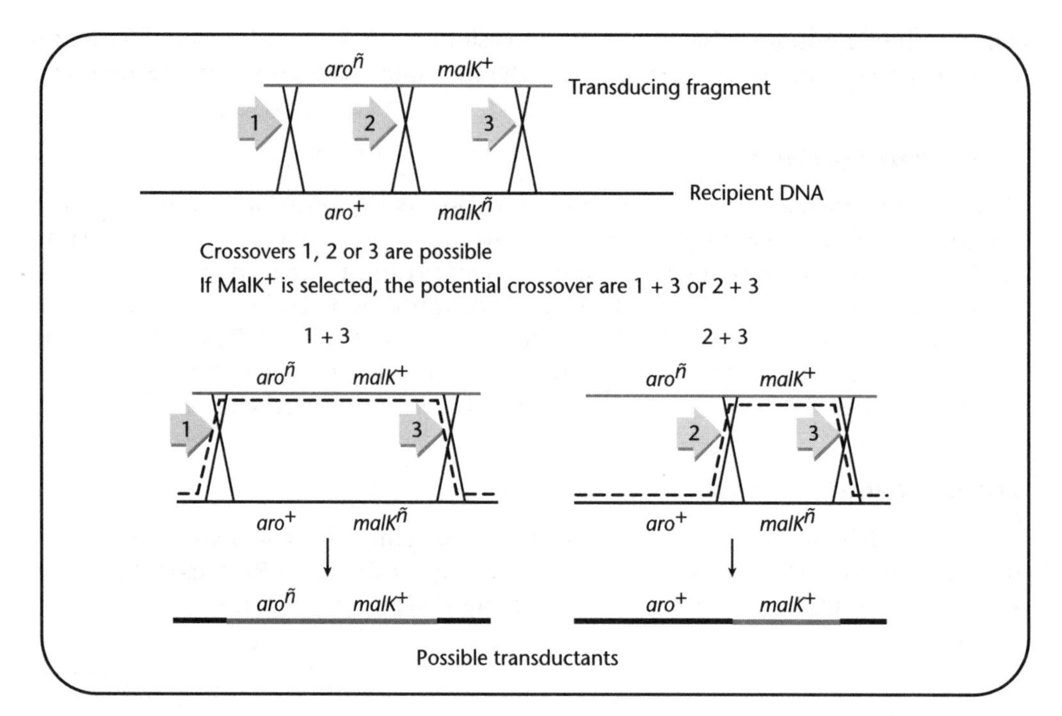

Fig. 8.7 Transducing linked genes by P1. Two types of transductants are possible, *aro*$^-$ *malK*$^+$ and *aro*$^+$ *malK*$^+$. The frequency of obtaining each type of transductant depends on how close the *aro* and *malK* genes are to each other.

1 has phenotypes that are difficult to score; or
2 mutations in many different genes give the same phenotype.

A marker gene that is closely linked to the gene of interest is used to select the transductants. A strain carrying this marker gene is mutagenized using either a chemical mutagen or a physical mutagen (see Chapter 3 for a discussion of mutagens). P1 is grown on the mutagenized strain and used to transduce another strain (the recipient) selecting for the marker gene. The transductants are then screened for the mutant phenotype of the gene of interest. Usually a second test is employed to prove that the mutation is in the gene of interest and not another gene close by that can also give the phenotype of interest.

Specialized transducing phage

Specialized transducing phage carry a defined region of the chromosome. Each specialized transducing phage and its descendants carry the same piece of chromosomal DNA. Unlike generalized transducing phage, specialized transducing phage carry chromosomal DNA in addition to phage DNA. One of the best-studied specialized transducing phage is λ. λ has a double-stranded DNA genome of 48,514 bp. The DNA is encased in a proteinaceous, icosohedron-shaped head, much like P1 (see Chapter 7).

To make a specialized transducing phage, a section of the λ DNA is removed and a gene of interest is added in its place. The *b* region of λ is nonessential for the growth of the phage in the laboratory and is usually the region that is replaced by the chromosomal DNA. The first specialized transducing phage were isolated using genetic techniques rather than DNA cloning techniques (Fig. 8.9). Normally, λ recombines with the bacterial chromosome using *attB*. *attB* is located between the *gal* and *bio* genes on the chromosome. Starting with a lysogen, λ can be induced out of the cell and usually the λ DNA recombines out of the chromosome by a site-specific recombination

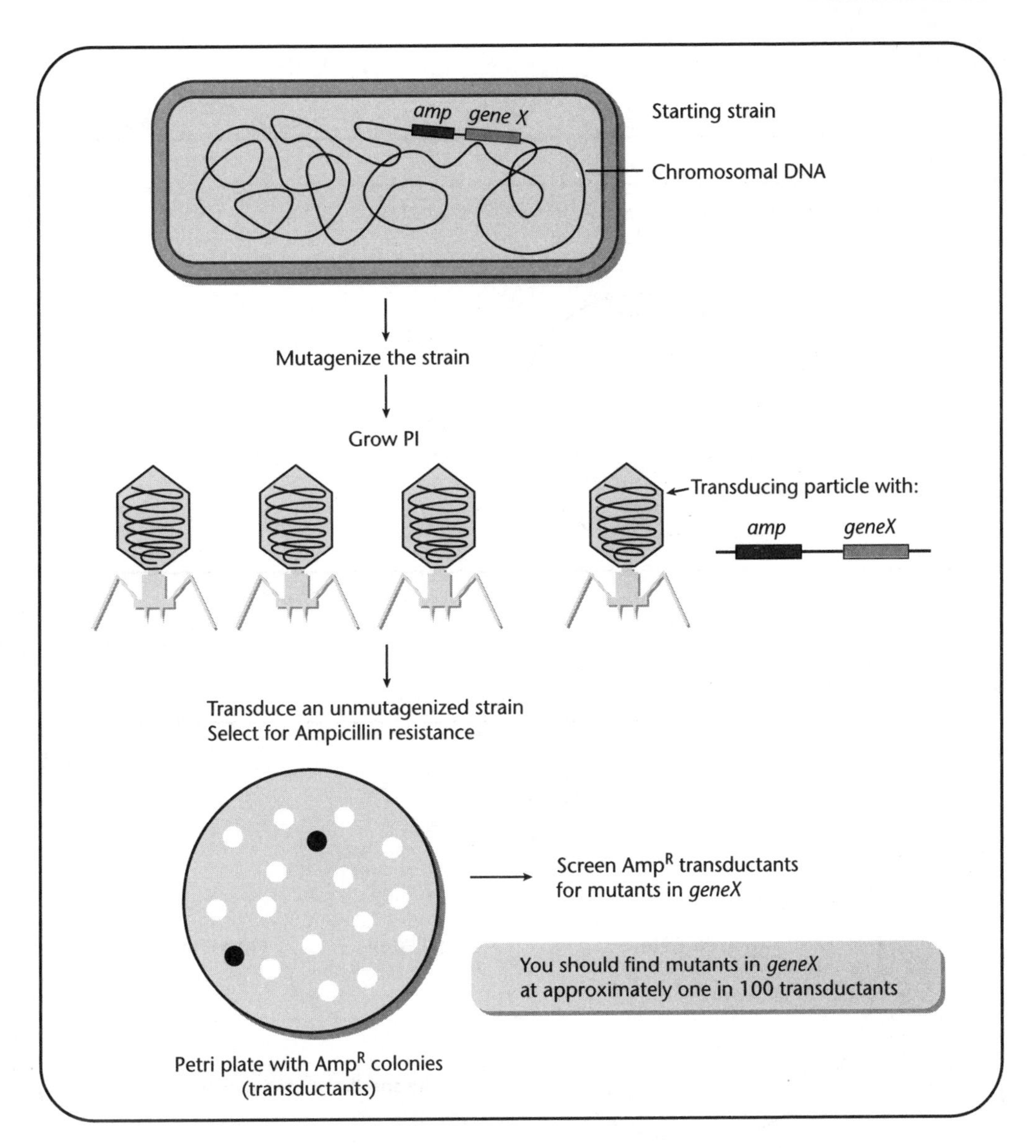

Fig. 8.8 The strategy for a localized mutagenesis experiment.

event (see Chapter 7). At a very low frequency, an illegitimate recombination event takes place and a λ phage carrying either *gal* or *bio* is formed. These phage can be identified genetically by their ability to confer a Gal$^+$ or Bio$^+$ phenotype on a *gal*$^-$ or *bio*$^-$ strain. If the *attB* site is removed from the chromosome, λ can integrate into other regions of the chromosome at a low frequency using secondary *att* sites. Genes located near the secondary *att* sites can also be isolated in a λ specialized transducing phage. With the advent of cloning, specialized transducing phage carrying any region of the chromosome can now be constructed.

Making merodiploids with specialized transducing phage

One very important use of λ specialized transducing phage is in constructing strains that have two copies of a specific gene or are **merodiploids** (Fig. 8.10). λ specialized transducing phage, if they carry *attP*, recombine into the chromosome at *attB* and not at the site of the cloned gene. If the bacterial copy of the gene is mutant and the copy

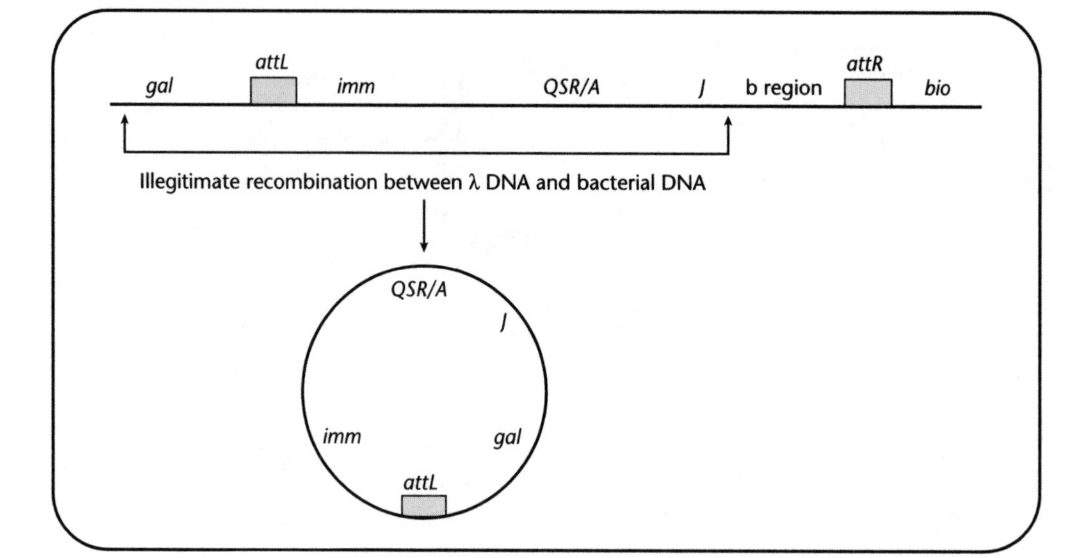

Fig. 8.9 Isolating a specialized transducing phage from λ integrated at *attB*. Illegitimate recombination between λ DNA and the bacterial chromosome will create a phage that is missing the b region of λ but carries the *gal* genes that are located close to *attB* in the bacterial chromosome.

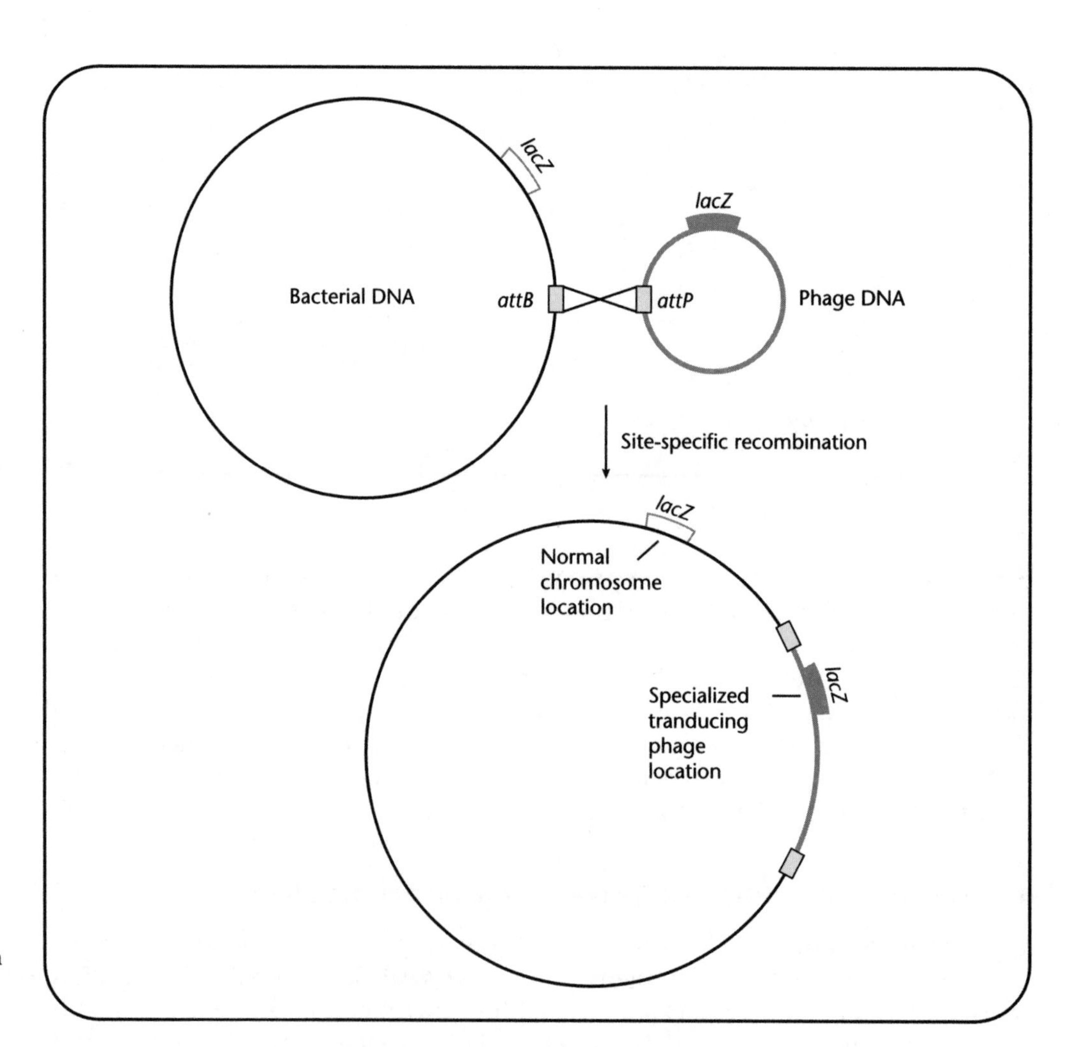

Fig. 8.10 Constructing a merodiploid using a λ specialized transducing phage. In this example, the phage carries *attP* and recombines with *attB* (see Chapter 7 for details of *attP* and *attB*).

of the cloned gene on the λ is wild type, then the merodiploid can be used to determine if the mutant copy of the gene is dominant or recessive (see Chapter 3). If the specialized transducing phage are missing *attP* or the bacterium is missing *attB*, then the phage will recombine into the chromosome by homology (see below).

Merodiploids can be very helpful in mutant isolation. If a specific selection always gives recessive mutations in one gene, then a cell can be made merodiploid for that gene and the selection repeated. Mutating two copies of one gene to give a certain phenotype is much harder to do. If the mutation appears at a frequency of 1 in 10^{-5} cells, then the frequency of mutating both genes in the merodiploid is 1×10^{-5} times 1×10^{-5} or 1×10^{-10}. Thus, using a merodiploid allows mutations in other genes to be isolated. A second strategy used in making mutations is to have a wild-type copy of the gene of interest in a cell and a recessive mutant copy of the same gene (Fig. 8.11).

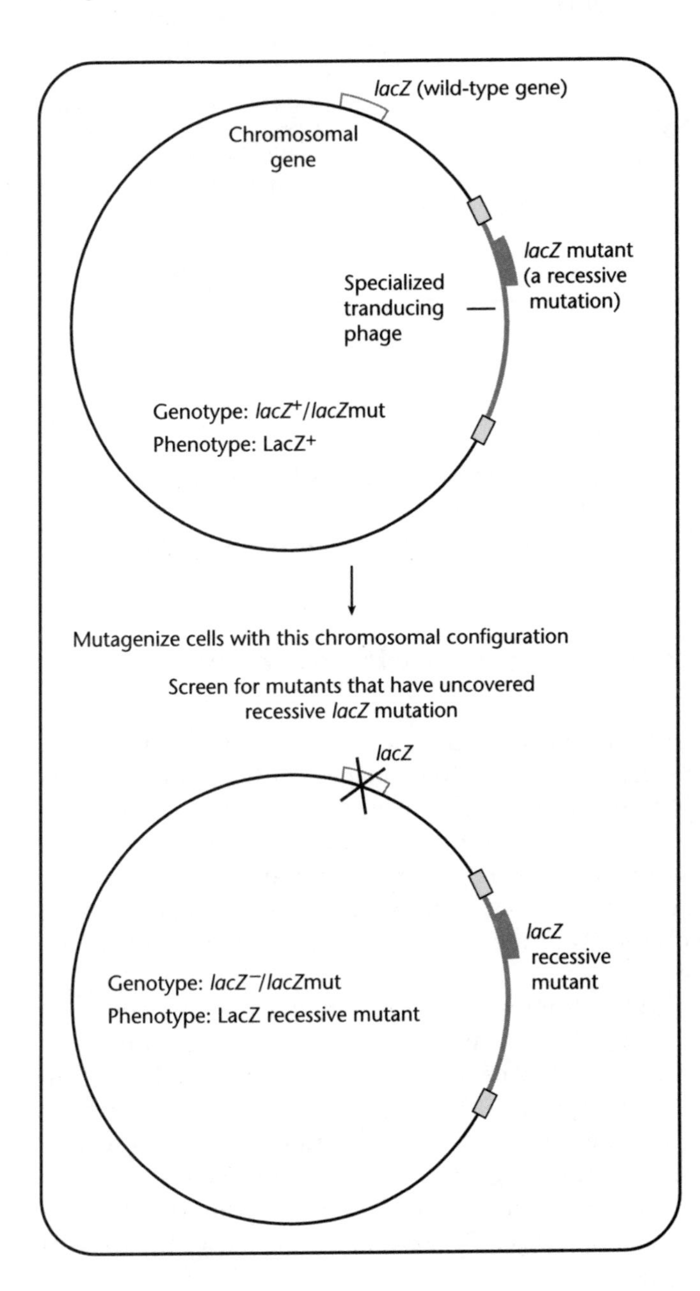

Fig. 8.11 Making mutations in a wild-type copy of a gene using a merodiploid. In this strategy, mutations in *lacZ* that lack any function can be isolated. This strategy is very useful if the gene being mutagenized is an essential gene where isolating lack-of-function mutations is lethal to the cells.

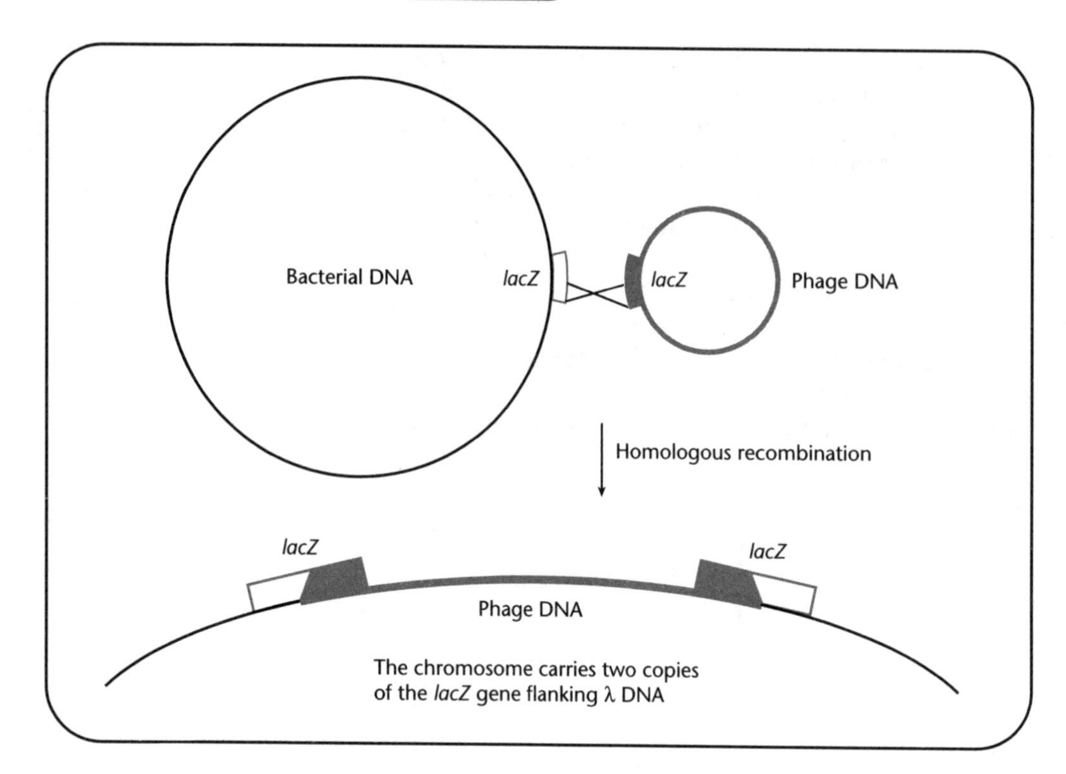

Fig. 8.12 Constructing a merodiploid using a λ specialized transducing phage. In this example, the phage does not carry *attP* and so it must recombine into the bacterial chromosome using the *lacZ* gene as homology. In this case, the phage is located in the chromosome at the position of *lacZ*.

Mutations are then isolated in the wild-type copy of the gene that render it nonfunctional and uncover the phenotype of the recessive mutant allele. This second strategy allows isolation of null or loss-of-function mutants in the wild-type copy of the gene of interest and can be used even if the gene is essential.

Moving mutations from plasmids to specialized transducing phage to the chromosome

λ specialized transducing phage that either do or do not carry *attP* can be constructed. If λ carries *attP* then the majority of the time that phage will lysogenize by site-specific recombination at *attB*. If the λ does not carry *attP* but does carry a cloned gene that is homologous to a gene in the *E. coli* chromosome, then the λ will recombine with the chromosome using the cloned gene (Fig. 8.12). A λ recombining at the site of the cloned gene requires all of the proteins needed for homologous recombination (see Chapter 5). The λ will be physically located on the chromosome at the site of the cloned gene.

Because the λ recombined into the chromosome by homology, it can come back out of the chromosome using homology. If both the phage and the chromosome carry wild-type copies of the cloned gene then all of the phage that come out of the chromosome will also carry a wild-type gene (Fig. 8.13a). If, however, the phage carries a wild-type copy of the cloned gene and the chromosome carries a mutant copy then where the recombination to remove the phage from the chromosome takes place is very important (Fig. 8.13b). Two different phage can be liberated from the chromosome. One is identical to the starting phage. The second phage carries the mutant

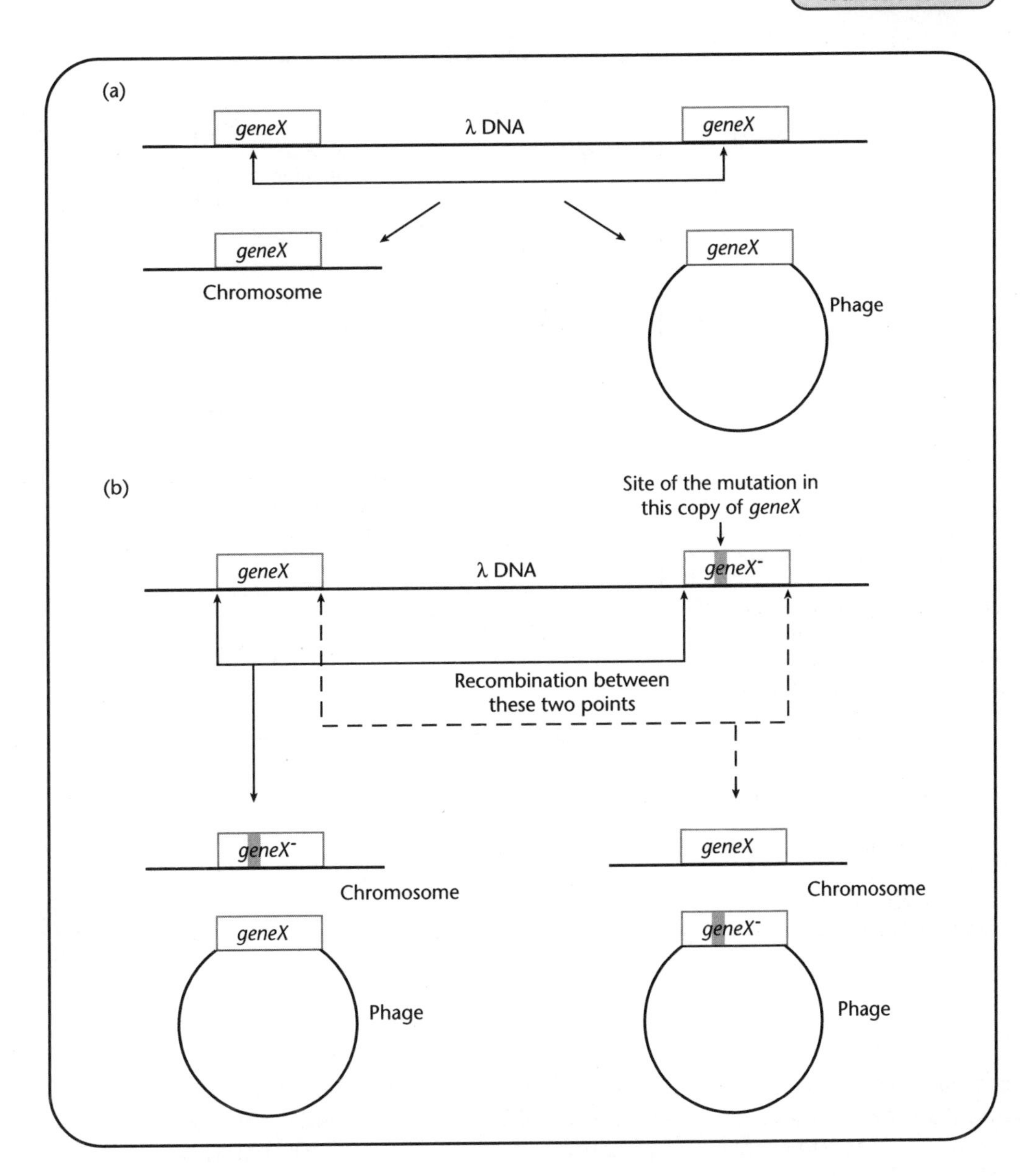

Fig. 8.13 Using a specialized transducing phage to exchange chromosomal alleles of the cloned gene. (a) Both genes are identical. (b) One copy of *gene X* is mutant and one copy is wild type. In example (b), two different outcomes are possible, depending on where the crossover takes place. The mutant gene can be left in the chromosome or the mutation can be recombined onto the phage.

copy of the gene. In this manner, a mutation can be moved from the chromosome to the specialized transducing phage.

A specialized transducing phage can also recombine with a plasmid that carries a copy of the same cloned gene. Usually, mutations are made on plasmids and then must be moved back into the chromosome to be studied. Using two separate crosses, a mutation can be recombined from the plasmid to the specialized transducing phage and then from the phage to the chromosome (Fig. 8.14). This is a very reliable method for moving mutations from plasmids, where they can be made and sequenced, to phage, where they can be tested for dominance, to the chromosome, where they can be studied in their normal copy number and chromosomal location.

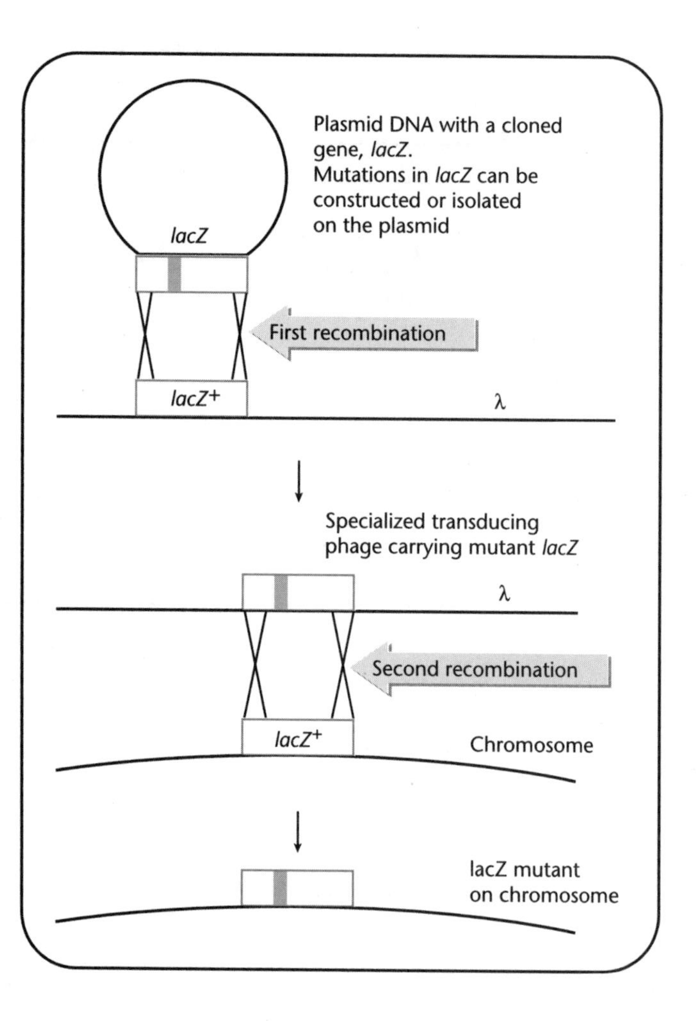

Fig. 8.14 Moving a mutation from a plasmid to λ and from λ to the bacterial chromosome. The first recombination event takes place between plasmid DNA and λ DNA and moves the mutation from the plasmid to the λ. The second recombination event takes place between the λ and the bacterial chromosome and moves the mutation from λ into the bacterial chromosome. The two recombination events are carried out in separate reactions.

Summary

In summary, generalized transducing phage move any piece of the chromosome at any given time. The phage particles carry chromosomal DNA instead of phage DNA. The amount of DNA a phage carries is determined by the amount that will fit inside its capsid. Because the transducing particles are a rare byproduct of phage development, movement of chromosomal DNA in this manner must be selected for. Moving chromosomal DNA by generalized transduction has many uses including building strains, mapping genes, determining linkage between two genes, determining the order of linked genes, and moving specific regions of the chromosome after mutagenesis. The techniques developed using generalized transduction provide one of the major advantages of working with bacteria.

Specialized transducing phage carry a piece of chromosomal DNA in place of part of their genome. The chromosomal DNA is inserted into the phage genome concomitant with a deletion of a nonessential part of the phage DNA. In general, specialized transducing phage carry a much smaller piece of chromosomal DNA than generalized transducing phage. Specialized transducing phage are very useful for constructing merodiploid strains, in making mutations, and moving them from plasmids to phage to the chromosome using genetic crosses. Specialized transducing phage also represent a good source of DNA for molecular experiments such as cloning and sequencing.

Study questions

1 What type of transducing particles carry the same piece of DNA in every phage particle? Different DNA in different transducing particles?
2 How does P1 package chromosomal DNA?
3 What is coinheritance?
4 What is the difference between a screen and a selection? How many cells can be surveyed in a screen? A selection? Which one is more powerful?
5 What are merodiploids and why are they useful?
6 What information can be obtained in a three-factor cross?
7 Can you order genes using a two-factor cross? Why or why not?
8 Devise a strategy for isolating an amber mutation in *ftsZ*, an essential gene required for cell division.
9 What type of transducing phage can be used to move a mutation from a plasmid to the chromosome? Diagram how this happens.

Further reading

Masters, M. 1996. Generalized transduction. In *Escherichia coli* and *Salmonella typhimurium: Cellular and Molecular Biology*, 2nd edn., eds. F.C. Neidhardt, R. Curtiss III, J.L. Ingraham, E.C.C. Lin, K.B. Low, B. Hagasanik, W.S. Rexnikoff, M. Riley, M. Schaechter, and H.E. Umbarger, pp. 2421–41. Washington, DC: ASM Press.

Weisberg, R. Specialized transduction. In *Escherichia coli* and *Salmonella typhimurium: Cellular and Molecular Biology*, 2nd edn., eds. F.C. Neidhardt, R. Curtiss III, J.L. Ingraham, E.C.C. Lin, K.B. Low, B. Hagasanik, W.S. Rexnikoff, M. Riley, M. Schaechter, and H.E. Umbarger, pp. 2442–8. Washington, DC: ASM Press.

Zinder, N.D. 1992. Forty years ago: the discovery of bacterial transduction. *Genetics*, **132**: 291–4.

Zinder, N.D. and Lederberg, J. 1952. Genetic exchange in *Salmonella*. *J. Bacteriol.*, **64**: 679–99.

Chapter 9

Natural plasmids

Plasmids are pieces of DNA that exist separate from the chromosome (Fig. 9.1). They contain an **origin for DNA replication** and, as such, replicate independently from the chromosome. Plasmids can be as small as a few hundred base pairs or, in a few cases, as large as one-third to one-half the size of a bacterial chromosome (in *E. coli* the chromosome is 4639 kilobase pairs). Any one plasmid has the same number and sequence of base pairs. There are families of plasmids whose members have very similar sequences. Plasmids are faithfully transmitted to daughter cells to ensure that they are stably maintained in the population. In addition to an origin for DNA replication, plasmids can contain a variety of other genes. These other genes include **antibiotic resistance determinants**, genes that allow the cells to use a variety of different carbon sources, genes to specify the production of a phage particle, or genes that are involved in causing diseases, to name a few.

FYI 9.1

Naming plasmids

Historically, the scientist who discovered a plasmid named it what they wanted, usually trying to give some indication as to the functions carried by the plasmid. Hence, we have the F factor (fertility factor), which allows transfer of a cell's chromosomal DNA from one cell to another, the R factor (resistance factor), which carries a large number of antibiotic resistance genes, and ColE1, which produces the antibiotic colicin. For historical reasons, these names have persisted. The convention now is to name plasmids beginning with a small p, followed by a designation unique to that plasmid. For example, in pBR322, the small p indicates we are discussing a plasmid, B and R are the initials of the person who constructed the plasmid and 322 is a unique designation among the BR plasmids.

Origins of replication

A plasmid must contain an origin for DNA replication (*ori*). Without an origin present, the plasmid cannot replicate and when the cells divide, one daughter cell would inherit a plasmid molecule and the other would not. At each subsequent cell division, only one daughter cell would receive a plasmid molecule (Fig. 9.2). The following scenario arises when a plasmid lacks an origin of replication:

Division I: 1/2 of the population has a plasmid molecule.
Division II: 1/4 of the population has a plasmid molecule.
Division III: 1/8 of the population has a plasmid molecule.
Division IV: 1/16 of the population has a plasmid molecule.

If *E. coli* are inoculated into growth media and incubated overnight, they will divide approximately 20 times. If one cell in the population contained one plasmid molecule without an origin of replication, at the end of this period of cell growth, one cell would have a plasmid molecule and 1,048,575 cells would not! Without an origin of replication, a plasmid is lost from the cell population very rapidly.

An origin consists of a small fragment of DNA that binds specific proteins and/or RNA molecules (Fig. 9.3a). These components must open the double helix of the DNA at or near the origin sequence, provide for the synthesis of a primer for DNA polymerase, and provide for the continued DNA synthesis of each strand of the plasmid DNA molecule. As can be imagined, not every origin of replication functions

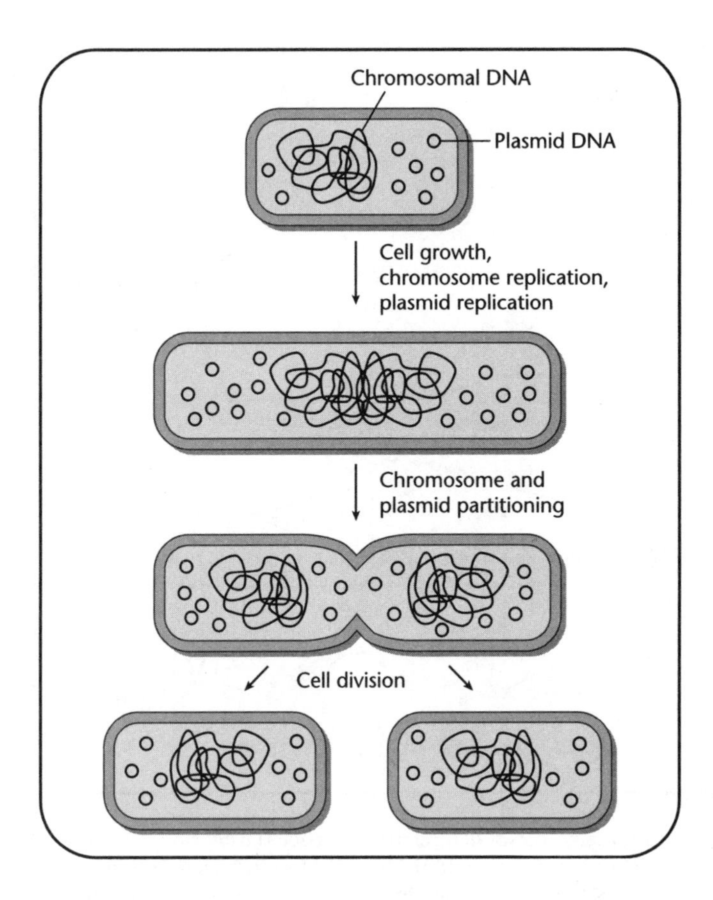

Fig. 9.1 Plasmids are extrachromosomal pieces of DNA that replicate independently from the chromosome and are faithfully transmitted to daughter cells.

identically. There are many different ways to accomplish the functions that an origin of replication is responsible for. Each plasmid has its own scheme for replication and many variations have been described.

Some plasmids encode a protein required for initiation of DNA replication and others do not (Fig. 9.3b). This plasmid-encoded **initiator protein** binds to the origin DNA and helps to open the double helix. Frequently, the host-encoded protein DnaA binds to the origin along with the plasmid-encoded initiator protein to facilitate the opening reaction. In other plasmids, the plasmid encodes several proteins required for double-helix opening and DnaA is not required (Fig. 9.4). Still other plasmids use an RNA molecule to open the double helix (Fig. 9.5). This RNA is also used as a primer for DNA polymerase.

Other steps in the DNA replication process that are carried out using several different mechanisms include:

- DNA synthesis from the origin proceeding in one direction at a time (**unidirectional DNA synthesis**) or both directions simultaneously (**bidirectional DNA synthesis**) (Fig. 9.6);
- requirements for different host proteins;
- differing requirements for RNA polymerase.

Because plasmids have a mechanism to replicate their DNA and ensure that they are stably maintained in a cell population, they are also known as **replicons**. Any DNA molecule that has these two properties, including the cell's chromosome, is a replicon.

FYI 9.2

Plasmids that carry transposons

Transposons (Chapter 6) can move themselves from one piece of DNA to another. We usually speak of transposons moving from one place in the chromosome to another. However, plasmids are also targets for transposons. Because plasmids are capable of moving between bacterial species, transposons that move into plasmid molecules can be widely disseminated in the bacterial kingdom. Transposons frequently carry antibiotic resistance genes. So, in exchange for a free ride, a transposon can provide a plasmid with an antibiotic resistance gene.

FYI 9.3

The proteins required for replicating plasmids

Most plasmids carry only a few if any of the proteins they need for replicating their DNA. Most of the replication proteins are supplied by the host bacterium. Almost all plasmids use host-encoded DNA polymerase for replication. Many plasmids, like the chromosome, use DNA Pol III and DNA Pol I for DNA replication. Plasmids can use host-encoded RNA polymerase to synthesize primers for DNA polymerase, DNA helicases to open the DNA, DNA primases to synthesize the primers needed by DNA polymerase III, or single-stranded DNA binding proteins to keep the strands of the DNA separate. Different plasmids utilize different combinations of these host enzymes to ensure that the plasmid DNA is replicated.

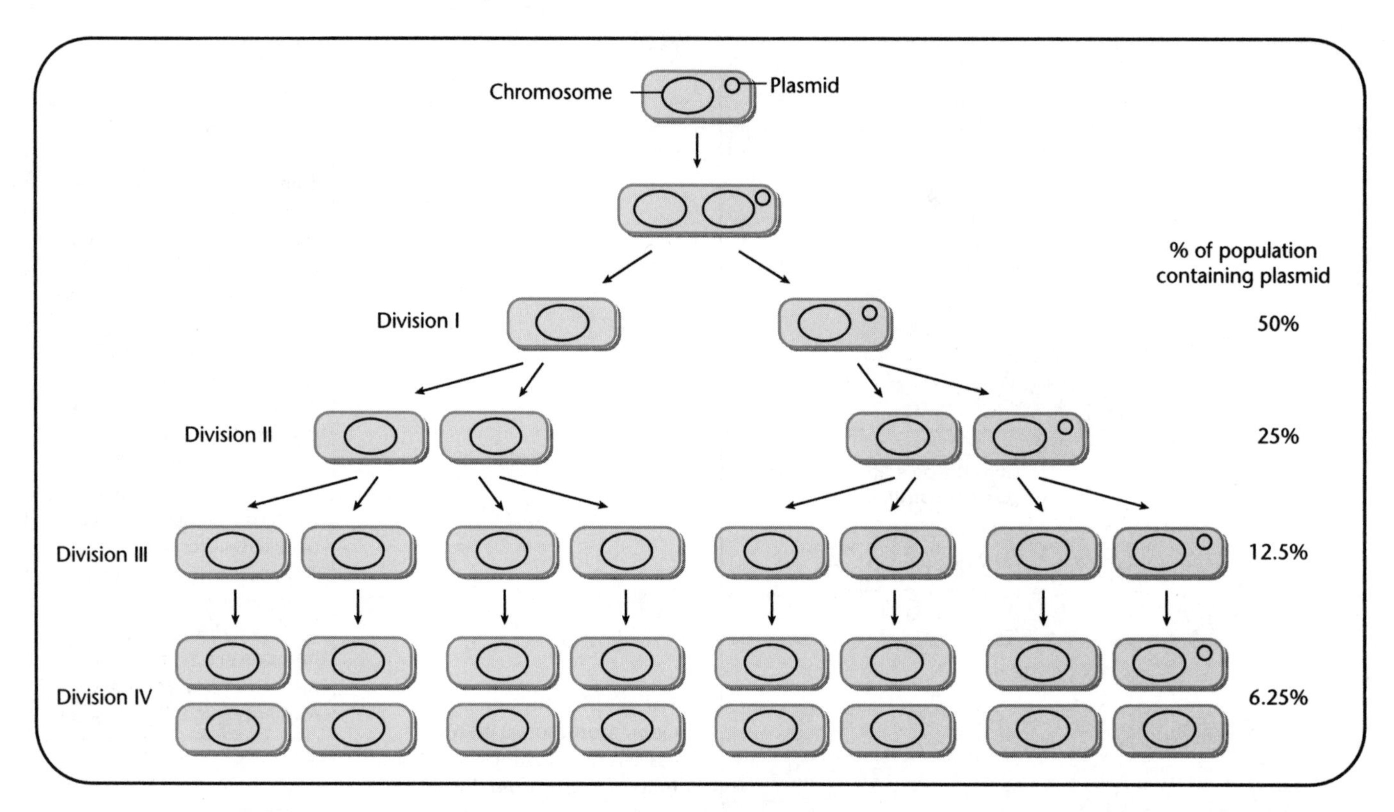

Fig. 9.2 The consequences of a plasmid not having an origin of DNA replication. If a plasmid present in one copy in a cell cannot replicate, then after the first cell division of the bacterium only 50% of the cells contain a plasmid. After the second bacterial division, 25% of the cells contain a plasmid. After the third bacterial division, 12.5% of the cells contain a plasmid and after the fourth bacterial division, only 6.25% of the cells contain a plasmid. Without an *ori*, plasmids are lost from the bacterial population very quickly.

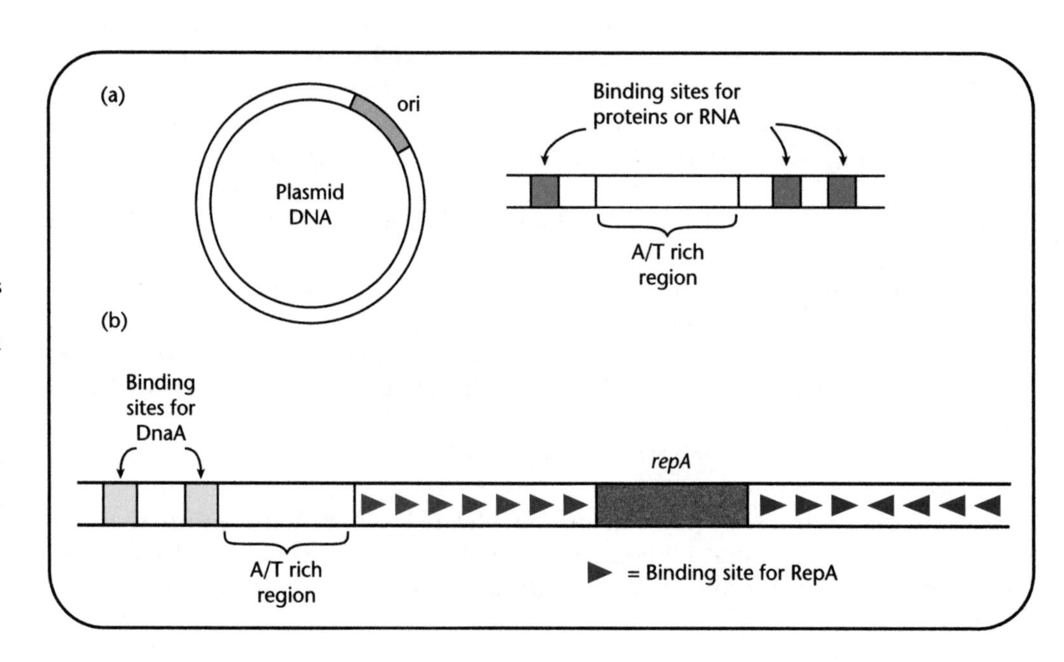

Fig. 9.3 Two examples of the structure of an origin of DNA replication (*ori*). (a) Origins frequently contain an A–T rich region of DNA and binding sites for an initiator protein or initiator RNA. (b) P1 encodes an initiator protein called RepA. RepA, in conjunction with the host-encoded protein DnaA, binds to the DNA and causes it to open at the A–T rich region. Replication forks proceed out of the origin and around the plasmid so that both strands of the plasmid molecule are replicated.

Plasmid copy number

A plasmid is present in a cell in a defined number of copies, depending upon its origin of replication. Plasmids such as F (see Chapter 10) or phage–plasmid hybrid P1 (see Chapter 7) are present in one to two copies per cell. Others such as pSC101 are present in 10–15 copies per cell. These are considered **low copy number plasmids**. The ColE1 plasmids are **high copy number plasmids** found in approximately 50 copies per cell. Other plasmids have been specifically engineered to be present in extremely high copy numbers, up to 100 copies per cell.

When a cell divides, the plasmid molecules are inherited by both daughter cells. For the high copy number plasmids, there are enough plasmid molecules for each daughter cell to receive at least some (Fig. 9.7a). The daughter cells do not have to inherit exactly half of the plasmid molecules. If there are 50 plasmid molecules before cell division, even if a daughter cell only receives a few copies of the plasmid rather than 25, this can be remedied during replication of the plasmid. For this reason, high copy number plasmids do not have to have any specific mechanisms to segregate or **partition** plasmid molecules into daughter cells. ColE1

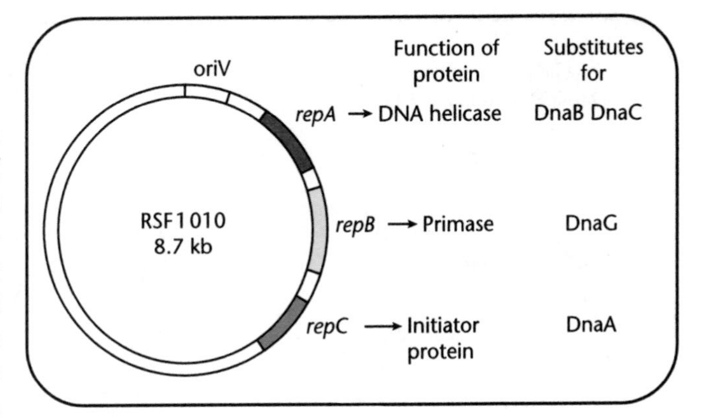

Fig. 9.4 Some plasmids encode all of their own initiator proteins and do not require host proteins for this step. RSF1010 encodes three proteins, RepA, RepB, and RepC, which substitute for the host proteins DnaA, DnaB, DnaC, and DnaG.

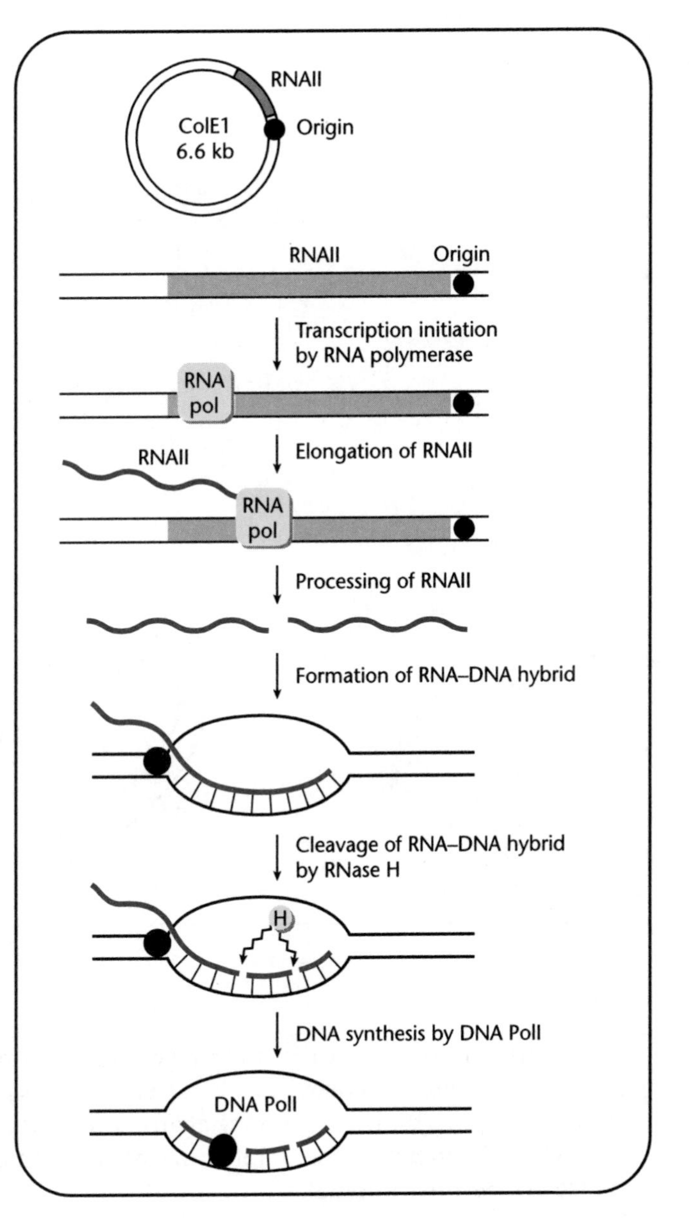

Fig. 9.5 ColE1 plasmids use an RNA molecule to open the DNA and initiate DNA synthesis.

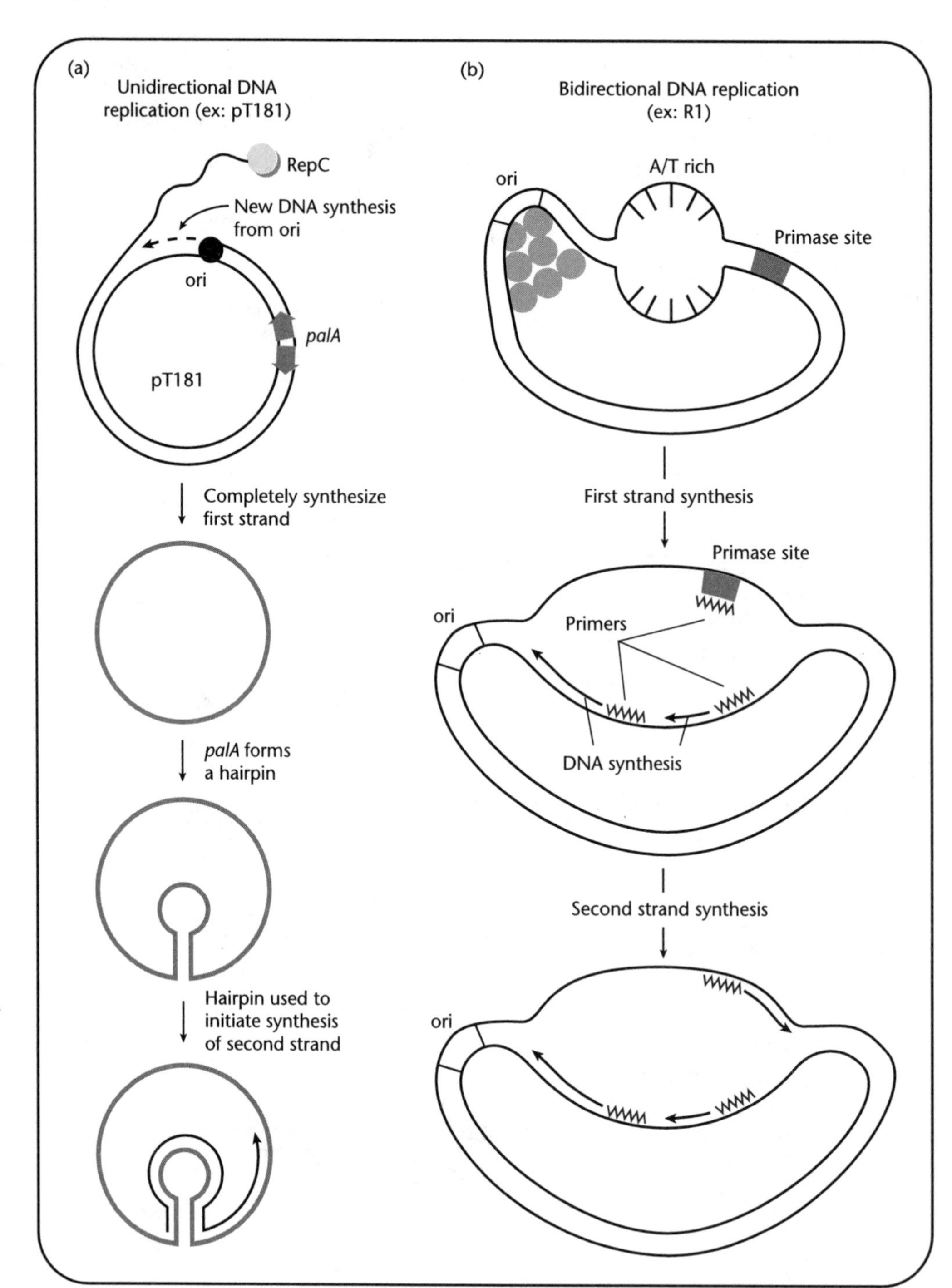

Fig. 9.6 Plasmids can replicate unidirectionally or bidirectionally. (a) pT181 synthesizes one strand of the DNA in one direction from the origin. After this strand is made, the second strand is synthesized. PT181 was isolated from the Gram-positive bacteria *Staphylococcus aureus*. (b) RI uses bidirectional DNA replication to synthesize its DNA. RI initiates synthesis of one strand and shortly thereafter, synthesis of the second strand. After initiation, both strands are simultaneously replicated.

is an example of a high copy plasmid that contains a function to aid in stability of the plasmid. The *cer* gene encodes a product that converts plasmid multimers to plasmid monomers (Fig. 9.7b). This ensures that all copies of the plasmid partition independently. While Cer increases the stability of ColE1 plasmids, it is not essential for maintenance of ColE1 in a population.

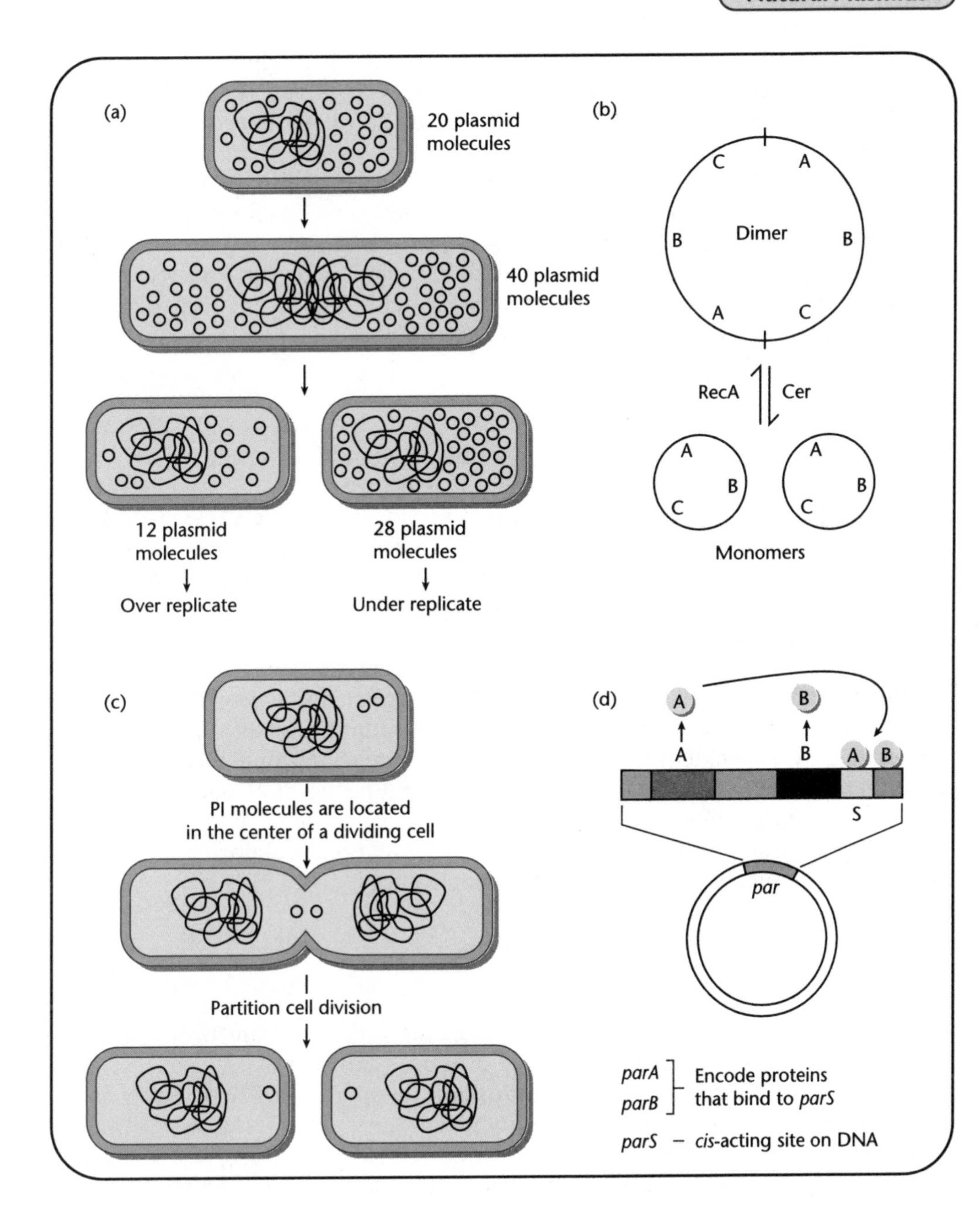

Fig. 9.7 Low and high copy number plasmids use different strategies to ensure efficient partitioning of plasmid molecules. (a) High copy number plasmids randomly partition plasmid molecules to daughter cells. If daughter cells do not receive approximately equal numbers of molecules, DNA replication will fix the problem. (b) Some high copy number plasmids contain a mechanism for resolving dimer plasmid molecules into monomer plasmid molecules. Monomer plasmid molecules can partition independently. (c) Low copy number plasmids must contain a partition system to ensure partitioning to daughter cells. (d) In P1, the *par* region encodes two proteins, ParA and ParB. The ParA–ParB complex binds to a cis-acting site called *parS*.

For the low copy number plasmids, partition of plasmid molecules into daughter cells is much more critical. Because there are very few plasmid copies, if one daughter inherits two molecules, the other daughter will not inherit any plasmid molecules (Fig. 9.7c). Thus, the low copy number plasmids must have a mechanism to ensure their proper partitioning (Fig. 9.7d). P1 has a specific site in the plasmid, called *parS*, which is required for segregation. The ParA and ParB proteins, which are also produced by the plasmid, bind to the *parS* sequence. It is thought that one function of the *parS*–ParA–ParB complex is to keep all of the molecules of the plasmid together in the middle of the cell until daughter cells are clearly distinguished. Then the plasmid molecules can be partitioned into daughter cells just before the final phases of septum formation and cell division take place. How the plasmids distinguish between different physical locations in the cell has not yet been determined. The result of the P1 partition system is that the low copy number of the P1 plasmid does not hinder the plasmid's ability to be inherited by daughter cells.

Computer simulations of random segregation of molecules in dividing cells have shown that a partition mechanism is required for plasmid stability once the copy number of the plasmid is approximately 5–10 per cell. Plasmids with copy numbers above this will be stably maintained in the population by random segregation. Plasmids with a copy number below this must have a specific mechanism for plasmid partitioning or they will be rapidly lost from the population.

Setting the copy number

Introducing either a low or a high copy number plasmid into cells devoid of any plasmid has shown that the copy number is usually fixed by controlling how often replication initiates. If a low copy number plasmid such as pSC101 is introduced, the plasmid replicates only enough so that daughter cells inherit a few copies of the plasmid. If a ColE1 plasmid is introduced, the plasmid undergoes much more DNA replication so that there are enough copies of the plasmid for each daughter cell to inherit approximately 50 plasmid molecules.

Several general strategies for maintaining copy number have been described for plasmids isolated from *E. coli* (Fig. 9.8). Both strategies employ negative regulation tactics. In one group of plasmids, the amount of initiator protein that is produced is regulated by a specific inhibitor. In a second strategy, multiple binding sites for the initiator protein compete with the origin for binding of a limited pool of initiator protein. In a third strategy, an RNA molecule that is required for initiation is prevented from binding to the origin by a second RNA and a protein. These strategies decrease the frequency of initiation of DNA replication.

Plasmid incompatibility

Cells can maintain more than one plasmid at a time, faithfully transmitting multiple plasmids to daughter cells. Maintenance of more than one plasmid in a cell can only occur if the plasmids carry different origins of replication. The inability of two plasmids with the same origin to be maintained in the same cell is known as **incompatibility**. Incompatibility is thought to be related to limiting concentrations of the initiator proteins and how cells pick which plasmid molecules are replicated (Fig. 9.9). When the concentration of initiator protein is high enough, one plasmid molecule from the population is randomly chosen and replicated. Both of the newly replicated plasmid molecules then become part of the pool of plasmid molecules. When the concentration of initiator protein is again high enough, a plasmid molecule is randomly picked from the pool and replicated. If two different plasmids that contain the same origin of replication are introduced into the same cell, one is lost after a few generations precisely because of this random picking of plasmid molecules for replication. If there is one copy of plasmid A and one of plasmid B, either plasmid A or B can be chosen to be replicated. If plasmid A is replicated then the cell will contain two molecules of plasmid A and one of plasmid B. This means that when the next plasmid molecule is chosen at random for replication, there is a greater probability that it will be a molecule of plasmid A. In this case, plasmid B will be lost from the cell and plasmid A maintained. Incompatibility is not only important for the maintenance of plasmids in a population of cells but also in cloning experiments where plasmids are used to carry specific pieces of chromosomal DNA.

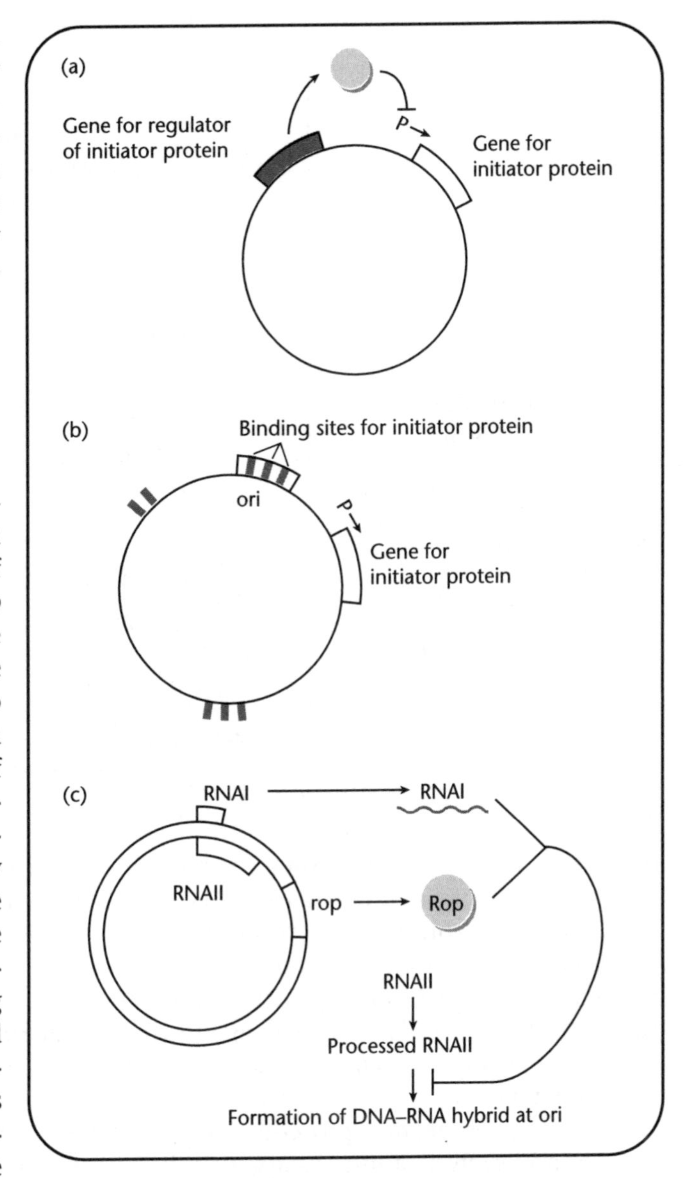

Fig. 9.8 Three strategies for maintaining the copy number of a plasmid. (a) Regulation of the initiator protein by a second protein. The second protein usually blocks the synthesis of the initiator protein. When the regulator protein is diluted out by cell division, the initiator protein is made and the plasmid is replicated. (b) Extra binding sites for the initiator protein on the plasmid. The binding sites outside the origin bind the initiator protein first. At high concentrations of initiator protein all of the binding site fill, including the origin sites, and the DNA is replicated. (c) An inhibitory RNA and protein prevent initiation. ColE1 requires an RNA molecule for initiation (RNA II). It also encodes RNA I and the Rop protein. RNA I and the Rop protein prevent RNA II from initiating DNA replication.

Plasmid amplification

Many of the known plasmids require new protein synthesis for each round of DNA replication (***de novo* protein synthesis**). The ColE1 plasmids are one exception to this. They do not require *de novo* protein synthesis for initiation of DNA replication. One consequence of this independence is that the copy number of ColE1 can be amplified relative to the copy number of the chromosome. If cells containing a ColE1 plasmid are treated with protein synthesis inhibiting antibiotics, such as chloramphenicol, the replication of the chromosome will stop but the replication of the ColE1 plasmid will continue. Many of the popular cloning plasmids, including the first and most famous cloning plasmid, pBR322, use a ColE1 origin of replication.

Other genes that can be carried by plasmids

In addition to an origin of replication, naturally occurring plasmids can carry a wide variety of other genes. From a medical standpoint, the most important genes that can be carried by plasmids are the antibiotic resistance determinants. This family of genes confers resistance to many of the medically important antibiotics. Each gene in the family is usually responsible for the resistance to only one antibiotic. However, a plasmid can carry many different antibiotic resistant genes. Given the widespread overuse of antibiotics, plasmids carrying antibiotic resistance genes have been selectively amplified in cellular populations, including many of the disease-causing micro-

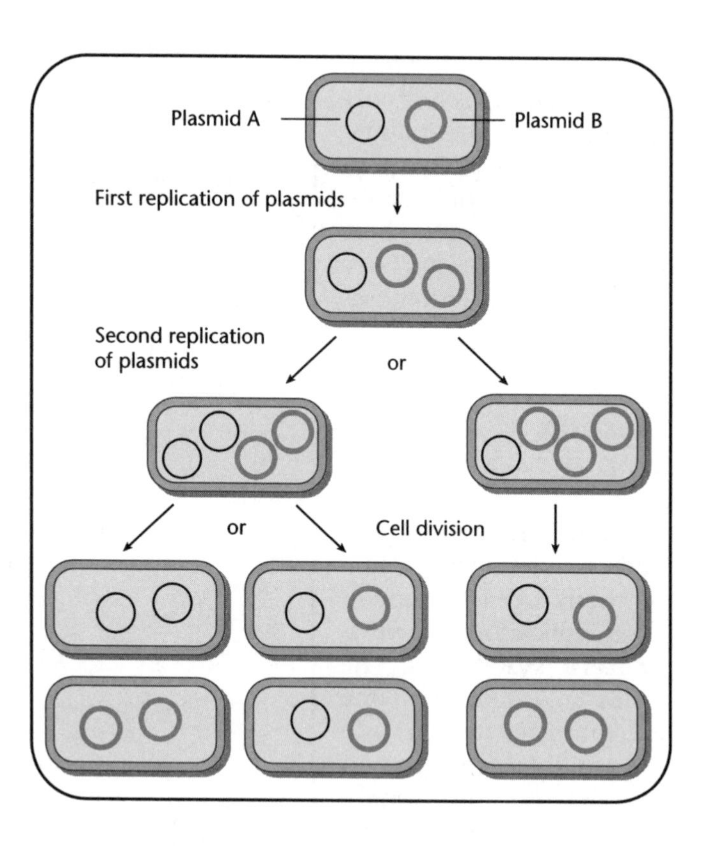

Fig. 9.9 Incompatibility can be explained by a limiting concentration of initiator protein and random replication of plasmid molecules. Plasmid molecules are randomly chosen from the population and undergo DNA replication each time the plasmid is replicated. Different plasmids that use the same or very similar origins and initiator proteins are rapidly separated from each other.

FYI 9.4

Pseudomonas aeruginosa camphor-resistant plasmid

Plasmids come in many sizes, including some that approach the size of the chromosome of the bacteria that carries them. For example, a plasmid from *Pseudomonas aeruginosa* allows the bacteria to use camphor as a carbon source. This plasmid is approximately one-third the size of the *Pseudomonas* chromosome, which points to several interesting facts. Carrying a plasmid this large can cost the bacteria a significant amount of energy to replicate and maintain this large amount of extra DNA. Metabolizing unusual compounds can require a significant number of enzymes and thus require a significant amount of DNA to encode all of these proteins. Plasmids can dramatically expand the compounds that can be used as food by bacteria. *Pseudomonas* can be considered one of the junk food eaters of the bacterial world. It is capable of eating camphor, and also phenol, pesticides, and a wide variety of other toxic compounds, but only in moderation! Many of the enzymes used to break down these toxin compounds are encoded by plasmids.

organisms. The increased resistance to antibiotics in disease-causing microorganisms can be easily explained by this selective amplification and the ability of plasmids to easily move from one cell to another across species. In addition to antibiotic resistance genes, plasmids can also carry genes that confer resistance to heavy metals, ultraviolet light, bacteriophages, enzymes that lyse bacteria cells, intercalating agents that deform DNA, and many other agents that damage or kill bacteria.

Naturally occurring plasmids can also carry many different metabolic enzymes. These enzymes can alter or increase the compounds that can be used as carbon or nitrogen sources, degrade proteins or nucleic acids, allow cells to synthesize specific compounds such as amino acids, metabolic cofactors, vitamins or pigments, or carry out other metabolic reactions. Plasmids can dramatically expand the metabolic capabilities of cells.

Plasmids that carry genes involved in pathogenicity and symbiosis are numerous. The genes encoded by such plasmids allow the cells that carry them to invade and colonize eukaryotic cells. The best-studied example of this kind of plasmid is the Ti plasmid of *Agrobacterium tumefaciens*.

The bacterium, *Agrobacterium tumefaciens*, induces the formation of large tumors, known as crown gall tumors, on plants. *Agrobacterium* contains plasmids, called **Ti plasmids** that are responsible for the tumor formation. The Ti plasmids, which are between 180 and 240 kilobases, contain a very large number of genes that have a variety of different functions (Fig. 9.10). The tumors in the plants are actually caused by the transfer of a 25 kb segment of the Ti plasmid, known as the **T-DNA**, from the bacterium into the plant cells (Fig. 9.11). The T-DNA goes to the nucleus of the plant cell where it is stably integrated into a plant chromosome. The T-DNA contains fully functional plant promoters that drive the synthesis of plant hormones and opines.

The plant hormones lead to the large tissue overgrowths known as crown gall tumors. The opines produced are used by the bacterium as a food source. In addition to the T-DNA, Ti plasmids contain the genes that encode the proteins needed for transfer of the plasmid from the bacterium to the plant and the genes that encode the proteins need to consume the opines. The opines produced by the T-DNA also promote movement of the Ti plasmid from a bacterium that harbors it to other *Agrobacterium* that are plasmid free. As can be seen, the Ti plasmids produce all of the functions necessary to ensure that they are maintained by the bacteria.

Interactions between the functions provided by two distinct plasmids can also be demonstrated using the Ti system. Certain strains of *Agrobacterium* grow on specific plants by harboring a second compatible plasmid that provides unique metabolic capabilities. An analysis of grapevines in California uncovered a strain of *Agrobacterium* that formed crown gall tumors specifically on grapevines. This *Agrobacterium* contained the Ti plasmid and a second plasmid that was named pTAR. The pTAR plasmid carries genes for the utilization of L-tartrate as a carbon source. L-tartrate is a compound that is very prevalent in grapevines. The pTAR plasmid was not found in the *Agrobacterium*

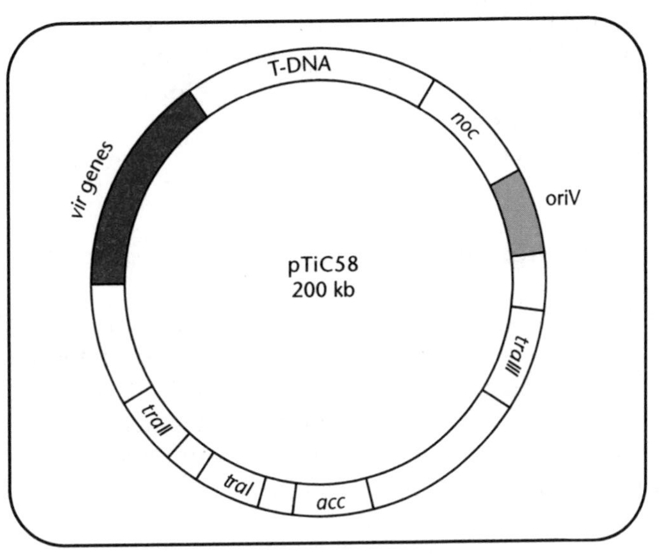

Fig. 9.10 The structure of a Ti plasmid that was isolated from *Agrobacterium tumifaciens* strain C58. Ti plasmids carry many of the same genes but they can differ slightly, depending on what strain of *Agrobacterium* they were isolated from. *traI*, *traII*, and *traIII* are needed for transfer of the T-DNA from the bacterium to the plant cell. *vir* genes encode products to sense the plant signals and process the T-DNA out of the plasmid. *acc* gene products catabolize the opine, agrocinopine. *noc* gene products take up and degrade nopaline.

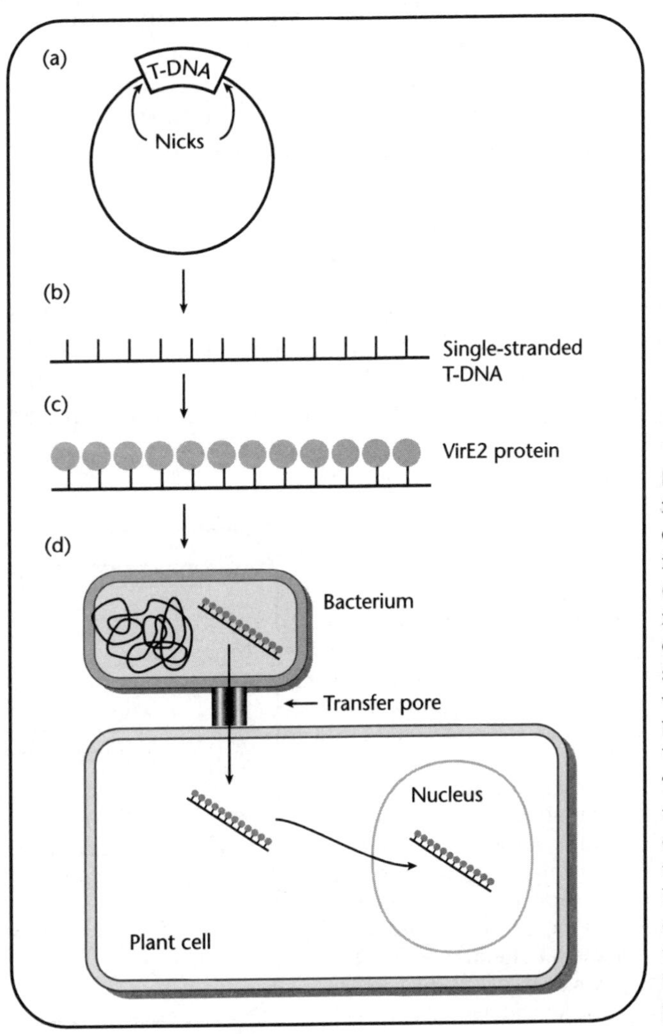

Fig. 9.11 Transfer of the T-DNA from the *Agrobacterium* to a plant cell. (a) The T-DNA is nicked at its borders on specific strands. (b) DNA synthesis leads to formation of a single-stranded copy of the T-DNA. (c) The single-stranded T-DNA is coated with a single-stranded DNA binding protein, VirE2. (d) The VirE2-coated single-stranded T-DNA is transferred through the transfer pore and into the cytoplasm of the plant cell. The transfer pore is composed of between 10 and 20 plasmid-encoded proteins. VirE2 directs the T-DNA to the nucleus where the T-DNA is integrated into a plant cell chromosome.

strains in the soil around the grapevines, only in those living on the grapevines. When the pTAR plasmid was removed from the *Agrobacterium* leaving only the Ti plasmid, the bacteria were still capable of forming crown gall tumors on the grapevines but the tumors were very small. Thus, this cooperation between plasmids confers a greater ecological fitness on the *Agrobacterium* carrying both plasmids.

Plasmids can be circular or linear DNA

The first plasmids described were circular double-stranded DNA molecules. In fact, all of the plasmids in the most widely studied bacterium, *E. coli*, are circular. However, circular plasmids are not the only possibility. Once people began looking for linear molecules, more and more linear plasmids and chromosomes were found. Linear plasmids have been characterized from *Streptomyces* species and from the bacterium that causes Lyme disease, *Borrelia burgdorferi*. Several other bacteria also contain linear chromosomes. Linear chromosomes or plasmids pose several specific challenges (Fig. 9.12). First, double-stranded DNA ends are the substrate for many exonucleases, and as such, are very unstable in cells. The linear plasmids must protect their ends. Second, double-stranded DNA cannot be replicated all the way to the end of the molecule because of DNA polymerase's requirement for a primer. Linear plasmids must be able to replicate their genomes completely. In *Borrelia*, the plasmids have covalently closed hairpins at the ends to protect them (Fig. 9.13). The hairpins are also used to replicate the ends of the plasmids. If the hairpin sequences are added to a circular plasmid, the circular plasmid is converted to a linear plasmid.

Broad host range plasmids

Most plasmids can only replicate and be maintained in one or a few closely related bacterial species. Several exceptions to this have been documented. Two classes of plasmids, exemplified by RK2 and RSF1010, can be maintained in almost all Gram-negative bacteria. Plasmids that can be replicated and maintained in many different

FYI 9.5

The chromosomes and plasmids of *Borrelia*

Borrelia burgdorferi is the causative agent of Lyme disease. This bacterium is carried by deer and mice, transmitted by ticks, and thus is not found free-living in the environment. *Borrelia* contains a chromosome and 21 plasmids, all of which have been sequenced. The plasmids total 613 kilobases of DNA, with nine of the plasmids being circular and 12 being linear. The chromosome contains a minimal set of genes needed for metabolic activities, while the plasmids contain numerous truncated and damaged versions of chromosomal genes as well as unique genes. The plasmids contain genes for such diverse functions as nucleotide metabolism, transporting small molecules, and DNA metabolism. Some of the plasmids are derivatives of other plasmids and some plasmids appear to be bacteriophage genomes. All of the plasmids are low copy and some of them are dispensable for growth of the bacteria, while other plasmids contribute to the infectivity of the bacteria. It is clear from these studies that *Borrelia* has a very unusual complement of DNA that it maintains, and that plasmids play an important role in its hereditary material.

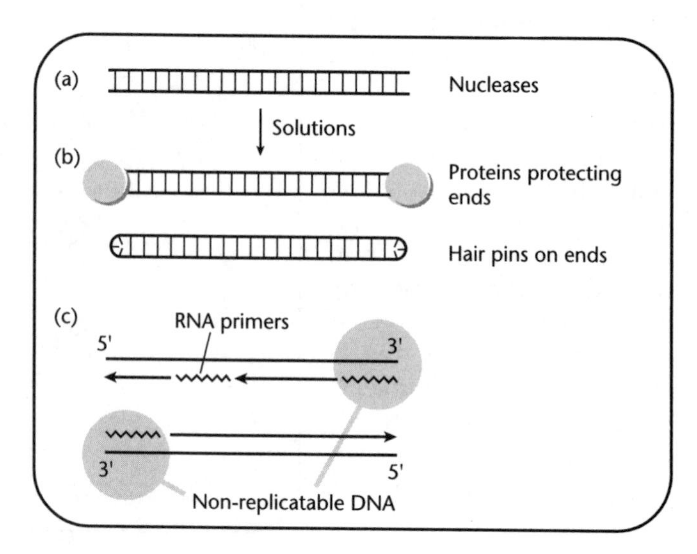

Fig. 9.12 The challenges posed by a linear DNA molecule. (a) The ends of a linear molecule are subject to degradation by nucleases. (b) Two solutions to this problem are to either protect the ends with proteins or have the DNA backbone be continuous and form a hairpin structure. (c) A second major problem with linear molecules is that the very last part of the molecule cannot be replicated because of DNA polymerase III's need for a primer.

species are known as **broad host range plasmids**. These plasmids' strategies for DNA replication and gene expression must have evolved so that they can flourish in a wide variety of diverse species.

RK2 is a large plasmid of approximately 60 kilobases that encodes multiple antibiotic resistant genes. Replication of RK2 in *E. coli* requires a DNA sequence on the plasmid (*oriV*) and a protein encoded by the plasmid (TrfA). Insertion mutations in *oriV* abolish its function and make RK2 unable to replicate in *E. coli*. However, these same *oriV* mutations do not effect replication in *Pseudomonas aeruginosa*. This result suggests that the plasmid may use different origins of replication in different bacterial species. RK2 depends mostly on host-encoded proteins to carry out the replication of the plasmid. These host-encoded proteins should be very different in different bacterial species. RK2 has found a way to utilize these very different host proteins to ensure it is replicated and transmitted to daughter cells.

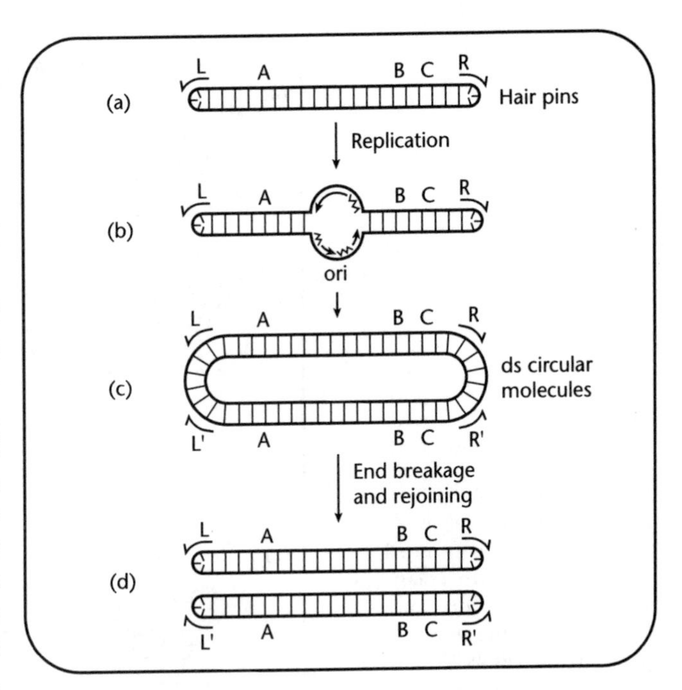

Fig. 9.13 Some *Borrelia* plasmids have covalently closed hairpins at the ends. It takes a minimum of three unpaired bases to make up a hairpin. Hairpins with more than three unpaired bases are also prevalent. To solve the problem of replicating a linear molecule, the model shown was proposed. (a) The linear plasmid has inverted repeat sequences at its ends (labeled L and R). (b) Replication begins from an internal origin and proceeds bidirectionally towards both ends. (c) Replication results in a double-stranded DNA molecule that is a dimer of the original plasmid. (d) End breakage and rejoining at L, L′, R, and R′ result in two copies if the original plasmid.

RSF1010 takes a different strategy for DNA replication in different bacterial species. This plasmid encodes many of the proteins needed for replication of the plasmid, making the plasmid much more independent of host-encoded proteins. In this strategy, the plasmid requires only a few host-encoded proteins, such as DNA polymerase, whose function is well conserved across different bacterial species.

Moving plasmids from cell to cell

The two broad host range plasmids discussed above, RK2 and RSF1010, also serve as examples of plasmids that can move from one cell to another. RK2 carries in its genome all of the functions necessary to transfer RK2 DNA from one cell to another in a process called **conjugation** (Fig. 9.14 and see Chapter 10). Plasmid transfer can take place between two bacterial cells of the same species or between two bacterial cells of different species. This ability to move between species has earned RK2 the name **shuttle vector**. The multiple antibiotic resistance genes carried by RK2 are moved between species as part of the plasmid DNA. The medical consequences of this are that many different bacterial species, including those responsible for diseases, can easily obtain antibiotic resistance genes. Experimentally, being able to move DNA into uncharacterized bacteria gives us a powerful tool to study novel species. Some shuttle vectors are naturally occurring plasmids and other are specifically constructed (see Chapter 14 for how to construct plasmids) to replicate in specific experimentally useful species.

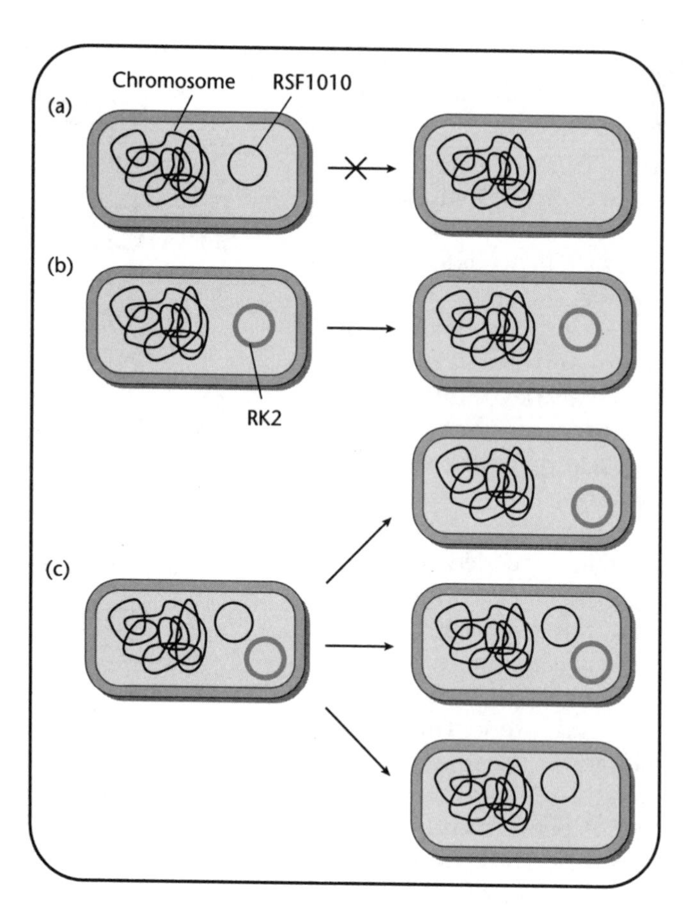

Fig. 9.14 Movement of plasmids from one cell to another by plasmid-encoded mechanisms. (a) Other plasmids, such as RSF1010, cannot move by themselves.(b) Some plasmids, such as RK2, can move themselves by conjugation. (c) However, if RK2 and RSF1010 are present in the same cell three outcomes are possible: RK2 can move itself; RK2 can move itself and RSF1010; RK2 can move only RSF1010.

RSF1010 does not carry all of the genes necessary for self-transmission of its DNA from one cell to another. However, if RSF1010 is put in the same cell with a plasmid that does carry the self-transmission genes (such as RK2), RSF1010 can be mobilized into another cell using the other plasmid's self-transmission genes (Fig. 9.14). Thus, RSF1010 can be **mobilized** or moved from one cell to another, while RK2 can mobilize itself and other plasmids. Not all plasmids can be mobilized. In fact, many plasmids cannot be mobilized under any known conditions.

Summary

Plasmids are extrachromosomal pieces of DNA that are capable of DNA replication and transmission to daughter cells. They have evolved many different strategies to ensure their survival in a bacterial population. Plasmids use host-encoded proteins for various parts of their lifecycle, including replication of their DNA and transcription and translation of their genes. Plasmids carry a wide variety of genes that increase the genetic diversity of their bacterial host. The genes carried by plasmids often provide a selective advantage to the bacteria that carry them. Because some plasmids are capable of moving between bacterial species, plasmids provide a mechanism for the horizontal transfer of large blocks of genes between species. Thus, plasmids are a very valuable addition to the genetic diversity of bacterial species.

Study questions

1 What functions must a plasmid carry in order to be maintained in a bacterial population?
2 What size(s) are plasmid molecules?
3 How do plasmids initiate DNA replication?
4 What is the major difference between unidirectional and bidirectional DNA replication?
5 Under what circumstances must a plasmid have a dedicated partition system?
6 How is the copy number of a plasmid determined?
7 What feature of a plasmid allows it to be amplified by treatment of plasmid-containing cells with chloramphenicol?
8 What are three general types of genes that can be found on plasmids?
9 What is the fate of the T-DNA after *Agrobacteria* infects a plant?
10 What is the mechanism for moving a broad host range plasmid from one DNA species to another?

Further reading

Das, A. 1998. DNA transfer from *Agrobacterium* to plant cells in crown gall tumor disease. *Subcellular Biochemistry*, **29**: 343–63.
Kado, C.I. 1998. Origin and evolution of plasmids. *Antonie van Leeuwenhoek*, **73**: 117–26.
Kornberg, A. and Baker, T. 1992. Plasmids and organelles. In *DNA replication*. pp. 637–87. New York: W.H. Freeman.
Novick, R.P. 1980. Plasmids. *Scientific American*, December: 102.

Chapter 10

Conjugation

Conjugation is the process of moving DNA from one cell to another using a specific type of plasmid to mediate the transfer of DNA and requiring direct cell-to-cell contact. This process was first discovered by Lederberg and Tatum in 1946. Not all *E. coli* strains and not all bacteria are capable of conjugation. To be able to carry out this process, a bacterium must carry an F factor or an R factor and be physically mixed with a bacterium that does not contain one of these plasmids. Through a series of steps, the F or R factor can move itself and, in some instances the chromosome, from one cell to another.

FYI10.1

The Lederberg and Tatum experiments

Normally *E. coli* does not require growth factors. Lederberg and Tatum, by mutagenizing *E. coli*, isolated two strains, one that required biotin and methionine to grow and one that required threonine and leucine. They mixed these two mutants together and plated them on agar without any supplements. Surprisingly, they were able to isolate some strains that were now able to grow without biotin, methionine, threonine, or leucine. A further characterization of how these wild-type strains were generated revealed that the process requires direct cell-to-cell contact. This was the first demonstration of conjugation and movement of a large piece of chromosomal DNA from one cell to another.

The F factor

The F factor or F for short is a 100 kilobase pair plasmid (Fig. 10.1). The name F factor stands for fertility factor and was given to the plasmid based on the ability of F to mediate conjugation. The F factor was described genetically long before plasmids were discovered, which is why it was first called a factor and not a plasmid. For historical reasons, the name F factor is still used.

The F factor contains origins for replicating the plasmid during cell growth (RepF1A and RepF1B). Because it is present at one to two copies per cell, F has a partition system to ensure that both daughter cells inherit a copy (see Chapter 9 for details). F contains an origin of transfer or *oriT* where the transfer of the F factor DNA from one cell to another begins. *oriT* is physically separated from the origins used to replicate the F DNA during normal growth. F also encodes all of the proteins needed to build the F pilus, a structure on the surface of the cell needed to bring the two mating cells into direct contact with each other. Other F-encoded gene products help stabilize the mating pair. F contains two insertion sequences, IS2 and IS3, and one transposon, Tn*1000*. The insertion sequences and transposon are used as portable regions of homology for chromosome mobilization, as described below.

The R factors

The R factors are a family of similar but not identical plasmids that share features with F. Like F, R factors are circular, double-stranded DNA. Unlike F, R factors carry resistances to antibiotics and heavy metals, hence R for Resistance factor (Fig. 10.2). Not all

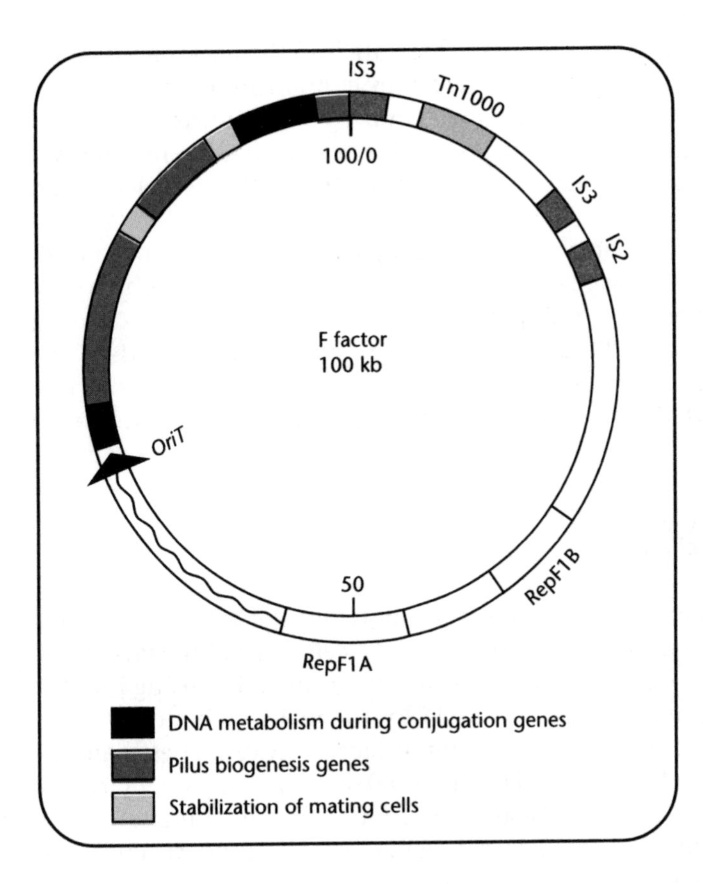

Fig. 10.1 A map of the F factor. F contains two copies of IS3, one copy of IS2, and one copy of Tn*1000*. RepF1A is the main replication region. It contains two origins of replication, *oriS* (specifies unidirectional replication) and *oriV* (specifies bidirectional replication). RepF1B is a backup origin that functions in the absence of RepF1A. *oriT* is the origin of transfer that is used during conjugation. The direction of transfer during conjugation is from *oriT* towards RepF1A. The region from ~65 kb to 100 kb contains genes that encode functions used during conjugation.

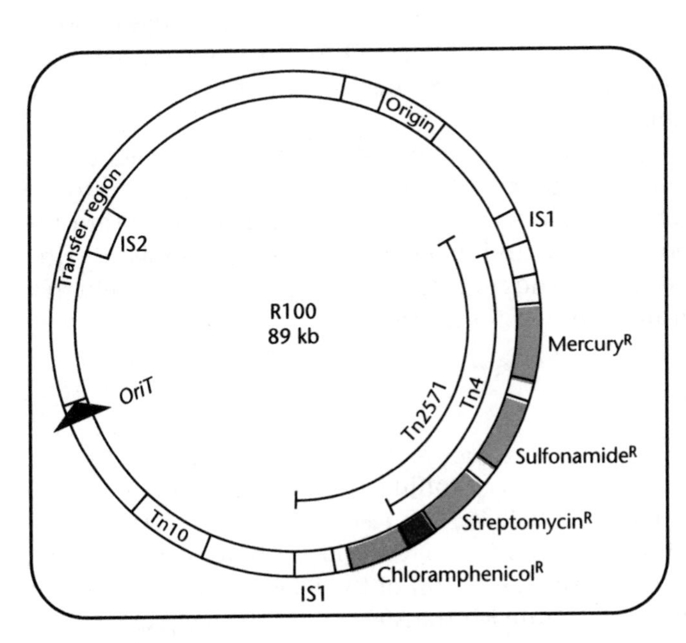

Fig. 10.2 A map of the R factor, R100. R100 contains an origin for DNA replication and an *oriT* for transfer of DNA during conjugation. The region carrying the transfer genes has extensive homology to the F factor. R100 has several resident IS and transposons including IS1, IS2, Tn*4*, Tn*10*, and Tn*2571*. R100 carries genes that encode resistance to mercury and the antibiotics sulfonamide, streptomycin, and chloramphenicol.

FYI10.2

The plasmid addiction system for maintaining F in a population

The F factor contains two genes that ensure that daughter cells inherit a copy of the F factor—or die! One gene encodes a protein that is very stable and very toxic to the cell. The other gene encodes a protein that is very unstable but is an antidote that prevents the toxic protein from working. Upon cell division, the toxin/antidote complex, which is found in the cytoplasm, is inherited by both daughter cells. If a cell inherits the F factor, then it makes both proteins and is fine. If a cell does not inherit the F factor and only gets the toxin/antidote complexes, then the trouble begins. The unstable antidote is degraded and the toxin protein is not. The toxin is free to kill the F factorless cell and eliminate it from the population, a pretty drastic way to insure that a population of bacteria keep the F factor!

FYI10.3

The initial discovery of the R factor in Japan

Shigella infections are one cause of bacterial dysentery in humans. For many years, a *Shigella* infection could be treated with the antibiotics, sulfonamide, streptomycin, chloramphenicol, or tetramycin. In the late 1950s in Japan, a strain of *Shigella* from a patient that had been treated with just one of these antibiotics was found to be resistant to all of them. By 1961, it had been shown that this multidrug resistance could be transferred from one bacterium to another and even from one species to another, including such species as *Haemophilus, Escherichia, Salmonella,* and *Shigella.* The R factor was shown to be responsible for the multidrug resistance. One important consequence of the conjugation of the R factor between species is that the normally harmless *Escherichia,* which is a minor component of the intestinal flora of humans and animals, can serve as a reservoir of these plasmids and transfer them to pathogenic species. This greatly diminishes the therapeutic benefits of large groups of antibiotics.

R factors are capable of conjugation. Some can integrate into the bacterial chromosome but at a much lower frequency than F. Like F they contain an origin of replication, essential replication proteins, and a copy number control locus. In some R plasmids, the replication control mechanism is very closely related to the mechanism used by F. In fact, the replication control can be so similar that the two factors can exhibit incompatibility.

R factors have been found that carry resistance for up to eight different antibiotics and three heavy metals. R factors were first discovered in clinical isolates where they are responsible for the resistance to multiple antibiotics that is found in increasing numbers of pathogenic bacteria. Because of the many resistance genes they can carry and their ability to move from one species of bacteria to another, R factors pose a serious medical threat. With the increased use of antibiotics, this threat continues to rise.

The conjugation machinery

F and R factors carry all of the necessary genes to promote transfer of DNA between a cell carrying the factor (the male cell or donor cell) and a cell without the factor (the female cell or recipient cell) (Fig. 10.3). We will examine conjugation using F and *E. coli* as examples. With proteins made from the F DNA, the male cell builds between one and three long hair-like structures on the outer surface of the cell. These hair-like structures are 2 to 3 μm in length and called F pili (plural) or an F pilus (singular). The F pilus is composed of a single protein and is anchored in the inner membrane (see Fig. 1.11). The tip of the F pilus comes in contact with a female cell and binds to its cell surface. The F pilus is depolymerized, with the protein subunits being incorporated into the membrane of either cell. Depolymerization serves to shorten the bridge between the two cells and bring them into direct contact. The cell-to-cell contact is stabilized by large regions of association between the outer membranes of the two cells and several proteins encoded by the F factor.

DNA is transferred from the male cell to the female cell. How this transfer occurs is not well understood. It has been suggested that the DNA transfers directly into the cytoplasm of the recipient cell. Alternatively, the DNA could first move into the periplasm and then be transported across the inner membrane to the cytoplasm. After DNA transfer, the two cells disassociate and continue growing and dividing. Cells that have undergone conjugation are called exconjugants.

Transfer of the DNA

Transfer of the F DNA begins at *oriT* in every conjugation. *oriT* is asymmetric and transfer is always in one direction (Fig. 10.4a,b). This results in the same bases always being transferred first. Next, the middle portion of the F factor that carries the DNA replication origins is transferred. Lastly, the region of the F factor that encodes the pilus genes is transferred. Only if the entire F is transferred will the recipient cell become an F^+ or male cell. If conjugation is interrupted, only part of the F will be transferred. Transfer of the F factor is very efficient. In *E. coli,* if male cells and female cells are mixed, virtually every female cell will end up with a copy of the F factor.

Transfer of the F DNA begins with a single-stranded nick in the DNA at *oriT* (Fig. 10.4a). Only the DNA strand that contains the nick is transferred to the female cell.

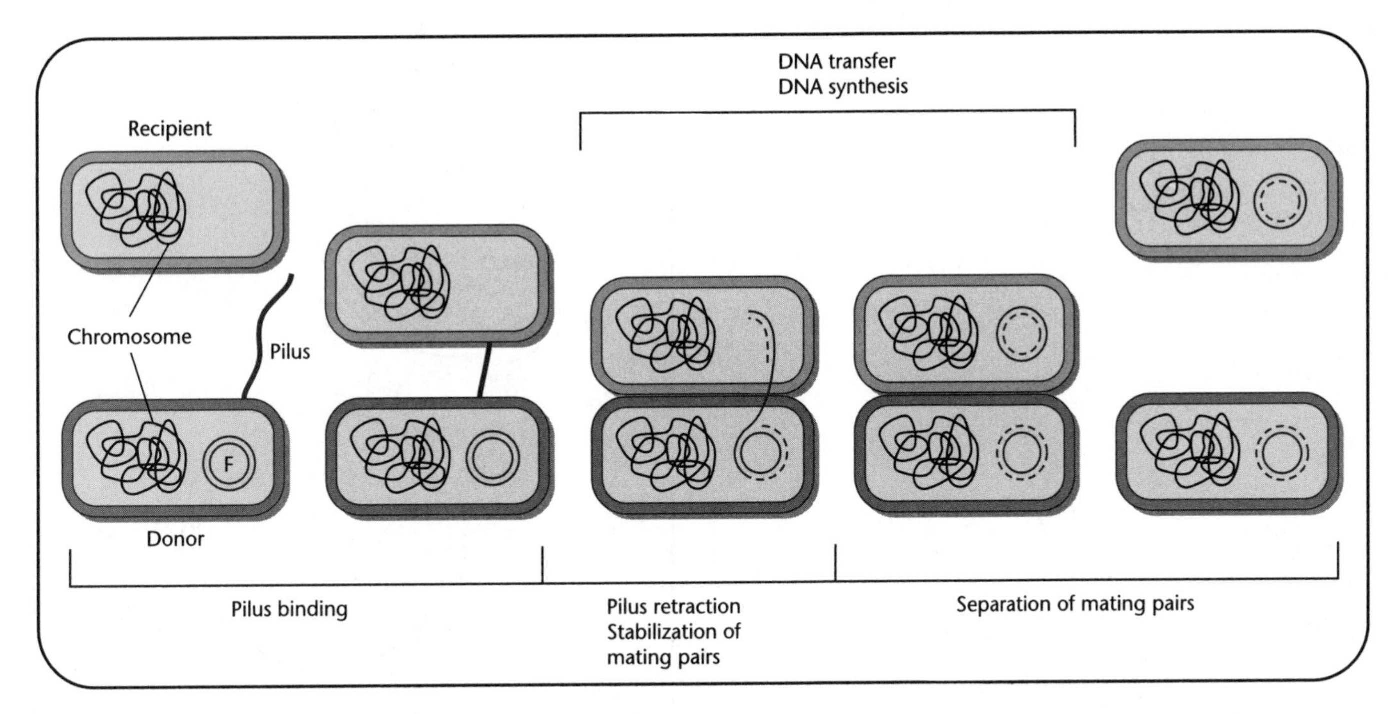

Fig. 10.3 The steps of conjugation. The F pilus from the donor cell binds to the surface of the recipient cell and brings the two cells into contact with each other. The mating pair is stabilized and transfer of the F DNA begins at *oriT*. Single-stranded DNA is transferred from the donor to the recipient. The F DNA is replicated as transfer is occurring. After transfer, the mating pair separates. After conjugation, both cells will contain an F factor.

Starting from the nick, the DNA is displaced and moved into the female cell in a 5′ to 3′ direction (Fig. 10.4b). At the same time the single strand of DNA is being moved, it is also replicated (Fig. 10.4c). DNA replication of the transferring strand in the female cell and the single-stranded DNA that is left behind in the male cell ensures that both cells end up with double-stranded DNA. DNA replication depends on the cell's DNA replication machinery. Lagging strand synthesis occurs only in the female or recipient cell, while leading strand synthesis takes place only in the male or donor cell. DNA transfer can take place even if replication is inhibited, although the single-stranded DNA is quickly degraded. Once the F factor DNA has been moved into the recipient cell and undergone replication to double-stranded DNA, it is established as a plasmid, the same as in the donor cell.

Surface exclusion

After the F DNA is transferred to the female cell, two of the genes that are transferred late are expressed. These two genes, *traS* and *traT* encode proteins that are responsible for surface exclusion. Surface exclusion greatly reduces transfer of multiple F factors into the same cell. TraS inhibits DNA transfer while TraT inhibits mating pair formation. Surface exclusion also works on closely related plasmids that move by conjugation.

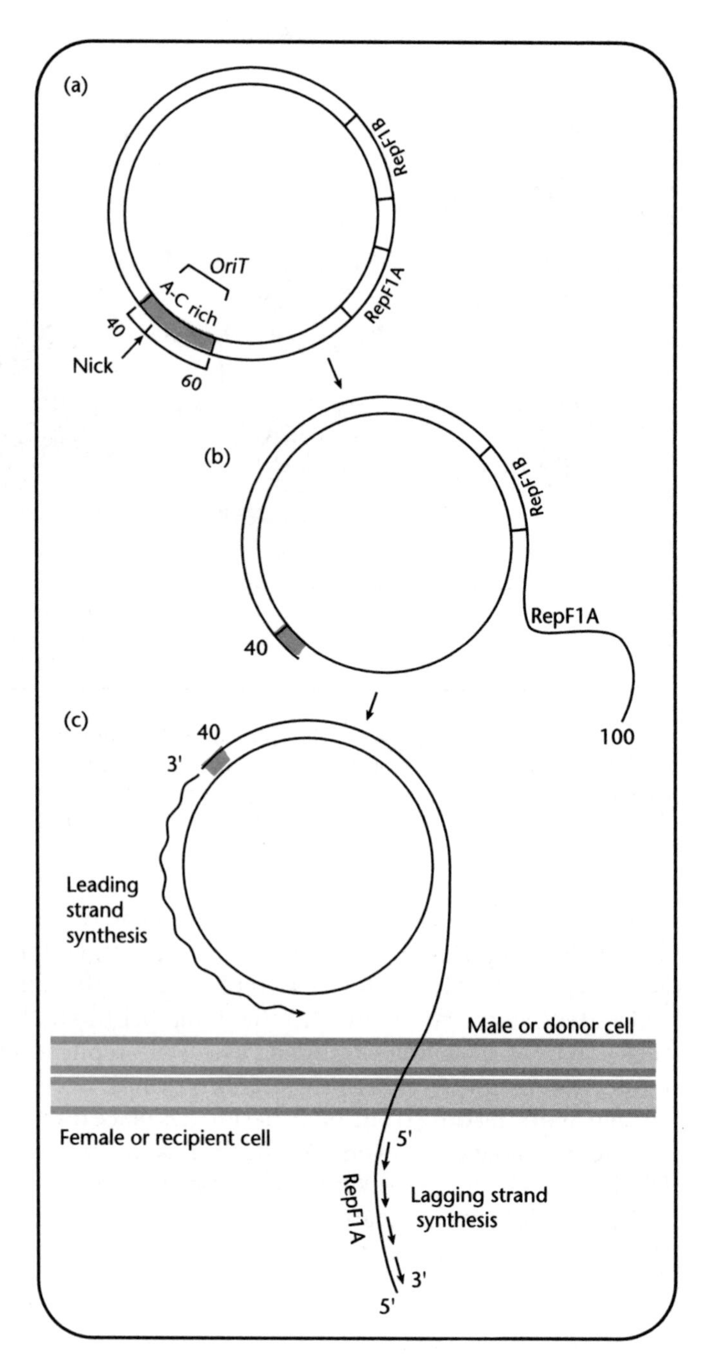

Fig. 10.4 Transfer of DNA from *oriT*. (a) *oriT* is a 100 bp sequence. It is A–C rich on one strand. A single-stranded nick occurs on the strand opposite the A–C rich strand 40 bases from one end of *oriT*. (b) The end containing 60 nucleotides of *oriT* is transferred first. One strand is transferred to the recipient cell. (c) As the strand is transferring, it is being replicated by host enzymes. Leading strand synthesis occurs in the donor cell and lagging strand synthesis occurs in the recipient cell.

F, Hfr, or F-prime

When the F factor is extrachromosomal and contains only F DNA, it is known as an F factor and cells that contain it are F^+ (Fig. 10.5a). Under certain circumstances, F can integrate into the chromosome of the host organism (Fig. 10.5b). A bacterium that contains an F factor in its chromosome is known as an Hfr. As an Hfr, the F factor is capable of mobilizing the chromosome from one cell to another and thus the name, High frequency transfer of chromosomal markers. F can also come back out of the chromosome of an Hfr strain. If it comes out and only carries F DNA, it is still called an F. If it comes out and has chromosomal sequences incorporated into the F DNA, then it is known as an F′ or F-prime (Fig. 10.5c). The size of the chromosomal sequences incorporated into an F-prime can be as small as a few hundred base pairs or as large as approximately a million base pairs (one-third to one-half the size of the *E. coli* chromosome).

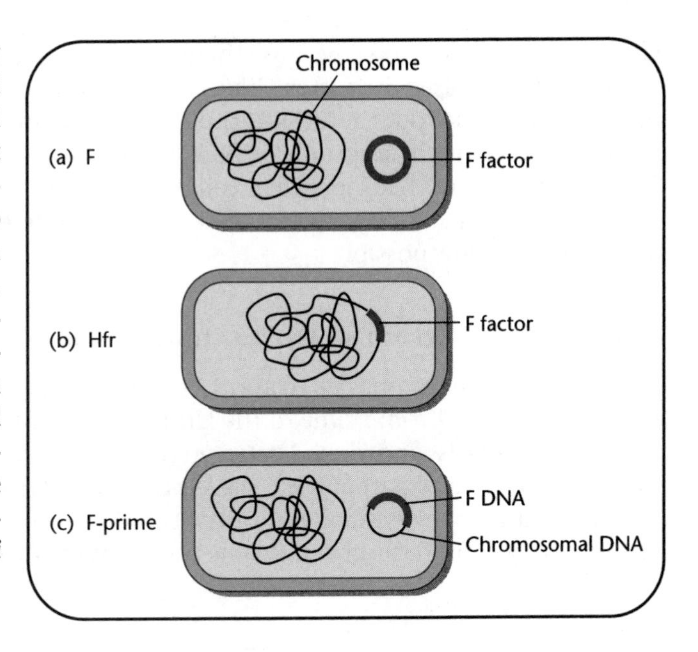

Fig. 10.5 The different places in the cell that an F factor can reside. (a) The F factor contains only F sequences and is free in the cytoplasm. (b) An Hfr contains the F factor in the bacterial chromosome. (c) An F-prime is free in the cytoplasm but contains chromosomal DNA in addition to F DNA.

Formation of the Hfr

The F factor has been shown to integrate into the *E. coli* chromosome in approximately 20 different locations. Several mechanisms have been proposed to explain Hfr formation. If both the F and the chromosome have any DNA in common, then this DNA can be used to recombine the F into the chromosome by homologous recombination. The F factor carries two copies of IS3, one copy of IS2, and one copy of Tn*1000*. The chromosome has about six copies of IS2 and approximately five copies of IS3. The number of copies of each IS varies from strain to strain. The chromosome also contains multiple copies of Tn*1000*. These can be used as a source of homology for recombination (Fig. 10.6 and Fig. 10.8). In other instances, how the F is incorporated into the

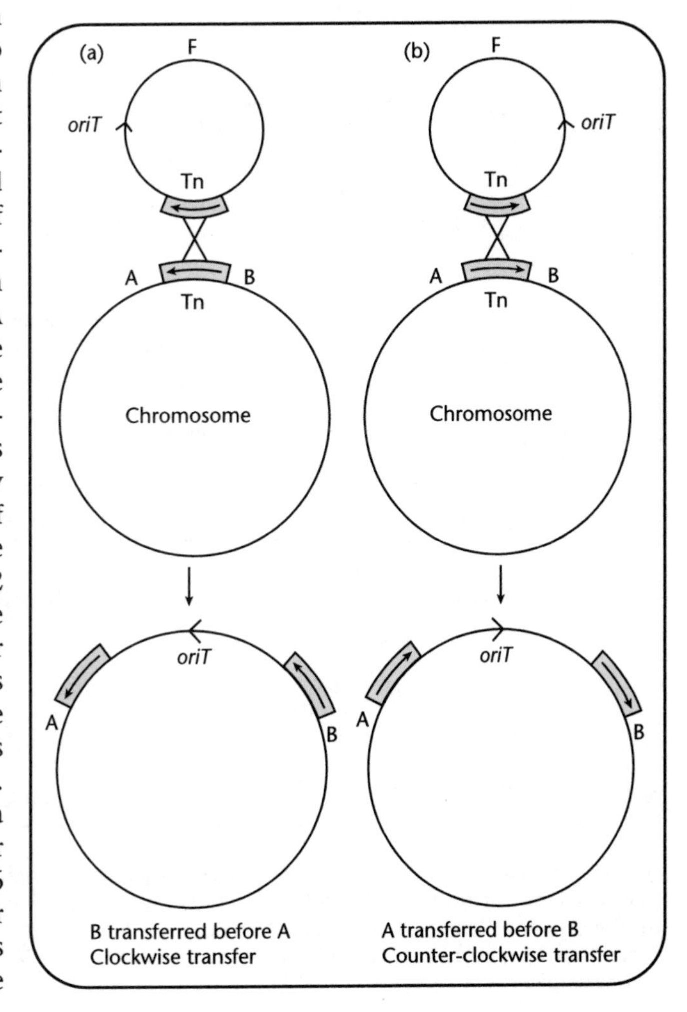

Fig. 10.6 Insertion sequences and transposons are used as portable regions of homology for recombination of the F DNA into the host chromosome. Depending on the orientation of the IS or Tn in the chromosome, the resulting Hfr will transfer chromosomal DNA in the clockwise (a) or counterclockwise (b) direction.

chromosome is not known because there is no evidence of homology at the insertion point. In these cases, it is most likely an illegitimate or nonhomologous recombination event that put the F DNA into the chromosome. In any case, once an F has integrated into the chromosome of a cell, it stays at that place and in that orientation. The descendants of that individual cell will also have the Hfr in the same place and orientation. This stability is what makes the genetic experiments using Hfrs that are described below possible.

Transfer of DNA from an Hfr to another cell

Transfer of DNA from one cell to another can be initiated from the F factor that is integrated into the chromosome in the Hfr (Fig. 10.7a,b). During the transfer process, the chromosome is sandwiched between the two ends of the F factor. Transfer begins as normal at *oriT* and part of the F DNA is transferred first. The chromosome is transferred next and in a very small minority of conjugation events, the transfer continues all of the way around the chromosome such that the other end of the F DNA is trans-

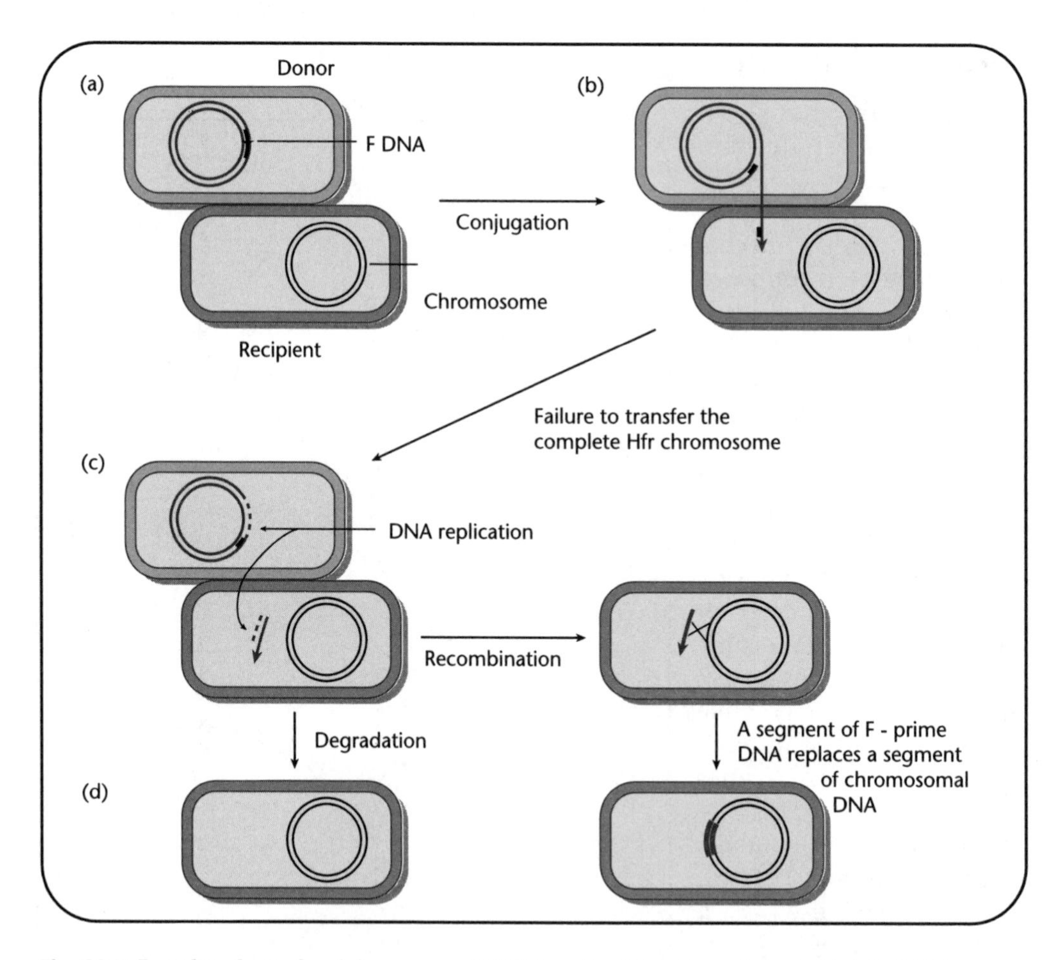

Fig. 10.7 F-mediated transfer of chromosomal DNA from the Hfr to the recipient cell. (a) F DNA is transferred through the normal reactions. (b) The chromosomal DNA comes along for the ride. (c) Most of the time, the entire chromosome will not be transferred. (d) One fate of the transferred piece of DNA is degradation by nucleases. (e) The other fate of the chromosomal DNA is to recombine with the host chromosome.

ferred. Even if the entire chromosome and F DNA is transferred into the recipient cell, it is not maintained as a plasmid. The reason for this is unknown but may have to do with how *E. coli* keeps track of the number of copies of the chromosome it contains. If the chromosome with F DNA was established as a plasmid, there would essentially be two complete chromosomes in the recipient cell. One would be the chromosome that was already present in the female cell and the second would be the chromosome with F DNA that was transferred from the male cell. *E. coli* can contain a chromosome with multiple replication forks. However, *E. coli* has never been shown to stably maintain, through many rounds of growth and cell division, two completely replicated and physically separated chromosomes.

In the majority of Hfr conjugation events, the first part of the F DNA is transferred, followed by some of the chromosomal DNA and then the mating pair is disrupted and DNA transfer stops. Once the transferred DNA is replicated, it becomes a double-stranded linear DNA molecule. At this point, one of two things usually happens to this DNA (Fig. 10.7c,d). The transferred DNA and the recipient's chromosomal DNA can recombine with each other by homologous recombination. This recombination requires two crossover events. It replaces a portion of the female cell's chromosome with DNA from the male cell's chromosome. Because the DNA moved from the male cell is a linear DNA molecule, a single crossover would leave a double-stranded break in the chromosome. Double-stranded chromosomal breaks are lethal for the cell. Alternately, the transferred DNA can be degraded. Degradation occurs because the transferred DNA has double-stranded ends that make it a target for nucleases. If the transferred DNA escapes degradation and is not recombined with chromosome, it will usually not be inherited from cell to cell. In order for the transferred DNA to be stably maintained and inherited by daughter cells, it would have to contain an origin of replication, a partition system, and it would have to circularize. The probability that a piece of transferred DNA will meet all three requirements is very low.

Formation of F-primes

F-primes are formed from Hfr strains. The F that is integrated into the chromosome can come back out. Depending on where the DNA is broken and rejoined, F-primes can be formed at this step (Fig. 10.8). The F factor can come back out with only F DNA, reforming an F. The F factor can come back out and carry chromosomal DNA from either the left or the right of the original insertion. These are known as Type I F-primes. If the F-prime carries genes that were transferred early by the starting Hfr strain then it is a Type IA F′. If the F-prime carries genes that were transferred late by the starting Hfr then it is known as a Type IB F′. It is also possible for the F′ to carry DNA from both sides of the F′ and these are called Type II F-primes. The Type II F-primes have created a novel joint in the DNA that is not present in the original chromosome.

An F-prime carries a defined part of the chromosome incorporated into the F factor (Fig. 10.8c,d). The chromosomal DNA carried by the F-prime depends on where the F factor was incorporated into the chromosome. For example, the F factor in one Hfr is located near the genes necessary for degrading the sugar lactose (the *lac* genes). F-primes derived from this Hfr will carry the *lac* genes at a high frequency and transfer them by conjugation from cell to cell. The F factor in a second Hfr is located near the *srl* genes, which are required to degrade the sugar sorbitol. F-primes derived from this second Hfr will carry the *srl* genes at a high frequency and transfer them from cell to cell. Because the *srl* genes and the *lac* genes are not close together on the chromosome, it would be very difficult to find an F-prime from the Hfr near *lac* that carries the *srl*

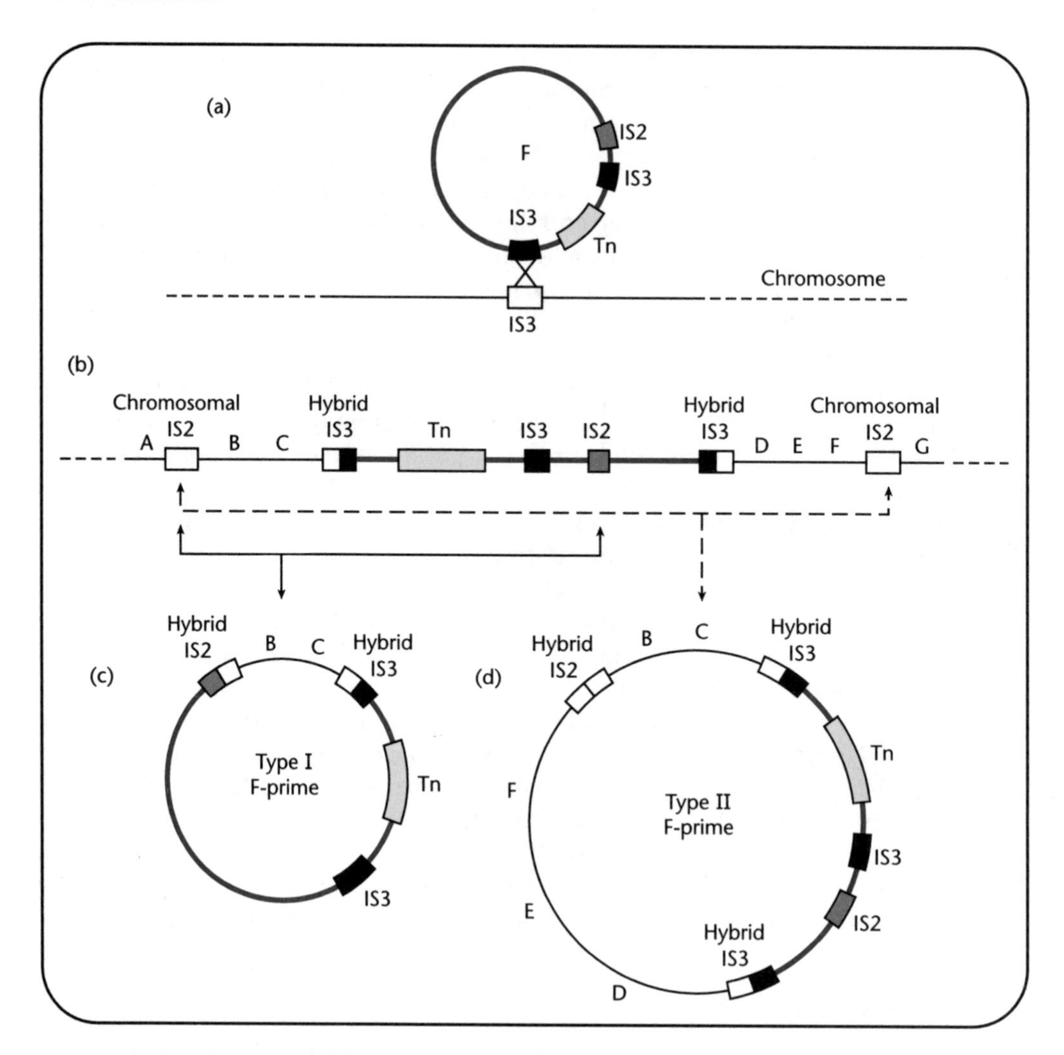

Fig. 10.8 Formation of F-primes. (a) One IS3 from the F DNA is used to recombine the F into the chromosome, forming the Hfr. (b) The F DNA in the chromosome is flanked by hybrid IS3 elements. Part of the hybrid IS3 is from a chromosomal copy of the IS3 and the other part is from the IS3 from F. The chromosome contains additional IS and Tn sequences. (c) Type I F-primes contain DNA from one side of the inserted F DNA or from the other side of the inserted F DNA, depending on which chromosomal IS element is used. Recombination uses one chromosomal IS and one IS in the F DNA. (d) Type II F-primes use two chromosomal IS or Tn elements for recombination. A novel chromosomal joint is created. In this example, gene *F* is placed next to gene *B*.

genes or one from the Hfr near *srl* that carries the *lac* genes. If such an F-prime was found, it would have to carry approximately 2250 kb of DNA. F-primes this large are usually very unstable and suffer deletions of the chromosomal DNA or excision of the F factor from the chromosomal DNA. The most stable F-primes contain approximately 5% to 20% of the *E. coli* chromosome, or 250 to 1000 kb of chromosomal DNA.

Transfer of F-primes from one cell to another

F-primes contain enough of the original F factor DNA so that they can be stably maintained as plasmids (Fig. 10.9a,b). F-primes contain the origin of DNA replication and partition systems of the F factor. During the breaking and joining of DNA molecules

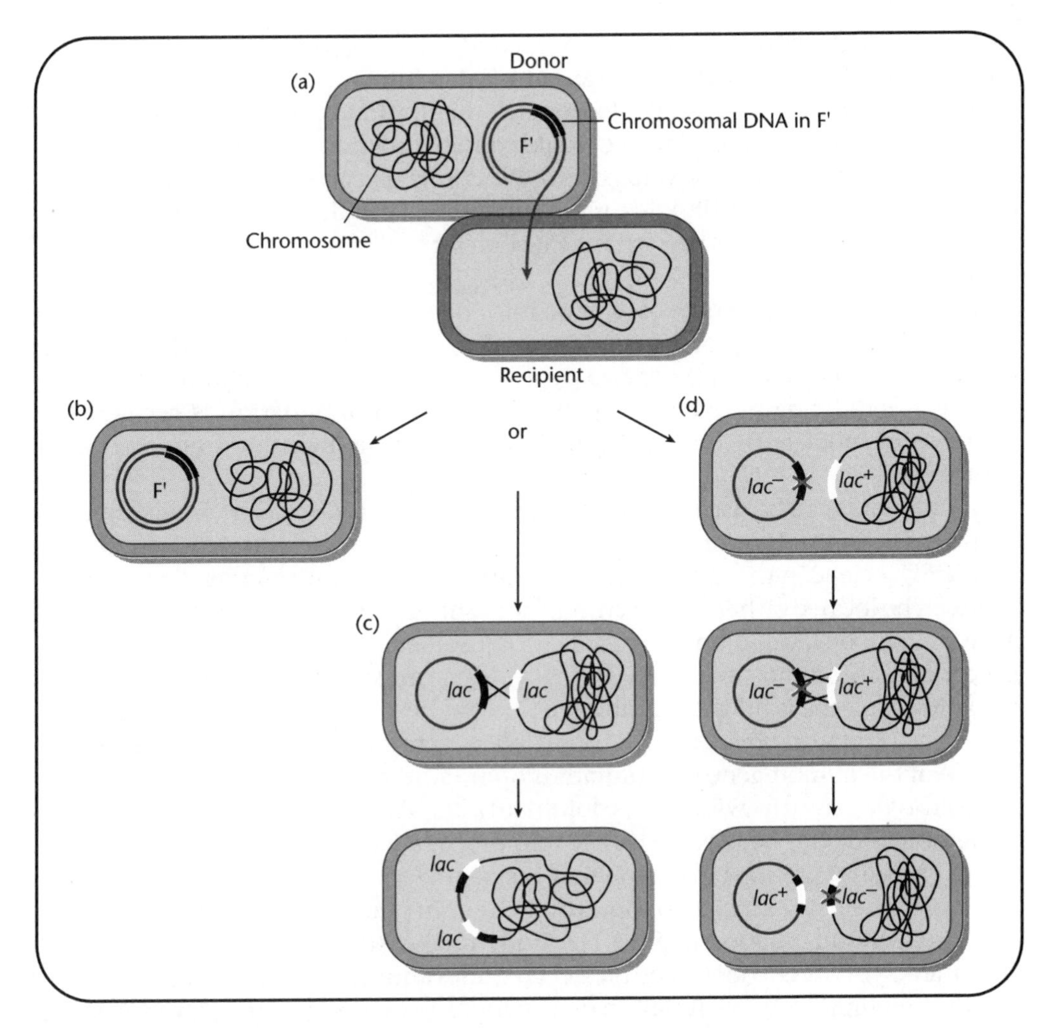

Fig. 10.9 The fate of the F-prime DNA after transfer. (a) The F-prime is transferred to the recipient cell. (b) The F-prime can establish itself as a plasmid. This is the most frequent event. (c) The F-prime can recombine with the chromosome through a single crossover and end up incorporated into the chromosome. (d) The chromosomal DNA in the F-prime can recombine with the host's chromosome. If the F-prime and the chromosome contain different alleles, the alleles can be exchanged.

that form the F-prime, the F-prime DNA is circularized. The replication origin, partition system, and circular DNA molecule are the minimal requirements for being a low copy plasmid. F-primes also contain *oriT* and all of the genes necessary for transfer by conjugation. Because F-primes contain chromosomal DNA, they can recombine into the chromosome by a single crossover event (Fig. 10.9c) or part of their chromosomal DNA can be incorporated into the bacterium's chromosome by a double crossover (Fig. 10.9d). This means that if the chromosomal DNA on the F-prime carries a mutation, then the mutation can be recombined into the host's chromosome (Fig. 10.9d).

Genetic uses of F-primes

There are several important laboratory uses for F-primes. All of them rely on the fact that an F-prime strain contains two copies of a defined region of the chromosome.

One copy is in the chromosome and the second copy is on the F-prime. For example, if the F-prime *lac* described above was conjugated into a wild-type strain, the strain would now carry two copies of the *lac* genes. If the Hfr that the F-prime *lac* was derived from was also Lac^+, then the F-prime would contain functional *lac* genes and the chromosome would contain a second copy of functional *lac* genes. The F-prime *lac* containing strain is said to be diploid for the *lac* region of the chromosome. Because only part of the chromosome is present in two copies, the strain is called a **merodiploid** or partial diploid (Fig. 10.10).

Being able to put two copies of a gene into the same cell is very useful for genetic studies. Making merodiploids with F-primes has the added advantage that the copy number of the F-prime is one or two per cell or very close to the chromosome copy number. Merodiploids have been used to determine if a mutation is dominant or recessive, to mutagenize essential genes, and to move a mutation from one cell to another cell.

Merodiploids can be used to determine if a mutation is dominant or recessive (see Chapter 3). For this type of study, the wild-type gene and the mutated gene must be physically located in the same cell and present in similar copy number. The wild-type gene can be located either on the chromosome or the F-prime. Likewise the mutated gene can also be located on either the chromosome or the F-prime. If the wild-type gene is on the chromosome, then the mutant gene must be located on the F-prime. If the wild-type gene is on the F-prime then the mutant must be on the chromosome. Once a cell contains both genes, it is possible to determine if the cell has the phenotype(s) of the mutant gene (the mutant is dominant) or if it has the phenotype(s) of the wild-type gene (the wild type is dominant).

Merodiploids can be used to isolate and propagate mutations in essential genes. Many types of mutations in a gene that is essential for cell growth will result in the death of the cell. By having two copies of the gene in one cell, one can be mutated and the other will still function (Fig. 10.11). Studying the lethal mutations to learn about the function of the encoded gene product requires another genetic trick. As described above, merodiploids allow isolation of mutations in essential genes. However, studying the mutation and the effect it has on the cell is not possible because a copy of the wild-type gene must be present to keep the cells alive. To get around this problem, what is needed is a conditional merodiploid or a cell that is merodiploid under one growth condition but haploid under another growth condition. A temperature-sensitive mutation in the F factor that allows the F DNA to replicate at 30°C but not at 42°C can be used to make a conditional merodiploid. For example, if the wild-type gene is carried on a temperature-sensitive F-prime and the mutated gene is located on the chromosome, the cells will grow at 30°C. When the temperature is raised to 42°C, the F DNA will not replicate and the existing F-prime molecules will be diluted out of the cells by growth and division. This will unmask the lethal mutation that is carried on the chromosome. After the shift to 42°C, the physiology of the cells can be studied to determine the defects caused by the lethal mutation.

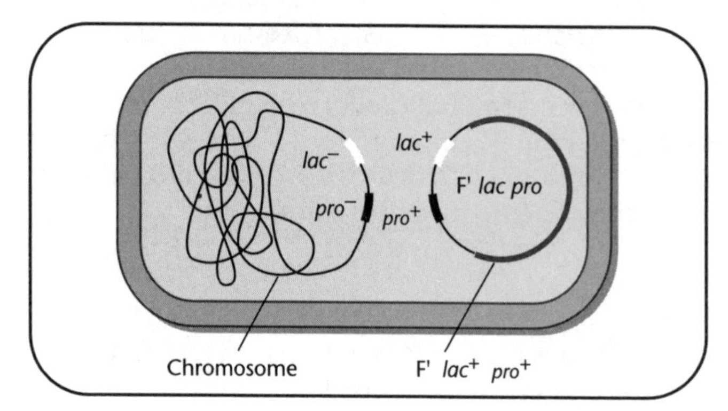

Fig. 10.10 Merodiploids contain two copies of a defined region of the chromosome. In this case, both the *lac* and *pro* genes are present in two copies. If the chromosome contains mutations in the *lac* and *pro* genes, the F-prime can supply these functions to the cell. The cell shown would be able to grow on lactose and in the absence of proline.

F-primes can be used to move mutations from one cell to another. The chromosomal DNA in the F-prime can recombine with the chromosome by homologous recombination. The practical consequences of this are that a mutation that is located on the chromosome can be recombined to an F-prime. The F-prime can be conjugated into a new cell where the mutation can be recombined into the chromosome of the new cell. If this strategy is carried out using the F factor that is temperature sensitive for the DNA replication described above, then the last step in moving the mutation into a new cell would be to raise the temperature and get rid of the F-prime DNA. While there are several ways to move mutations from one *E. coli* strain to another, in many medically relevant and other interesting bacteria there are fewer options. Using F-primes or more frequently R factors that carry chromosomal DNA to move mutations from one cell to another is a very powerful and frequently used technique in many bacteria.

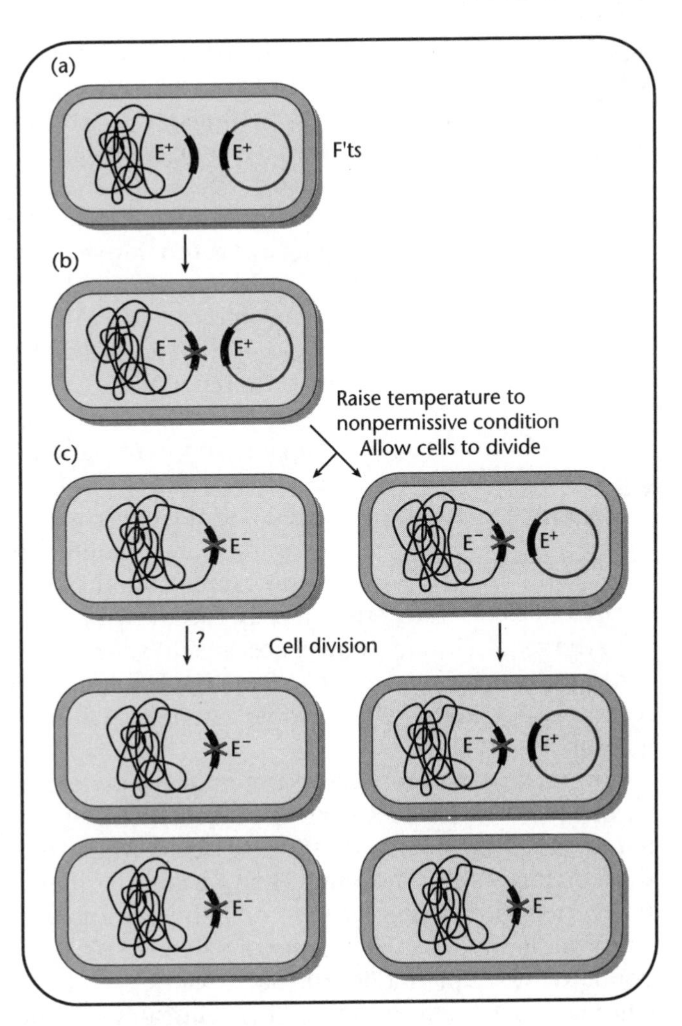

Fig. 10.11 Merodiploids can be used to study mutations in essential genes. (a) In this example of a merodiploid, an essential gene, *E*, is present in two copies. (b) Mutations in the chromosomal copy of *E* can be identified or constructed. (c) If the F-prime is temperature sensitive for replication, raising the temperature will hold the number of copies of the F-prime static. Cell division results in one cell with an F-prime and one without. Depending on the nature of the *E* gene product, the cells might be able to undergo a few more cell divisions, each division producing more cells without the F-prime. The cells without the F-prime can be examined to determine the effects of the mutation in *E*.

Genetic uses of Hfr strains—mapping genes on the *E. coli* chromosome using Hfr crosses

E. coli contains a single circular chromosome that takes approximately 90 to 100 minutes to completely transfer from a male cell to a female cell in an Hfr cross (Fig. 10.12). For this reason, the chromosome was divided into 100 minutes and, before the sequence of the chromosome was completed, the placement of a gene on the chromosome was described in terms of minutes. For historical reasons, the genes required for biosynthesis of threonine (*thrABC*) are located at 0 minutes. The *lac* genes (*lacIZYA*) are located at 8 minutes. From DNA sequence analysis, the chromosome contains approximately 4000 potential genes. Only about 2000 of the potential genes have been shown to have a biological function and the large majority of these genes were mapped using genetic techniques. One of the first questions that is usually asked about a newly discovered gene is where on the chromosome does it reside. It is very useful to determine the position of a new gene relative to other known genes. This allows the classification of the number of different genes involved in a specific process

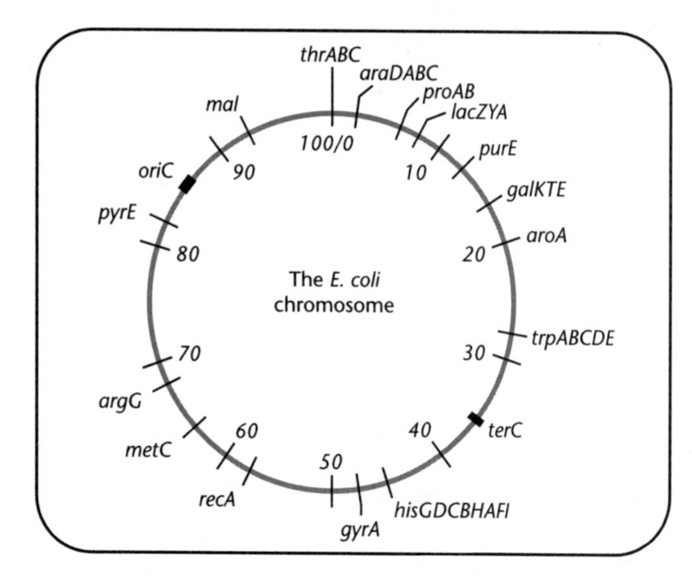

Fig. 10.12 A partial map of the *E. coli* chromosome. Only a few of the more than 2000 known genes are shown. The 0 minute position of the chromosome is the site of the threonine biosynthesis genes. These were the first genes transferred in the original report on conjugation, *oriC* is located at 84 min, and *terC* is directly opposite *oriC*.

as well as determining whether newly isolated mutations are in a known or novel gene. Using conjugation, there are two methods for determining map position of a gene. One of these, time of entry mapping, has more of a historical importance, while the second, Hfr mapping relative to known markers, is still widely used. We will consider mapping of a gene relative to known genes in detail.

The basic goal of an Hfr cross is to move a defined piece of chromosomal DNA from an Hfr cell into a female cell (Fig. 10.13). The piece of Hfr DNA that was transferred is recombined with the chromosome of the female cell so that is can be stably inherited. Then it is determined if a gene of interest and a known gene were transferred together on the Hfr DNA or if the known gene was transferred and the other was not. If the two genes are transferred and recombined together into the recipient's chromosome then they must both be in the defined piece of DNA that was transferred. If only the known gene is transferred and recombined, then the gene of interest must be located in a part of the chromosome that is not present in the defined piece of DNA from the Hfr. Using several different Hfr strains, different regions of the chromosome can be tested for the presence or absence of the gene of interest.

How each part of an Hfr cross is monitored or determined is examined below. First, we must know what piece of chromosomal DNA was transferred from the Hfr to the female cell. Next, we must be able to detect when the Hfr DNA has recombined into the female cell's chromosome. Then we must be able to tell if our gene of interest was transferred with the known gene or if only the known gene was transferred.

Hfr mapping relies on transfer of a defined piece of chromosomal DNA from the donor to the recipient (Fig. 10.13a). Because each Hfr has an F factor located in a specific place on the chromosome, this delineates one end of the transferred DNA molecule. For example, HfrH has an F factor located at 98 minutes on the chromosome and it transfers chromosomal DNA in the clockwise direction. If HfrH is used for conjugation, the chromosomal DNA from 98 minutes will be transferred first, followed by the DNA at minute 99, minute 100, minute 0, minute 1, etc. The known gene that is to be used in the cross marks the other end of the transferred fragment of the Hfr chromosome. For example, if HfrH is used as the starting Hfr and *lacZ* is used as the known gene, then the transferred piece of Hfr chromosome will start at 98 minutes and end at 8 minutes which corresponds to the position of the *lac* genes on the chromosome. We can then determine if the unknown gene maps between 98 and 8 minutes on the chromosome. By choosing appropriate Hfrs and known genes we can predetermine the region of the chromosome that we want to test for the presence of our unknown gene.

It is not enough that a defined piece of DNA is transferred into the recipient cell. The defined piece of DNA must also be recombined into the recipient cell's chromosome so that it is stably inherited. The recombination step can be ensured by using a mutant allele of the known gene and including this mutation in the recipient cell (Fig. 10.13b). For example, if *lacZ* is used as the known marker, then the recipient cell should contain a mutant allele of *lacZ* and be phenotypically Lac^-. The Hfr should contain the wild-type *lacZ* gene and be Lac^+. The Hfr cells and the recipient cells are mixed together to allow mating and then plated on agar containing lactose as a sole

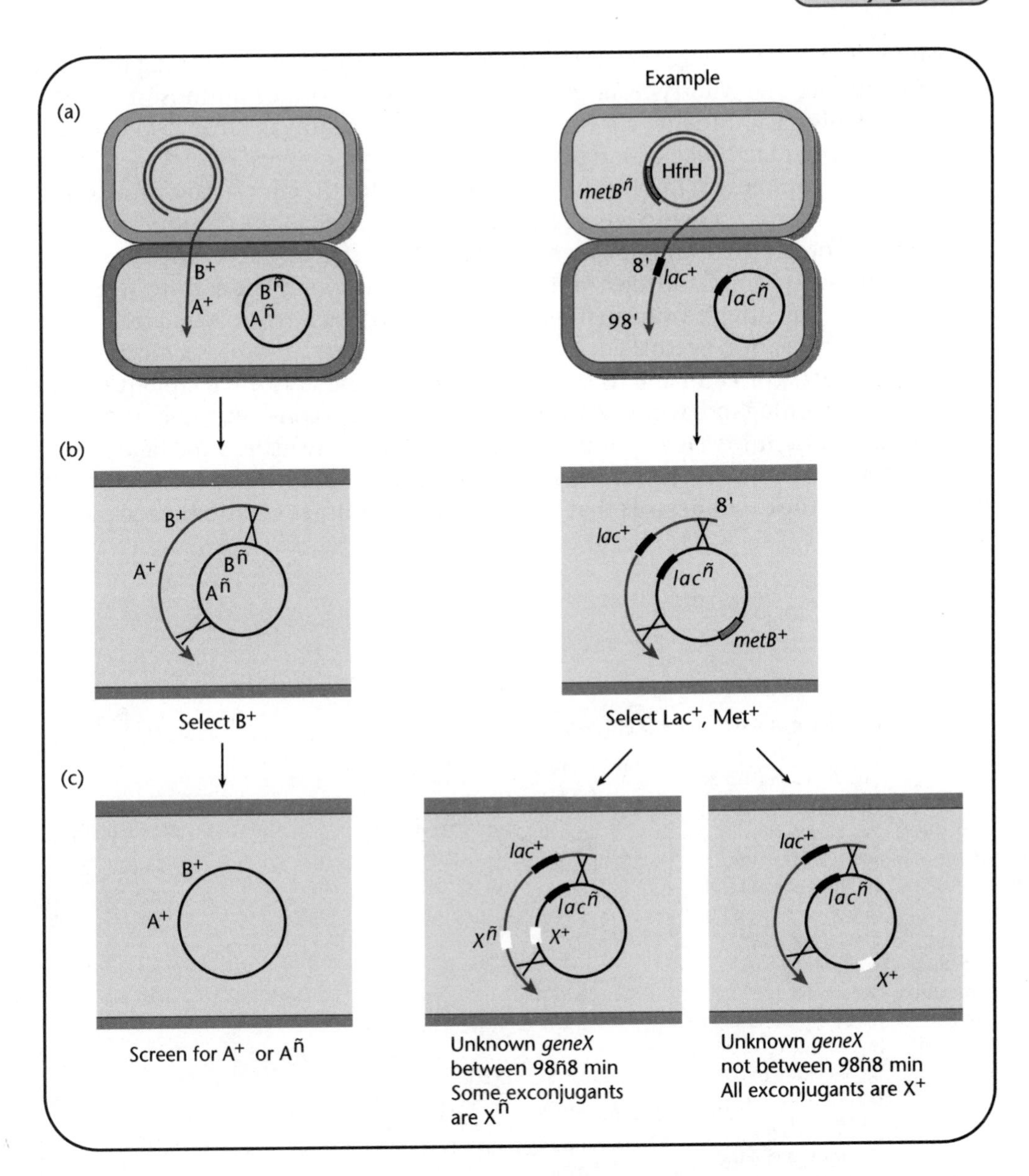

Fig. 10.13 Hfr mapping requires three basic steps. (a) A defined piece of chromosomal DNA is moved from the donor into the recipient. In the example, HfrH, which starts transfer at 98 minutes and proceeds clockwise is used. The other end of the transferred DNA is marked by the *lac* genes. The donor contains wild-type *lac* genes (lac^+) and the recipient, mutant *lac* genes (lac^-). (b) The defined piece of transferred DNA must recombine with the chromosome. In the example, recombinational replacement of the lac^- genes from the recipient with the lac^+ genes from the donor is selected for. *metB* is used as the counterselection gene. The Hfr is $metB^-$ and the recipient is $metB^+$. By omitting methionine from the selection agar, the Hfr cannot grow. (c) The recombined exconjugants are screened for the presence or absence of the unknown gene. In the example, the Hfr carries a mutation in the unknown gene (X^-) and the recipient carries the wild-type allele of the unknown gene (X^+). The recombined exconjugants are screened for the phenotype associated with X^-. If all of the exconjugants are X^+ like the recipient, then *X* does not reside between 98 and 8 minutes on the chromosome. If 50% of the exconjugants are X^+ and 50% are X^- then *X* does reside between 98 and 8 minutes on the chromosome.

carbon source where only Lac$^+$ cells can grow. The Hfr is Lac$^+$ and exconjugants that have recombined the wild-type *lac* genes from the Hfr into the chromosome of the female cell will be Lac$^+$. To prevent the Hfr strain from forming colonies or as a counterselection on the lactose selection plates, another mutation must be included in the Hfr. The main requirement for the mutation to be used for the counterselection is that the gene is not located between where the F factor is inserted in the chromosome and the known gene. In the example given above, the counterselection gene cannot be located between 98 and 8 minutes on the *E. coli* chromosome. The counterselection gene should be far enough away on the chromosome so that it is not transferred in the majority of conjugation events. For example, if the HfrH strain carries a mutation in *metB* (90 minutes) or is unable to synthesize methionine and the recipient carries *metB*$^+$, only exconjugants whose chromosome had undergone recombination can grow on agar containing lactose and lacking methionine. By using two alleles of the known gene and two alleles of the counterselection gene we can design an Hfr cross in order to isolate the exconjugants that have recombined Hfr DNA into the recipient's chromosome.

FYI10.4

Time of entry mapping and determining that the chromosome is circular

Time of entry mapping takes advantage of the fact that for a single Hfr, DNA from the same location always enters the recipient or female cell first. The mutation to be mapped is included in the chromosome of the Hfr strain. The Hfr strain carrying the mutation is simply mixed with the female strain and every minute a sample is removed. The cells in the sample are examined for the phenotype of the mutation. The time of appearance of the mutant phenotypes in the recipient cell indicates how far from the origin of transfer of the Hfr that the mutation is located. In a time of entry experiment, other known genes are included as controls. For example, if it is known from previous experiments that the gene order is gene *A*, *B*, *C*, and *D*, then any new gene can be mapped relative to these. If the phenotype of the new gene appears between the phenotypes of gene *A* and gene *B*, then the new gene must be physically located on the chromosome between *A* and *B*. Time of entry mapping was used to show that the *E. coli* chromosome is circular. The options were either a circular molecule or a linear molecule. If it was circular, then by time of entry mapping, using an Hfr between *thr* and *lac* that transfers DNA in the clockwise direction, *lac* should enter first, then *bio*, *trp*, *his*, *cys*, *mal*, *arg*, and *thr*. Using an Hfr between *mal* and *cys* that transfers clockwise, *mal* should enter first, then *arg*, *thr*, *lac*, *bio*, *trp*, *his*, and *cys*. By using enough different Hfrs, even if one is not sure where they reside and what direction they transfer, it should be possible to demonstrate that there are no gaps in transfer of the chromosome. If the chromosome was linear, a different result would be obtained. Using the same Hfr between *thr* and *lac*, *lac*, *bio*, *trp*, *his*, *cys*, *mal*, and *arg* would be transferred in the same order but *thr* would not be transferred. Using the second Hfr between *mal* and *cys*, *mal* and *arg* would be transferred but not any of the other genes. Because the first result was obtained, early *E. coli* geneticists were able to conclude that *E. coli* has a single circular chromosome. How did they know there was only one chromosome?

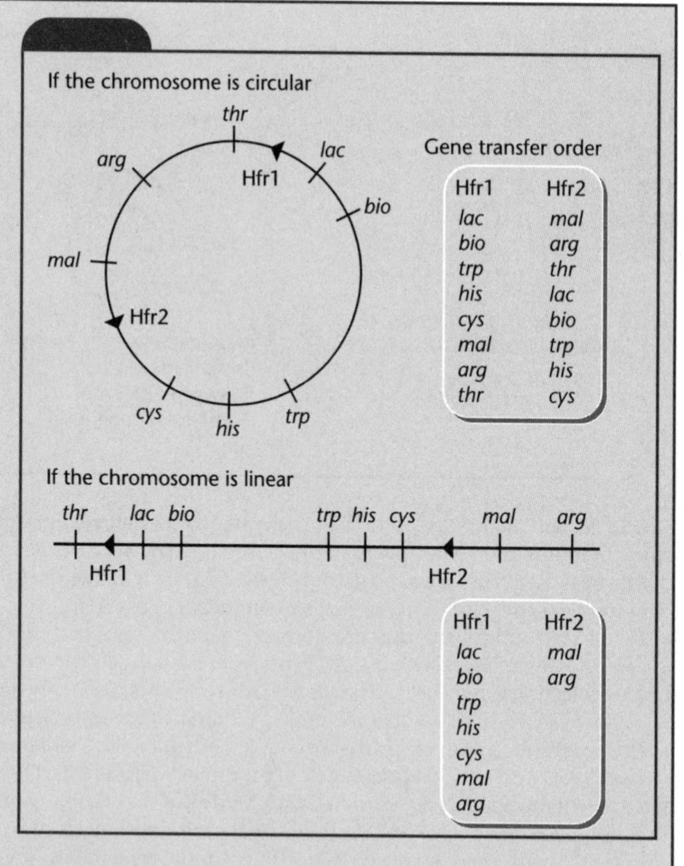

The last step in an Hfr cross is to determine if the exconjugants that have undergone recombination between Hfr DNA and recipient cell chromosomal DNA have also inherited the unknown gene (Fig. 10.13c). Again, a mutant allele and a wild-type allele of the unknown gene are required. Each allele must have a distinct phenotype that can be easily determined. The Hfr must contain one allele and the recipient cell a different allele. In general, it does not matter if the Hfr contains the mutant allele or the wild-type allele as long as the Hfr and the recipient are different. If the recipient cell carries the wild-type allele and the Hfr the mutant allele, then the exconjugants are screened for the presence of the mutant allele. If the exconjugants contain the mutant allele then they must have received it from the Hfr strain. This indicates that the unknown gene maps in the region of the chromosome covered by the DNA that was transferred from the Hfr.

The 50% rule

Inheritance of the unknown gene in an Hfr cross does not occur in 100% of the exconjugants. Because the recombinational crossover can take place on either side of the unselected gene, the maximum number of exconjugants that can gain the unknown gene is 50% (Fig. 10.14a). If the unknown gene is inherited significantly less than 50% (10% for example) of the time by the exconjugants, it is not in the interval between the F DNA and the selected gene (Fig. 10.14b). These 10% represent conjugations where DNA transfer has gone past the selected marker. Occasionally a marker will be exchanged greater than 50% (Fig. 10.14c). These cases are rare. They indicate that the unknown gene is very close to the selected gene.

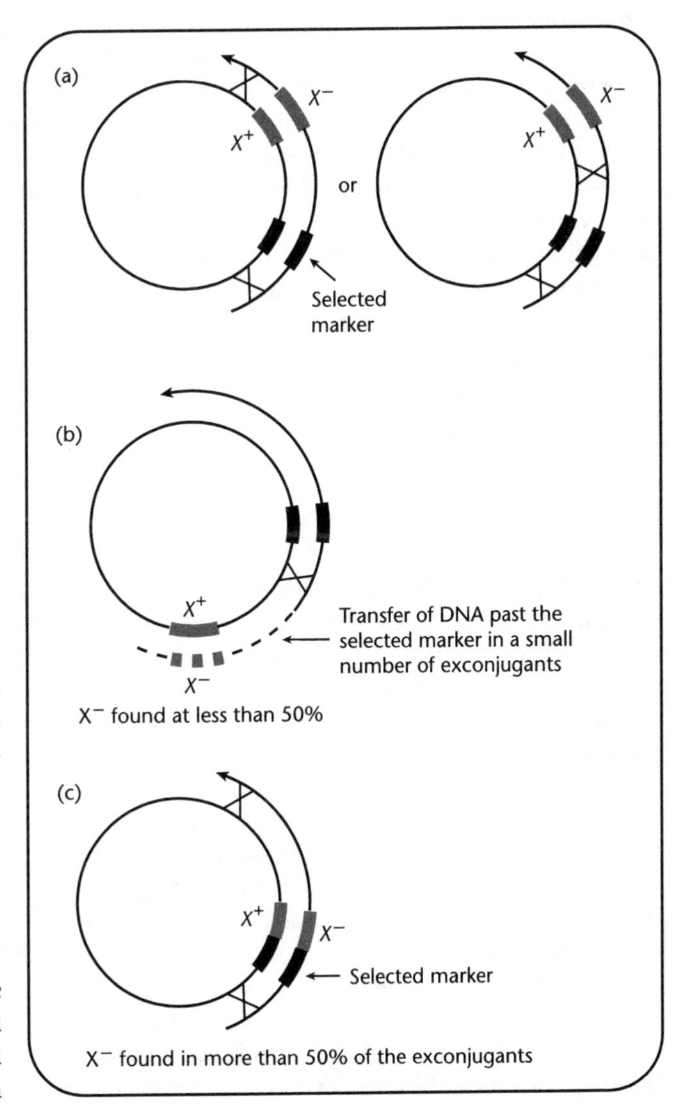

Fig. 10.14 The fate of the unknown gene in an Hfr cross. (a) If the unknown gene *X* is located in the interval of the chromosome that is being tested, 50% of the exconjugants will be X^+ and 50% will be X^-. This is because there are two places that the recombinational crossover can take place. (b) If gene *X* is located outside of the interval, then a much lower percentage of exconjugants will be X^-. Ideally there would be no X^- exconjugants but in practice there are usually a few. These few have transferred chromosomal DNA beyond the selected gene. (c) If more than 50% of the exconjugants have inherited X^-, then the *X* gene is very close to the selected gene.

Using several Hfr strains to cover the chromosome

Different Hfr strains have the F DNA incorporated into the chromosome in a different place and in a specific orientation. Using several different Hfrs whose location and direction of DNA transfer is known as well as appropriately chosen genes for selection and counterscoring, it is possible to determine where in the chromosome that the gene of interest is located. In fact, a series of 10 Hfrs has been developed so that a novel gene can be easily mapped to a region that is approximately 10% of the *E. coli* chromosome. Once a gene has been localized to a region, it can be more precisely mapped by transduction.

Transfer of DNA from the Hfr can result in a large piece of chromosomal DNA being moved. It is not unusual for the selected marker to be 25% of the chromosome or about 1000 kb away from the incorporated F DNA. If two genes are both in this 1000 kb, they will be transferred together and can recombine into the exconjugants chromosome together. Mapping using P1 transduction results in about 90 kb of chromosomal DNA being transferred from one cell to another. Experimentally, Hfr mapping allows determination of a general chromosome location and transductional mapping results in a much more specific chromosomal position. Usually Hfr mapping is conducted first to localize a gene to a region of the chromosome that corresponds to about 10–15% of the chromosomal DNA. Transductional mapping can further refine this position to a region that is 1% or less of the chromosome.

Why map a gene rather than just determine its DNA sequence? When a gene is first identified by a mutant phenotype, nothing is known about where it is located on the chromosome. The mutation could be anywhere in the several thousand kilobase pairs that make up a bacterial chromosome. It is too time consuming and costly to determine the DNA sequence of the entire wild-type bacterial chromosome and the entire chromosome carrying the mutation. Even if the sequence of all of this DNA was determined, it is very unlikely that only a single base change would be found. Due to the error rate of the DNA replication machinery, several changes would be present. Sequencing alone will not tell you which change is responsible for the mutant phenotype. By mapping the new mutation first, the amount of DNA that must be sequenced is dramatically reduced and a specific sequence change can be correlated with the mutant phenotype.

Mobilization of non-conjugatible plasmids by R and F

In addition to mobilizing themselves and the chromosome, both R and F factors can mobilize other plasmids. The hallmarks of these other plasmids are that they are not capable of conjugation themselves and they must carry a region of homology to the F or R. Most frequently the homology is a small region of the chromosome, often only a single gene. Using this homology, it is thought that the non-conjugatible plasmid is recombined into the F or R and simply transferred as part of the F or R (Fig. 10.15). In nature, this feature allows the movement of a non-conjugatible plasmid from one cell to another and from one species to another. In the laboratory, this process allows the movement of DNA, cloned genes, or mutations in the cloned genes into a wide array of bacterial species. In some cases, this is the only known way to easily introduce DNA into these organisms. Mobilization of non-conjugatible plasmids is used to study marine bacteria, pathogenetic bacteria, soil bacteria, and many other important and interesting bacterial species. Because many of the genetic techniques developed for *E. coli* are specific to it, they are not useful for studying anything but *E. coli* and very closely related bacteria. Plasmid mobilization has proved to be a much more general technique that can be adapted to many different vastly unrelated species.

Conjugation from prokaryotes to eukaryotes

Conjugation has been demonstrated to occur between prokaryotes and eukaryotes. In the known cases the DNA moves from the prokaryote to the eukaryote. In one case, F-primes from *E. coli* can be conjugated into the yeast, *Saccharomyces cerevisiae*. In the most well-studied case, DNA is conjugated from the bacterium *Agrobacterium tumifasciens* into certain plants. As described in Chapter 9, *A. tumifasciens* is a Gram-negative soil bacterium that causes crown gall tumor disease on plants. It contains a plasmid that varies in size from 180 to 200 kb, depending on the strain the plasmid is isolated from. A 10 to 25 kb piece of the plasmid, termed the T-DNA, is transferred from the bacterium to the plant cell nucleus and incorporated into a plant chromosome. The T-DNA is transferred by conjugation. Transfer of the T-DNA is not restricted to plant cells, it can also be conjugated into yeast and *Streptomyces*. In these other organisms, the T-DNA is also incorporated into the host's chromosome. Thus, the T-DNA contains the necessary information to cross kingdoms.

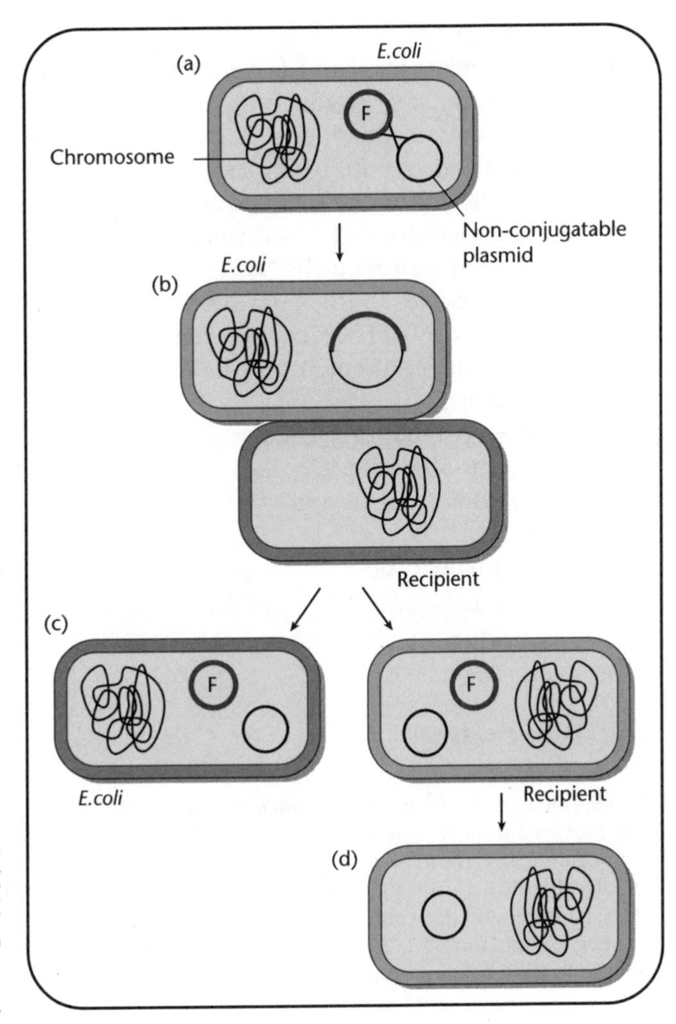

Fig. 10.15 Mobilization of non-conjugatible plasmids. (a) The F DNA and the non-conjugatible plasmid recombine. (b) F directs the transfer of both plasmids. (c) The F factor and the non-conjugatible plasmid can recombine into separate molecules. (d) If the recipient cell cannot support the replication of F then the F factor can be lost. Alternatively, the temperature-sensitive F factor can be used and in the final step the temperature can be raised to select for loss of the F.

Summary

Movement of DNA by conjugation has proven to be very important to both organisms and scientists. For the organisms, it represents a significant way to inherit large segments of foreign DNA. It also allows for movement of large amounts of DNA from one cell to another and thus through the population. For scientists, it has provided a way to map the genes on the chromosomes and carry out sophisticated genetic experiments to understand how genes and gene products interact with one another.

Study questions

1 What elements must an F factor contain in order to be stably maintained in a cell? To be transferred from one cell to another?
2 What characteristics do the F factor and the R factor have in common? What characteristics do they not share?
3 What is the difference between an F, an Hfr, and an F-prime?
4 What happens to the F DNA after it is transferred? Hfr DNA? F-prime DNA?
5 During conjugation, is the DNA transferred as a single-stranded or double-stranded molecule? If single stranded, then how did the DNA become single stranded and why did conjugation require single-stranded DNA for transfer?
6 What are the requirements for moving a mutation from one cell to another by F or R factors?
7 How does Hfr mapping work?
8 Explain the 50% rule. What feature(s) of conjugation are responsible for it?
9 What elements are necessary for mobilization of a plasmid from one cell to another?
10 What are merodiploid strains and what can they be used for?

Further reading

Clewell, D.B. ed. 1993. *Bacterial Conjugation*. New York: Plenum Press.
Firth, N., Ippen-Ihler, K., and Skurray, R.A. 1996. Structure and function of the factor and mechanisms of conjugation. In *Escherichia coli* and *Salmonella typhimurium: Cellular and Molecular Biology*, 2nd edn., eds. F.C. Neidhardt, R. Curtiss III, J.L. Ingraham, E.C.C. Lin, K.B. Low, B. Hagasanik, W.S. Rexnikoff, M. Riley, M. Schaechter, and H.E. Umbarger, pp. 2377–401. Washington, DC: ASM Press.
Lederberg, J. and Tatum, E.L. 1946. Gene recombination in *E. coli*. *Nature* (London), **158**: 558.
Smith, G.R. 1991. Conjugational recombination in *E. coli*: myths and mechanisms. *Cell*, **64**: 19–27.

Chapter 11

Transformation

Chapters 8 (Transduction) and 10 (Conjugation) describe ways in which DNA can be exchanged among bacteria. Transformation represents a third way. **Transformation** is the process that allows bacteria to take up free or naked DNA from their surrounding environment. Bacteria that have undergone transformation are called **transformants**. Transformation does not require cell-to-cell contact like conjugation nor does it require a bacteriophage-like transduction. Transformation was discovered by Fred Griffith in 1928; he called the unknown agent responsible for transformation, the transforming principle. Sixteen years later, Avery, Macleod, and McCarty determined that the transforming principle is naked DNA.

Transformation was the first form of genetic exchange to be identified in bacteria. It requires that a bacterium can become naturally **competent** to take up DNA from its environment or that, through a series of physical manipulations, it can be made artificially competent to take up DNA. Some of the bacteria that can become naturally competent are shown in Table 11.1. For those bacterial, plant, or animal cells unable to become naturally competent, modern molecular biological techniques have developed artificial means to transform DNA into these cells. As we will learn at the end of this chapter, artificial transformation methods have fueled the progress in biotech-

FYI 11.1

The transforming principle

In 1928 Frederick Griffith was studying a bacterial-induced pneumonia caused by *Streptococcus pneumoniae*. Virulent or pathogenic strains and avirulent or nonpathogenic strains of *S. pneumoniae* have very distinctive phenotypes. The virulent strain, which is called the S-strain, has a polysaccharide capsule that is required for infection. The presence of the capsule leads to a smooth colony morphology when the cells are grown on solid media. This capsule protects the bacterium from being destroyed by our immune system. Specifically, the bacterial capsule prevents human macrophages from eating the virulent bacterium. The avirulent strain of *S. pneumoniae*, called the R-strain, does not have the smooth polysaccharide capsule. R-strains cannot infect humans and do not cause disease. R-strains produce colonies that phenotypically appear rough on solid media. If mice are injected with a live S-strain, they die within a few days of pneumonia (positive experimental control), whereas mice injected with a live R-strain continue to live (negative experimental control). If an S-strain is treated with heat and then injected in to mice, the mice continue to live. This indicates that the virulent S-strain can be killed by heat (negative experimental control). Griffith co-injected into the same mouse a heat-killed S-strain and a live R-strain and observed that the mouse died (the experiment). The *S. pneumoniae* isolated from the dead mouse had a smooth phenotype. This result suggested that the *S. pneumoniae* phenotype was transformed from rough to smooth and that the change was accompanied by transformation of an avirulent to virulent phenotype. Griffith put forth the hypothesis that something of a chemical nature was passing from the heat-killed S-strain to the live R-strain. He called this something the "Transforming Principle".

nology, such that many types of cells are capable of being transformed with naked DNA.

Natural competency

When do bacterial cells become competent to take up DNA? The physiological state of the cell influences its ability to become competent. For some bacteria, establishment of the competent state is based on cell density. Competency is induced when the number of cells in a given volume increases above a certain level. For other bacteria, competency is established by the impact of shifting cells from optimal growth conditions to poor growth conditions.

In the Gram-positive, spore-forming bacterium, *Bacillus subtilis*, establishment of natural competency occurs when the cells reach a high density towards the end of **exponential growth**. At this point, vital nutrients such as carbon, nitrogen, or phosphorous have been depleted. It is during this period that two separate pathways can activate the expression of a specific set of genes whose products are needed for the es-

FYI 11.2

The chemical nature of the transforming principle

What was the chemical nature of the transforming principle? Was it the capsular polysaccharide? Was it a protein? Lipids? DNA? It wasn't until 16 years after Griffith's reported observations that Oswald Avery, Colin McCleod, and Maclyn McCarty, reported that DNA from a heat-killed *S. pneumoniae* S-strain caused the transformation of a nonvirulent R-strain into a virulent S-strain. This was the first demonstration that DNA transmitted genetic information and was the hereditary material. Yet Avery's report was met with skepticism. In the words of W.M. Stanley, "(there were) some aspects of the scientific climate existing at the time which probably were responsible for the fact that the discovery was really undiscovered, generally unappreciated, and failed, for several years, to affect the trend of science." Why didn't the scientific community acknowledge this discovery for its obvious impact and profoundness?

Avery, one of the founders of the field of immunochemistry, and his coworkers used an experimental approach that biochemically separated and purified the chemical components (capsule, lipid, protein, DNA) of the heat-killed S-strain. Each purified S-strain component was mixed with a live intact R-strain and injected into a mouse. The capsule, lipids, or protein of the heat-killed S-strain did not transform the R-strain into a virulent S-strain, and the mouse lived. Only when the DNA of the heat-killed S-strain was mixed with the live R-strain, did R-strain transform into a virulent S-strain. Their critics would argue that protein was the transforming principle because protein is found bound to DNA and they may not have made a preparation of DNA that was completely protein free. Avery and coworkers were as meticulous as possible in generating their proof that the transforming material was DNA. They used many different techniques to make DNA preparations and determine the purity of the DNA. They also demonstrated that the transforming material could be inactivated by DNases and not proteases. This result should have convinced even their harshest critics, because DNases degrade only DNA whereas proteases degrade protein. If the transforming material was preferentially inactivated by DNases and not proteases, then the transforming material had to be DNA.

Even with the results reported by Avery and coworkers, the hypothesis persisted that it was proteins and not DNA that played the prevailing role in the transmission of genetic properties. At the time of Avery's report, the scientific community, including Avery and his coworkers, were strongly influenced by the "Levene school of thought." Levene and coworkers considered DNA's structure to be a tetranucleotide containing repeating units of the bases adenine, guanine, cytosine, and thymine. Because of DNA's apparent chemical and structural simplicity, and because of the apparent chemical and enzymatic complexity and diversity displayed by proteins, no attempt was made to relate biological activity or genetic information to DNA. The research focus was entirely on proteins. It was generally assumed and vigorously argued that DNA did not carry biological information and that protein was responsible for genetic transmission. Apparently, even Avery and his coworkers were astonished to learn from their experiments that DNA was the transforming material. Another potential factor had nothing to do with hypotheses and experimental data, but quite possibly had every thing to do with personality and timing. At the time of his discovery, Dr. Avery was close to retirement, and since World War II was intensifying, Dr. Avery's younger coworkers had to turn their attention to matters related to the war. Dr. Avery was known as a friendly, yet cautious and timid man. He apparently did not have any desire to argue the merits of his research in front of the scientific community. Thus, no one was available to promote this research, argue its significance and present defenses to vocal critics. As a result, many years passed before the significance and impact of Avery's work was realized, and it was accepted that DNA is the hereditary material. The experiments of Avery, McCleod, and McCarty are one of the most important discoveries in biology that never received a Nobel Prize.

Table 11.1 The transformable state of some bacterial species.

Some naturally transformable Gram-negative bacteria	
Acinetobacter calcoaceticus	Member of the Neisseriaceae family, generally do not cause disease
Azotobacter agilis	Free living bacteria
Butyrivibrio	Cellulose digesting bacteria that live in the rumen
Campylobacter	Found worldwide in the soil and intestinal tracts of wild and domestic animals; causes infectious diarrhea
Haemophilus influenzae	Causative agent of pneumonia, meningitis, and other acute infections
Haemophilus parainfluenzae	Normal flora of the mouth of some individuals (carriers); opportunistic infection of bloodstream
Helicobacter pylori	Causative agent of gastric ulcers
Moraxella	Some species cause upper respiratory infections in humans
Neisseria meningitides	Causative agent of meningococcal meningitis
Neisseria gonorrhoeae	Causative agent of gonorrhea
Pseudomonas stutzeri	Widely distributed soil bacterium
Synechococcus	Grows at very high temperatures
Some naturally transformable Gram-positive bacteria	
Bacillus subtilis	Normal soil bacterium
Bacillus cereus	Normal soil bacterium frequently mistaken for *B. anthracis*; associated with emesis- and diarrheal-related food borne gastro-enteritis
Bacillus stearothermophilus	Grows at very high temperatures
Bacillus anthracis	Causative agent of anthrax
Clostridium perfringens	Commonly found in soil and animal intestinal tracts, produces toxins that cause muscle necrosis in gas gangrene
Clostridium botulinum	Causative agent of botulism poisoning
Micrococcus	Normal flora of human skin
Streptococcus pneumoniae	Causative agent of pneumonia; normal flora of nose and throat for some individuals who are classified as carriers
Some naturally transformable acid-fast bacteria	
Mycobacterium	Specific *Mycobacterium* species are the causative agents of tuberculosis, leprosy, and other diseases; some species do not cause disease in humans
Streptomyces	Normally found in soil and water
Not naturally transformable laboratory workhorses	
Escherichia coli	Soil bacterium; minor component of intestinal tract
Salmonella enterica serovar *typhimurium*	Soil bacterium; minor component of the intestinal tract

tablishment of competency (Fig. 11.1). These genes are known as the *com* genes. Both pathways for inducing the expression of the *com* genes sense the cell density. One of these pathways is also used to induce a second developmental program, sporulation. The induction of competency in *B. subtilis* elegantly demonstrates how several seemingly unrelated processes are integrated to provide the most benefits to the bacterium. For this reason, we will consider induction of competency in *B. subtilis* in detail.

One pathway in *B. subtilis* that senses the density of cells uses a small peptide or **competence factor** called ComX that is excreted by actively growing cells (Fig. 11.2). ComX is synthesized in the cytoplasm as a large precursor molecule. The precursor molecule is cleaved and modified by the ComQ protein. The mature ComX is transported out of the cell. The concentration of ComX in the environment only reaches high levels when cells are at a high density (Fig. 11.3).

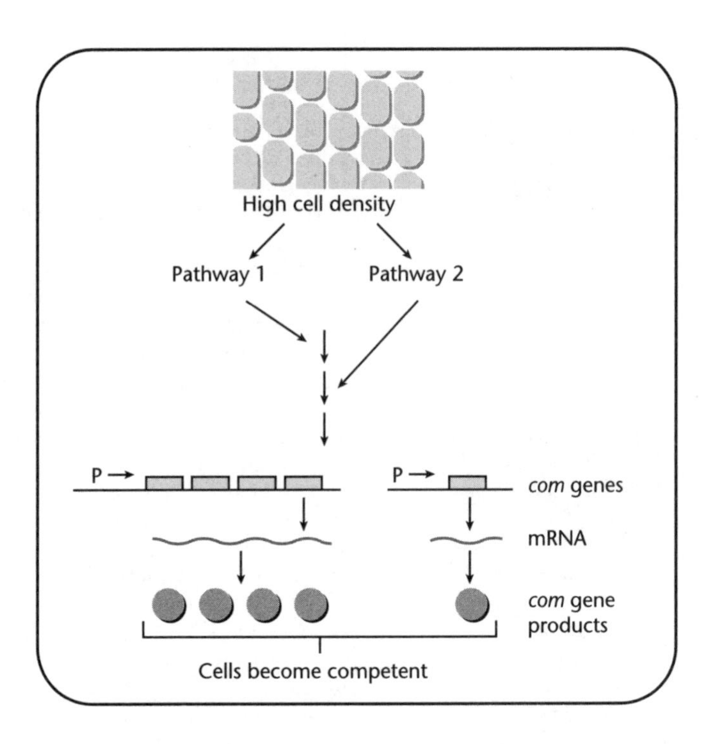

Fig. 11.1 Two pathways activate the competence genes in *B. subtilis*. Pathway 1 senses the cell density and when enough cells are present, signals the cells to induce the *com* genes. Pathway 2 monitors the available nutrients and when they become limiting, signals the cells to induce the *com* genes.

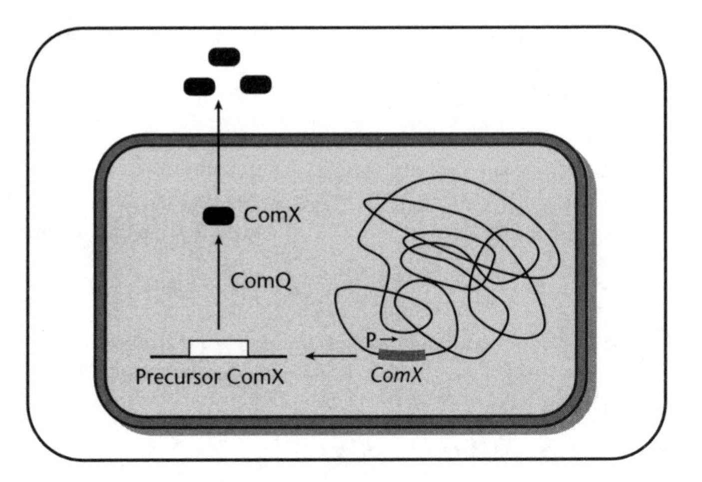

Fig. 11.2 One pathway used by *B. subtilis* relies on a small peptide, ComX, to sense cell density and induce the *com* genes. ComX is produced in the cytoplasm as a larger precursor molecule. The ComQ protein processes and modifies precursor ComX to make a smaller active peptide. The active ComX is excreted from the cell.

The high levels of ComX in the environment are sensed by way of a molecular communication system referred to as a **two-component signal transduction system** (Fig. 11.4 and see Chapter 12). A two-component signal transduction system consists of at least two proteins, one is called the **sensor-kinase protein** and the other is called the **response-regulator protein**. The sensor-kinase spans the membrane, with a portion of the protein exposed to the outer environment and another portion of the protein exposed to the cytoplasm. The externally oriented portion of the sensor-kinase senses signals in the environment and relays this information to the response-regulator. The **kinase** portion of the protein adds phosphates to a specific place on a molecule or substrate. The relay is carried out by the kinase part of the sensor-kinase through a series of **phosphorylation** (adding phosphates) and **dephosphorylation** (removing phosphates) steps. First, in response to the environ-

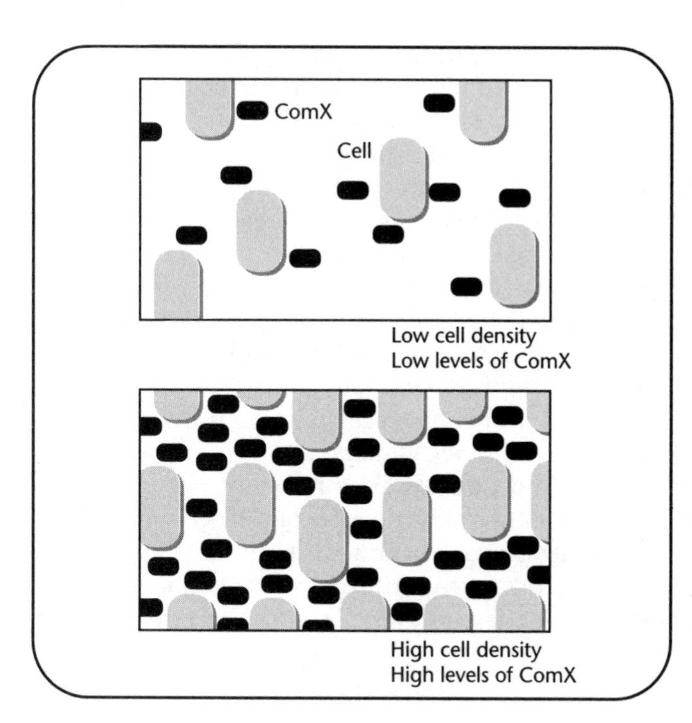

Fig. 11.3 ComX is produced by all of the cells in the population. At a low cell density, low amounts of ComX are present in the population. At high cell densities, high levels of ComX are present in the population. It is the change in the concentration of ComX that the cells respond to.

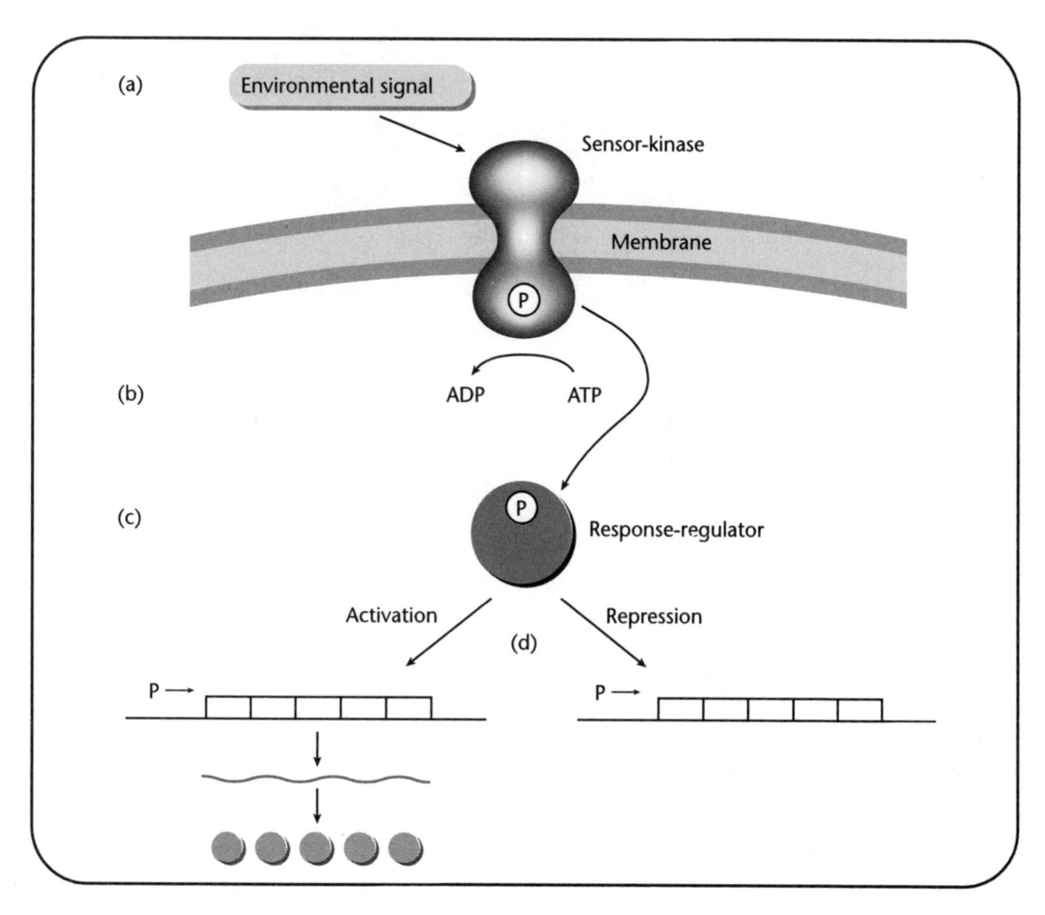

Fig. 11.4 Two-component signal transduction systems use a membrane protein called a sensor-kinase to sense an environmental signal (a). The sensor-kinase autophosphorylates on a specific histidine residue in the protein (b). The sensor-kinase transfers this phosphate to the response-regulator (c). This activates the response-regulator so that it can induce and/or repress specific genes (d). These genes allow a cell to respond to the original environmental signal in an appropriate way.

mental signal the sensor-kinase phosphorylates itself (**autophosphorylation**) on a specific histidine that is conserved among all sensor-kinase proteins (Fig. 11.4a,b). This phosphate is then transferred by the sensor-kinase to an aspartate in the response-regulator protein (Fig. 11.4c). Again, this aspartate is conserved among all response-regulator proteins. The phosphorylated response-regulator is able to either turn on or turn off genes (Fig. 11.4d). The regulation of these genes allows the cells to respond to the specific change in the environment.

The *B. subtilis* two-component signal transduction system that senses the levels of ComX is composed of ComP and ComA (Fig. 11.5). The sensor-kinase is ComP and the response-regulator is ComA. When the levels of ComX are high enough, ComX binds to ComP, causing ComP to autophosphorylate. The phosphate is then transferred from ComP to ComA. Once ComA is phosphorylated, it functions as a regulatory protein to activate the transcription of *comS*, which leads to expression of the competence genes. Because the ComP and ComA system responds to the levels of ComX and the levels of ComX are dependent on the number of cells present, ComX, ComP, and ComA are also known as a **quorum sensing system**. Quorum sensing systems detect a change in cell density and use this information to influence which genes are expressed and ultimately what behaviors the cell is capable of carrying out.

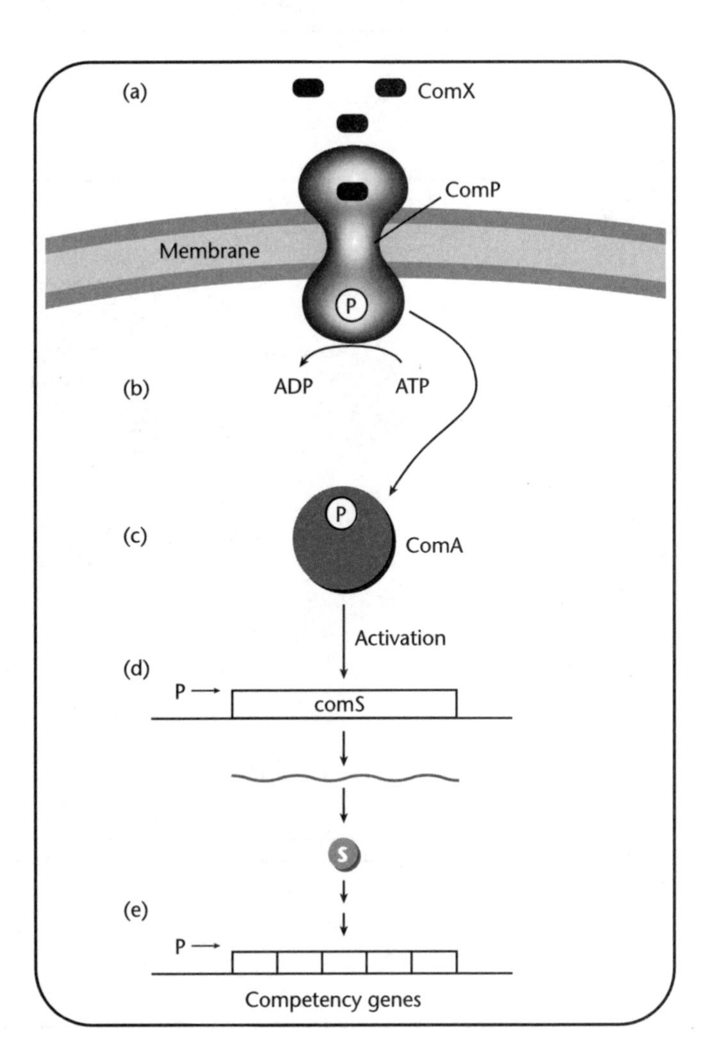

Fig. 11.5 ComP and ComA comprise a two-component signal transduction system in *B. subtilis* for the induction of the competency genes. (a) ComX binds to ComP. (b) This binding induces autophosphorylation of ComP. (c) The phosphate is relayed to ComA. (d) The phosphorylated ComA activates the *comS* gene. (e) ComS leads to induction of the competency genes.

Quorum sensing is used by many different bacteria to control such diverse processes as virulence, conjugation, antibiotic production, sporulation, and motility.

The second pathway for induction of the *com* genes also affects a second developmental process, the formation of spores. For *B. subtilis* and a few other bacteria (i.e. *Clostridium*), high cell density, depletion of nutrients, and other signals activate **sporulation** (Fig. 11.6). Sporulation begins when normal, vegetative bacterial cell growth is arrested and continues as the bacterial cell undergoes progressive changes that result in the production of a spore. A **spore** contains a chromosome completely surrounded by a series of protective layers. A spore is the dormant form of a bacterium; it does not actively grow or metabolize nutrients. The spore can survive many harsh conditions that a bacterial cell cannot, including high levels of radiation, lack of food, lack of water, high temperatures, and increased levels of toxic chemicals. Spores can survive for decades and even centuries in some cases. Once a spore finds itself in a

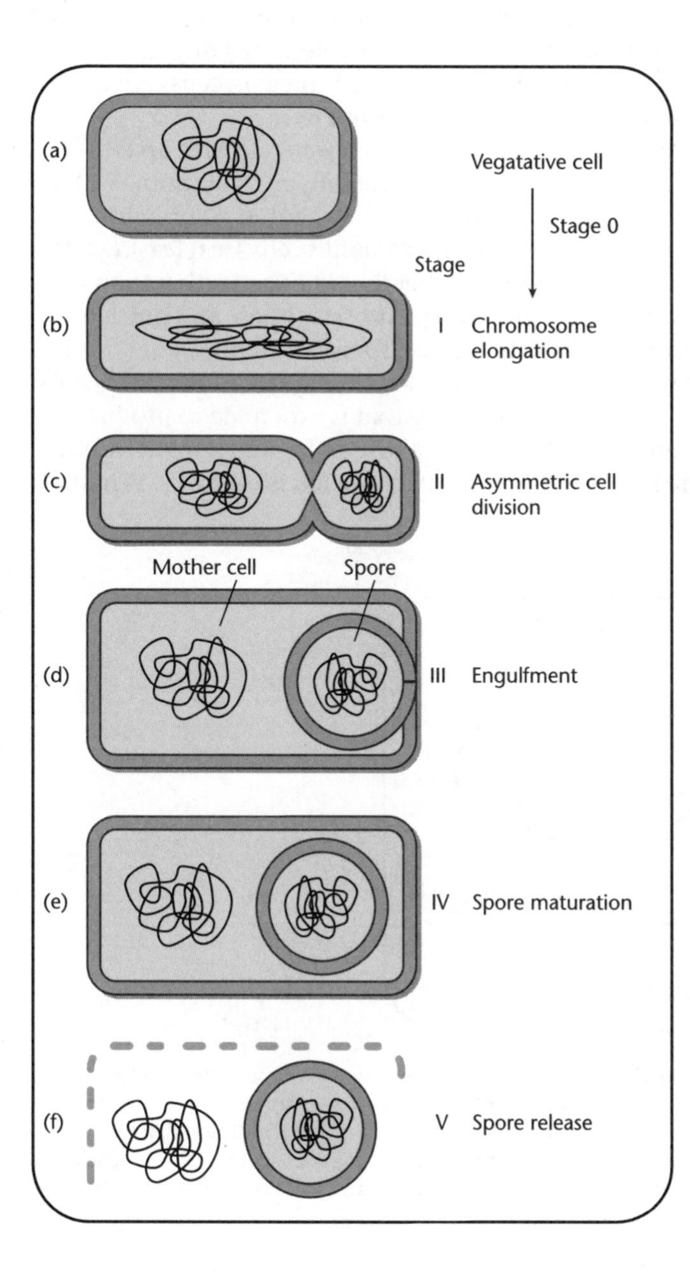

Fig. 11.6 Specific bacteria can undergo a developmental program called sporulation. (a) Actively growing and dividing bacterial cells are known as vegetative cells. The transition from vegetative growth to recognizable morphology changes associated with sporulation is known as Stage 0. (b) Stage I is defined by the elongation of the chromosome. (c) During Stage II, the cells divide asymmetrically to produce a polar septum. One chromosome is left in the mother cell and the other is moved into the spore. (d) During Stage III, the polar septum bulges out to encompasses the spore in a process called engulfment. (e) The spore is matured by addition of protective layers to surround the dormant chromosome (Stage IV). (f) The mother cell lyses and releases the spore into the environment (Stage V).

nutrient-rich environment, it can germinate back into an actively growing bacterium.

Activation of competency and sporulation occurs through a second pathway that bypasses ComX and ComP but feeds into ComA (Fig. 11.7). A second competence peptide, CSF (competence and sporulation factor), interacts with a second membrane protein called Spo0K. Spo0K is an oligopeptide permease or a hole through the membrane allowing small peptides to pass from the outside of the cell to the inside of the cell. CSF passes through Spo0K and into the cytoplasm.

The cytoplasmic levels of CSF tell the cell when to activate competency and when to activate sporulation. In fact, when CSF activates sporulation it actually represses competency. How does CSF carry out its dual roles? At low cytoplasmic levels, CSF inhibits a protein called RapC. Normally RapC removes the phosphate from ComA and keeps it from activating *comS* (Fig. 11.8a). The more RapC that is inactivated, the more phosphorylated ComA is available for *comS* activation and competency is induced (Fig. 11.8b). At high cytoplasmic levels, CSF inhibits competency by directly inhibiting ComS (Fig. 11.8c). High cytoplasmic CSF levels also inhibit a protein called RapB and this leads to production of the genes needed for sporulation.

What these overlapping pathways tell us is that the cell wants to take up DNA but only at specific times. Late in exponential growth as the cells enter stationary phase some cells will lyse and release their DNA into the environment. It is thought that if cells become competent at the same time that some *B. subtilis* cells are releasing DNA, they will be more likely to take up DNA from another *B. subtilis* cell rather than DNA from an unrelated species. As the cells progress into stationary phase and nutrient depletion worsens, the cells become committed to the sporulation program and shut down competency. The reason for turning off the competency genes during sporulation is unknown. Perhaps it is the amount of energy that is expended to produce the *com* gene products that is the problem. Or perhaps taking up foreign DNA at the same time as a dormant form of the bacterium is being constructed is too risky. Whatever

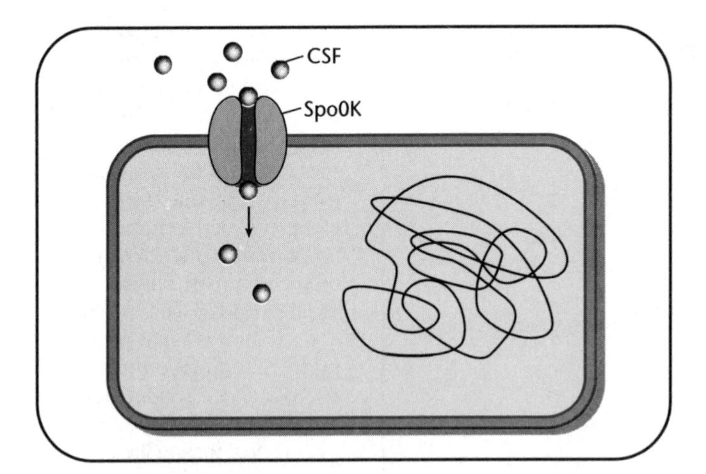

Fig. 11.7 The second pathway to induce the competency genes in *B. subtilis* uses another small peptide called CSF for competence and sporulation factor. CSF, unlike ComX is transported into the cell via a membrane protein called Spo0K. The "zero" in front of the K in the name Spo0K indicates that this protein is found in the cells during Stage 0, as defined in the caption to Fig. 11.6. Because there are so many genes and gene products required for sporulation, the *spo* genes and proteins are named this way to help keep track of when they are expressed.

the reason, the cells keep very tight control over what genes are expressed under what conditions to maximize their abilities as needed.

Even though *B. subtilis* appears to have two pathways to activate the competent state, only 20% of the cells can become competent. These cells, however, can stay competent for several hours. Establishment of competency in the Gram-positive, non-spore forming bacterium, *Streptococcus pneumoniae*, is also dependent on high cell density and a critical concentration of a competence factor, much like *B. subtilis*. However, in contrast to *B. subtilis*, 100% of *Streptococcus pneumoniae* cells can become competent, yet they only stay competent for a very short period of time.

Establishment of competency in the Gram-negative bacterium, *Haemophilus influenzae*, appears to be dependent on the growth media. A competence factor, such as ComX, has not been identified. *H. influenzae* can achieve a low level of competency when cells enter the late exponential phase of growth. However, if cells are transferred from normal growth media to a nutritionally deficient media (a nutritional downshift), or if cells are exposed to transient anaerobic conditions, then a high level of competency is achieved. *H. influenzae* presents an example of media dependence rather than cell density dependence in the establishment of high-level competency.

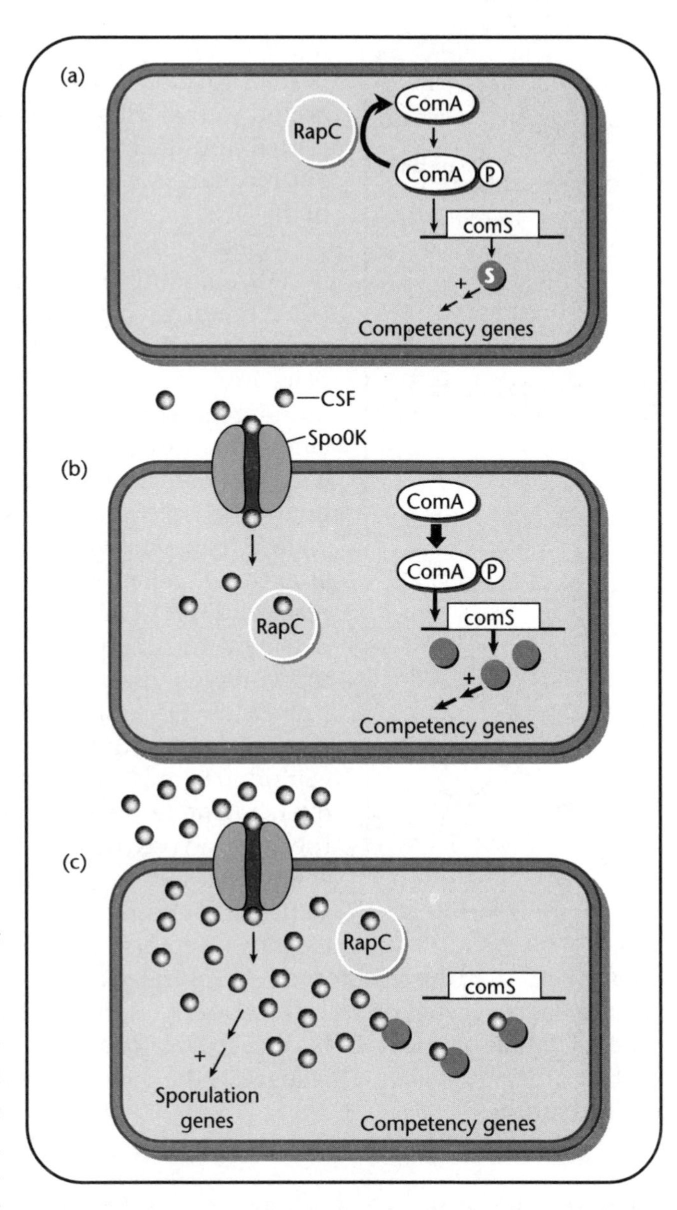

Fig. 11.8 Regulation of the competency genes and the sporulation genes of *B. subtilis* by CSF. (a) Phosphorylation of ComA is needed to induce synthesis of the competency genes as described in the text. The RapC protein is a phosphatase (phosphate-removing enzyme) that removes the phosphate from ComA and prevents it from activating *comS*, thus preventing the activation of the competency genes. (b) At low intracellular levels of CSF, CSF binds to RapC and inhibits it. This allows the ComA to stay phosphorylated and activate *comS* and the competency genes. (c) At high intracellular levels of CSF, CSF binds directly to the ComS protein and inactivates it, preventing the expression of the competency genes. These same high levels of CSF activate the sporulation genes.

The process of natural transformation

So far we have considered when a cell induces the competency genes. But what functions do the *com* gene products carry out? The first step in transformation is the binding of free DNA to the cell (Fig. 11.9). For the Gram-positive and Gram-negative bacteria that can be naturally transformed, free DNA first appears to

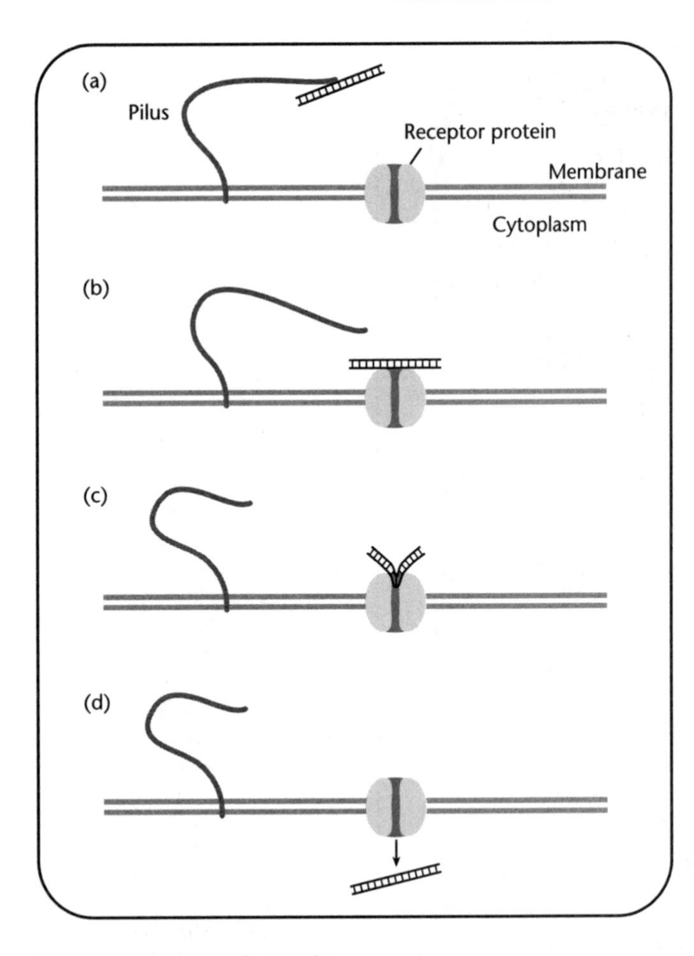

Fig. 11.9 The steps of natural transformation. (a) Double-stranded DNA binds to the pilus. (b) The DNA is transferred to a membrane receptor. The initial binding is reversible. (c) The DNA becomes irreversibly bound to the receptor. (d) The DNA is transported into the cytoplasm.

make contact with the bacterial cell surface by a **pilus** structure. The free DNA must then be irreversibly bound to the cell and, subsequently, the DNA is transported into the cytoplasm. Different bacteria use different mechanisms for irreversible binding and transport of the DNA.

B. subtilis and *S. pneumoniae* use similar mechanisms for DNA binding and transport. Proteins that bind DNA (DNA binding receptors) are encoded by some of the *com* genes. These DNA binding receptors are assembled and placed on the cell surface. It is thought that the DNA binding receptors pluck the double-stranded DNA molecules from the pilus. The DNA bound to the receptor undergoes a conversion from loosely bound to irreversibly bound. Loosely bound DNA can be removed from the cell surface by a phenol-detergent treatment. A short time after the DNA has bound to the receptor, a change takes place and the DNA becomes impervious to the phenol-detergent treatment. What happens to the DNA–receptor complex during the change from reversible to irreversible DNA binding is not fully understood. Any piece of double-stranded DNA can be bound by these receptors. There is no requirement for a specific base sequence. One consequence of this lack of specificity is that the DNA can originate from any source, including another bacterial species, plasmids, or bacteriophage chromosomes. It has been estimated that approximately 50 DNA binding receptors can be found on the surface of each *B. subtilis* cell. All of the receptors can be saturated or bind DNA at the same time. *S. pneumoniae* has fewer sites on its cell surface that can bind DNA. Approximately 10 molecules of double-stranded DNA of 15 to 20 kilobase pairs in length can bind to the cell surface of *S. pneumoniae*. This suggests the presence of 10 DNA binding receptors on the cell surface.

The machinery of naturally transformable cells

In *B. subtilis*, the pili that first bind the DNA are only one of the many different pili found on the cell surface and are called type IV pili (Fig. 11.10a). The DNA binding receptor is the ComEA protein. It prefers to bind double-stranded DNA but has no sequence specificity (Fig. 11.10b). The ComG proteins are involved in making the ComEA receptor accessible to DNA. DNA bound by the ComEA receptor is susceptible to degradation by endonucleases that cut the bound DNA into smaller pieces (Fig. 11.10c). Once the DNA is irreversibly bound by the ComEA receptor, a DNA entry nuclease degrades one of the bound DNA strands. The ComI protein is a membrane-bound nuclease used for this purpose. Either strand of the bound DNA molecule can be degraded (Fig. 11.10d). The remaining DNA strand is transported into the cell (Fig. 11.10e). The ComEA receptor and the ComEC protein are required for DNA transport. Once inside the cell, the single-stranded DNA is protected by a *com* encoded DNA binding protein. This DNA–protein complex can undergo homologous recombination with the chromosome of the recipient cell if homology exists and if the cells are *recA*⁺. A similar transformation machinery with similar proteins exists for *S. pneumoniae*.

Interestingly, a gene sequence similar to the *B. subtilis comEA* receptor gene has been identified in chromosomes of the Gram-positive bacterium, *S. pneumoniae*, but not in the chromosomes of the Gram-negative bacteria, *H. influenzae*, *Neisseria meningitidis*, and *Helicobacter pylori*. A gene sequence similar to the *B. subtilis comEC* gene, which encodes a DNA transport protein, has been identified in both Gram-negative and Gram-positive bacteria including *S. pneumoniae*, *H. influenzae*, *Neisseria meningitidis*, and *Neisseria gonorrhoeae*, but not in *Helicobacter pylori*. This comparative sequence information suggests there may be differences in the binding of DNA to the receptor but similarities in how bound DNA is transported into the cell.

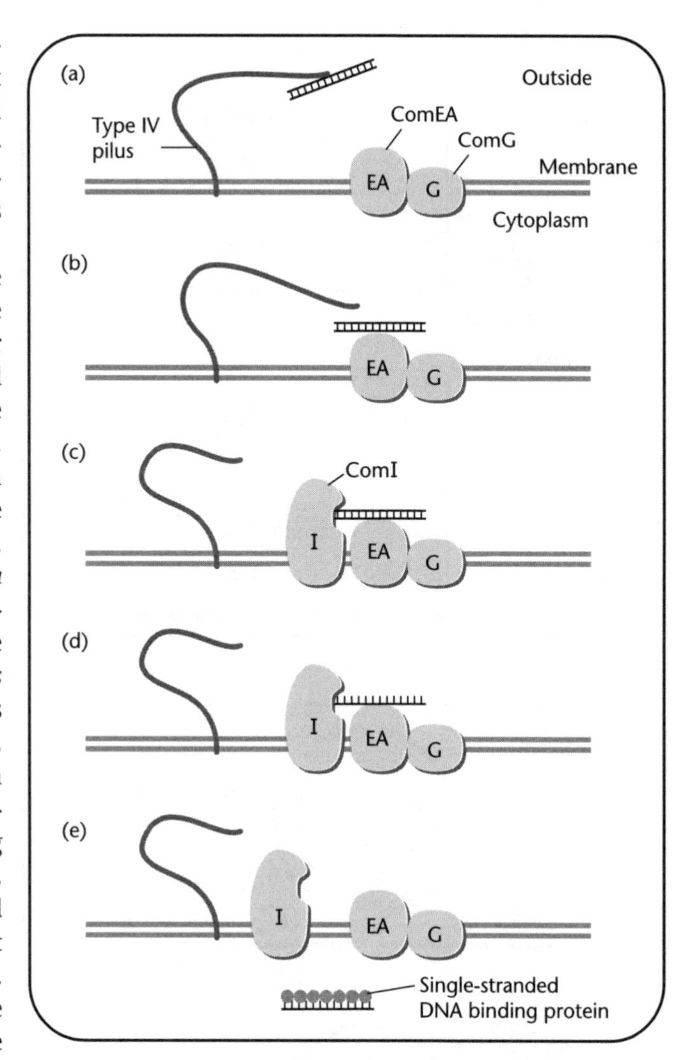

Fig. 11.10 Transfomation in *B. subtilis*. (a) A type IV pilus binds the double-stranded DNA first. (b) The DNA is transferred to the ComEA receptor. ComG helps make ComEA accessible to the DNA. (c) An endonuclease like ComI recognizes the receptor with bound DNA. (d) ComI degrades one strand of the DNA. (e) The single-stranded DNA is transported into the cell and protected by a single-stranded DNA binding protein.

The suggestion from the comparative sequence analysis is supported by the data that *B. subtilis* and *S. pneumoniae* DNA binding receptors can bind any sequence of DNA, yet *H. influenzae* and *N. meningitidis* can only take up DNA that contains a specific base pair sequence (Fig. 11.11). This specific sequence is called an **uptake signal sequence** (USS) and must be present in the naked DNA in order for it to be bound and transported into the cell. For *H. influenzae*, the core of the USS contains the sequence 5′AAGTGCGGT3′. The core USS is present 1471 times in the *H. influenzae* chromosome. For *N. meningitidis*, the USS is 5′GCCGTCTGAA3′, and this sequence is present 1891 times in the *N. meningitidis* chromosome. The USS are often found as part of the base pair sequences located at the end of a gene that are used as transcription stop signals. The strategic location of the USS may increase the probability that the DNA taken up contains an intact gene that could benefit the cell. The second benefit to using an USS is that the bacterium will only take up DNA from its own species or a very closely related species. In contrast to *H. influenzae* and *N. meningitidis*, another naturally transformable Gram-negative bacterium, *H. pylori* does not contain uptake signal sequences in its chromosome. The lack of a USS presents the interesting situa-

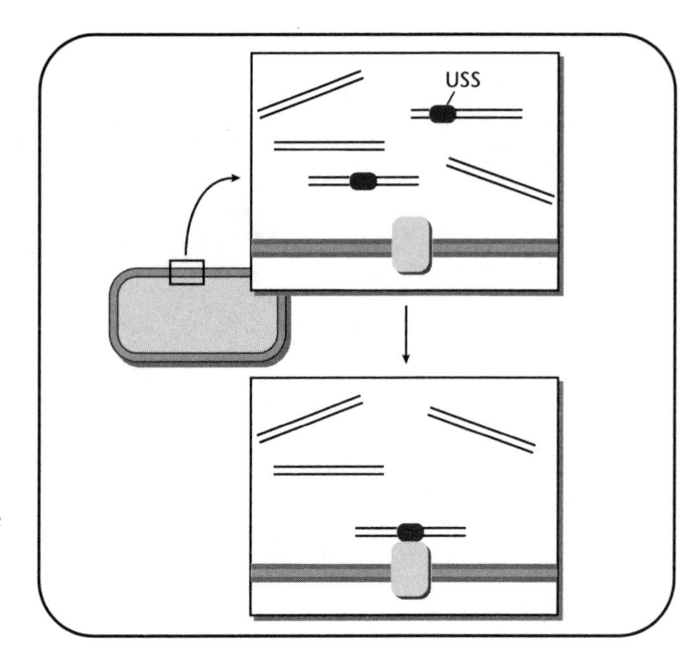

Fig. 11.11 Some species that are naturally competent can only take up DNA that contains a specific base sequence called a USS or uptake signal sequence.

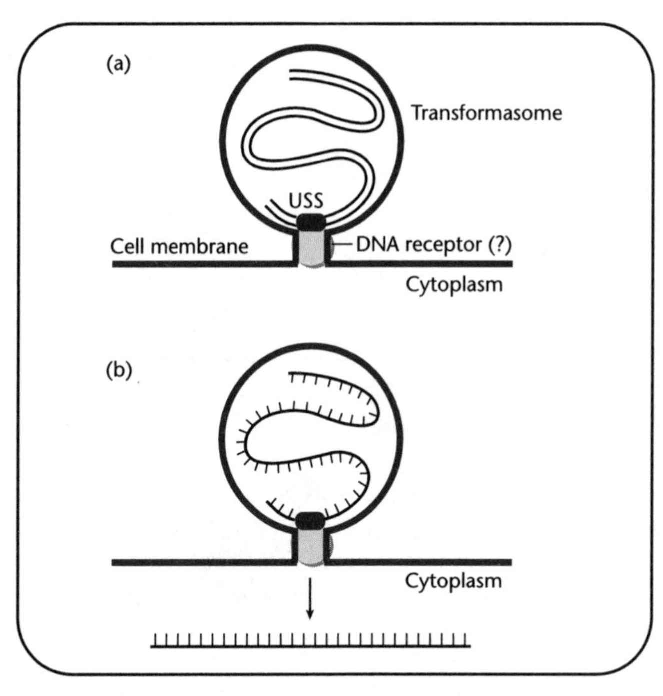

Fig. 11.12 Some species use a specialized structure called a transformasome to import DNA. The DNA must contain a USS and initially is double stranded. During the process of translocation, the DNA is degraded to a single-stranded molecule.

tion that similar to *B. subtilis* and *S. pneumoniae*, *H. pylori* might be able to uptake DNA from any species.

Additional differences in the molecular mechanisms of natural transformation among bacterial species are found. For example, the first step in *H. influenzae* is to take up USS-containing DNA into a membrane-enclosed compartment, referred to as a **transformasome** (Fig. 11.12). Within the transformasome, a 30–50 kilobase pair double-stranded DNA fragment containing a USS sequence can be retained. It is unclear if a true DNA binding receptor exists within the transformasome, however, it is known that a *B. subtilis* like *comEA* receptor gene sequence does not exist in *H. influenzae*. Once the double-stranded DNA is in the transformasome, it is slowly translocated from the outside of the *H. influenzae* cell into its cytoplasm. During this translocation process, one strand is completely degraded, while the other strand remains mostly intact. It is the intact strand that will be available for homologous recombination with the chromosome.

Where does free DNA come from? For *B. subtilis*, chromosomal DNA is released into the environment during sporulation. To release a newly made spore, the bacterial cell must lyse, thus releasing its chromosomal DNA. A bacterial cell might have an unfortunate encounter with a lytic bacteriophage (see Chapter 7), thus liberating bacterial chromosomal DNA into the environment when the phage lyses the bacterial cell. Alternatively, a bacterial cell might become an unwilling donor of free DNA if it were to encounter hostile environmental conditions that cause cell lysis, such as inappropriate salt or pH concentrations, harmful chemicals, or a very aggressive immune system.

What might be the biological rational for the natural transformation process? Desperate times call for desperate measures. Cells able to take up DNA and use it to modify their genetic information are at a definite advantage, especially when environmental conditions are not optimal. It is easy to imagine hostile environments, where the bacterium can no longer find appropriate nutrients, and might have to adjust its enzymatic repertoire, by acquiring new gene sequences. It is possible that the bacterium uses the carbon and nitrogen atoms of the transformed DNA molecule as its actual source of nutrients. If a bacterium sustains a large amount of DNA damage, importing free DNA could allow the bacterium to repair its chromosome and survive. The free DNA can be used to fill in the gaps in the damaged chromosome through a process such as recombinational repair (see Chapter 4).

Artificial transformation

Many bacterial species, especially those exploited for biotechnology purposes, are not naturally transformable. This is because they do not contain genes that encode products needed in the uptake and transport of DNA. Yet both naturally transformable and non-transformable bacteria can be rendered transformable if they are exposed to a chemical or physical agent that perturbs their membrane.

Many bacteria can be made competent by exposure to a divalent or multivalent cation, such as calcium chloride, manganese chloride, or a rubidium chloride/calcium chloride mix. The cation exposure must take place at a very cold temperature (0°C) and is followed by the addition of free DNA and a heat shock at 42°C. This approach results in up to one out of 100 cells becoming capable of taking up DNA (~1% efficiency). It is the combination of cations and low temperature, followed by heat shock that encourages DNA association with the cell and transport of DNA into the cell.

FYI 11.3

Transformation as a pathogenic problem

The discovery of the transformation process and the role DNA plays in this process was due to research efforts focused on pathogenic bacteria. *Streptococcus pneumoniae* causes pneumonia. Griffith and his colleagues were analyzing those features of *S. pneumoniae* responsible for its pathogenic behavior (FYI11.1). In 1928, when Griffith reported his transformation observations, antibiotics, such as penicillin, had not yet been perfected for human consumption. The impact of bacterial-induced pneumonia on society was enormous during this time. People died of simple bacterial infections because of the lack of antibiotics. Research efforts in the United States concentrated on understanding how pathogenic bacteria caused disease in humans and what could be done to prevent the bacteria's devastating effects. With the evolution of effective purification procedures for antibiotics and the discoveries of antibiotics less toxic to humans, society was lulled into a sense of false security. An assumption was made that people would no longer die from simple bacterial infections because the antibiotic arsenal available to kill bacteria was effective and extensive. What was not predicted was that bacteria would develop resistance to antibiotics and that this resistance could be transferred to other bacteria by genetic mechanisms such as transformation, transduction, or conjugation. Today, transformation as a pathogenic problem casts a different light on this intriguing mechanism. Naturally transformable bacteria such as *Bacillus subtilis*, pathogenic members of the *Bacilli* family (i.e. *B. anthracis*), *Streptococcus pneumoniae*, or *Helicobacter pylori* have the potential to pick up genes encoding antibiotic resistances because they are capable of being transformed with any sequence of DNA. In contrast, naturally transformable pathogenic bacteria such as *Haemophilus influenzae*, *H. parainfluenzae*, or *Neisseria meningitidis* have very specific sequence requirements for the DNA they can take up. It could be predicted that the likelihood of these pathogenic bacteria picking up genes encoding antibiotic resistances or virulence factors would be very small given the assumption that they can only take up DNA from their own species. Recently, however, interspecies exchange of DNA has been demonstrated between *Haemophilus* and *Neisseria*, suggesting that these unrelated bacteria can exchange DNA sequences through transformation. Not only does intergeneric transfer have implications for antibiotic resistance development, but also may present a way for *Neisseria* genes encoding deadly toxins or other virulence factors to be transferred to *Haemophilus* thus creating a *Haemophilus* transformant that is highly invasive, toxin producing, and very deadly.

Electroporation, a process that subjects bacteria to a brief pulse of high voltage electricity (~1–5 kilovolts), has been found to be effective in transforming both prokaryotic and eukaryotic cells. It is thought that the high jolt of electricity creates temporary pores in the cell's membrane so macromolecules, such as DNA or proteins, gain entry. Electroporation is very efficient at creating cells capable of taking up DNA. This efficiency is thought to approach 10%.

Transformation as a genetic tool: gene mapping

Mapping genes requires that a piece of chromosomal DNA from one bacterium be introduced to another bacterium of the same species. Transduction (Chapter 8), conjugation (Chapter 10), and transformation can be used to introduce DNA from one cell to another. Once the DNA has been introduced into the cell, the principles of mapping are the same. Transformation has most often been used for mapping in those species where natural competency occurs at a high frequency.

Choosing a method to introduce DNA into a cell depends on the bacterial species being studied. Natural transformation, conjugation, and transduction all require more than just the bacterium. Natural transformation needs the cell-encoded competency machinery, conjugation requires an F or R factor, and transduction requires a transducing phage. Depending on what is available for the bacterium in question limits the choices. If multiple choices are available, the next consideration is to choose the method that is the most efficient at getting DNA into a cell. In the case of *E. coli*, transduction and conjugation are both very efficient at introducing DNA. In contrast, for naturally transformable bacteria, such as *B. subtilis*, transformation is very efficient at introducing DNA. Choosing a method to introduce DNA into a cell for gene mapping comes down to what is available for the bacterial species being used and how finely the gene location needs to be determined.

The rules for determining gene linkage are essentially the same whether one uses transformation, conjugation, or transduction to introduce the DNA. If two genes are inherited together 98% of the time, then the genes must be physically located very close together on the chromosome. If they are only inherited together 10% of the time, then the genes are physically located much further apart. Thus, the frequency of co-inheritance gives a relative measure of the distance between two genes.

Transformation as a molecular tool

As we have seen, not all bacteria are naturally transformable, however, many bacterial species can be transformed with DNA by artificial treatments. This apparently simple feat fueled the biotechnology revolution, making it possible to introduce new genes into many different bacteria. However, there are many factors that can affect the efficiency of artificial transformation, and at times, when one encounters difficulty in a particular DNA transformation protocol, it would seem that transformation is not such a simple feat.

One of the major factors affecting artificial transformation is the physiological makeup of the bacterial cell. For example, in *E. coli*, the long O side chains of the lipopolysaccharide (LPS) are believed to interfere with the association of DNA and the cell. An *E. coli* Gal$^-$ mutant does not synthesize the galactose component of the long O side chains of the LPS, and does not have the O side chain on its cell surface LPS. Gal$^-$

strains have an increased efficiency in the uptake of DNA. Another example in *E. coli* involves *deoR*, which encodes a transcriptional repressor protein involved in regulating nucleoside catabolism. An *E. coli deoR*⁻ mutant strain shows increased plasmid transformation efficiency.

Certain *E. coli* strains are used for very specific transformation goals. *E. coli* strains mutant in *recB*, *recC*, *recD*, and *sbcB* can be transformed with linear DNA molecules. *recB*, *recC*, and *recD* encode subunits of the DNA degrading enzyme Exonuclease V and *sbcB* encodes Exonuclease I. These endonucleases degrade linear DNA molecules. The *E. coli* strains mutant in *recB*, *recC*, *recD*, and *sbcB* have been constructed to get around the requirement of only being able to transform circular plasmid DNA molecules in to *E. coli*. However, the *E. coli recBCD*⁻ *sbc*⁻ mutant strains can still undergo homologous recombination. This is because these strains express both functional RecA and nucleases that can substitute for RecBCD, so that strand assimilation and phosphodiester bond breakage needed during homologous recombination will still occur.

E. coli recA⁻ mutant strains are used to ensure that transformed plasmids will not rearrange or will not recombine with the chromosome, thus maintaining their independently replicating state. *recA*⁻ mutants are also useful when the transforming DNA sequence contains direct repeats of DNA base sequences.

E. coli strains mutant for certain DNA restricting (endonuclease: *hsdR*) or modifying (methylase: *hsdM*⁻ or *hsdS*⁻) activities can take up many foreign DNA sequences without fear that the DNA will be degraded by HsdR because it is not correctly modified by methylation by HsdM and HsdS (see Chapters 7 and 14 for more details on the Hsd system). DNA isolated from *E. coli* strains mutant for the nonspecific DNA nuclease, Endonuclease I (*endA*), has been reported to be of higher quality for transformation and other molecular biology procedures (see Chapter 14). *E. coli* strains mutant for certain protein degrading enzymes or **proteases** (*lon*⁻ or *prc*⁻) may produce increased amounts of protein when the transformed genes are expressed. As can be seen, choosing the appropriate bacterial mutant can greatly affect the outcome of a transformation experiment.

Summary

The ability of a bacterial species to be transformed is either a natural part of the bacterial lifestyle or is inducible by artificial treatments. Naturally competent bacteria are equipped to pick up DNA sequences that may be of benefit to them. The transformed DNA can be used as a carbon or nitrogen source. It can be used to repair gaps in a recipient bacterium's damaged chromosome, or to increase the genetic, and thus enzymatic diversity of the recipient bacterium. Early research in the transformation process yielded an unexpected yet profound discovery: DNA, not protein, was the transforming principle or hereditary material.

Bacteria that can become naturally competent contain *com* genes whose products are responsible for the binding, processing, and transporting of DNA into the bacterial cell. Some bacteria can be transformed by DNA of any base pair sequence. Other bacteria are preferentially transformed by DNA that contains an uptake signal sequence (USS). We are at the beginning stages of understanding the transformation process at the molecular level. This elegant process not only requires the correct machinery, but also relies on environmental signaling and signal processing to inform the naturally transformable bacterium to produce the transformation machinery.

Study questions

1 How does transformation differ from transduction and conjugation?
2 Can all bacteria become naturally competent? Why or why not? Devise a strategy for determining if a newly discovered bacterial species is naturally competent.
3 Is DNA ever free of proteins? Why or why not? If DNA is never found free of protein, then how did Avery and his colleagues convince the scientific community that DNA was the transforming principle?
4 What happens to the transformed DNA?
5 What are the steps in natural transformation?
6 How does *B. subtilis* know when to express its *com* genes?
7 Do all naturally transformable bacteria have the same *com* genes as *B. subtilis*? What does this tell you about how different bacteria bind and take up free DNA?
8 What is a transformasome?
9 How does free DNA arise?
10 If an *E. coli* strain is *recBCD*$^-$, *sbcB*$^-$, and *recA*$^-$ what will happen to a piece of linear DNA that it takes up by transformation? What will happen to a circular DNA molecule that contains an origin of replication?

Further reading

Avery, O.T., MacLeod, C.M., and McCarty, M. 1944. Studies on the chemical nature of the substance inducing transformation of pneumococcal types: Induction of transformation by a deoxyribonucleic acid fraction isolated from pneumococcus type III. *Journal of Experimental Medicine*, **79**: 137–58.

Dubnau, D. 1993. Genetic exchange and homologous recombination. In *Bacillus subtilis and other Gram-positive Bacteria: Biochemistry, Physiology and Molecular Genetics*, eds. A.L. Sonenshein, J.A. Hoch, and R. Losick, pp. 555–84. Washington, DC: ASM Press.

Hanahan, D. and Bloom, F.R. 1996. Mechanisms of DNA transformation. In *Escherichia coli and Salmonella typhimurium: Cellular and Molecular Biology*, 2nd edn., eds. F.C. Neidhardt, R. Curtiss III, J.L. Ingraham, E.C.C. Lin, K.B. Low, B. Hagasanik, W.S. Rexnikoff, M. Riley, M. Schaechter, and H.E. Umbarger, pp. 2449–59. Washington, DC: ASM Press.

Provvedi, R. and Dubnau, D. 1999. ComEA is a DNA receptor for transformation of competent *Bacillus subtilis*. *Molecular Microbiology*, **31**: 271–80.

Stanley, W.M. 1970. The "undiscovered" discovery. *Archives of Environmental Health*, **21**: 256–62.

Chapter 12

Gene expression and regulation

Bacterial genomes usually contain several thousand different genes. Some of the gene products are required by the cell under all growth conditions and are called **housekeeping genes**. These include the genes that encode such proteins as DNA polymerase, RNA polymerase, and DNA gyrase. Many other gene products are required under specific growth conditions. These include enzymes that synthesize amino acids, break down specific sugars, or respond to a specific environmental condition such as DNA damage.

Housekeeping genes must be expressed at some level all of the time. Frequently, as the cell grows faster, more of the housekeeping gene products are needed. Even under very slow growth, some of each housekeeping gene product is made. The gene products required for specific growth conditions are not needed all of the time. These genes are frequently expressed at extremely low levels, or not expressed at all when they are not needed and yet made when they are needed. This chapter will examine **gene regulation** or how bacteria regulate the expression of their genes so that the genes that are being expressed meet the needs of the cell for a specific growth condition.

Gene regulation can occur at three possible places in the production of an active gene product. First, the transcription of the gene can be regulated. This is known as **transcriptional regulation**. When the gene is transcribed and how much it is transcribed influences the amount of gene product that is made. Second, if the gene encodes a protein, it can be regulated at the translational level. This is known as **translational regulation**. How often the mRNA is translated influences the amount of gene product that is made. Third, gene products can be regulated after they are completely synthesized by either **post-transcriptional** or **post-translational regulation** mechanisms. Both RNA and protein can be regulated by degradation to control how much active gene product is present. Both can also be subjected to modifications such as the methylation of nucleosides in rRNA, the extensive modifications made to tRNAs (over 80 modified nucleosides have been described), or the phosphorylation of response-regulator proteins (see below). These modifications can play a major role in the function of the gene product.

In general, every step that is required to make an active gene product can be the focus of a regulatory event. In practice, most bacterial regulation occurs at the transcriptional level. Transcriptional regulation is thought to be more frequent because it would be a waste to make the RNA if neither the RNA nor its encoded protein is needed. The first three systems discussed—repression of the *lac* operon by Lac repressor, activation of the *lac* operon by cAMP-CAP, and control of the *trp* operon by attenuation—describe classic transcriptional regulation mechanisms. The last three systems—regulation of the heat shock genes, regulation of the SOS response, and regulation of capsule synthesis—rely on different combinations of transcriptional, post-transcriptional, and post-translational regulatory mechanisms and will be used to demonstrate how several regulatory mechanisms can be integrated to fine-tune gene expression.

The players in the regulation game

Ribonucleic acid, or RNA, exists as messenger RNA (mRNA), transfer RNA (tRNA), and ribosomal RNA (rRNA). In most cases, RNA is a single-stranded, rather than a double-stranded, molecule. RNA participates in both genetic and functional activities. As mRNA, it allows the genetic information stored in the DNA molecule to be transmitted into proteins. As tRNA, it transfers amino acids during translation. As rRNA, it maintains the structure of the ribosome and helps carry out translation.

RNA is chemically synthesized by the action of RNA polymerase in a process called **transcription**. There are four identifiable steps during transcription: **promoter recognition**; **chain initiation**; **chain elongation**; and **chain termination** (Fig. 12.1). RNA polymerase catalyzes the formation of phosphodiester bonds between ribonucleotides using DNA as a template. Unlike DNA polymerase, RNA polymerase does not require a primer to begin synthesis of the RNA molecule (see Chapter 2 for details of DNA synthesis). Growth of the RNA chain, like the growth of a DNA chain, is in the 5′ to 3′ direction because RNA polymerase can only add a new nucleotide to a free 3′ OH group. The order in which the different ribonucleotides are added to a free 3′ OH is determined by a double-stranded DNA molecule in which one strand acts as the template (Fig. 12.2). The **template strand** is complementary to the RNA. The non-template or **coding strand** contains the same sequence as the RNA except for the substitution of uracil for thymine.

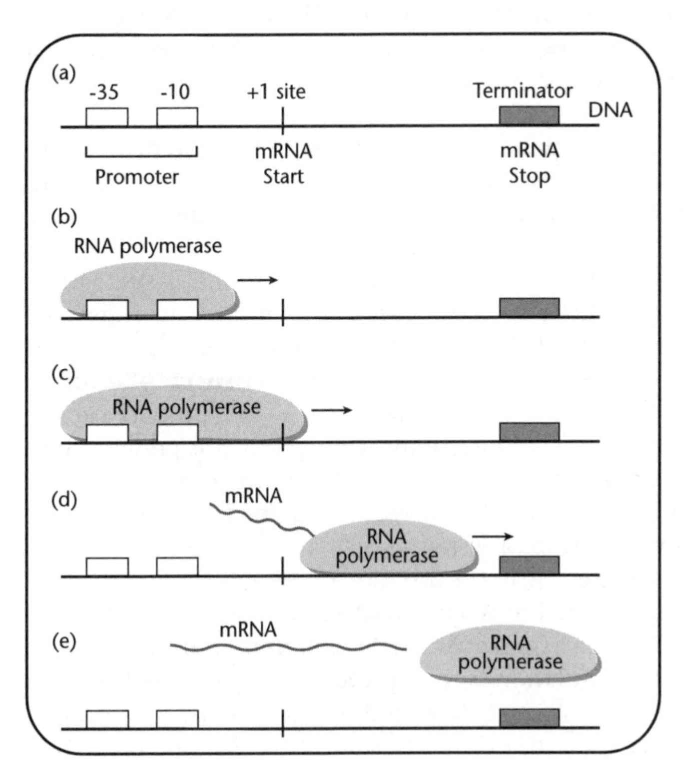

Fig. 12.1 The four major steps of transcription. (a) RNA polymerase recognizes the promoter. (b) RNA polymerase moves to the start site and begins polymerizing RNA. (c) RNA polymerase moves along the DNA template, elongating the RNA. (d) RNA polymerase stops RNA synthesis. The newly synthesized RNA disassociates from RNA polymerase and RNA polymerase disassociates from the DNA. RNA polymerase transcribes mRNA, rRNA, and tRNA.

Core RNA polymerase is a complex composed of the proteins β, β′ and two subunits of α (Fig. 12.3a). Core RNA polymerase is responsible for the synthesis of RNA, but it imparts no specificity as to where the start of the RNA occurs. Core RNA polymerase is converted to **holoenzyme** when one additional protein, **sigma factor** (σ) is associated with it (Fig.

12.3a). σ factor directs RNA polymerase to specific sequences in the DNA called **promoters** so that transcription initiates at the proper place.

Bacteria can contain more than one σ factor. In *E. coli*, the major σ factor present under normal growth conditions is called sigma 70 or σ^{70}. σ^{70} recognizes promoters that have a specific DNA sequence and directs the RNA polymerase molecule that it is part of to begin transcription near these specific sequences. σ^{70} was named for its molecular weight, which is 70 kilodaltons. Different σ factors recognize different sequences as promoters. For example, when cells are exposed to an increase in temperature or heat shock a group of genes is induced to cope with this stress (see below). The expression of the heat-shock genes is controlled by an alternative σ factor and all of the heat-shock genes have a common promoter sequence that is recognized by the alternative σ factor. The heat-shock promoter sequence is different from the promoter sequence recognized by σ^{70}.

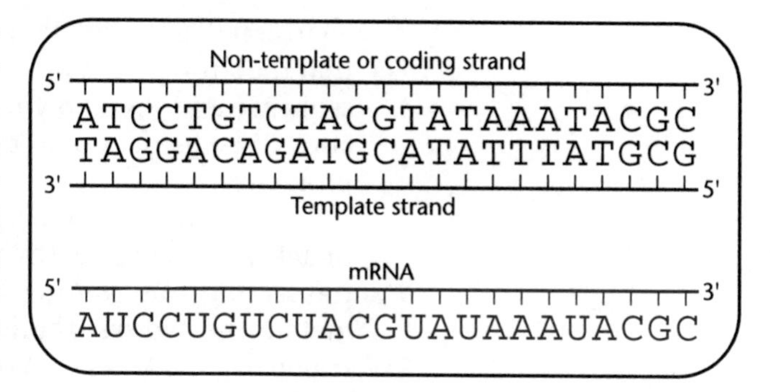

Fig. 12.2 The mRNA is synthesized using one strand of the DNA as a template. This makes the mRNA complementary to the template strand. When the DNA sequence encoding a gene is shown, by convention the DNA strand that is the same sequence as the mRNA (except for T to U) is usually shown.

Promoters contain two distinct sequence motifs that reside ~10 bases and ~35 bases upstream of the transcriptional start site or first base of the RNA. The transcriptional start site is known as the **+1 site**. All of the bases following the +1 site are transcribed into RNA and are numbered consecutively with positive numbers (Fig. 12.3b). The bases prior to the +1 site are numbered consecutively with negative numbers. The motif at ~10 bases upstream of the +1 site is called the **–10 region** and the motif at ~35 bases upstream of the +1 is called the **–35 region**. σ^{70} recognizes promoters with a **consensus sequence** consisting of TAATAT at the –10 region and TTGACA at the –

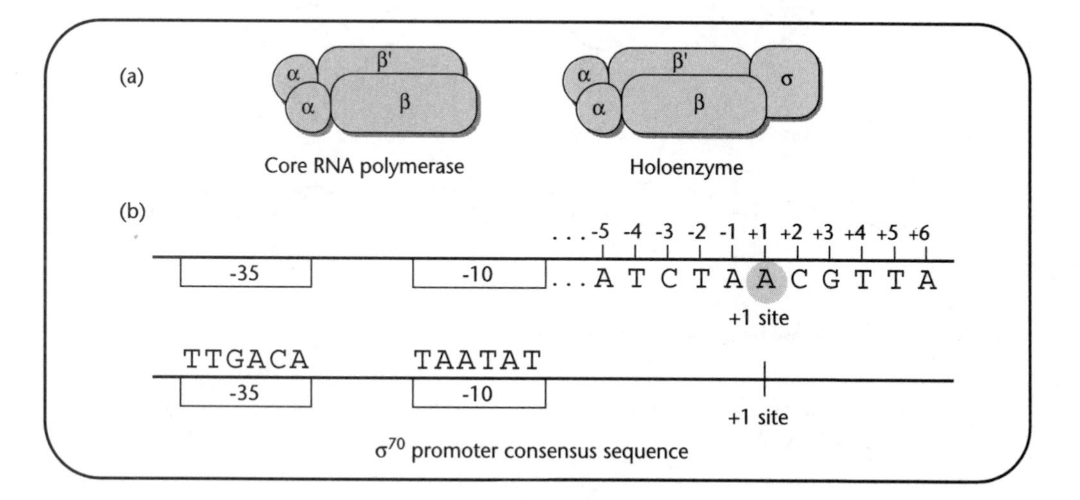

Fig. 12.3 The structure of RNA polymerase and the regions in the promoter it binds to. (a) RNA polymerase has two forms. Core RNA polymerase synthesizes RNA in vitro but starts at many places on the DNA. Holoenzyme also synthesizes RNA in vitro but it starts only at promoters. (b) The general organization of the signals used by RNA polymerase. RNA polymerase recognizes promoters, which consist of a –10 region and a –35 region. RNA polymerase begins synthesis of RNA at the +1 base. RNA polymerase containing σ^{70} recognizes specific sequences at the –10 and –35 regions. The closer a sequence is to the consensus, the better σ^{70} is at recognizing it as a promoter and the more times it will be transcribed into RNA. If any of the bases are changed, σ^{70} may still recognize the sequence as a promoter but it will do so with a lower affinity.

35 region (Fig. 12.3b). Promoters are defined according to their strength. This means that the stronger the promoter, the stronger the interaction between that promoter sequence and RNA polymerase. A general rule of thumb is that the closer the –10 and –35 sequences of a promoter are to the consensus sequence, the stronger the promoter.

RNA polymerase holoenzyme binds to promoter sequences and covers approximately 75 bases of the DNA from –55 to +20. Once bound, RNA polymerase initiates transcription by causing the double-stranded DNA template to open, effectively melting the hydrogen bonds that hold the two DNA strands together in the promoter region (Fig. 12.4). As RNA polymerase starts transcribing at the +1 site, it continues to open the double-stranded DNA molecule, creating a short region of single-stranded DNA. After RNA polymerase has passed through the opened DNA, this region will reform hydrogen bonds to give a closed DNA molecule. Approximately two turns of the double helix or 17 base pairs are unwound at any given time during the elongation phase of transcription.

Although σ factor is needed for RNA polymerase to bind to the right promoter sequences, it is not needed during the elongation phase of transcription. After a short RNA transcript is synthesized, σ factor dissociates from RNA polymerase and core RNA polymerase continues to elongate the RNA transcript. Transcription continues until core RNA polymerase encounters a transcription termination signal or

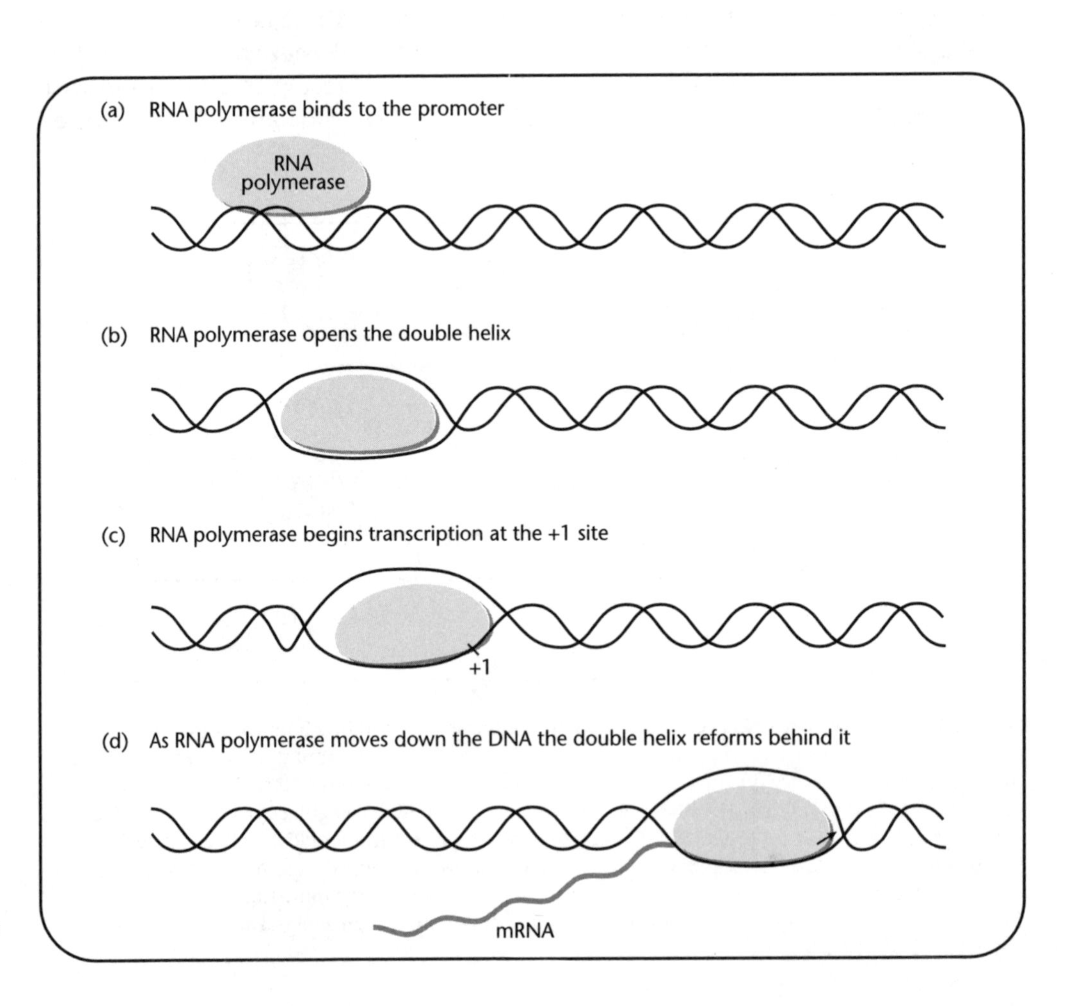

Fig. 12.4 The beginning steps of transcription. RNA polymerase opens a 17 base pair bubble in the DNA to gain access to the DNA template.

terminator where RNA polymerase and the newly synthesized RNA dissociate from the DNA to end transcription.

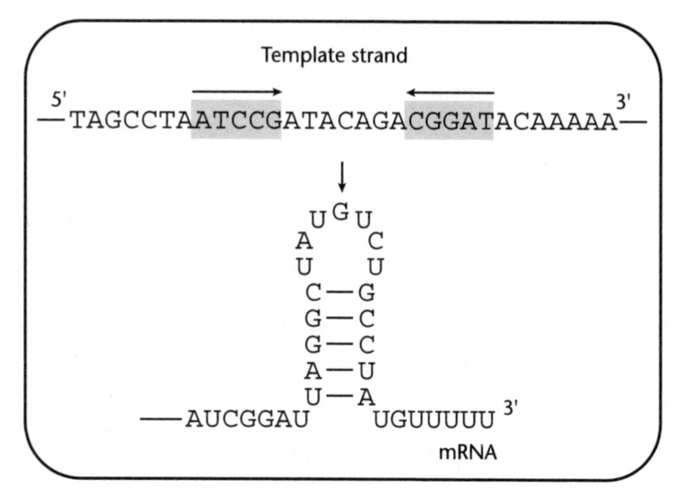

Fig. 12.5 Termination of transcription takes place at specific sites. Frequently, a hairpin structure is part of the termination site. Inverted repeats form the base of the hairpin. The hairpin is frequently followed by a stretch of uracils.

Inverted repeat sequences are often found as part of a terminator along with a stretch of uracils found at the end of the RNA transcript (Fig. 12.5). The inverted repeat sequences are able to hydrogen bond with each other and create a stretch of double-stranded RNA. When this secondary structure forms in the RNA, it takes on the appearance of a **stem loop structure**. Terminating transcription can either involve a protein called Rho or be Rho independent, depending on the sequence comprising the terminator. Terminators that do not involve Rho rely on the stem loop structure that forms at the end of the RNA to halt RNA polymerase's forward motion along the DNA. In contrast, Rho-dependent terminators rely on both Rho and the formation of the stem loop structure at the end of the RNA to halt transcription.

Initiation of a new round of transcription does not require that the previous round be terminated. This means that there can be several RNA polymerase complexes transcribing the same template DNA at the same time (Fig. 12.6). Multiple RNA polymerases, because of the size of RNA polymerase, must be spread at least 75 nucleotides apart. In bacteria, translation of an mRNA into its corresponding polypeptide does not require that the entire mRNA be synthesized or transcription be terminated before translation is initiated. Ribosomes are able to bind the **ribosome binding site** (RBS) or **Shine–Delgarno sequence** in the mRNA and initiate translation at the

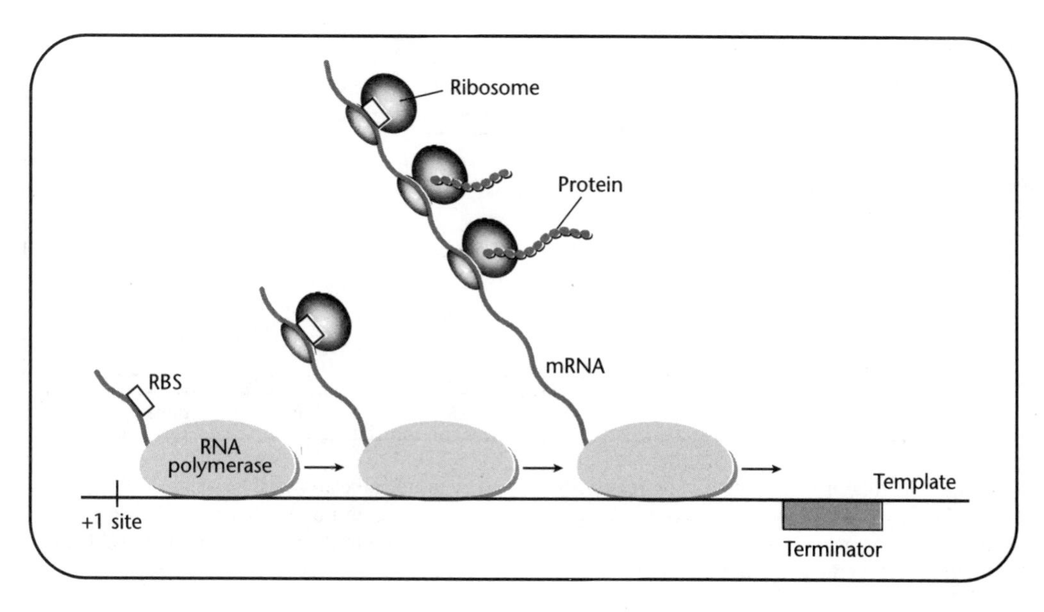

Fig. 12.6 Multiple rounds of transcription and translation take place at the same time. As RNA polymerase moves along the DNA template, the mRNA becomes accessible to ribosomes. The ribosomes bind to the RBS and initiate translation at the AUG start codon.

FYI 12.1

Coupled transcription and translation in different species

In bacteria, transcription and translation can take place on the same mRNA molecule at the same time. This is possible because bacteria do not have a membrane surrounding their DNA and all of the proteins needed for both processes have access to the DNA. The situation is quite different in eukaryotic cells that do have a membrane, called the nuclear membrane, surrounding their DNA. The proteins needed for transcription are inside the nuclear membrane but the proteins needed for translation are located outside of the nuclear membrane. Thus, coupled transcription and translation cannot happen in eukaryotic cells. A second difference between bacteria and eukaryotes is that while bacteria make many polycistronic mRNAs, eukaryotes only encode one protein per mRNA.

starting methionine codon (AUG) even if RNA polymerase is still transcribing the mRNA.

In transcriptional regulation, DNA sequences called **control regions** or **operators** are found adjacent to or overlapping the –35 and –10 regions of the promoter. Specific DNA binding proteins recognize these control regions and exert an effect on holoenzyme's ability to initiate transcription. Regulatory proteins can bind to their control region and prevent transcription. This is known as **repression**. Regulatory proteins can also bind to their control regions and promote RNA polymerase binding to the promoter. This effect is known as **activation**. The regulatory mechanisms controlling gene expression are typically discovered by mutational analysis. Mutations that eliminate the regulation or make the regulation even better usually localize to one of two places (Fig. 12.7). The mutations can be in the genes encoding the repressors or activators. The mutations can also be in the nucleotide sequences that these regulatory proteins bind to.

Operons and regulons

An adaptive response, metabolic pathway, or developmental program usually requires more than one gene. Frequently, the genes for a given process are regulated in a similar manner. Genes that need to be coordinately regulated can be located right next to each other or spread around the chromosome. An **operon** is a group of genes physically linked on the chromosome and under the control of the same promoter(s) (Fig. 12.8a). In an operon, the linked genes give rise to a single mRNA that is trans-

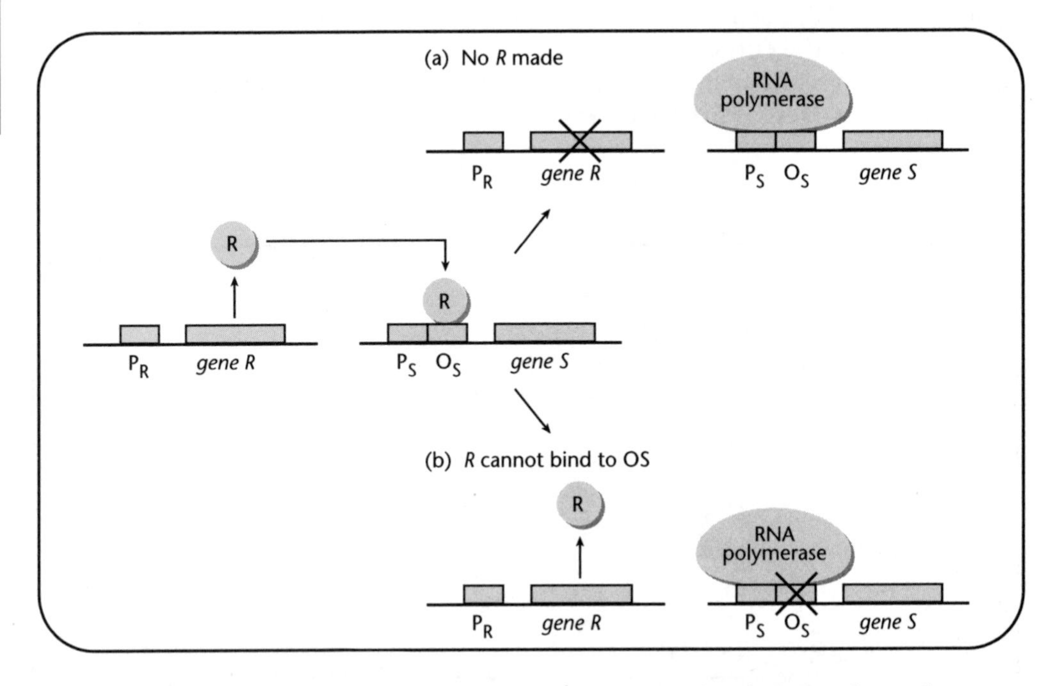

Fig. 12.7 Regulatory mechanisms are frequently studied by mutational analysis. Two classes of mutations are possible. (a) Mutations in the *R gene* or *R gene* promoter result in no R protein being produced. (b) Mutations in the operator (O) DNA sequence can result in the R protein not being able to bind to the DNA and carry out regulation.

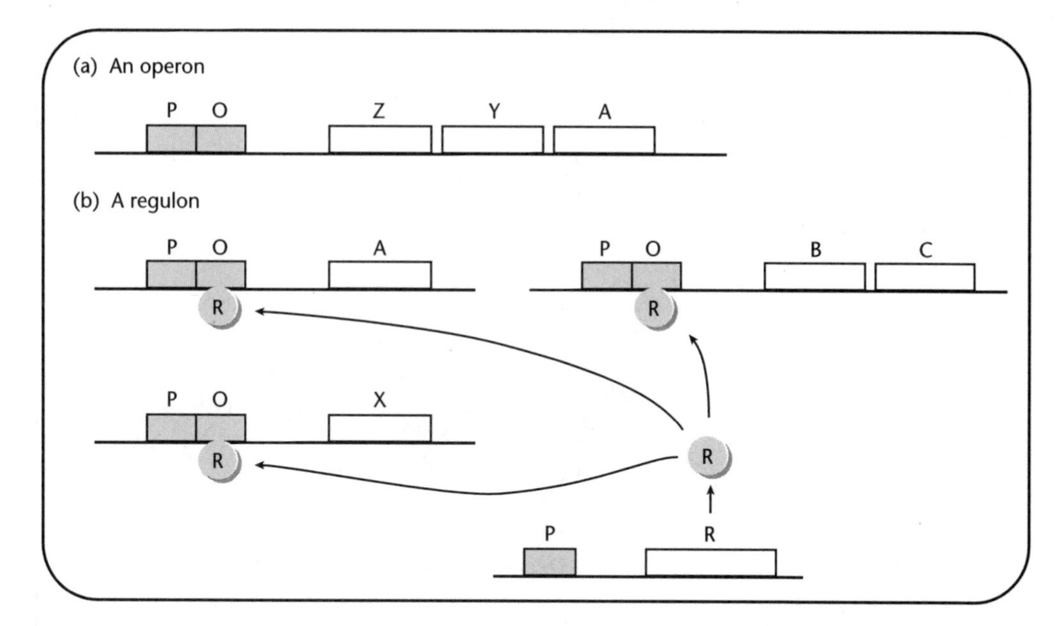

Fig. 12.8 There are two widely used organizational schemes in bacteria for multiple genes that are coordinately regulated. (a) In an operon, all of the genes are physically linked and under the control of one promoter. (b) In a regulon, the genes are in different places on the chromosome but they all have promoters that respond to the same regulators.

lated into the different gene products. This type of mRNA is called a **polycistronic mRNA** (Fig. 12.9). A **regulon** is a group of genes all needed for the same process but physically located in different parts of the chromosome and containing their own promoter(s) (Fig. 12.8b). In a regulon, the promoters are all regulated in the same fashion and allow for coordinate expression of the necessary genes.

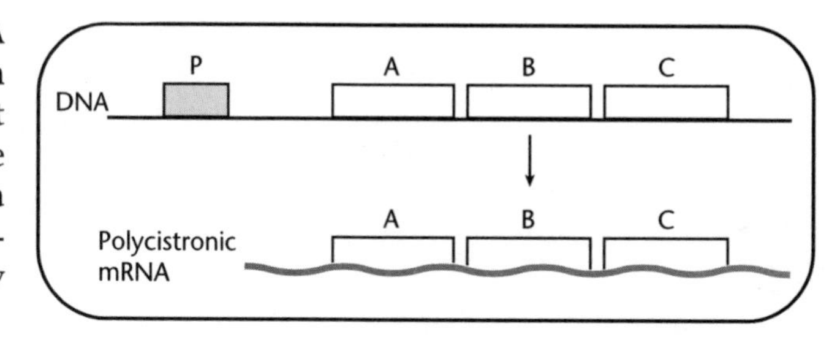

Fig. 12.9 Polycistronic mRNAs are translated into more than one protein. In this example, *gene A*, *gene B*, and *gene C* are encoded on one mRNA and will all be translated from this one mRNA.

Repression of the *lac* operon

The metabolism of the sugar, lactose, requires at least two proteins in *E. coli*. The lactose permease, encoded by the *lacY* gene, transports lactose into the cell. The enzyme β-galactosidase, called β-gal for short and encoded by the *lacZ* gene, catalyzes the cleavage of the β1,4-glycosidic linkage in lactose resulting in the monosaccharides, galactose and glucose (Fig. 12.10a). When a compound is broken down, the process is called **catabolic**. The *lac* operon expresses genes whose products catabolize lactose. Analogs are available that mimic lactose's ability to either be a substrate for β-gal (X-gal, ONPG) or converted to an **inducer** of the *lac* operon (IPTG) (Fig. 12.10b). This has made the study of the *lac* operon and its regulation amenable to examination in the laboratory. These analogs are frequently used in DNA cloning experiments as described in Chapters 13 and 14.

lacZ, *lacY*, and one other gene, *lacA*, are linked on the *E. coli* chromosome (Fig. 12.11). *lacA* encodes a detoxifying enzyme called β-galactoside transacetylase. The intracellular concentrations of β-gal and Lac permease are very low in *E. coli* cells grown

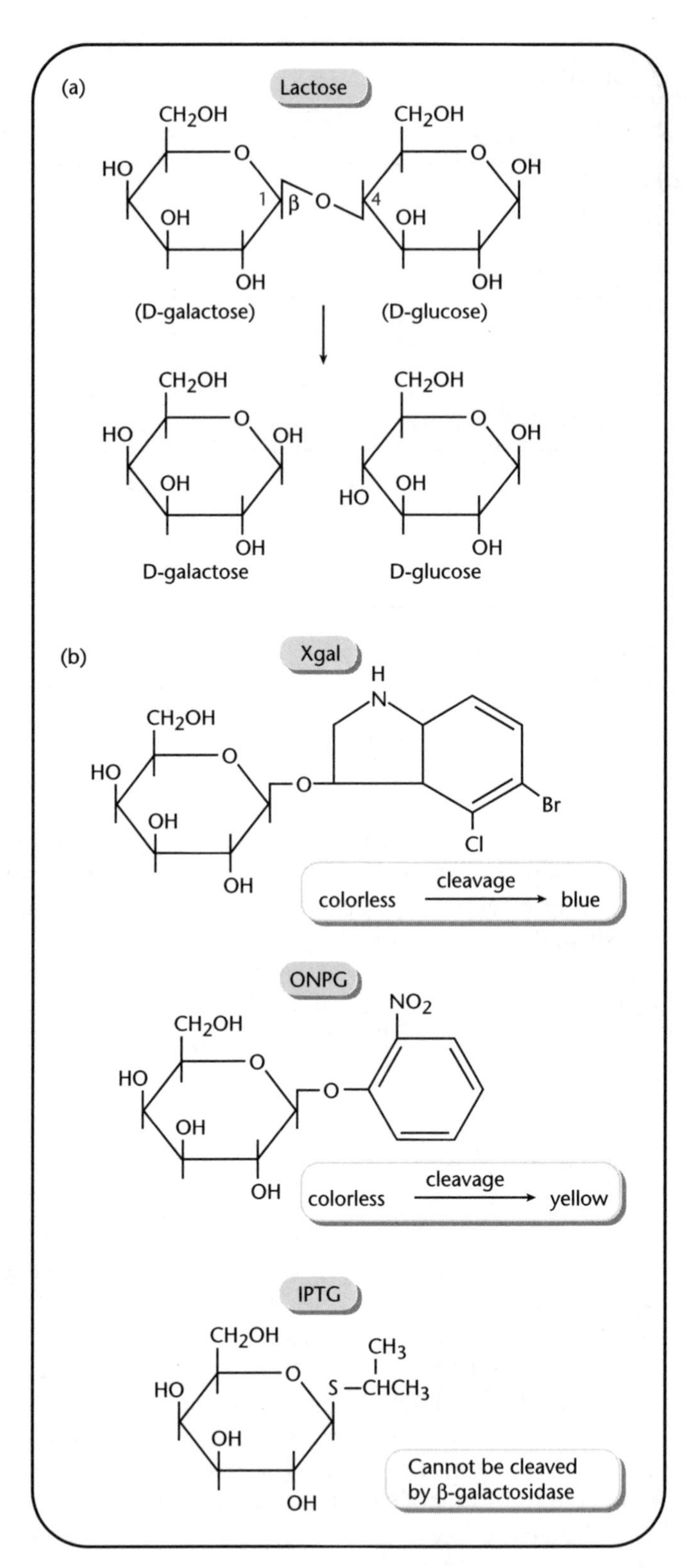

Fig. 12.10 (a) Cleavage of lactose by β-galactosidase. (b) Lactose analogs that are used to study the *lac* operon. X-gal is converted from a colorless compound to a dark blue compound when it is cleaved by β-galactosidase. ONPG is converted from a colorless compound to a yellow compound. IPTG cannot be cleaved by β-galactosidase but it can bind to LacI and induce the *lac* operon.

in medium that does not contain lactose (Table 12.1). If lactose is added to the medium, β-gal, Lac permease, and transacetylase appear at the same time in *E. coli* cells. After lactose is transported into the cell, some of it is converted to allolactose (1–6-O-β-galactopyranosyl-D-glucose) by the few molecules of β-gal found in uninduced cells. Allolactose is a rearranged lactose molecule and an inducer of the *lac* operon.

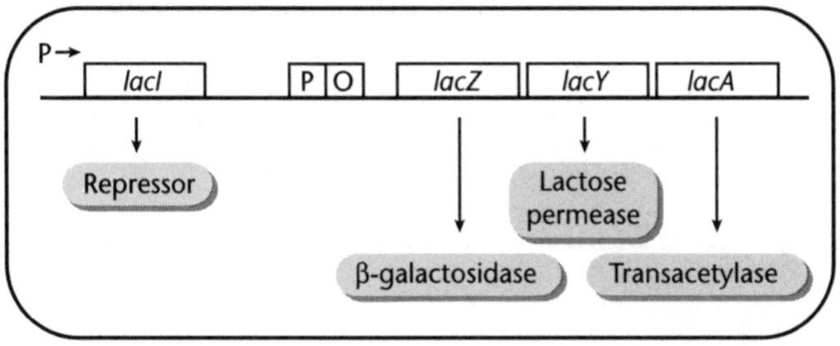

Fig. 12.11 The genes that encode the proteins used to transport and degrade lactose form an operon. The operon is regulated by LacI. The *lacI* gene is located just upstream of *lacZ*, *lacY*, and *lacA*. LacI is expressed from its own promoter that is on all of the time.

lacZ, *lacY*, and *lacA* are coordinately expressed from one promoter called *lacP* that directs expression of a polycistronic mRNA (Fig. 12.11). *lacP* is located upstream of *lacZ*. *lacI* encodes the lactose repressor protein and is located upstream of *lacP*. *lacI* has its own promoter and is constitutively expressed. Lac Repressor or LacI binds to the DNA at the *lac* operator site called *lacO*. *lacO* is located between *lacI* and *lacZ*. In the absence of inducer, Lac repressor is bound to *lacO*, preventing expression of the operon (Fig. 12.12a). In the presence of inducer, inducer binds to Lac repressor and prevents repressor from binding to *lacO*, leading to transcription of the operon (Fig. 12.12b).

How *lac* operon induction takes place and the players that control this induction were elucidated in the extensive genetic analyses undertaken by Francois Jacob, Jacques Monod, and their coworkers. Table 12.1 describes a few of the *lac* mutants isolated by Jacob and Monod and the effect the inducer has on the activity of β-gal and transacetylase. For example, Class I mutations eliminate the inducibility of the *lac* operon resulting in β-gal and transacetylase activity being detectable in the absence of lactose (uninduced). This class of mutation results in a constitutive Lac$^+$ phenotype and maps to either the *lacI* gene or to *lacO*. *lacI*$^-$ mutations are recessive or have no phenotype in the presence of a wild-type *lac* operon. Enough LacI is made from the wild-type *lacI* gene to repress both copies of the *lac* operon. Class II mutations result in a super-repressed Lac$^-$ phenotype. Super-repressed mutants have no β-gal or transacetylase activity in the presence of inducer. These mutations map to the *lacI*

Table 12.1 Some classes of *E. coli lac* mutant isolated by Jacob and Monod and the effect inducer has on the relative concentrations of ß-galactosidase and transacetylase in the mutant backgrounds. The mutations were examined when they resided in the chromosome in one copy and in merodiploids containing a second copy of the *lac* operon on an F′.

Class	Genotype	ß-galactosidase (*lacZ*)		Transacetylase (*lacA*)	
		Uninduced	Induced	Uninduced	Induced
WT	*lacZ A I*	0.1	100	1	100
I	*lacZ A* I^-	100	100	90	90
	lacZ A I^-/F′*lacZ A I*	1	240	1	270
II	*lacZ A* I^s	0.1	1	1	1
	lacZ A I^s/F′*lacZ A I*	0.1	2	1	3
III	*lacZ A* O^c	25	95	15	100
	lacZ$^-$ *A O*/F′*lacZ A*$^-$$O^c$	180	440	1	220
II + III	*lacZ A O* I^s F′*lacZ A* O^c *I*	190	219	150	200

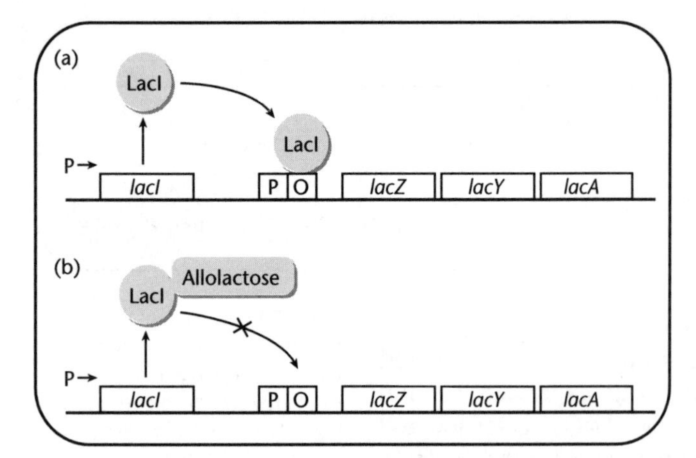

Fig. 12.12 A model for the regulation of the *lac* operon. The Lac repressor or LacI binds to the *lac* operator DNA (*lacO*) and prevents transcription of *lacZ*, *lacY*, and *lacA*. If lactose is present inside the cell, it is converted to allolactose. Allolactose binds to LacI and prevents its interaction with *lacO*. This allows expression of the *lacZ*, *lacY*, and *lacA* genes.

gene and result in a repressor that can no longer be inactivated by inducer. Super-repressor mutants are dominant to wild type because the mutant LacI can bind to both copies of the *lac* operon and turn them off. Class III mutations are in *lacO* and prevent Lac repressor from binding and turning off the operon. *lacO*^{c} mutants are dominant to wild type because wild-type LacI from either copy of the operon cannot repress the *lac* operon containing the *lacO*^{c} mutant. When super-repressor mutants are combined with *lacO*^{c} mutants, *lacO*^{c} mutants are dominant to super-repressor mutants because the super-repressor still cannot bind the mutant *lacO*.

Activation of the *lac* operon by cyclic AMP and the CAP protein

If lactose and glucose are added to a culture of wild-type *E. coli* cells, the *lac* operon is not induced. No *lac* mRNA or gene products are made. This effect of glucose is the result of a second regulatory mechanism, **catabolite repression**. Catabolite repression affects not only *lac* gene expression but also other operons that catabolize specific sugars such as galactose, arabinose, and maltose. Contrary to its name, catabolite repression describes an activating mechanism, involving a complex between cAMP (cyclic adenosine monophosphate) and the catabolite activator protein (CAP; also called cyclic AMP receptor protein or CRP) (Fig. 12.13).

If given the choice of sugars to metabolize, *E. coli* will use glucose first and then other sugars. When levels of glucose are high, there is no need to express high levels of the enzymes needed for the metabolism of lactose or other sugars, which yield less energy than glucose. High levels of intracellular glucose result in low levels of cAMP. Low levels of cAMP mean very few cAMP–CAP complexes. When intracellular levels of glucose drop and other sugars must be metabolized, levels of cAMP increase. Increased cAMP levels mean there are more cAMP–CAP complexes. cAMP–CAP binds to a specific site on the DNA that is located adjacent to the promoter for the *lac* genes (Fig. 12.14) and adjacent to promoters controlling the expression of other sugar metabolizing operons affected by cAMP–CAP. The –10 and –35 motifs in the promoter sequence for *lac* are not a perfect match to the consensus sequence for σ^{70} promoters. cAMP–CAP bound to the CAP binding site increases the binding of σ^{70} RNA polymerase to the *lac* promoter. By having two mechanisms controlling the expression of the *lac* operon, cells ensure that the *lac* gene products are only made when needed.

Regulation of the tryptophan biosynthesis operon by attenuation

The biosynthesis of the aromatic amino acid tryptophan requires the activity of three enzymes, anthranilate synthetase (*trpE*) complexed with phosphoribosyl-anthranilate transferase (*trpD*), phosphoribosyl-anthranilate isomerase (*trpC*), and tryptophan synthetase (*trpA* and *trpB*) to convert chorismate to tryptophan. The five *trp* structural genes encoding the three tryptophan biosynthetic enzymes are physically linked and coordinately regulated as an operon. The *trp* genes are arranged in the

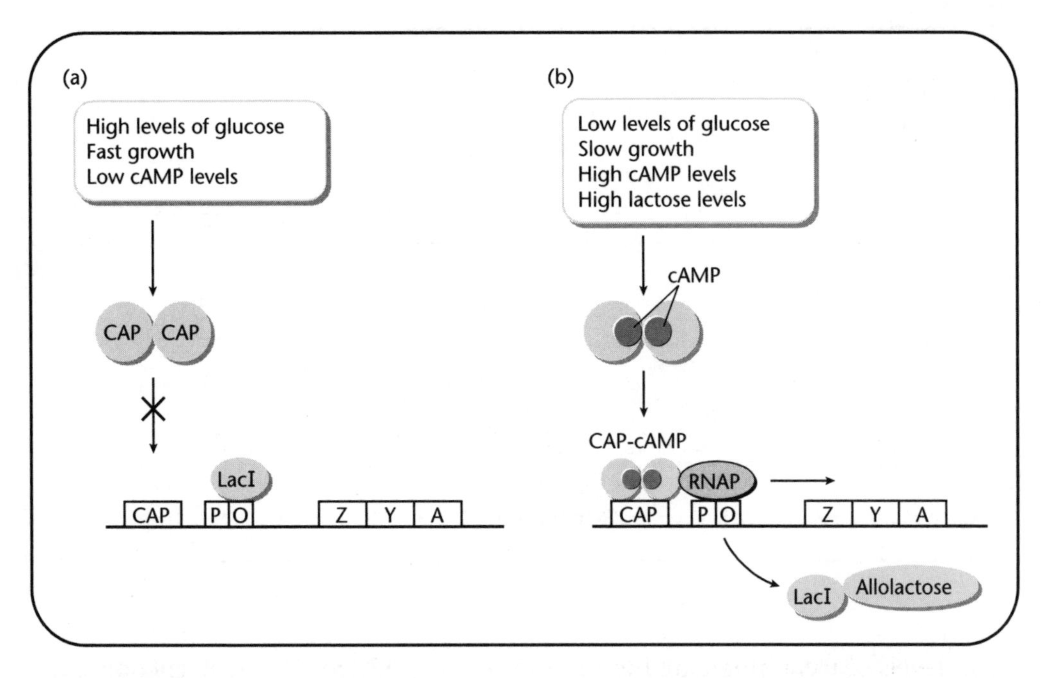

Fig. 12.13 Catabolite repression of the lactose operon. (a) High levels of glucose mean low levels of cAMP. There is not enough cAMP to complex with CAP and no induction of the *lac* operon takes place. LacI remains bound to *lacO*. (b) Low levels of glucose lead to high levels of cAMP. cAMP complexes with CAP and binds in front of *lacP*. If there is lactose present, it is converted to allolactose, allolactose complexes with LacI and prevents LacI from binding to lacO. This leads to induction of the *lac* operon.

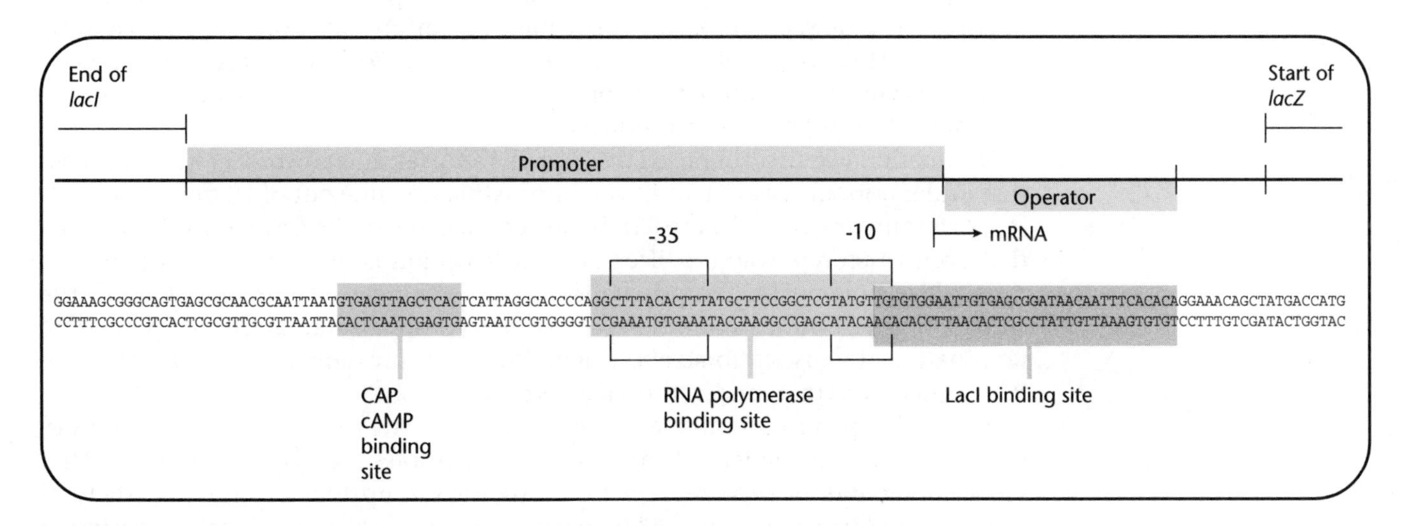

Fig. 12.14 The binding sites for CAP–cAMP, RNA polymerase, and LacI in the *lac* operon of *E. coli*.

order that their encoded enzymes function in the biosynthetic pathway (Fig. 12.15). The goal of a biosynthetic or **anabolic** pathway is to synthesize the end product only when it is needed. In 1959, Cohen and Jacob made the observation that addition of tryptophan to an actively dividing *E. coli* culture repressed the synthesis of tryptophan synthetase. The *trp* operon is repressed by TrpR, a transcriptional repressor,

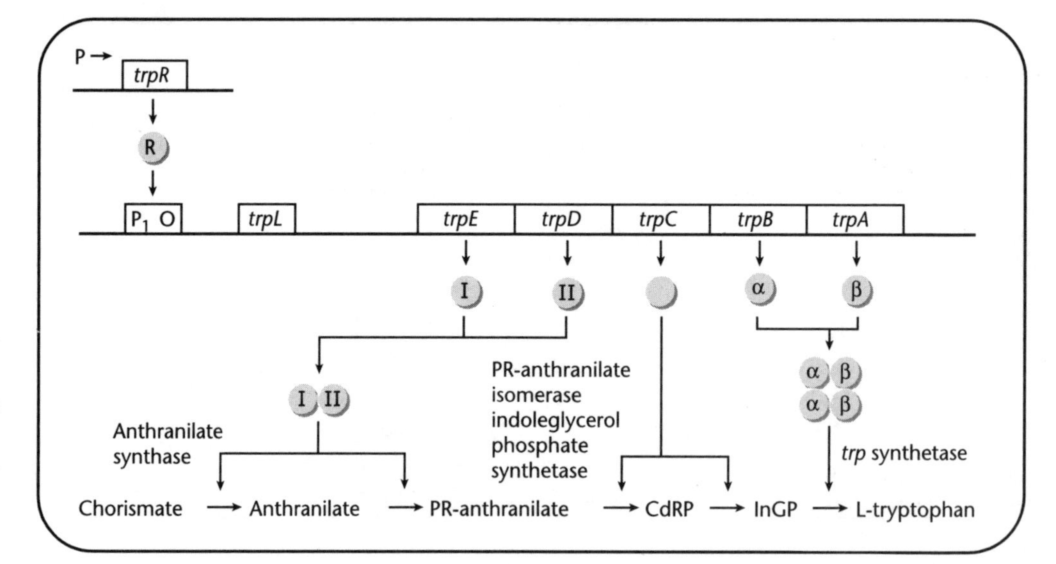

Fig. 12.15 The structure of the *trp* operon and the enzymatic activities encoded by the *trp* genes. The regulator of the *trp* operon, TrpR, is encoded by a gene that is not linked to the rest of the operon. The structural genes, *trp EDCBA*, are positioned in the operon in the same order as the enzymes they encode are used in tryptophan synthesis.

when levels of tryptophan are high in the cell (Fig. 12.15). The gene encoding the TrpR repressor is not located near the *trp* operon.

Even though expression of both the *lac* and *trp* operons are regulated by repression, the mechanism by which the LacI and TrpR repressors work is very different. The LacI repressor when bound to its effector molecule (i.e. lactose) cannot repress transcription. In contrast, the TrpR repressor must be bound by an effector molecule, in this case tryptophan, for it to repress transcription. The repression mechanism for the *trp* operon can be viewed as a coarse control mechanism; an all or none approach to gene expression. However, another regulatory mechanism, **attenuation**, is available to fine-tune gene expression of some biosynthetic operons, especially during times of extremely short supply of end products.

Attenuation was first proposed by Charles Yanofsky and coworkers who observed that under normal growth conditions approximately nine out of 10 *trp* mRNA transcripts terminate before they reach the structural genes of the *trp* operon. This means that even when repression is lifted and transcription is initiated from the *trp* promoter, RNA polymerase abruptly halts, failing to elongate the *trp* mRNA across the structural genes. During conditions of severe tryptophan starvation, the premature termination of transcription is abolished, allowing expression of the *trp* operon at 10-fold higher levels than under normal growth conditions.

What is happening to cause early termination of the *trp* mRNA? The answer lies in the regulatory sequence in front of the *trp* structural genes. Downstream from the *trp* promoter and the TrpR repressor binding site but upstream of the *trp* structural genes, are 162 base pairs. The 162 base pair sequence is called the **leader sequence** or *trpL* (Fig. 12.16a). *trpL* contains a ribosome binding site adjacent to an AUG translational start codon. This means there are two RBS sites and two AUG translational start codons located after the *trp* promoter but in front of the first *trp* structural gene (*trpE*).

The *trp* leader sequence, after it is transcribed into mRNA, can be translated into a short polypeptide consisting of 14 amino acids. Of these 14 amino acids, two are *trp* residues that are located next to each other. The presence of two tandem *trp* residues is very rare. Most *E. coli* proteins contain one *trp* per every 100 amino acids. At the end of *trpL* is a Rho-independent terminator. This signal resides approximately 40 base pairs

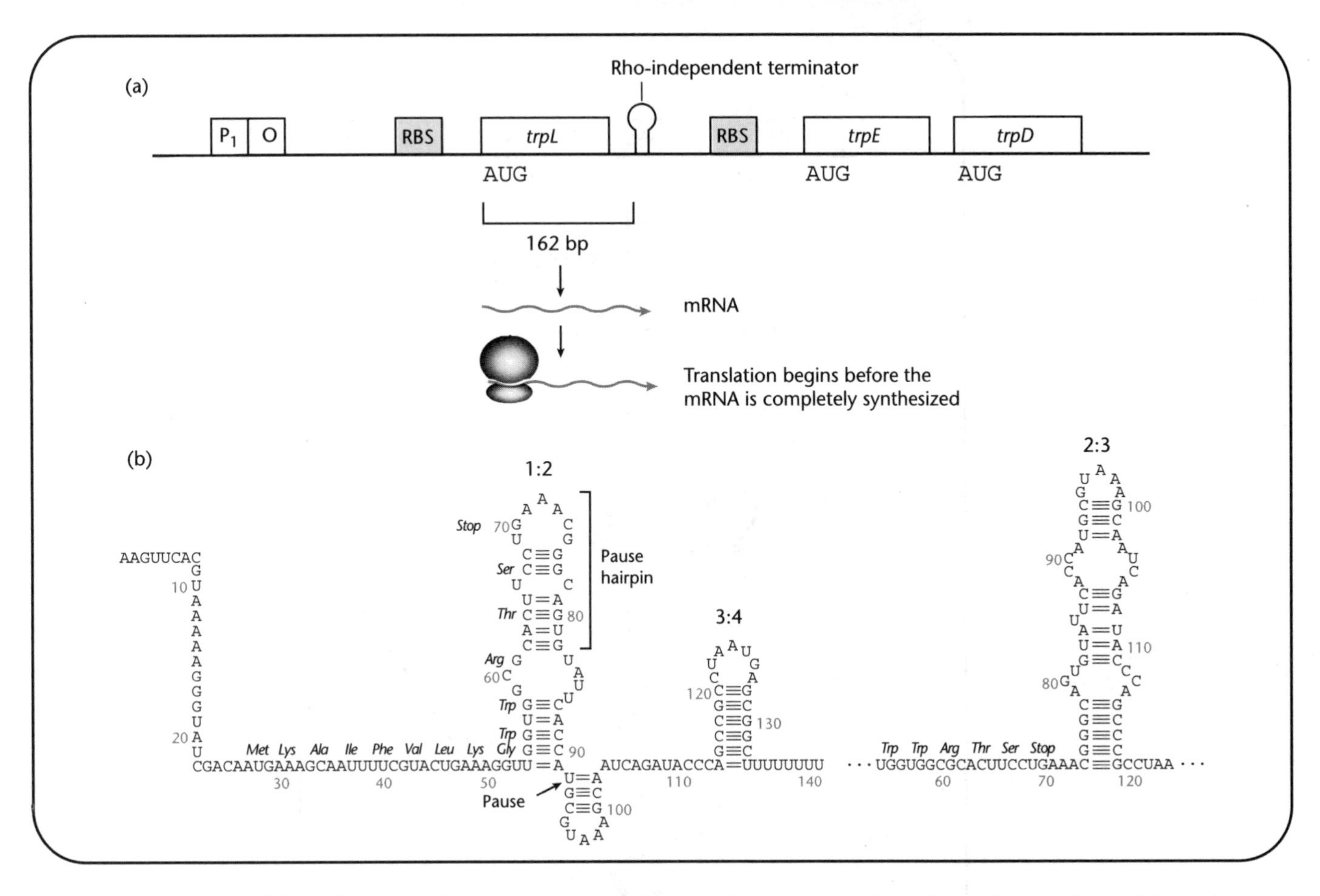

Fig. 12.16 The features of the leader region of the *trp* operon. (a) *trpL* is transcribed into mRNA. (b) The *trpL* mRNA can form three different stem loop structures depending on the prevalence of charged trp–tRNAs inside the cell (see text and Fig. 12.17 for details).

upstream of the signals that control the translation of the *trp* structural genes. The DNA sequence in this region suggests that the encoded mRNA has a very high probability of forming stem loop secondary structures. There are three combinations of stem loops that can form. They have been named the pause loop (1:2), the terminator loop (3:4), and the anti-terminator loop (2:3) (Fig. 12.16b).

Attenuation of the *trp* operon uses the secondary structures in the mRNA to determine the amount of tryptophan in the cell (Fig. 12.17). RNA polymerase initiates transcription at the *trp* promoter after the TrpR repressor has been removed. Because translation can be initiated before the mRNA is completed, a ribosome loads at the *trpL* RBS and initiates translation of TrpL (Fig. 12.17 step 1). As RNA polymerase elongates the *trpL* mRNA, stem loop structures will form in the mRNA. The first secondary structure that can form is the 1:2 pause loop. This large hairpin structure is called the pause loop because it forms behind the transcribing RNA polymerase and causes RNA polymerase's forward movement along the DNA template to pause (Fig. 12.17 step 2). Pausing transcription allows the translating ribosome to stay close to the transcribing RNA polymerase. This ensures that these two complexes stay in close communication during crucial times of abruptly changing levels of tryptophan. As the translating ribosome advances on the mRNA, the ribosome breaks apart the pause loop freeing RNA polymerase to continue transcribing *trpL* (Fig. 12.17 step 3). At this time, the ribosome

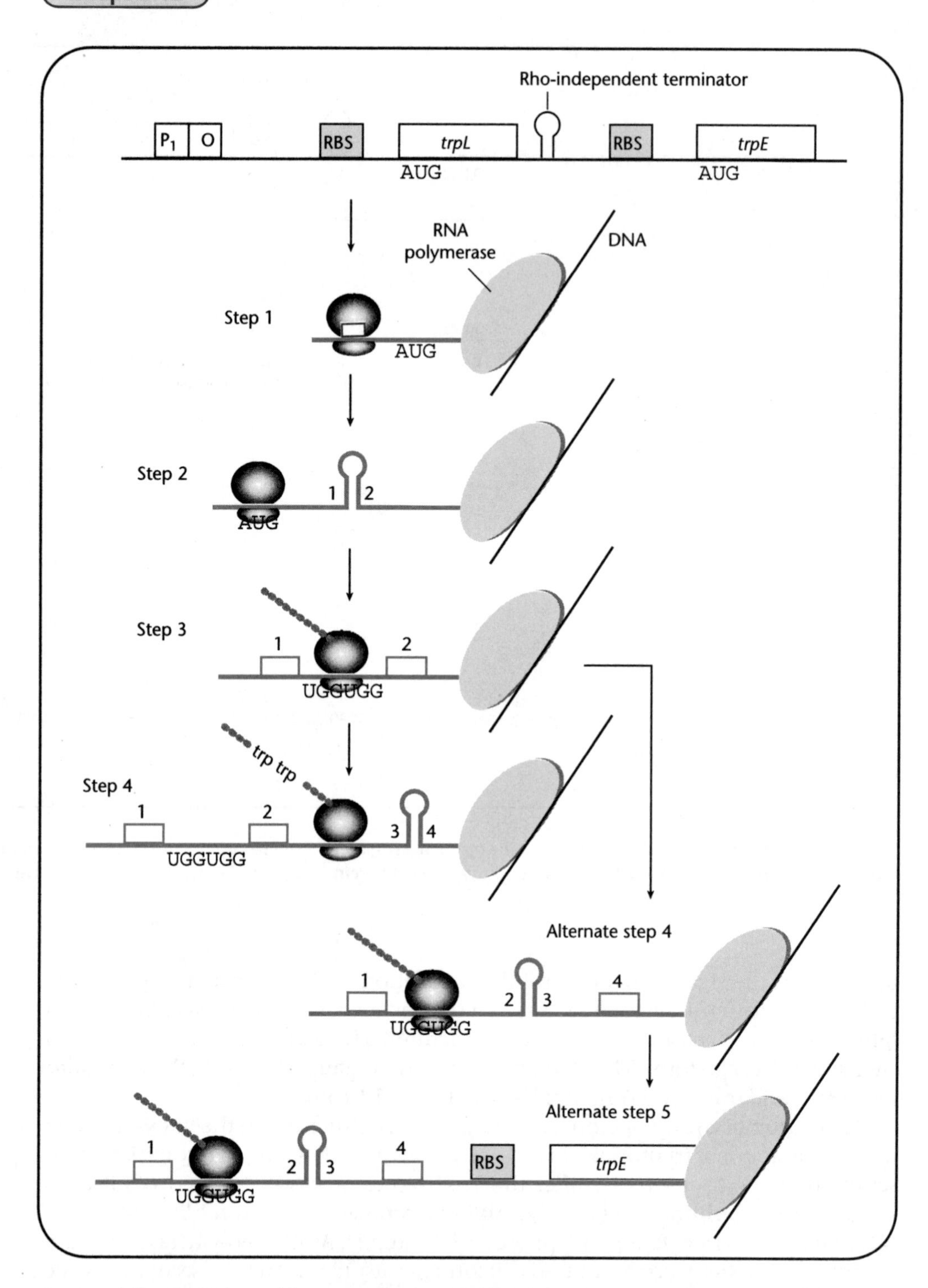

Fig. 12.17 The predictions of the attenuation model. (Step 1) Coupled transcription/translation takes place as for any bacterial gene. (Step 2) RNA polymerase pauses and 1 : 2 stem loop forms. (Step 3) The ribosome disrupts the 1 : 2 stem loop and encounters the two *trp* codons. (Step 4) If enough tryptophan is present, charged trp–tRNAs will be present and the ribosomes will translate *trpL*. This causes RNA polymerase to stop at the terminator composed of a 3 : 4 stem loop. (Alternate Step 4) If not enough *trp* is found in the cell, the ribosome stalls at the two *trp* codons, while RNA polymerase continues. The 2 : 3 stem loop forms. (Alternate Step 5) The 3 : 4 terminator cannot form and RNA polymerase continues transcribing into the *trp* structural genes. This exposes the RBS upstream of *trpE*, allowing translation of *trpE*.

encounters the two *trp* codons. If there are two charged tryptophanyl–tRNAs to pass tryptophan to the ribosome to add to the growing TrpL polypeptide, then the ribosome does not stall, and it is able to stay close to the transcribing RNA polymerase (Fig. 12.17 step 4). If they stay close, very few bases of the mRNA transcript will separate the ribosome from the RNA polymerase and only the 3:4 terminator loop structure is predicted to form. The 3:4 loop is a Rho-independent transcription terminator.

If two charged tryptophanyl–tRNAs are not available to pass tryptophan to the translating ribosome, then the ribosome stalls in its attempt to synthesize the TrpL polypeptide (Fig. 12.17 alternate step 4). The transcribing RNA polymerase continues to move away from the stalled ribosome, creating a longer stretch of bases that separate these two complexes. Because of the longer distance between RNA polymerase and the ribosome, the larger 2:3 anti-terminator stem loop structure forms. The 2:3 loop is not a transcription termination signal and RNA polymerase continues its forward motion, transcribing the *trp* structural genes (Fig. 12.17 alternate step 5). Thus, the leader peptide plays a decisive role in determining if transcription of the *trp* operon will terminate early or will continue into the structural genes. Other *E. coli* biosynthetic operons, including those that synthesize *phe*, *his*, *leu*, *thr*, and *ilv*, are controlled by attenuation. Of this group of operons, only *trp* has a repressor in addition to attenuation.

Regulation of the heat-shock regulon by an alternate sigma factor, mRNA stability, and proteolysis

The coordinately regulated genes of an operon are physically linked and controlled by the same promoter. In contrast, the coordinately regulated genes of a regulon are physically located in different parts of the chromosome and are controlled by their own promoters yet are regulated by the same mechanisms. Regulons are also called multigene systems or global regulatory systems. Table 12.2 describes a few of the regulons identified in bacteria. Three bacterial regulons, heat shock, SOS and eps, will be used to illustrate the regulatory mechanisms and components associated with regulons.

Cells respond to an abrupt increase in temperature by inducing synthesis of a specific group of proteins to cope with this stress. This heat-shock response has been documented in many different cell types and extensively studied at the molecular level in prokaryotic cells such as *E. coli*. *E. coli* is a mesophile, which means its normal growth temperature is between 20°C and 37°C. Shifts in temperature within this range result in very little stress for the bacterium and there is no noticeable adjustment in its gene expression. *E. coli* responds to an abrupt increase in temperature from 30°C to 42°C by producing a set of ~30 different proteins (Table 12.3). These proteins are collectively called the heat-shock proteins (Hsps). Within 5 minutes after a 30°C to 42°C temperature shift, the Hsp proteins increase by 10-fold. After this transient increase, the amount of the Hsp proteins decreases slightly to a steady-state level. The steady-state level is maintained as long as the cells remain at the elevated temperature. Once the temperature is decreased from 42°C to 30°C, the levels of Hsp proteins decrease abruptly, approximately a 10-fold decrease within 5 minutes. Besides a change in temperature, other agents such as ethanol, organic solvents, or certain DNA damaging agents can induce heat-shock gene expression. This suggests that the members of the heat shock regulon deal with cellular damage that can occur in several different ways.

Many of the *hsp* genes encode either proteases or chaperones. Proteases are proteins that degrade proteins especially those that may be abnormal. Abnormal proteins

Table 12.2 Examples of regulons identified in *E. coli*.

Multigene system	Stimulus	Regulator	Regulated genes
Heat shock	Shift to high temperature	σ^{32} (*htpR* = *rpoH*)	~30 heat-shock proteins (Hsps) including proteases and chaperones
pH shock	Acid tolerance	CadC	Several gene products needed for protection from acid damage
SOS response	Extensive DNA damage	LexA repressor	~20 gene products needed for DNA repair
Oxidation response	H_2O_2	OxyR repressor	~12 gene products needed for protection from H_2O_2 and other oxidants
Porin response	High osmolarity	Sensor EnvZ regulator	Porin genes OmpR
Catabolite repression	Carbon utilization	Crp-cAMP	Many different sugar catabolizing operons

Table 12.3 Heat-shock proteins in *E. coli* and their corresponding functions.

Regulon	Hsp protein	Function
σ^{32}	DnaK	Chaperone
	DnaJ	Chaperone
	GroEL	Chaperone
	GroES	Chaperone
	GrpE	Nucleotide exchange factor
	Lon	Protease
	ClpX	Protease
	ClpP	Protease
	HflB	Protease
	RpoD	Sigma factor σ^{70}
	HtpR	Heat-shock sigma factor σ^{32}
	HtpG	Chaperone
	GapA	Dehydrogenase
	ClpB	Chaperone
	HtrM	Epimerase
	IbpB	Chaperone
	IbpA	Chaperone
	ClpY	Protease
σ^E	DegP	Protease
	σ^E	Sigma factor
	σ^{32}	Sigma factor
unknown	LysU	Lysyl-tRNA synthetase

σ^{32} regulated Hsps whose function is unknown: HtpY, HslA, HslC, HslK, HtpX, FtsJ, HslO, HslP, HslV, HtrC, HslW, HslX, HslY, HslZ, HtpK, HtpT.
σ^{32} regulated Hsps whose function and regulation are unknown: PspA, HslE, HslF, HslG, HslI, HslJ, HslM, HslQ, HslR.

include incompletely synthesized proteins and proteins that are not folded correctly. Chaperones are proteins that bind to abnormal proteins, unfold them, and try to let them refold into an active configuration. Frequently, if a chaperone cannot fix an abnormal protein, it will target the abnormal protein for degradation by a protease. From looking at the genes that are part of the heat-shock regulon, it is interesting that they mainly deal with misfolded proteins. This suggests that misfolding of proteins is the predominant damage to cells at high temperatures.

The genes encoding Hsps are scattered around the chromosome but are coordinately regulated. Thus, the *hsp* genes constitute a regulon. The regulator of the heat-shock response is an alternative sigma factor, named sigma 32 or σ^{32}. σ^{32} is a 32 kilodalton protein and the gene encoding σ^{32} is *rpoH* (for RNA polymerase subunit heat shock). σ^{32} directed RNA polymerase recognizes the distinctive sequences of heat-shock promoters. The heat shock promoters have different –10 (CCCCAT) and –35 (CTTGAAA) consensus sequences than promoters recognized by σ^{70}. All of the *hsp* genes have a heat-shock promoter. In fact, once the *E. coli* genome was sequenced, heat-shock promoter sequences were used to identify novel potential Hsps.

σ^{32} is made at all temperatures yet the *hsp* gene products are not. σ^{32} is an unstable protein and at low temperatures it has a half-life of about one minute. At high temperatures, 10 times more σ^{32} protein is made and the σ^{32} that is made is five times more stable than the σ^{32} made at low temperatures. The regulation of the *rpoH* gene is at the translational level. There is a significant amount of *rpoH* mRNA in cells at low temperatures yet translation of this mRNA is inhibited. At high temperatures, translational inhibition is relieved and σ^{32} protein is made. Two regions of the *rpoH* mRNA are required for translational inhibition, a region near the +1 site and a region internal to the *rpoH* gene between nucleotides 150 and 250 in the mRNA. How these two regions of the mRNA inhibit translation at low temperatures is not well understood. The two regions are capable of forming a stem loop structure and this may prevent the ribosome from loading at the RBS and inhibit translation in addition to possibly increasing the stability of this mRNA.

σ^{32} is degraded by a specific protease called HflB. Degradation of σ^{32} at 30°C also requires a chaperone composed of three proteins, DnaK, DnaJ, and GrpE. Degradation of σ^{32} at 30°C is decreased ~ 10-fold by mutations in any one of the genes that encode HflB, DnaK, DnaJ, or GrpE. All four of these proteins are part of the heat-shock regulon and so their expression is induced by heat shock. If more chaperone proteins are present at higher temperatures to help degrade σ^{32}, then how does σ^{32} ever interact with core RNA polymerase to activate the heat-shock genes? While the precise answer to this question is unknown, recent experiments suggest that the interactions between DnaK and σ^{32} are temperature dependent and only occur at low temperatures. When the temperature is raised, DnaK and σ^{32} no longer interact yet interaction between σ^{32} and core RNA polymerase remains unchanged. It is possible that this temperature-dependent interaction of DnaK and σ^{32} plays a role in σ^{32} stability at high temperatures. If the temperature is lowered from 42°C to 30°C, translational inhibition of the mRNA returns and σ^{32} again becomes sensitive to degradation. The heat-shock response can be turned off as quickly as it was turned on.

A second heat-induced regulon is controlled by a second alternative sigma factor, σ^E. At temperatures around 50°C, σ^E controlled promoters are very active. Deletions of the gene encoding σ^E are temperature sensitive at 42°C, whereas deletions of the gene encoding σ^{32} are temperature sensitive at 20°C. σ^E responds to misfolded outer membrane proteins and σ^{32} responds to misfolded cytoplasmic proteins. The σ^{32} gene also has a σ^E promoter so that all of the Hsps are induced by a cascade effect when σ^E regu-

FYI 12.2

Weigle reactivation

If a lysate of lambda is UV irradiated, the DNA in many of the phage is damaged and these phage are not capable of growth. If this irradiated lysate is used to infect a population of *E. coli* that have also been irradiated, many of the damaged phage now survive. This basic observation is known as the Weigle effect or Weigle reactivation. At first it was not clear why two doses of irradiation (one to the phage and one to the bacterium) should fix the damage caused by one dose of irradiation. We now know that irradiating the bacteria induces the SOS response and induces the synthesis of DNA repair enzymes. When the damaged λ DNA is introduced into the SOS induced bacteria, the λ DNA is repaired and is used to produce more phage.

lated genes are expressed. It is thought that the σ^E regulon provides proteins that are needed at more extreme conditions. Given what we know, it is possible to induce the σ^{32} regulon by itself or the σ^{32} and σ^E regulons together. Having several overlapping regulons that respond to different degrees of one condition gives the cell more flexibility in how it responds. Cells can make large amounts of very specific gene products without wasting resources by making gene products that they don't need.

Regulation of the SOS regulon by proteolytic cleavage of the repressor

The SOS regulon permits a rapid response to severe or lethal levels of DNA damage. The goal is to quickly activate DNA repair proteins and deal with the DNA damage in a swift manner. Speed is of the essence to save the cell by restoring its chromosome to an intact double-stranded state. The DNA repair enzymes that are induced result in a higher level of mistakes than normally seen but this is balanced by their speed in repairing the majority of the chromosome. During the SOS response, cell division is halted so that damaged chromosomes are not segregated into daughter cells. Thus, in addition to DNA repair enzymes, a cell division inhibitor protein is also expressed at high levels during the SOS response. Table 12.4 describes the genes and proteins of the SOS regulon. The SOS genes are expressed at low levels during normal cell growth. They are induced to high-level expression upon extensive DNA damage.

Weigle and coworkers first reported from their observations of reactivated UV-irradiated λ that some DNA repair systems are inducible (see FYI12.2). Many of these repair mechanisms are activated as part of the SOS regulon. Like the heat-shock regulon, the SOS regulon has a mechanism for signaling when the regulon should be on or off. The SOS genes contain a common sequence that in some cases overlaps the promoter region and in other cases is adjacent to the promoter region. This common sequence is bound by a repressor protein called LexA. SOS promoters bound by LexA cannot be used to initiate transcription. As a result of significant DNA damage, the LexA repressor is inactivated and removed so that the SOS genes are expressed.

How is the LexA repressor inactivated and removed from the promoters of SOS genes? Upon exposure to DNA damaging agents, large amounts of single-stranded DNA accumulate (Fig. 12.18a). Single-stranded DNA is bound by the RecA protein (Fig. 12.18b). RecA participates in homologous recombination (see Chapter 5) and in

Table 12.4 Known SOS genes, proteins, and corresponding functions in *E. coli*.

Gene	Protein	Function
lexA	LexA	Represses transcription of SOS genes
recA	RecA	LexA activator and major homologous recombination protein
recN	RecN	Binds ATP; RecBCD substitute
ruvAB	Ruv	Branch migration and resolution
sulA (*sfiA*)	SulA	Cell division inhibitor
umuC	UmuC	Mutagenesis
umuD	UmuD	Mutagenesis
uvrA	UvrA	Excision repair
uvrB	UvrB	Excision repair
uvrD	UvrD	Excision repair

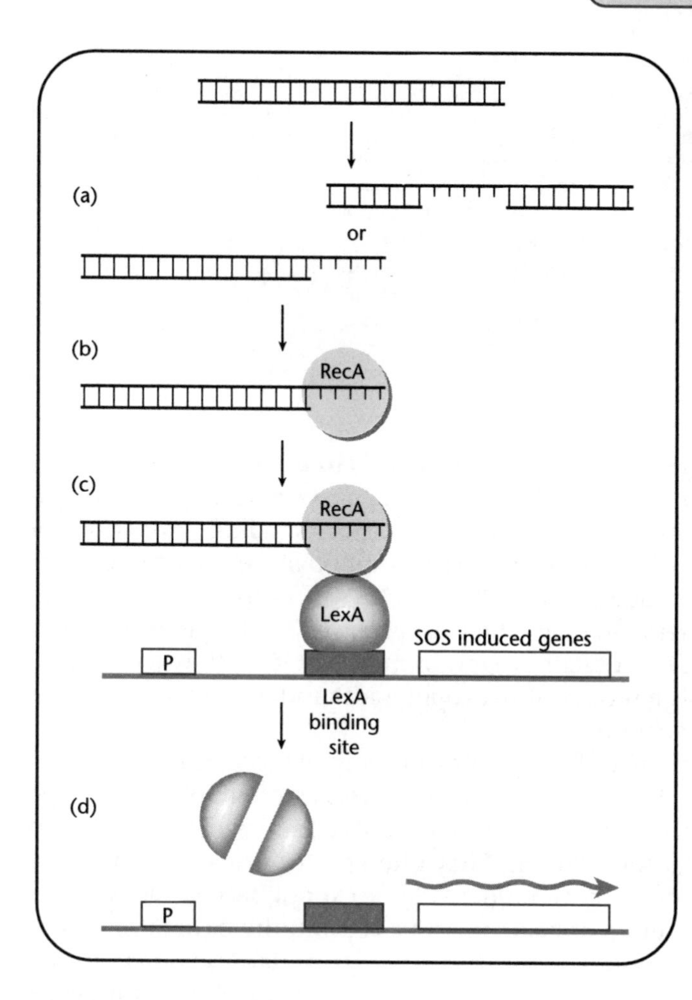

Fig. 12.18 The SOS response. (a) DNA damage leads to single-stranded DNA. (b) RecA binds to the ssDNA. (c) RecA–ssDNA binds to LexA. (d) Lex A undergoes autocleavage, relieving repression of the SOS genes.

post replication DNA repair (see Chapter 4). In the case of extensive DNA damage, RecA complexed to single-stranded DNA binds to LexA and induces LexA to cleave itself (Fig. 12.18c). Since LexA, with help from the RecA–ssDNA complex, cleaves itself, LexA is said to autocleave. Autocleavage of LexA takes place between two specific amino acids that separate the repressor into two domains, the DNA binding domain and the dimerization domain. Two LexA molecules bind to the sequence found in SOS regulon promoters. The disruption of dimerization results in the removal of LexA from the SOS promoters (Fig. 12.18d). Once LexA is removed, SOS genes are expressed at high levels. Eventually the induction signal, RecA complexed to single-stranded DNA, drops because of DNA repair, and LexA is no longer able to undergo autocleavage. This returns the regulon to its prestimulus or uninduced state.

Two-component regulatory systems: signal transduction and the *cps* regulon

Bacteria continually monitor their environment so as to set in motion a proper response to detected changes. There are molecular mechanisms that pick up signals from the environment and transmit this information to the appropriate regulators. This is the concept of **signal transduction**. The molecular mechanism associated with this process in bacteria is the two-component system. The **two-component**

Table 12.5 Examples of two component systems in *E. coli*.

System	Two-component response-regulator/sensor-kinase
Nitrogen limitation	NtrC regulator/NtrB sensor
Phophorus limitation	PhoR regulator/PhoB sensor
Oxygen limitation	ArcA regulator/ArcB sensor
Osmotic upshift	KdpD regulator/KdpE sensor
Capsule expression	RcsB regulator/RcsC sensor
Porin regulation	OmpR regulator/EnvZ sensor

signal transduction system consists of a **sensor-kinase protein** and a **response-regulator protein** (see Fig. 11.4). This system has been identified as a regulatory mechanism for many operons and regulons, including induction of competency in *B. subtilis* (see Chapter 11). The basic components of two-component signal transduction systems are described in Chapter 11. Two-component systems are found in many different bacteria. In *E. coli* there are at least 50 different examples of this type of regulation. Table 12.5 describes a few of these systems. We will use the *E. coli cps* system, which controls synthesis of an exopolysaccharide capsule, to illustrate two-component signal transduction.

E. coli produces a number of different types of exopolysaccharides. Exopolysaccharides are polymers consisting of sugars linked together, often in a lattice-type structure. They are located on the outside of the bacterial cell and while some remain attached to the bacterial cell, others do not. They often protect bacteria from harsh environments, toxic chemicals, or bacteriophage. Some exopolysaccharides serve as matrices to link a number of different bacteria into a community or biofilm. Other bacterial exopolysaccharides cause health problems in humans.

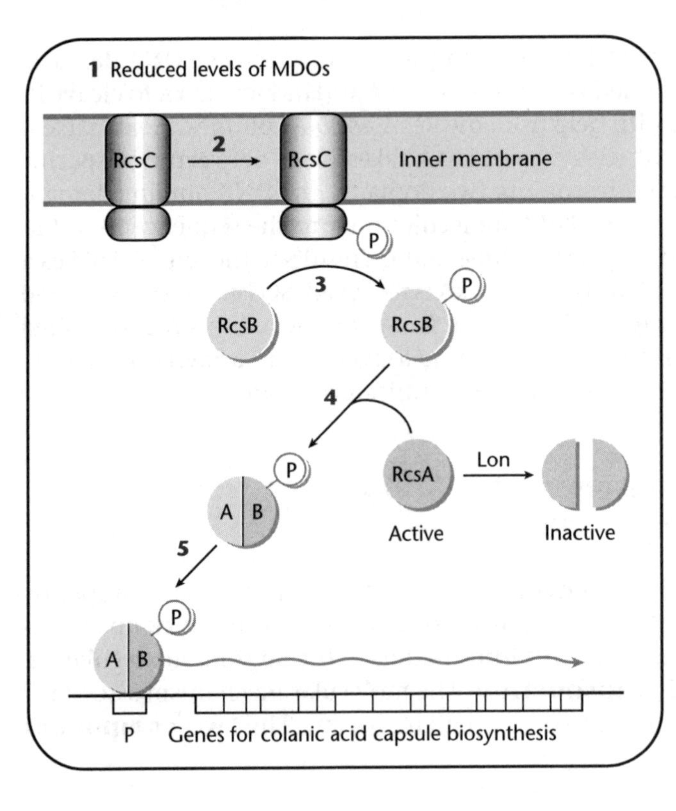

Fig. 12.19 Regulation of the *cps* genes. (Step 1) Reduced levels of MDOs (membrane-derived oligosaccharides) result in autophosphorylation of RcsC (Step 2). (Step 3) The phosphate group is transferred to RcsB. (Step 4) RcsB binds to RcsA. (Step 5) This complex acts on the *cps* promoter and activates transcription of the *cps* genes. RcsA is also an unstable protein that is degraded by Lon protease. When RcsA is in a complex with RcsB, it is protected from degradation. Expression of *cps* genes is controlled by several different mechanisms to ensure that the capsule in only made when needed.

An extensively studied *E. coli* exopolysaccharide is the colanic acid capsule. The genes encoding the enzymes that make this polymer are named *cps* (colanic acid polysaccharide). The majority of the *cps* genes are arranged in a single operon. The regulation of the *cps* genes involves several different regulatory mechanisms (Fig. 12.19). One such mechanism is a two-component system involving the sensor-kinase protein, RcsC (regulator of capsule synthesis), and the response-regulator protein, RcsB. This two-component system can be activated by osmotic shock or by cell desiccation. The signal that RcsC responds to during periods of high osmolarity is the level of membrane-derived oligosaccharides (MDOs). Levels of MDOs are low during periods of high osmolarity, and the reduced levels of MDOs are thought to act as a signal to trigger RcsC autophosphorylation. Once RcsC is phosphorylated it transfers the phosphate to the response-regulator, RcsB. Phosphorylated RcsB interacts with a second positive regulatory protein, RcsA, to activate the expression of the *cps* genes and lead to biosynthesis of the colanic acid capsule.

Interestingly, the heat-shock protease, Lon, also plays a role in regulating the expression of the *cps* genes. RcsA is an unstable protein and is degraded by Lon. This instability keeps the levels of RcsA in the cell very low and the capsule genes turned off. Removing Lon activity increases levels of RcsA and together with phosphorylated RcsB, this complex turns on *cps* genes expression. It is estimated that there are approximately 40 *cps* genes or about 1% of the genes in *E. coli*. Thus, making the colanic acid capsule requires a significant amount of energy and cellular resources. The cell uses several mechanisms to ensure that the capsule is only made when needed.

Summary

Many mechanisms regulate gene activity. The goal is to express genes only when needed. The catabolic machinery needed to metabolize lactose only needs be available when lactose is present. Likewise, tryptophan only needs to be made when the cell runs out of this amino acid. Transcriptional regulation is frequently observed because transcription is the first step in expressing a gene. Transcriptional regulation for the *lac* and *trp* operons involves a repressor, LacI or TrpR, respectively. LacI represses transcription when not bound by the inducer lactose. In contrast, TrpR represses transcription only when bound by tryptophan. Both of these operons have at least one additional regulatory feature. For the *lac* operon, its transcription is positively regulated by cAMP–CAP. For the *trp* operon, its transcription is fine-tuned by the process of attenuation. Attenuation is the mechanism by which the transcribing RNA polymerase pays attention to the translating ribosome, and bases its decision as to whether to transcribe the *trp* structural genes on if the translating ribosome stalls in its movement along the leader mRNA.

Regulons involve large networks of genes, often arranged in multiple operons, yet regulated by a common global regulatory mechanism. The heat-shock, SOS, and *cps* capsule regulons are controlled by several different mechanisms. The heat-shock alternate sigma factor, σ^{32}, directs RNA polymerase to transcribe only heat-shock promoters. σ^{32} levels are controlled by translational repression and protein stability. The SOS regulon is induced upon the removal of the LexA repressor. RecA complexed to single-stranded DNA facilitates LexA autocleavage, effectively removing LexA from SOS promoters. Signal transduction, through the transfer of phosphate groups, from RcsC to RcsB regulates the expression of genes needed to make colanic acid capsular polysaccharide. The second regulator of the capsule genes, RcsA, is regulated by protein stability.

Many examples of operons and regulons are known in *E. coli* and other prokaryotes. Conjugation under the control of the *tra* operon (Chapter 10) and λ's lytic/lysogenic decision-making process controlled by multiple operons (Chapter 7) represent examples of groups of genes coordinately regulated. Areas of intense research today center on the regulation of virulence gene activity in bacterial pathogenicity. The products encoded by virulence genes allow pathogenic bacteria to cause disease. Some virulence genes are organized and controlled as global regulons. Examples include regulation of the toxin genes of *Vibrio cholerae*, the causative agent of cholera and the regulation of the toxin genes of *Corynebacterium diphtheriae*, the causative agent of diphtheria. A sensor regulator pair involved in signal transduction regulates expression of the virulence genes of *Bordetella pertussis*, the causative agent of whooping cough.

Study questions

1 Give two examples each of positive and negative regulation.
2 Is it guaranteed that once mRNA is expressed that the encoded gene product will also be expressed? Why or why not?
3 What steps in the production of a protein can be regulated? What steps in the production of RNA can be regulated?
4 How does the presence of glucose in the medium in which *E. coli* is grown affect the synthesis of *E. coli*'s enzymes involved in the utilization of lactose? What happens at the molecular level to explain this effect?
5 Name a compound that only acts as an inducer of the *lac* operon and explain why it cannot be a substrate. Name a compound that is a substrate for β-galactosidase and explain why it cannot act as an inducer.
6 Describe transcriptional attenuation. Why hasn't this regulatory mechanism been described in eukaryotic cells?
7 Suppose you isolated a new bacterial species in the genus, *Finalococcus* and you wish to learn whether it has a multigene response (e.g. regulon) to the stress of taking a final exam. How would you experimentally demonstrate that this bacterium had a multigene response to the stress of taking a final?
8 What would happen to the regulation of the SOS regulon if the LexA binding regions were deleted? How would this impact the cell?
9 Contrast and compare repression in the *lac* and *trp* operons.
10 Why are super-repressor mutants dominant to wild-type repressor?

Further reading

Gottesman, S. 1995. Regulation of capsule synthesis: modification of the two-component paradigm by an accessory unstable regulator, In *Two-Component Signal Transduction*, eds. J. Hoch and T.J. Silhavy, pp. 253–62. Washington, DC: ASM Press.

Jacob, F. and Monod, J. 1961. Genetic regulatory mechanisms in the synthesis of proteins. *Journal of Molecular Biology*, **3**: 318–56.

Schwartz, D. and Beckwith, J.R. 1970. Mutants missing a factor necessary for the expression of catabolite sensitive operons in *E. coli*. In *The Lactose Operon*, eds. J.R. Beckwith and D. Zipser, pp. 417–22. Cold Spring Harbor, NY: Cold Spring Harbor Laboratory Press.

Walker, G.C. 1996. SOS regulon. In *Escherichia coli* and *Salmonella typhimurium: Cellular and Molecular Biology*, 2nd edn., eds. F.C. Neidhardt, R. Curtiss III, J.L. Ingraham, E.C.C. Lin, K.B. Low, B. Hagasanik, W.S. Rexnikoff, M. Riley, M. Schaechter, and H.E. Umbarger. Washington, DC: ASM Press.

Weigle, J. 1953. Induction of mutations in a bacterial virus. *Proceedings of the National Academy of Science, USA*, **39**: 628–36.

Yanofsky, C. 1981. Attenuation in the control of expression of bacterial operons. *Nature*, **289**: 751–8.

Chapter 13

Plasmids, bacteriophage, and transposons as tools

Genes reside in chromosomes where they have their own promoters and regulatory signals to direct their expression. A gene is usually present in a chromosome in a single copy. To study a specific gene, and its RNA or protein product, it is useful to take the gene out of the chromosome and place it in a situation where its expression and copy number can be controlled and manipulated. The gene is removed from the chromosome and placed in a piece of DNA, called a **vector**. A vector is capable of replicating and being transmitted to daughter cells (Fig. 13.1). The process of moving the gene from the chromosome to the autonomously replicating vector DNA is called **cloning**. In this chapter we will examine a number of different vectors and their uses. The enzymes and techniques used in cloning will be examined in Chapter 14.

Genes are cloned for several reasons. First, moving a gene from the chromosome to a vector allows the isolation of many more copies of the gene. The vector can contain a number of different promoters or biological tags that can be used for such diverse experiments as changing the expression of the gene, tagging the resulting gene product, sequencing the gene, aiding in mutagenizing the gene, or altering the cellular location of the gene product, to name a few. While cloning a gene opens up many experimental possibilities, the appropriate vector must be chosen to meet the requirements of the planned experiment. It is not unusual to clone a gene of interest into several different vectors, depending on the experiments to be carried out.

FYI 13.1

The frequency of restriction endonuclease cleavage sites

Given a restriction endonuclease that recognizes a 6 base pair sequence, the probability that this sequence will occur randomly in any piece of DNA is 4^6. The 4 represents the four bases that occur in DNA and the sixth power is the length of the sequence of interest. This means that any given 6 base pair sequence will occur once in every 4096 base pairs of DNA if the sequence of the DNA is random. It is not unusual for the DNA sequence to be nonrandom and restriction sites can occur more or less frequently than predicted.

What is a cloning vector?

A vector is the piece of DNA that the cloned gene is put into (Fig. 13.2). It must have an origin of DNA replication so that it can be replicated and transmitted to daughter cells. It is useful if the vector also contains a **selectable marker**, such as a gene for an **antibiotic resistance determinant**. By selecting for the appropriate antibiotic resistance, the presence of the vector in cells can be easily detected. Vectors are

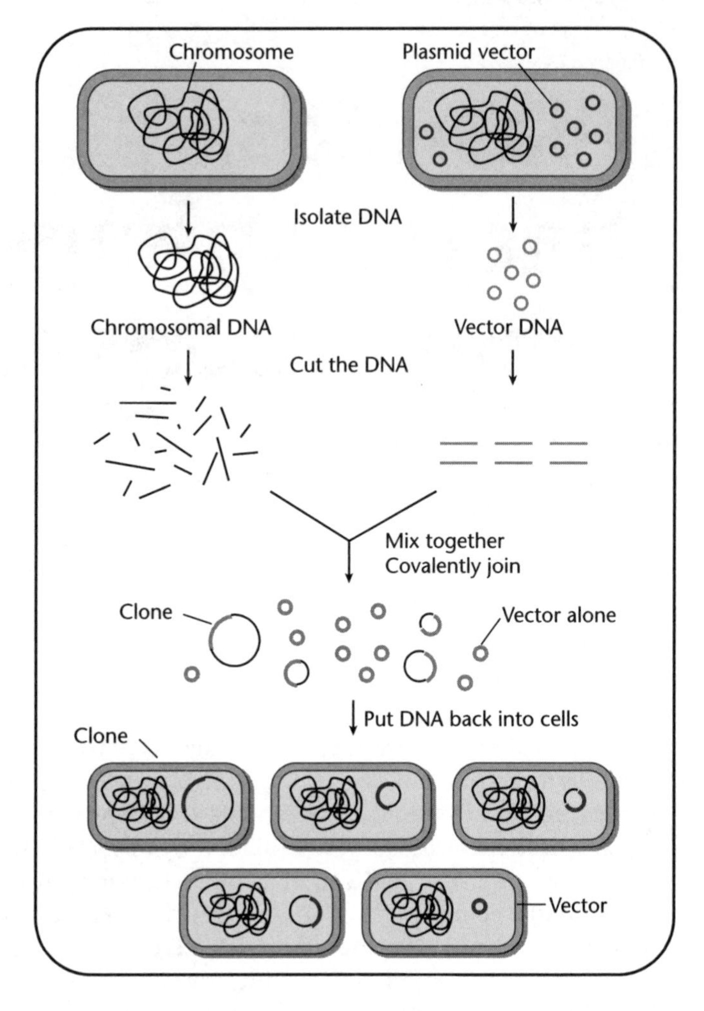

Fig. 13.1 Cloning involves isolating chromosomal and vector DNA, digesting the DNAs with restriction enzymes, and covalently joining pieces of the chromosomal DNA to the vector DNA. The rejoined chromosomal and vector DNA is put back into living cells and is propagated using signals on the vector DNA.

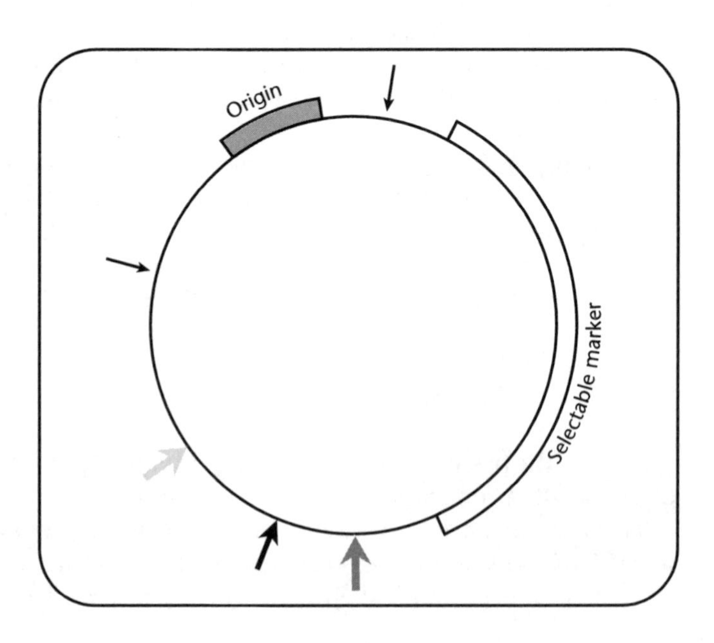

Fig. 13.2 A vector must contain an origin of replication. A selectable marker and unique restriction enzyme cleavage sites (arrows) make the vector more useful.

usually small pieces of DNA, under 10 kilobases for plasmids. The small size is useful for manipulations in vitro as well as for maintaining the clone in cells. Lastly, vectors need to contain **restriction sites** or DNA sequences that are recognized and cleaved by **restriction endonucleases** (see Chapter 14). The restriction sites provide places for the cloned gene to be inserted into the vector.

All of the characteristics either required or desirable for a vector can be found in plasmids and phage. Plasmids contain an origin of replication and are faithfully partitioned to daughter cells. They frequently contain antibiotic resistance determinants, are small in size, and contain the DNA sequences for restriction endonucleases. Phage genomes contain origins of replication and packaging signals for stuffing DNA into a phage capsid so that the DNA can be easily moved from cell to cell. Phage genomes can be designed to contain selectable markers and convenient restriction enzyme sites. We will examine plasmids as vectors first, then examine the unique uses for phage as vectors.

Why not use naturally occurring plasmids as vectors?

Naturally occurring plasmids have many of the determinants needed to make a good cloning vector, but rarely do they occur in optimal combinations. A naturally occurring plasmid's origin of replication forms the backbone of every plasmid cloning vector. To the origin of replication a variety of selectable markers are added. Sometimes the selectable marker used in the cloning vector comes from the naturally occurring plasmid, other times the selectable marker is from the chromosome, a bacteriophage, or a transposon. Naturally occurring plasmids frequently contain DNA sequences that are not essential for the replication and transmission of the plasmid in cells grown in the laboratory. These nonessential sequences can be deleted to make the vector smaller in size. And lastly, naturally occurring plasmids rarely contain more than a few useful restriction sites and even more rarely do they contain useful biological tags. These features of cloning vectors are always added to the basic backbone of the naturally occurring plasmid.

The importance of copy number

In general, the higher the copy number of the plasmid, the easier it is to obtain large quantities of pure DNA for *in vitro* manipulations. In many cases, high copy number plasmids work very well and the cells will tolerate a cloned gene in many more copies per cell than normal. However, in some cases, when a gene is present in a large number of copies, it is detrimental or even lethal to the cell. For this reason, plasmids with different copy numbers are used to construct cloning vectors. If a gene is detrimental or lethal to cells when present in 100 copies per cell then a lower copy number vector can be tested. Vectors with copy numbers of 1–2, 10–15, ~50, and ~200 copies per cell have been developed.

An example of how a cloning vector works—pBR322

To demonstrate how cloning vectors are used, it is helpful to look at the first plasmid vector constructed, pBR322 (Fig. 13.3). pBR322 replicates in Gram-negative bacteria

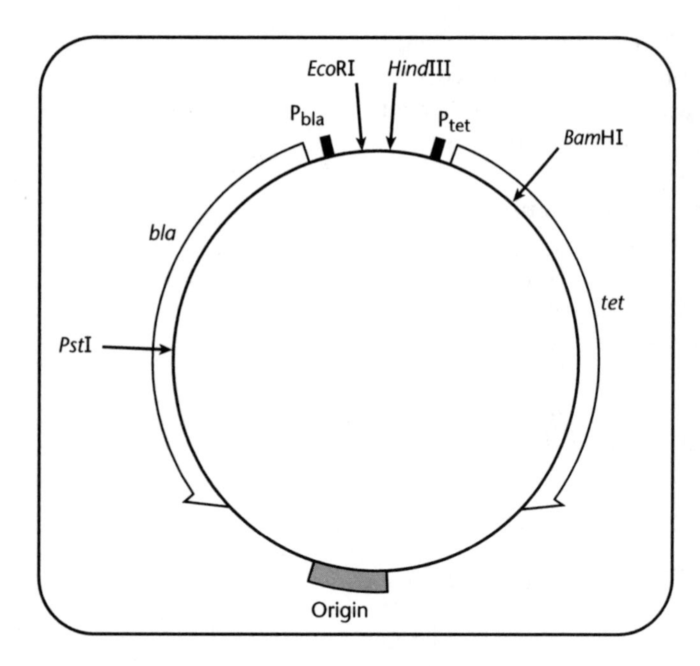

Fig. 13.3 A map of the first cloning vector, pBR322. *PstI*, *Eco*RI, *Hind*III, and *Bam*HI are restriction sites.

such as *E. coli* and *Salmonella enterica* serovar *typhimurium*. pBR322 contains the replication origin from ColE1 and, as such, is present in approximately 30 to 50 copies per cell. It contains two antibiotic resistance genes that produce proteins that impart a selectable phenotype on the cells. One of these makes the cells resistant to the antibiotic ampicillin by producing the enzyme β-lactamase (the *bla* gene from the transposon, Tn*3*) and the other makes the cells resistant to the antibiotic tetracycline (the *tet* gene from the plasmid, pSC101). β-lactamase actually cleaves the ampicillin to inactivate it. The tetracycline resistance protein is a pump that resides in the membrane and pumps the tetracycline out of the cell. pBR322 contains several useful restriction enzyme sites. One of these is for the enzyme *Bam*HI. The *Bam*HI site is in the *tet* gene. If a gene is cloned into the *Bam*HI site in pBR322 then the cells containing such a plasmid will no longer be resistant to tetracycline. However, because the *bla* gene is intact, cells containing such a plasmid will still be resistant to ampicillin. By cloning genes into one of the two selectable markers, it is easy to identify a plasmid containing an insert.

Multiple cloning sites

In the example given above, pBR332 has only a few useful restriction sites for cloning purposes. In order for a restriction site to be useful for cloning, it must occur only once in the plasmid vector. It must also not be in a region of the DNA that is essential for the plasmid, such as the origin of replication. Because plasmids usually have a small number of useful restriction sites, **multiple cloning sites** (MCS) have been synthesized *in vitro* and inserted into plasmid vectors. Multiple cloning sites are simply a string of base pairs, usually 50 to 75 base pairs in length that specify restriction enzyme sites for many different restriction enzymes (Fig. 13.4). The restriction enzymes sites are usually chosen based on two criteria. First, the site does not occur anywhere else in the vector and, second, the restriction enzyme is readily available. Most cloning vectors contain one MCS.

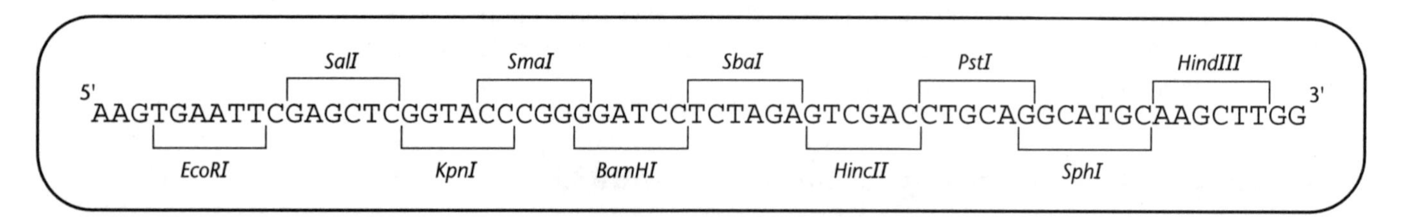

Fig. 13.4 The multiple cloning site (MCS) from the vector pUC19. This piece of DNA was synthesized in vitro and cloned into pUC19. Each of the restriction enzyme sites in the MCS cleaves pUC19 only once, making each of them useful for cloning.

Determining which plasmids contain an insert

As described in detail in Chapter 14, cloning is a very inefficient process. Isolating chromosomal DNA, digesting it with restriction enzymes, and joining it to a vector that itself has been digested with a restriction enzyme requires many experimental manipulations on a small amount of DNA. Consequently, when this mixture of chromosomal and vector DNA is transformed back into cells (see Chapter 11 for a discussion of transformation), many of the vector plasmid molecules will not contain a chromosomal DNA insert. A group of vectors (pUC plasmids) has been constructed so that plasmid molecules that do have a chromosomal insert can be distinguished from those molecules that do not contain an insert (Fig. 13.5). The pUC vectors contain a fragment of the *lacZ* DNA that encodes the α-complementing fragment of β-galactosidase (see Chapter 12 for a discussion of the *lac* operon). β-galactosidase can physically be divided into two parts. The first ~60 amino acids are known as the **α fragment** and the last ~960 amino acids are known as the **ϖ fragment**. If the α and ϖ fragments are expressed from two physically distinct partial *lacZ* genes, the resulting two protein fragments will associate and form an active enzyme. The pUC vectors contain the α fragment of the *lacZ* gene. If the ϖ fragment is present in the cells containing pUC, then the cells are $LacZ^+$. The ϖ fragment can be carried on an F′, a lysogenized phage, or another plasmid that can be maintained in the cell with the pUC vector. A derivative of lactose, 5-bromo-4-chloro-3-indolyl-β-D-galactoside (Xgal, Fig. 12.10b), can be used to indicate if the cells are $LacZ^+$ or $LacZ^-$. The MCS is located in the α fragment of the *lacZ* gene in pUC. pUC vectors without an insert produce blue-colored colonies in the presence of the ϖ fragment when plated on agar containing XGal. If a piece of chromosomal DNA is inserted into the MCS of the pUC vector, then only colorless or "white" colonies result. This is the basis of the **blue/white screen** used in many vectors.

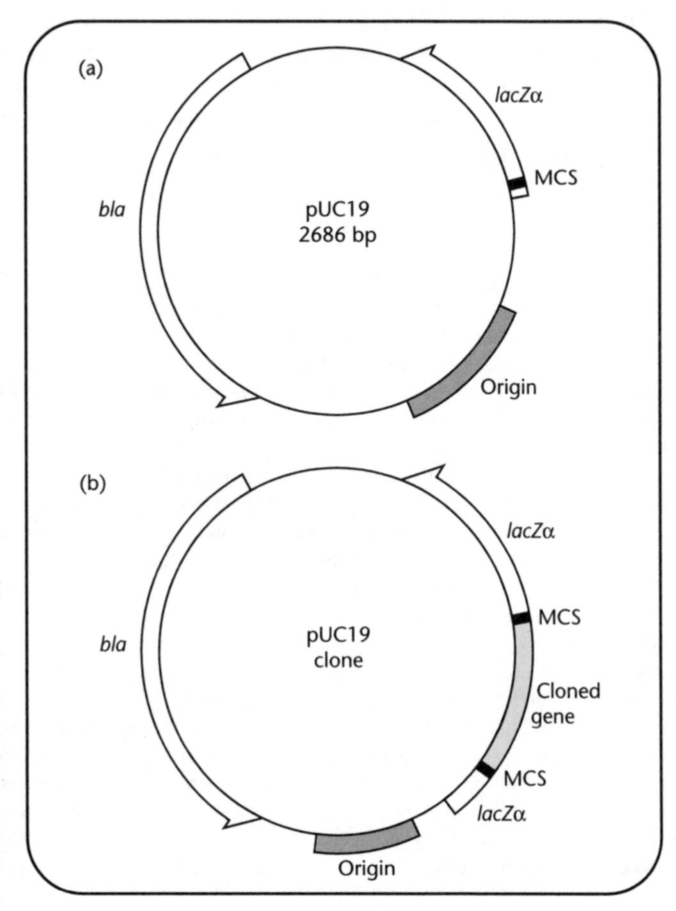

Fig. 13.5 (a) pUC19 forms blue colonies when put in a cell containing the ω fragment of *lacZ*. (b) pUC19 with an insert forms white colonies in the presence of the ω fragment.

Expression vectors

Many different types of cloning vectors have been developed to specifically direct the transcription and translation of a cloned gene. Collectively, vectors designed for these purposes are known as expression vectors (Fig. 13.6). The promoters used in expression

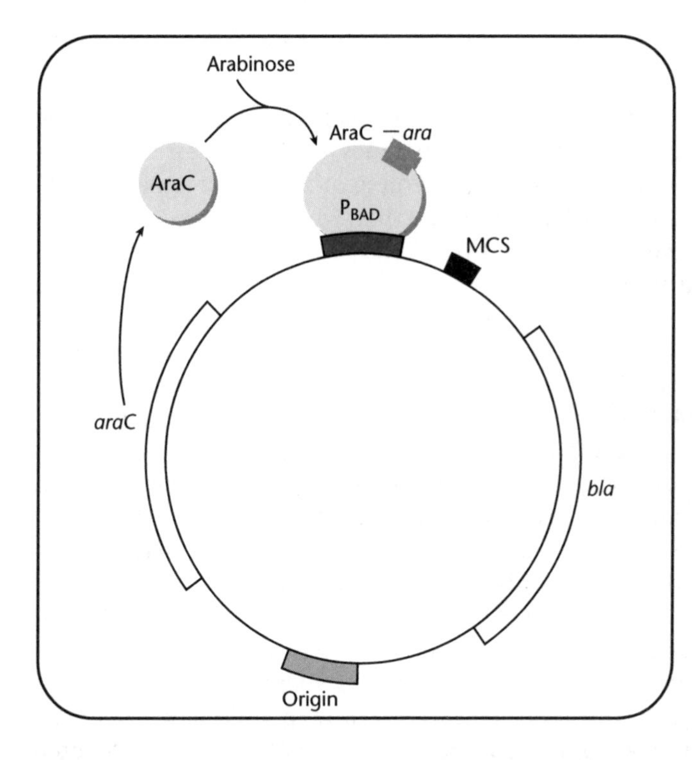

Fig. 13.6 An example of an expression vector. P_{BAD} is a promoter from the arabinose operon. If no arabinose is present, the P_{BAD} promoter is not active. If arabinose is added, it complexes with the protein encoded by the *araC* gene. The arabinose-AraC complex binds to P_{BAD} and activates it.

vectors are chosen with care. It is most desirable if the cloned gene is not expressed under one growth condition and yet is expressed under another. Two promoters, one from the *lac* operon and one from the *ara* operon, have been used successfully for the expression of cloned genes in bacteria. The *lac* promoter is induced by the addition to the growth media of either lactose or a non-metabolizable analog of lactose called isopropyl-β-D-thiogalactoside (IPTG, Fig. 12.10b). The *ara* promoter is induced by the addition of arabinose. For promoters used in expression vectors, the major concern is the level of transcription when no inducer is present, which is described as the basal level of transcription. The *lac* promoter has a higher basal level of transcription than the more tightly regulated *ara* promoter. If the cloned gene is to be expressed in eukaryotic cells then a promoter that is active in eukaryotic cells is built into the expression vector. Like bacterial promoters, there are constitutively expressed and regulated eukaryotic promoters. Unlike bacteria, some

FYI 13.2

Components that can be included in vectors

Many other useful elements have been incorporated into specific vectors. It would be nearly impossible to catalog each and every element that has been used in the thousands of vectors that have been constructed. Many of these elements function independently and can be used in a mix and match fashion. Some of these elements include:

- Lox and Cre—Cre is a protein that recognizes and recombines two identical sequences of DNA called *lox* sites. If one *lox* site is incorporated into the chromosome and the other in a plasmid vector, then the vector will be recombined into the chromosome. This strategy is used to recombine plasmid clones into a specific place on a eukaryotic chromosome.
- M13 origin—M13 is a single-stranded DNA phage of *E. coli*. During its lifecycle it replicates one strand of its DNA from a specific origin without replicating the other strand. If this M13 origin is incorporated into a vector then a single strand of the vector is produced.
- CMV promoter—CMV is a human virus (cytomegalovirus) that contains very strong promoters that function in mammalian cells (much like T7 promoters in *E. coli*). The CMV immediate early promoter results in high-level constitutive expression of any downstream gene in mammalian cells.
- Poly A sites—Eukaryotic mRNAs contain a string of adenines that are added to the 3′ end of the transcript (poly-A-tail) at a specific sequence in the mRNA. The sequence for the poly-A-tail can be inserted into a vector after the MCS so that the mRNA from the clone can be poly-A-tailed and subsequently translated in eukaryotic cells.
- G418 resistance—Most antibiotics that function in bacterial cells do not function in mammalian cells. One antibiotic resistance marker that does function in mammalian cells is for resistance to the antibiotic known simply as G418. For vectors used in mammalian cells, the gene for G418 resistance is frequently used as the selectable marker.
- SV40 origin—SV40 is a mammalian virus. In order for a plasmid to replicate in mammalian cells, it needs a mammalian origin and the SV40 origin fits the bill.
- Localization signals—Both eukaryotic and prokaryotic cells utilize specific amino acid sequences to tell them where a protein belongs in the cell. In all cells, membrane proteins frequently have an amino terminal signal sequence. In eukaryotes, nuclear proteins have a nuclear localization signal. The codons that specify localization signals can be incorporated into plasmid vectors to direct cloned proteins to specific places in the cell.

eukaryotic promoters are expressed in specific tissues or at specific times in development.

Vectors for purifying the cloned gene product

A second group of expression vectors has been developed for a completely different purpose. These vectors use regulated bacteriophage promoters to drive the expression of as much of the cloned gene product as is possible. Vectors containing either lambda or T7 promoters have been developed. These phage promoters are some of the strongest known promoters in biological systems. The phage promoter vectors are used to purify the cloned gene product. In the case of the T7 vectors, the cloned gene product can be overexpressed until it represents up to 10% of the total cell protein. At this level, the cloned gene product is frequently lethal to the cells. As long as the cells do not lyse and the main objective remains purification of the gene product, it does not matter if the cells are alive or dead.

The T7 and lambda promoter vectors are designed to express large amounts of a wild-type protein. Several other vectors have been constructed to make purification of proteins easier. Several histidines in a row (usually 6–8) will bind reversibly to metal ions, especially nickel (Ni^{++}) and zinc (Zn^{++}). If the codons specifying a histidine tag are added to a gene to make a gene fusion, the resulting protein will contain the **histidine tag**. This his-tagged protein will usually bind to nickel ions, a property that allows purification of the his-tagged protein in one step. To further aid in characterizing his-tagged proteins, several antibodies to the his-tag have been developed and characterized. The antibodies will recognize almost any his-tagged fusion protein.

Two other types of fusion vectors have proven extremely useful for purifying proteins. In one, the coding sequences for a 26 kilodalton domain from glutathione S-transferase (GST) can be fused to the gene of interest. This portion of GST binds to glutathione. The resulting **GST-fusion protein** will usually bind to glutathione that is coupled to a solid support. Free glutathione will release the fusion protein from the glutathione coupled to the solid support. Like his-tags, GST tags allow purification in a single step. **Maltose binding protein (MBP)**, a protein normally exported to the periplasm of *E. coli*, can also be used to make fusion proteins for purifying gene products (Fig. 13.7). MBP binds to maltose. The fusion proteins usually can attach to the solid-support-coupled maltose and be released by a solution of maltose. Several

FYI 13.3

Expression from a T7 promoter

The T7 promoter used in cloning vectors is one of the strongest known promoters in bacteria. T7 promoters require their own RNA polymerase for transcription. The T7 RNA polymerase is cloned under the *lac* promoter and integrated into the chromosome of specific strains of *E. coli*. Once a gene is cloned behind the T7 promoter in a vector, the clone is moved into the strain containing the T7 RNA polymerase. Once synthesis of the T7 RNA polymerase is induced, transcription from the T7 promoter will begin.

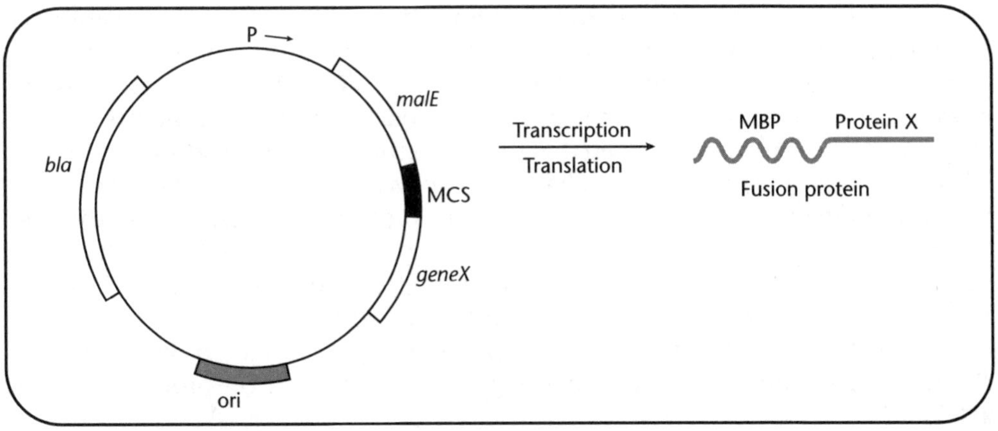

Fig. 13.7 Vectors to make maltose binding protein fusions contain the coding sequences for *malE* (encodes maltose binding proteins) and an MCS. The gene of interest is cloned into the MCS in frame with *malE*, resulting in a protein fusion between *malE* and the gene of interest.

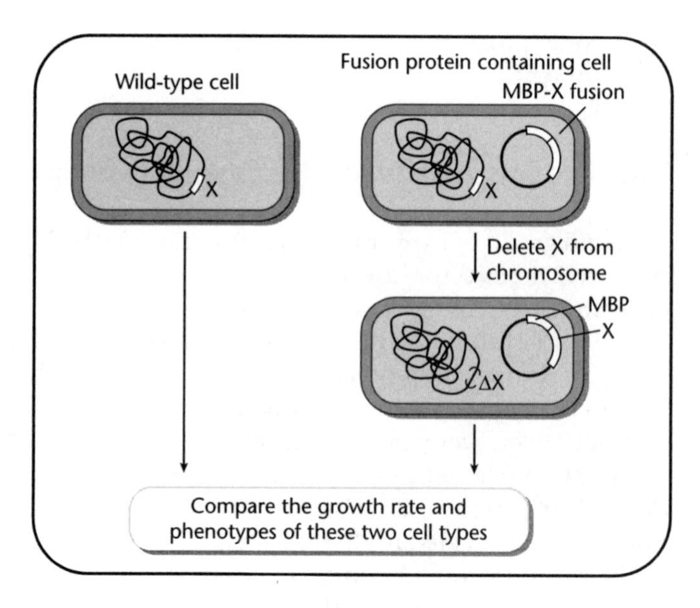

Fig. 13.8 To determine if a fusion protein has the same properties as the wild-type protein, the gene encoding the fusion protein is put in a cell and the gene for the corresponding wild-type protein is deleted. The phenotypes of a cell containing the wild-type gene and the fusion protein gene are compared in many different assays. If these two strains behave the same, then the fusion protein does not have any mutant properties and is studied as a representative of the wild-type protein.

other tags that bind to specific substrates have been used to make vectors similar to the ones described above. There is no guarantee that any specific tag will work with any specific gene product. Sometimes several different tags must be tried before one is found that has the desired properties.

One experiment that must be carried out on a fusion protein before a large amount of time is invested in studying it is to determine if the fusion protein has the same activities as the wild-type protein. The most reliable strategy for this experiment is shown in Fig. 13.8. First, the wild-type gene is deleted from the cell. Next, the gene for the fusion protein is put in the cell in its place. The phenotypes of the cell when it is running exclusively on the fusion protein are studied and it is determined if the cell has any growth defects or other deficiencies. If the wild-type gene is essential for the cell, the gene for the fusion protein can be put in the cell and then a deletion of the wild-type gene can be attempted. Regardless of how the experiment is carried out, assays to determine the functional status of the fusion protein must be conducted to ensure that the addition of the amino acid tag has not destroyed the protein's function.

One disadvantage of the his, GST, or MBP tagged fusion proteins is that the amino acids that comprise the tag remain attached to the protein after it has been purified. A strategy that some vectors employ to remove the tag after the fusion protein has been purified is exemplified by the vector shown in Fig. 13.9. A protease known as thrombin has a specific sequence of amino acids (LVPRGS) that it recognizes and cleaves. Cleavage occurs between R and G. The codons for these amino acids are incorporated into the vector between the codons for the tag and the codons for the gene of interest. Once the fusion protein has been expressed and purified, it can be treated with thrombin to cleave between the tag and the protein of interest. The tag can then be removed from the protein of interest and the protein of interest studied. This strategy sometimes leaves a few amino acids from the thrombin cleavage site but the number of foreign amino acids on the purified protein are fewer than without the thrombin cleavage site. Several different proteases can be used, not just thrombin. The requirements are that the cleavage sequence is known and the protease is readily available.

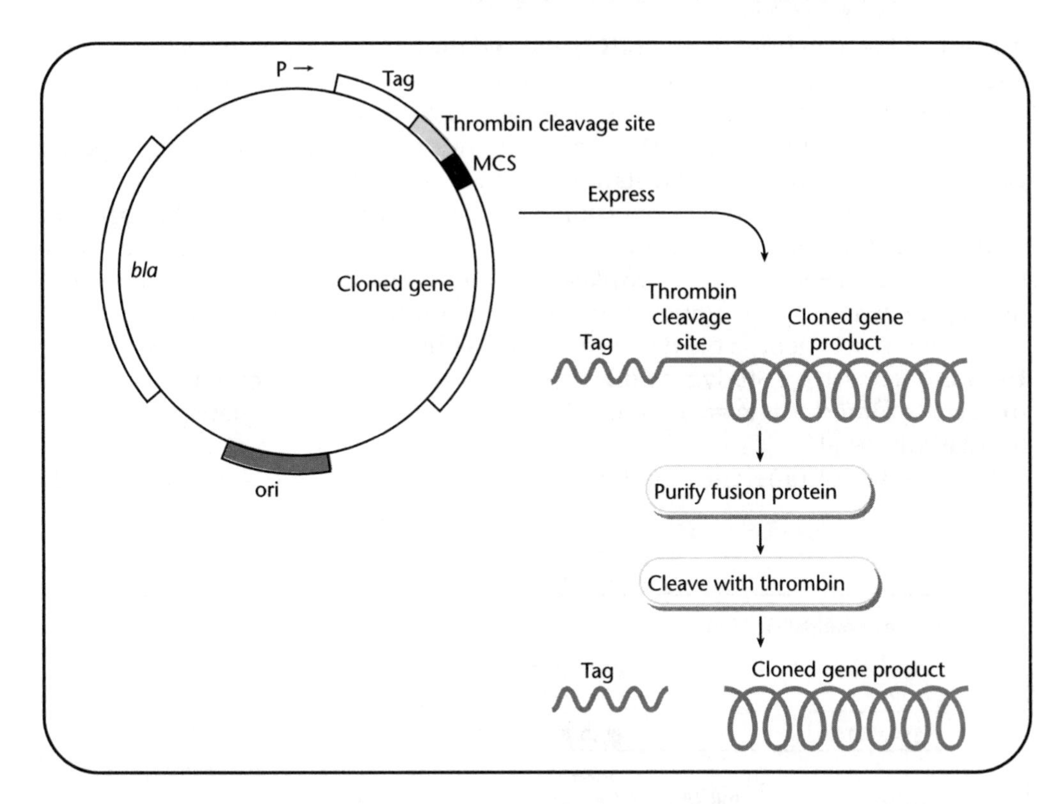

Fig. 13.9 A vector useful for protein purification is shown. The base pairs encoding a tag useful for protein purification are fused to the base pairs encoding a protease cleavage site and an MCS. The gene of interest is cloned behind this construct. When the gene fusion is expressed, the protein of interest will contain the purification tag and the protease cleavage site. The protein is purified using the tag amino acids and after purification, they are removed by cleaving the purified protein with the protease. In this example, the protease used is thrombin, which is readily available.

Vectors for localizing the gene product

For many proteins, it is desirable to know where in the cell the protein resides. One type of fusion vector that has been developed relies on the **green fluorescent protein** (GFP) isolated from the jellyfish, *Aequoria victoria*. GFP is a 35 kilodalton protein that accepts energy from blue-green light at an optimum wavelength of 400 nm and emits green light at a wavelength of 509 nm. Practically, this means that GFP fluoresces and this fluorescence can be detected by a microscope equipped with a low light camera and the appropriate filters. Fusions between the *gfp* gene and a gene of interest can be generated using *gfp*-containing vectors. If the fusion protein is active, then the location of the gene product in vivo can be determined by the location of the GFP. A major advantage to GFP is that localization of GFP can be carried out in live cells.

Vectors for studying gene expression

Determining how a gene is regulated can be carried out using a vector containing a reporter gene. A **reporter gene** is a well-characterized gene whose product can be quantitatively measured in an easy assay. The promoter region from a gene of interest is cloned in front of a reporter gene and the amount of reporter gene product and the conditions under which it is produced are examined. Two types of reporter gene constructs can be made. In the first type, the signals needed for transcription (the −35, −10, and +1 site) from the gene of interest are placed in front of the reporter gene. The signals needed for translation (the RBS and starting met codon) are already present in the vector. This type of fusion is known as a **transcriptional** or **operon fusion** (Fig. 13.10b). In the second type, all of the signals for transcription and translation from the gene of interest are placed in front of the reporter gene. This creates a novel gene whose 5′ end comes from the gene of interest and whose 3′ end comes from the reporter gene. This type of fusion is known as a **translational** or **gene fusion** (Fig. 13.10a). Studying both types of fusions gives information about how the gene of interest is regulated and if the regulation is at the transcriptional or translational level.

Any number of different reporter genes can be utilized. Many of the currently

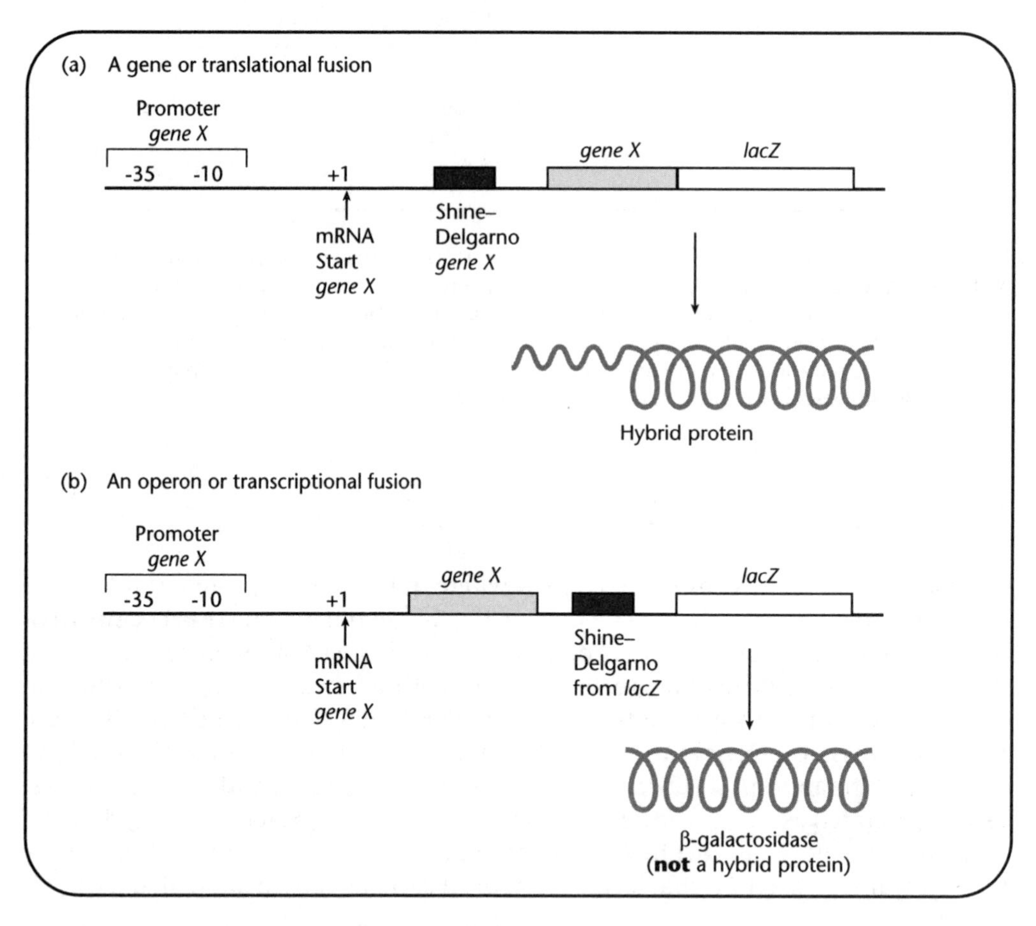

Fig. 13.10 Making gene fusions with β-galactosidase. Amino-terminal extensions on β-galactosidase do not usually interfere with the functioning of the enzyme. (a) In a gene fusion, the gene of interest is cloned in front of *lacZ* and in the same reading frame as *lacZ*. A fusion protein that contains some amino acids from the gene of interest followed by amino acids from β-gal is produced. (b) In an operon fusion, the *lacZ* gene in the vector has its own ribosome binding site. No protein fusion is created but the transcription of the *lacZ* gene is directed by the cloned promoter.

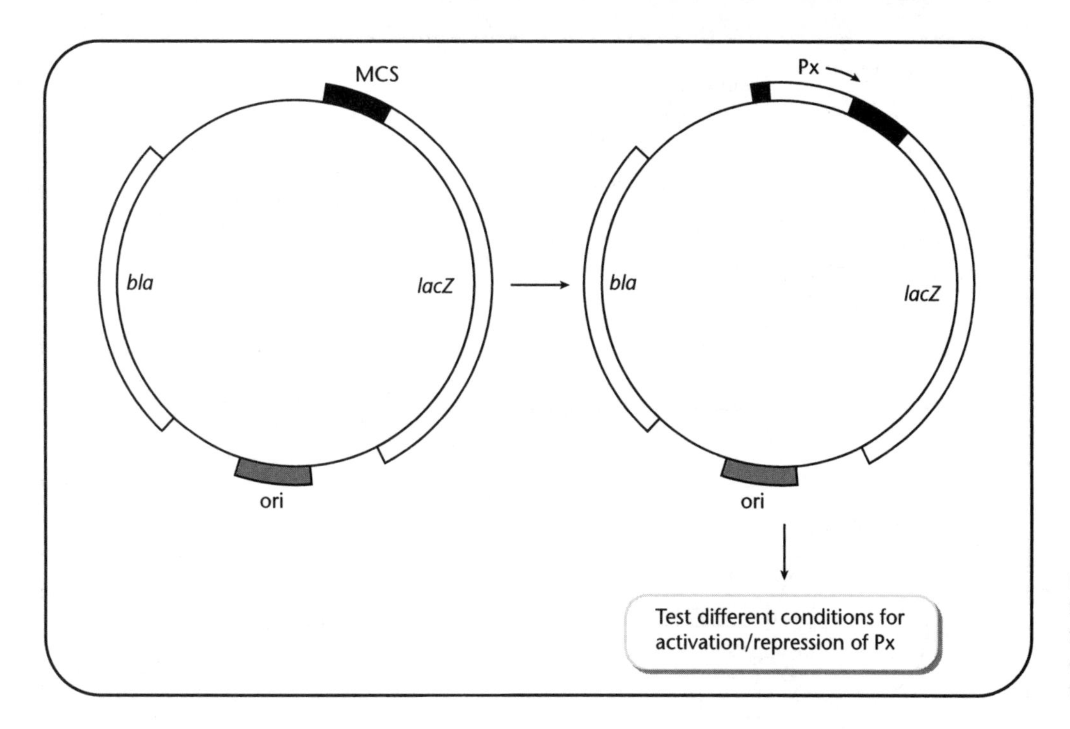

Fig. 13.11 A promoter of interest Px is cloned in front of the reporter gene, *lacZ*. Many different conditions can be tested for how they affect Px.

available reporter gene vectors use *lacZ*. *lacZ* possesses an unusual property that makes it amenable to making translational fusions. Amino acids can be added to the amino terminus of β-galactosidase, frequently without affecting the function of β-galactosidase. This means that both transcriptional and translational fusions can be made using the same reporter gene.

Two different strategies are used with the fusions vectors. In the first strategy, the promoter region from a gene of interest is cloned into a vector containing the *lacZ* reporter gene (Fig. 13.11). In the second strategy, random pieces of chromosomal DNA are inserted into the *lacZ* reporter gene vector (Fig. 13.12). These random clones are then examined for promoter activity or for promoters that respond to given stimuli, such as a change in pH, presence of a specific sugar, or growth at a specific temperature.

One problem with measuring promoter activities on plasmids is that the copy number of the plasmid can impact the measurement. A promoter present in 100 copies per cell is frequently not regulated as it is in one copy per cell. If specific protein molecules bind to the promoter to regulate it and the genes for the regulatory proteins are present on the chromosome in only one copy, there will not be enough of the regulatory proteins to physically bind to each copy of the promoter on every plasmid. For this reason, reporter gene vectors are most useful if they are present in one or a few copies per cell. Several plasmid vector-λ systems have been developed for recombining the fusion back into the chromosome (Fig. 13.13). In these systems, the fusion is constructed on the plasmid. The plasmid is recombined with a specific λ so that the fusion is moved into the λ chromosome. A single copy of the λ is lysogenized into the chromosome of the host bacterium. This approach takes advantage of the strengths of both vectors, cloning in plasmids and genetics in λ.

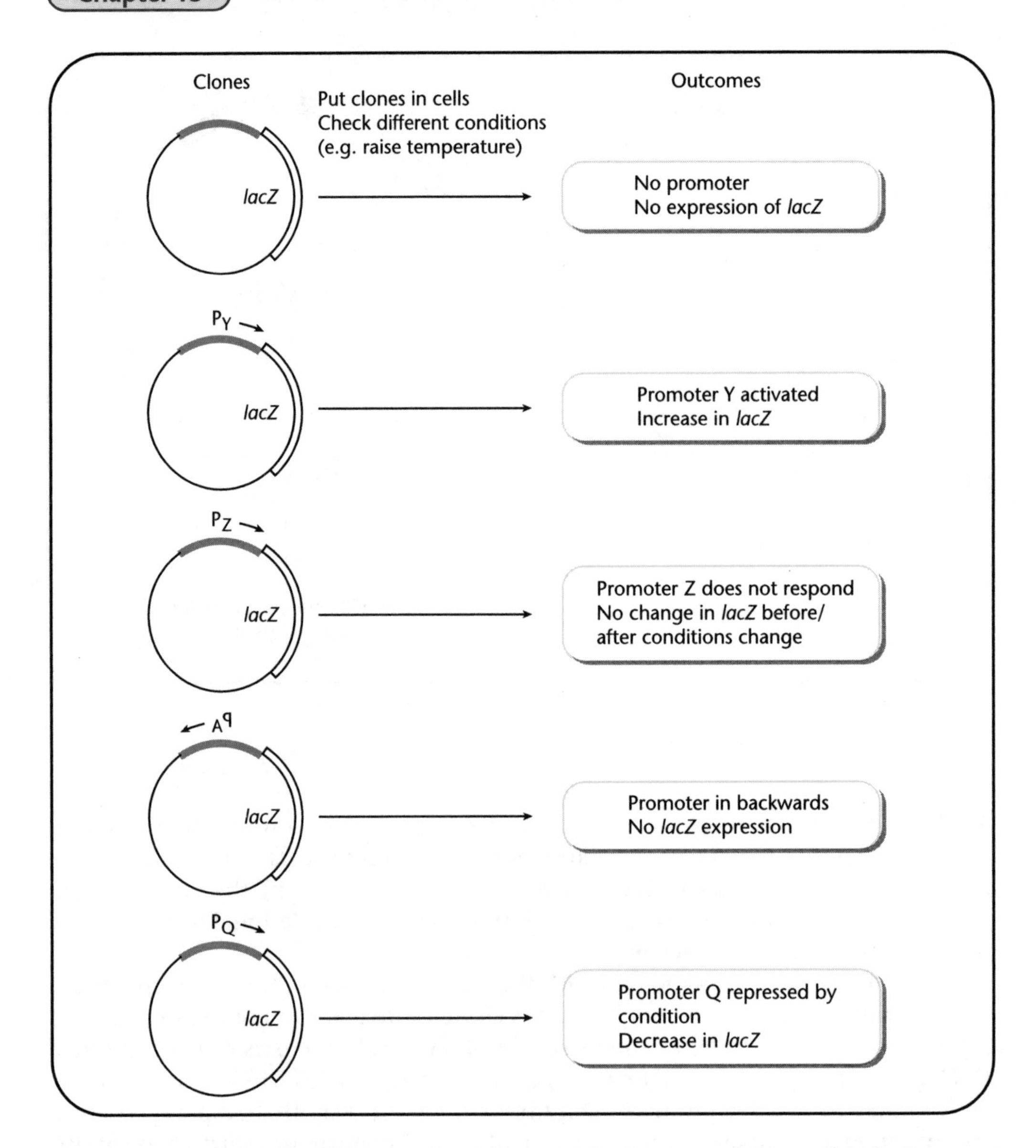

Fig. 13.12 Random fragments of chromosomal DNA (wavy line) are inserted in front of the *lacZ* gene. The clones are tested for β-galactosidase activity before and after a change in growth conditions such as a temperature change or a change in carbon source. Promoters that respond to the condition change can be identified.

Shuttle vectors

Shuttle vectors are plasmids that contain origins of replication and selectable markers for two different species (Fig. 13.14). One of the origins and one of the selectable markers function in one species and the other origin and selectable marker function in the second species. One of the species the shuttle vector replicates in is almost always *E. coli*. The other species can be as diverse as *Bacillus subtilis*, the yeast *Saccharomyces cerevisae*, rodent cell lines, plants, or human cell lines. The advantages of the shuttle vectors are that the plasmid can be isolated and manipulated in *E. coli* and the experiments can be conducted in cells from the species of interest. It can be from difficult to almost impossible to isolate large quantities of plasmid DNA from many species. Almost all of the DNA manipulation techniques were developed for *E. coli*. So it is logical to use *E. coli* for its strengths. Likewise, much of the research taking place today is on eukaryotic cells so it is very useful to be able to take finished DNA constructs back into the species of interest.

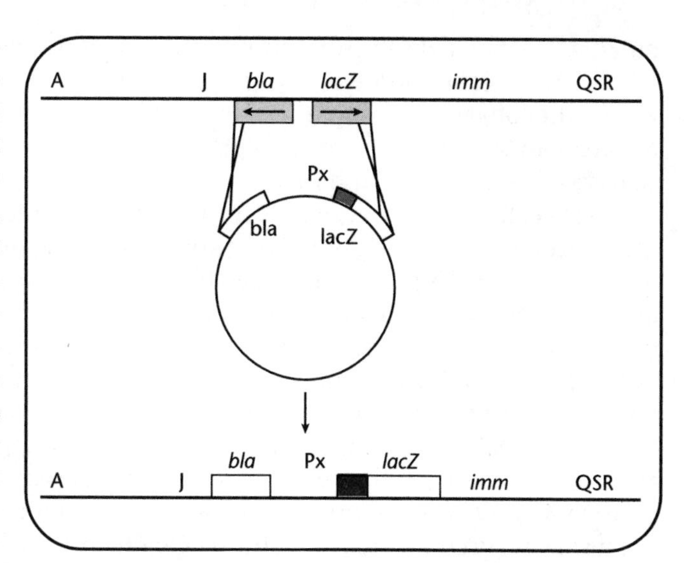

Fig. 13.13 Plasmid/λ systems have been designed so that the *lacZ* fusion can be constructed on a plasmid. Using homologous recombination, the specially designed λ vector can recombine the fusion from the plasmid onto the phage. The fusion-containing phage can be recombined into the host chromosome in a single copy.

Artificial chromosomes

With the increased frequency of sequencing whole chromosomes and the advent of bioinformatics (see Chapter 15), vectors that can accept large inserts have been developed. Most of the vectors described above were designed for inserts of ~10 kilobases or less. The artificial chromosome vectors are designed for fragments in the 40 to 1000 kilobase range. Artificial chromosome vectors have been constructed that are based on the chromosomes of λ (cosmids), P1 (PACs), bacteria (BACs), and yeast (YACs). These vectors, along with their applications are discussed in Chapter 15.

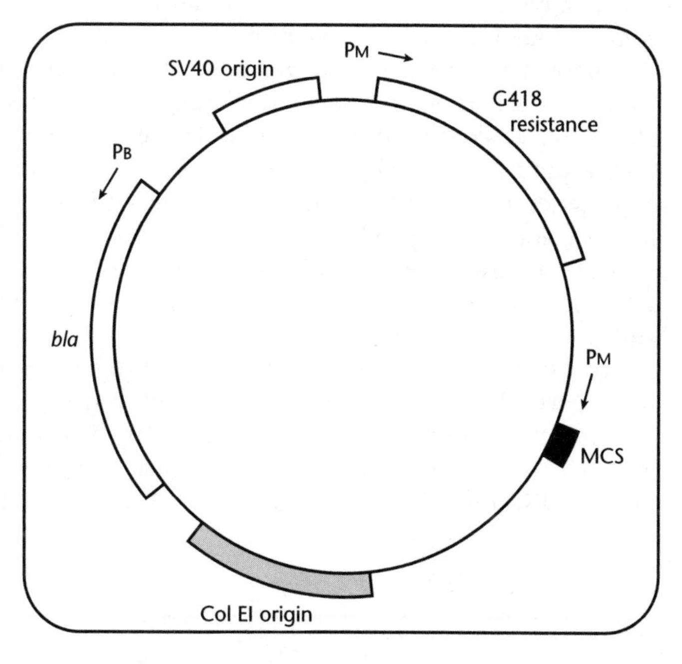

Fig. 13.14 A shuttle vector is designed to replicate in two species. The example shown above contains *bla* with a bacterial promoter and a ColEI origin for replication in bacteria. The SV40 origin works in mammalian cells and not bacterial cells. The G418 resistance determinant has a mammalian promoter. The other mammalian promoter Pm and the MCS can be used for cloning and expression of a gene of interest.

Constructing phage vectors

Many of the features described for plasmid vectors can also be incorporated into phage vectors. The reasons to use a phage vector over a plasmid vector depend on the specifics of a phage's lifecycle and those features of the lifecycle that can be exploited for experimental purposes. Phage are very good at moving DNA into cells. Upon infection by a recombinant phage, virtually every cell will receive a copy of the recombinant DNA molecule. By isolating phage, a very pure and highly concentrated source of DNA is available. Some phage are maintained as lysogens in low copy numbers, while other phage produce particles with single-stranded DNA in them. Each of these features can be exploited to make a phage vector. Phage vectors have been constructed based on λ, M13, Mu, and P1.

λ lytic vectors retain the genetic information to undergo lytic development while the information encoding lysogenic processes is removed. In λ lytic vectors, the DNA between the *J* and *N* genes can be deleted, omitting the *b* region, the *att* site, and the *int* and *xis* genes (Fig. 13.15). Unique restriction sites are created by first determining how many times a specific restriction enzyme cuts the λ DNA. Next, the sites that are not useful for cloning are removed by mutation. Sometimes, all of the sites for a given restriction enzyme are removed and one site is added back to the vector in a convenient place. DNA can be cloned into this type of vector, packaged into λ phage heads and introduced into *E. coli* cells. Lytic vectors are used to produce large quantities of pure DNA. Most lytic vectors can accept DNA fragments up to about 10 kilobases.

λ lysogenic vector systems retain the genetic information for both lytic and lysogenic development. This type of vector has a much different goal from that of a λ lytic vector. λ lysogenic vectors are used to deliver the cloned DNA to the *E. coli* chromosome through site-specific recombination between *attP* and *attB* (see Chapters 5 and 7). These types of vectors are generally not used to produce large amounts of the cloned gene. Rather, they are used to construct *E. coli* recombinants that contain one chromosomal copy of the cloned gene. Lysogenic vectors are particularly useful when a gene and its expressed protein product need to be studied in a single copy. Single-copy genes mimic physiological conditions and can frequently overcome the problem of a cloned gene product being toxic to *E. coli*.

There are several features of M13 that make it useful as a vector for recombinant DNA. Several recombinant DNA techniques require the use of single-stranded DNA and M13 specifically packages ssDNA into its phage particle (see Chapter 7). The phage particles can be easily isolated and the ssDNA recovered in a pure, concentrated form. The double-stranded replicative form (RF) of M13's genome also serves an important function. Cloned DNA can only be inserted into a dsDNA vector and, for M13, the RF DNA is used for this purpose. Adding a multiple cloning site to the M13 RF DNA provides unique restriction sites that can be used for cloning. Another

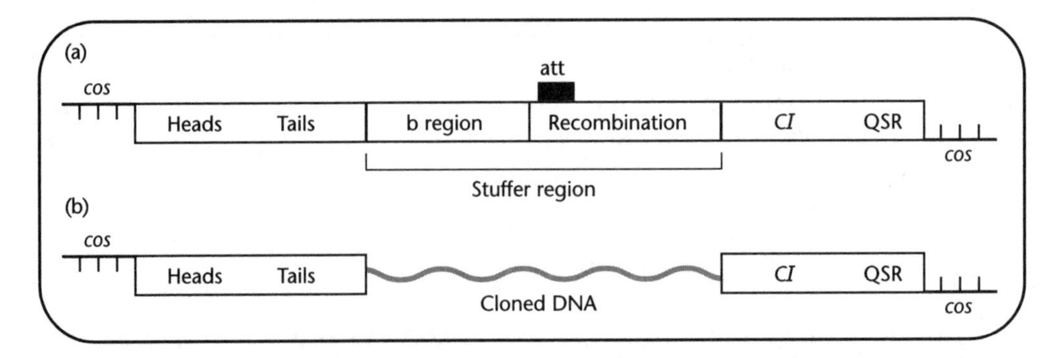

Fig. 13.15 The *b* region and recombination region of λ are nonessential for growth of the phage in the lab. They can be deleted and replaced with DNA containing genes of interest. Because λ has a size requirement for packaging its DNA into phage heads, λ vectors are usually built with a stuffer region that is removed and replaced by the cloned DNA.

positive feature of M13 is the length of the foreign DNA that can be cloned into M13 genome is not limited because filamentous phage, such as M13, do not have a fixed size. The intact phage particle will be as long as the enclosed DNA molecule. In practice, small M13 phage replicate quicker than large M13 phage. In a mixture of several different sized M13 phage, the smaller phage end up outgrowing the larger phage, even if the smaller phage start out as a minority in the population.

Suicide vectors

Suicide vectors have been developed using derivatives of λ. This type of vector is called a suicide vector because once the phage DNA enters the cell, it is not capable of replicating on its own (Fig. 13.16a). In the case of the λ suicide vectors, they have been made incapable of replicating in certain *E. coli* strains due to conditional mutations

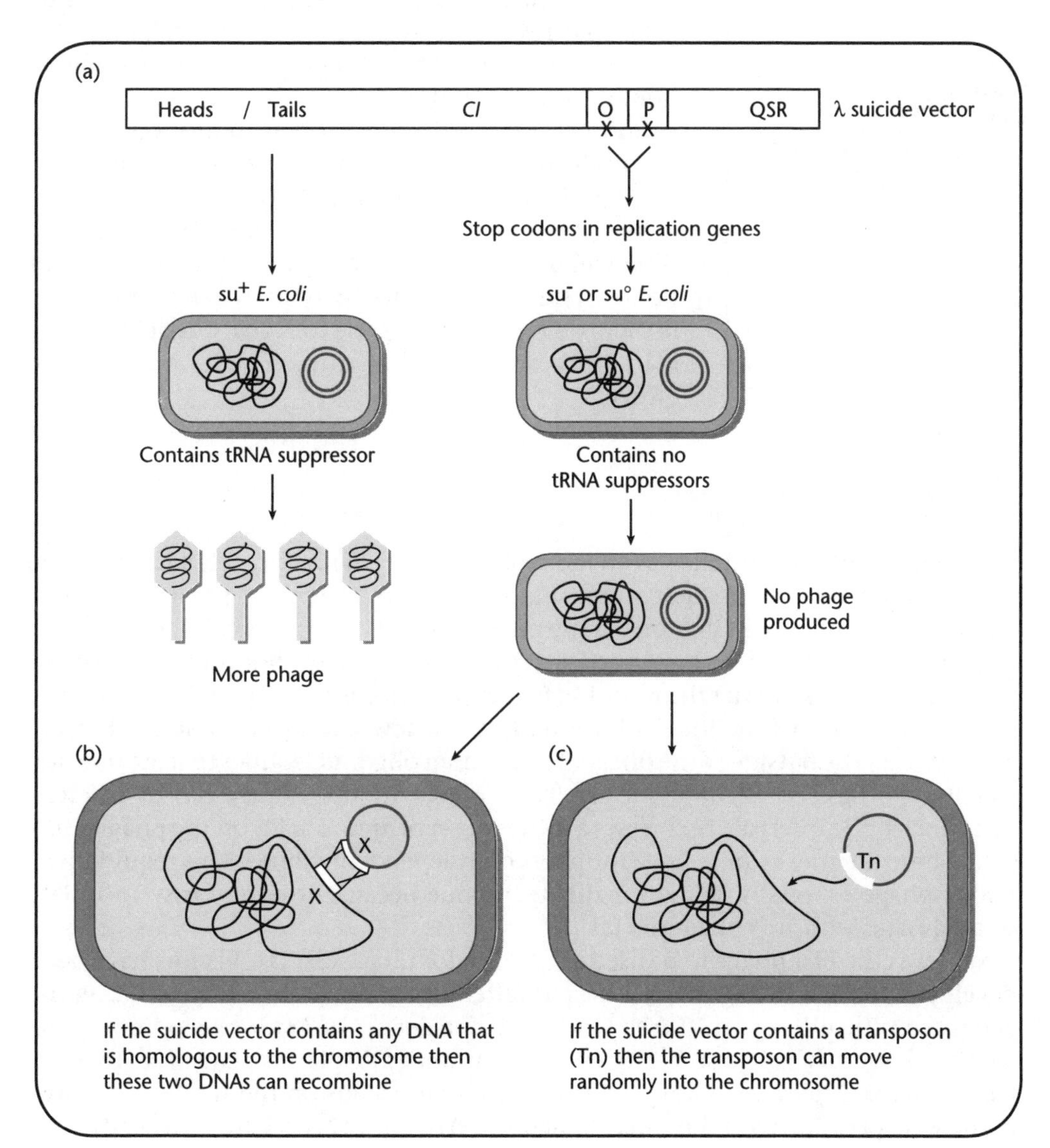

Fig. 13.16 (a) Suicide vectors grow under one set of condition and fail to grow under a different set of conditions. Growth under one set of conditions allows the vector to be propagated. Under the nonpermissive conditions, suicide vectors can be used to deliver cloned genes (b) or transposons (c) to the chromosome.

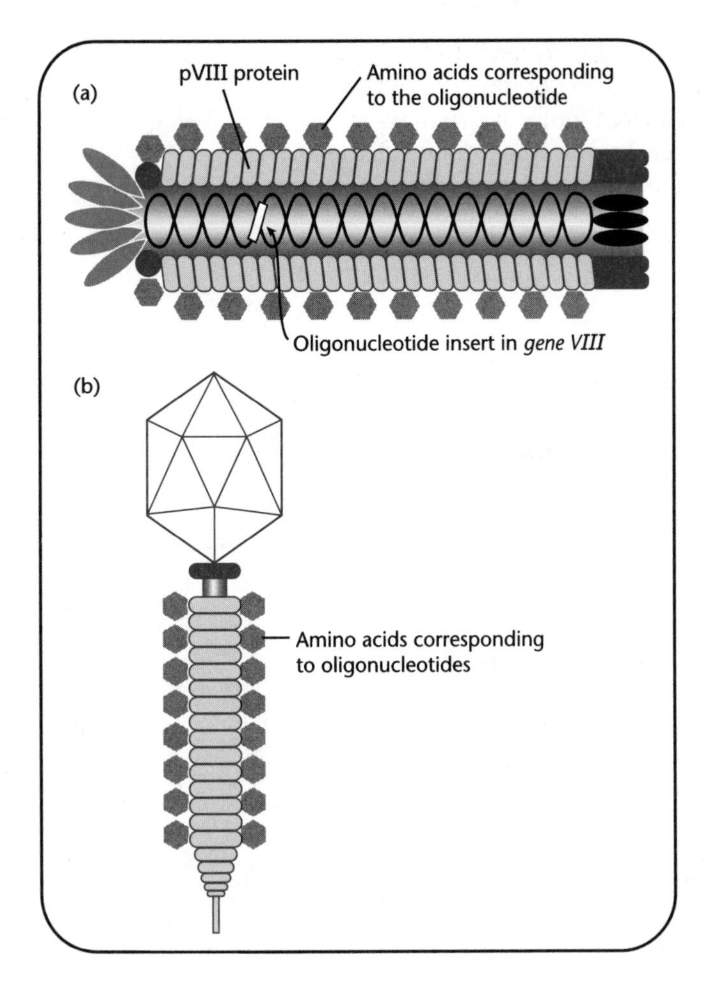

Fig. 13.17 Phage display vectors express a portion of a cloned gene as a gene fusion to a protein that appears on the outside of the phage. (a) An example of an M13 phage display vector. (b) An example of a λ phage display vector.

in the genes encoding the λ DNA replication proteins, O and P. Amber mutations in O and P prevent the phage from replicating in a wild-type cell. If the phage is propagated on a bacterial host containing an amber suppressing tRNA (see Chapter 3), the phage will grow. This conditional growth allows for the propagation of the suicide vector under one condition and the use of it as a suicide vector under another. Suicide vectors have also been rendered incapable of lysogeny at *attB* due to the removal of their *attP* site and/or the *int* gene.

If a gene is cloned into a λ suicide vector, this recombinant DNA molecule can be packaged into λ phage heads. Once packaged, the recombinant DNA molecule can be delivered to *E. coli* cells by the normal infection process. After the recombinant DNA molecule is inside *E. coli,* it will circularize but not replicate. The cloned gene that resides on the recombinant DNA molecule can be delivered to the *E. coli* chromosome through a homologous recombination event (Fig. 13.16b). An additional use of suicide vectors is to introduce transposons into a cell. If a transposon has been included in the λ suicide vector, then a transposition event can deliver the transposon to the chromosome while the λ DNA will be lost from the cell (Fig. 13.16c). This type of vector system is particularly useful when you want to use λ to deliver a cloned gene to *E. coli*, but do not want λ to hang around either through lytic or lysogenic growth.

Phage display vectors

The phage display technique has an incredible number of practical applications. The basics of the technique are that a large number of short stretches of nucleotides (usually 18–30 bp) are cloned into one of the phage genes whose product is located on an outer surface of the final phage particle. Once cloned, these short oligonucleotides create gene fusions so that the resulting fusion protein contains enough of the phage protein to make a functional phage and also a few extra amino acids that are accessible on the outside of the phage. The random oligonucleotides that are used are usually synthesized *in vitro*. Ideally, each phage in the library has a different oligonucleotide insert and expresses a different set of amino acids on the phage's surface. These libraries require large numbers of independent clones. This requirement makes phage especially suited for this technique because it is very easy to isolate phage lysates with 10^{10} phage per ml.

M13 was the first phage to be used as a vector for phage display. Vectors have been developed for fusing oligonucleotides to different M13 coat proteins including the genes for pIII, pVI, or pVIII (see Chapter 7 for details of M13 development, Fig. 13.17a). λ vectors have also been developed as phage display systems (Fig. 13.17b). One λ phage display vector expresses foreign amino acids on the surface of λ virus particles via a fusion with the C-terminal portion of one of λ's tail proteins. The major feature of all phage display vectors is that the foreign amino acids are accessible on the outer surface of the phage.

What are the phage in a phage display library used for? The library can be screened for peptides that bind to a specific antibody, protein, enzyme, receptor, any other molecule of interest, or even whole cells (Fig. 13.18). For example, a protein of interest can be immobilized on a solid support such as latex beads. The immobilized protein is mixed with the phage display library. Any phage that carry a peptide sequence that interacts with the protein of interest will bind to it. The immobilized protein-phage complex is isolated from the mix, the phage are released from the complex, and used to infect *E. coli* so that the potentially interesting phage are amplified. Several cycles of this basic experiment are usually carried out, a technique known as **biopanning**. Once a lysate that is enriched for potentially interesting phage has been generated, individual phage are purified and tested in individual binding assays. Phage display has been used for such varied experiments as investigation of host–parasite interactions, identification of novel enzyme substrates, identification of enzyme inhibitors, determination of the substrates for cell adhesion molecules, production of antibodies, and determination of substrates for specific proteins.

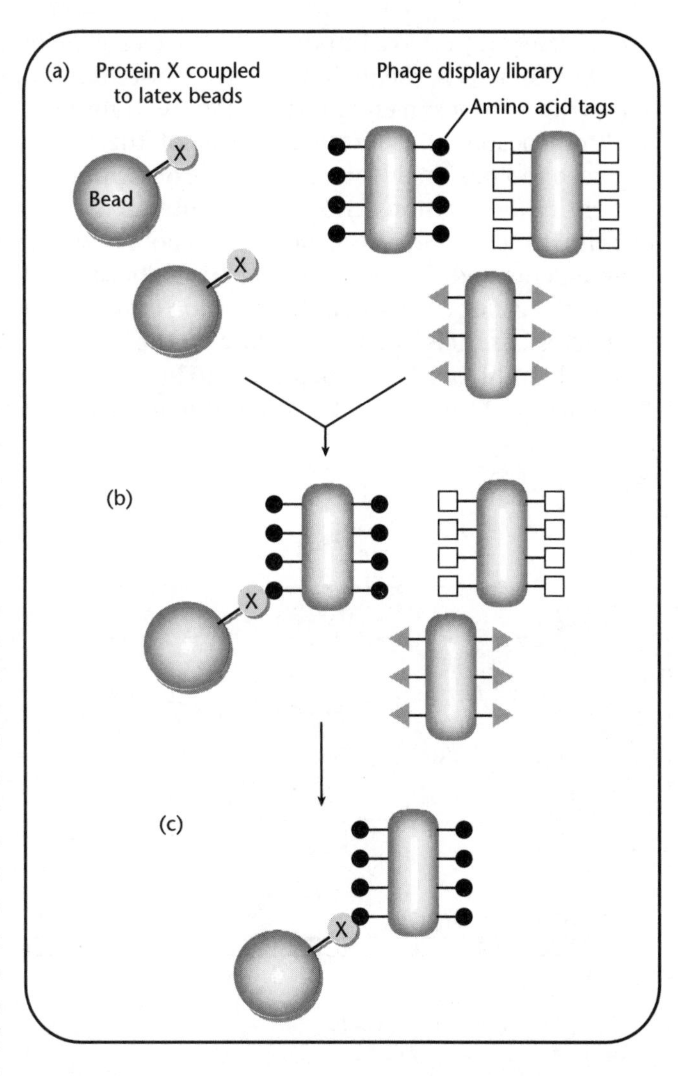

Fig. 13.18 Phage display libraries can be used to identify interactions between the amino acids expressed on the surface of the phage and any other molecule of interest (X) in a process called biopanning. (a) The molecule of interest is coupled to a solid support such as a large latex bead and mixed with the phage display library. (b) Any phage whose tag can interact with X will bind to it. (c) Properties of the solid support (usually its size) are used to remove X with any attached phage from the mixture. The phage can be purified and used in individual binding assays to assess the strength of the interaction between X and the amino acid tag.

Combining phage vectors and transposons

Gene fusions can be constructed using transposition. If the *lacZ* gene or any other reporter gene were inserted between the ends of a transposon and in the correct orientation, then transposition could be used to deliver *lacZ* to almost any place on the chromosome (Fig. 13.19). Some of the insertions of the *lacZ*-transposon would result in the formation of a fusion between a chromosomal gene and *lacZ*. Using transposition, fusions can be made without ever isolating DNA.

Several considerations must be taken into account when constructing the *lacZ* gene into a transposon. Most importantly, the *lacZ* gene must be very close to one end of the element and be facing in the correct orientation (Fig. 13.19). The reading frame across the transposon end must contain no translational stop codons. In most of the *lacZ*-transposons, the terminal repeat of the transposon abutting *lacZ* has been trimmed to a minimal sequence capable of supporting transposition.

Identifying insertions into a gene of interest is as simple as knowing the phenotype of a null mutation in that gene. A large pool of independent *lacZ*-transposon insertions is generated. The pool is screened for those insertions that have the null phenotype of the gene of interest. If the insertion is null for the gene of interest and expresses β-galactosidase, it is a good candidate for the correct fusion.

A slightly different approach can also be taken to identifying fusions of interest. If the goal is to study all genes that respond to a particular stimulus, fusion technology

FYI 13.4

Isolating *lacZ* fusions—the numbers game

When using a *lacZ*-transposon to generate fusions, how many transposition events must be screened to find ones of interest? If the antibiotic resistance determinant is used to select transposition events, a consideration of a few numbers can provide an estimate. Using *E. coli* as an example, because 87.8% of the *E. coli* genome encodes proteins, the number of insertions between genes is small. Half of all insertions into the chromosome will be facing the same direction as a promoter and half will be facing away from the promoter. For protein fusions, the transposon must insert in the correct frame. Because a codon consists of three base pairs, only one-third of the fusions in the correct orientation are in the correct frame. Frame does not figure in operon fusions.

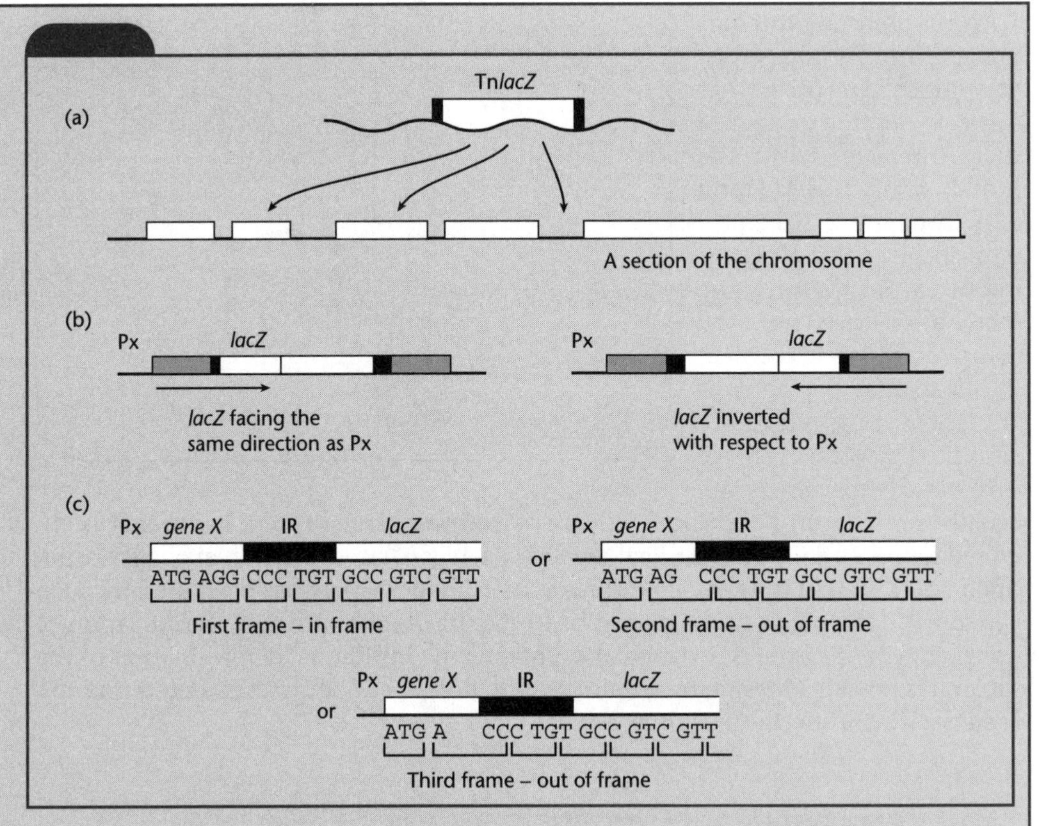

If we start with 1000 transposition events, then, for a protein fusion:
878 of the insertions will be in a protein-encoding gene.
439 of the insertions will be in the correct orientation.
146 of the insertions will be in the correct frame.

For operon fusions, 439/1000 transposition events should lead to a fusion. For protein fusion, 146/1000 transposition events should lead to a fusion. If every protein-encoding gene was the same size and the transposon showed no target site preferences, then 29,369 insertions would produce 4288 protein fusions or one protein fusion in each *E. coli* gene. 9767 insertions would produce 4288 operon fusions. These numbers show that to identify a fusion in any one gene, at least ~30,000 protein fusion insertions or ~50,000 operon fusions must be examined. The good news is that 500 to 1000 independent insertions can be examined on a single agar plate. Bacterial genetics is very powerful!

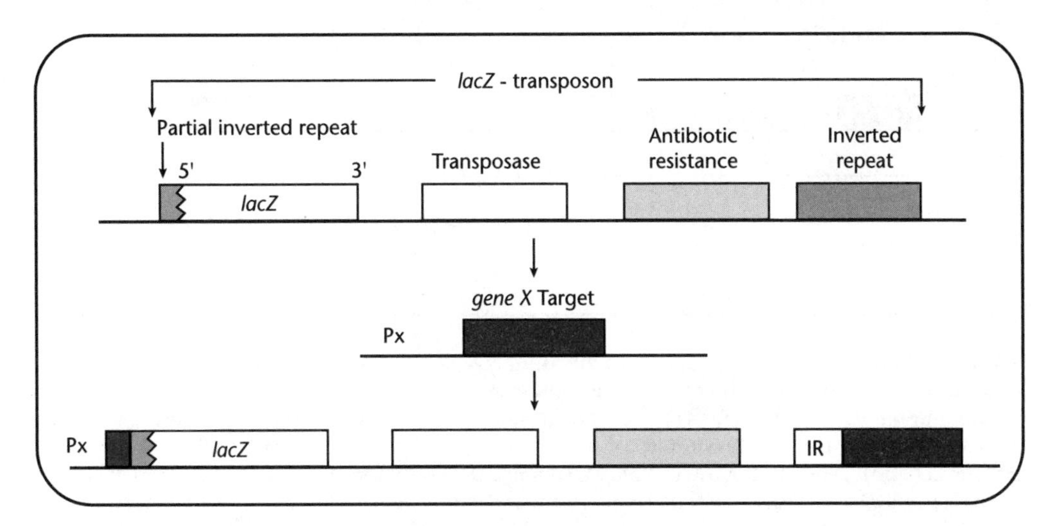

Fig. 13.19 Using transposition to make reporter gene fusions. A transposon with *lacZ* or any other reporter gene inserted between the inverted repeats is used. This constructed transposon is allowed to transpose randomly into the chromosome. Active *lacZ* fusions are isolated and the fusion to a gene of interest is identified by its phenotypes. The requirements for a functional transposable *lacZ* gene include inverted repeats bracketing the element. *lacZ* is very close to one of the ends of the element. The transposon also carries a selectable marker and the gene encoding transposase.

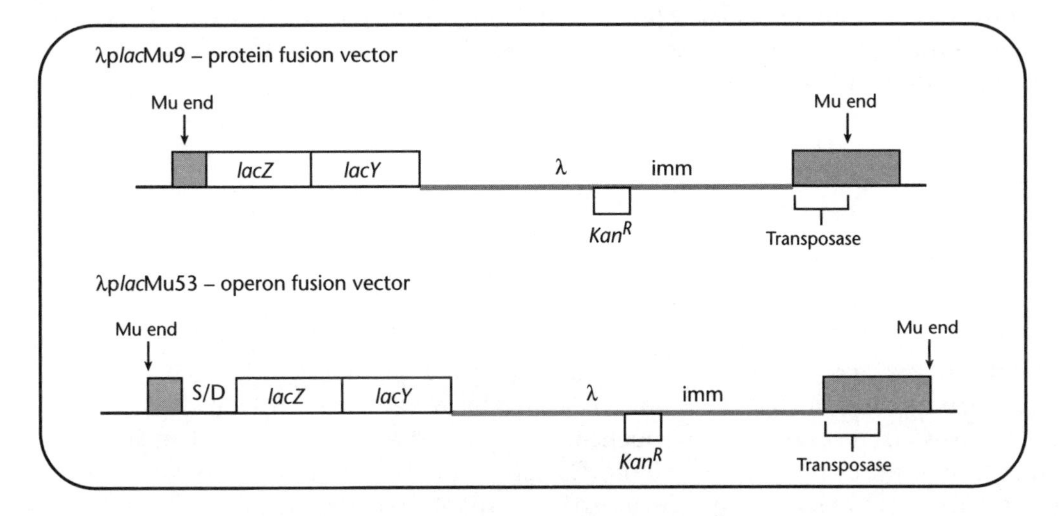

Fig. 13.20 λ-Mu-*lac* hybrid phage/transposons for making *lacZ* fusions. Two versions of this hybrid phage exist: one to isolate gene fusions and one to isolate operon fusions.

can be used to identify the gene set. A large pool of *lacZ*-transposon insertions is made. The pool is then screened for those fusions that respond to the stimulus of interest (i.e. heat shock, cold shock, changes in pH). The genes that respond to the stimulus can be identified by mapping and sequencing the fusions.

A type of transposon fusion vector has been constructed from bacteriophage Mu, the *lac* operon, and bacteriophage λ (Fig. 13.20). Mu phage are actually very large transposons. The phage uses transposition to replicate its DNA. The hybrid fusion vector transposes in the same way as Mu to deliver *lacZ* to random locations on the

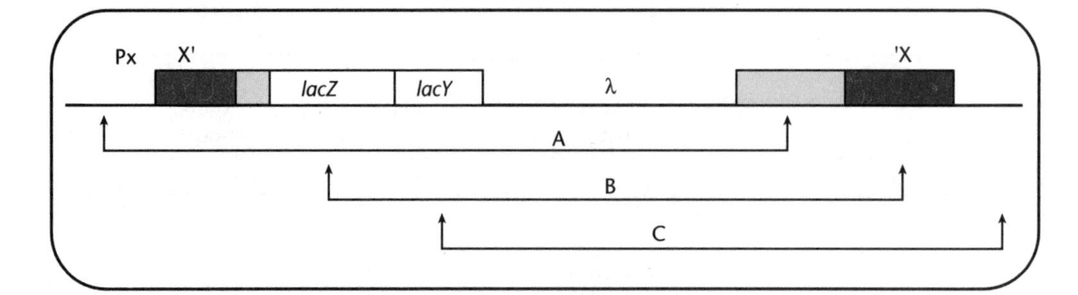

Fig. 13.21 Inducing a λ*plac*Mu phage out of the chromosome using ultraviolet light. Because the λ*plac*Mu went into the chromosome by a transposition event, the only way it can come back out of the chromosome is by illegitimate recombination (Chapter 5). Three different types of phage can be produced. Phage A carries the fusion and is usually the phage that is isolated and used. Phage B carries DNA from λ and is generally not useful. Phage C contains DNA from the other side of the fusion. Phage A can be identified by its β-gal activity that is expressed under the same conditions as the original fusion.

chromosome but grows and produces more phage like λ. Once the phage has been inserted into the chromosome, it behaves like a λ lysogen. All of the genetic tricks that have been developed for λ can be used for this transposable Mu-λ hybrid. One of the most valuable of these is inducing the phage out of the chromosome using ultraviolet light (Fig. 13.21). The phage can be induced out of the chromosome such that it will form a specialized transducing phage that carries the fusion between *gene X* and *lacZ*. The λ-Mu-*lacZ* hybrid phage combines knowledge of all three systems to make available more and better genetic techniques for the study of a gene of interest.

Summary

Cloning genes is a very powerful tool for studying the expression and function of the gene product. In many cases, it is the starting point for investigating a gene or process. The vector that the genes are put into must have three basic elements. It must have an origin of replication, a selectable marker, and unique sites for restriction endonucleases. Many vectors have other useful elements, depending on what cell types and uses the vector has been built for. These other elements are modular and can usually be mixed and matched to create many different types of vectors.

While designing cloning experiments on paper can be quite easy, actually building the correct construct or a useful library can be very difficult. In reality, many *in vitro* manipulations are carried out and each manipulation has a low success rate. When the correct clone has been identified and its DNA sequence determined to ensure that it is correct, it still may not function as desired. Many of the vectors described in this chapter have been designed to carry out specific experiments. For example, the GFP fusion vectors are designed to localize the gene product in live cells. The success of this experiment relies on the fusion protein having biological activity and not having a detrimental effect on the cells. The experiment becomes impossible if, for example, the fusion protein is toxic. Most of the time, the only way to determine if the construct will work is by trial and error!

Study questions

1 What element(s) are essential for a plasmid cloning vector and what element(s) are highly desirable?
2 Can the *lacZ* gene be used as a selectable marker for a plasmid? Why or why not?
3 If you want to measure the expression from a promoter of interest that is cloned onto a plasmid, what copy number vector should be used?
4 What vector would you use to determine the location of a gene of interest?
5 Which type of fusion requires both transcriptional and translational signals to be provided by the gene under study? Which requires only the transcriptional signals? What signals are needed for transcription? Translation?
6 If you were to design a hybrid cloning vector that was self-replicating and could present itself as either single-stranded DNA or double-stranded DNA, which components would you need for your hybrid vector? Why these components?
7 What are the unique qualities of phage vectors that are not shared with plasmid vectors?
8 Design a vector for purifying a large amount of Protein X in a single step.

Further reading

Bolivar, F., Rodriguez, R.L., Greene, P.J., Betlach, M.C., Heyneker, H.L., and Boyer, H.W. 1977. Construction and characterization of new cloning vehicles. II. A multipurpose cloning system. *Gene*, **2**(2): 95–113.
Chauthaiwale, V., Therwath, A., and Deshpande, V. 1992. Bacteriophage lambda as a cloning vector. *Microbiological Reviews*, **56**: 577.
Kleckner, N., Bender, J., and Gottesman, S. 1991. Uses of transposons with emphasis on Tn*10*. *Methods in Enzymology*, **204**: 139–80.
Mikawa, Y.G., Maruyama, I.N., and Brenner, S. 1996. Surface display of proteins on bacteriophage lambda heads. *Journal of Molecular Biology*, **262**: 21–30.
Yanisch-Perron, C., Vieira, J., and Messing, J. 1985. Improved M13 phage cloning vectors and host strains: nucleotide sequences of the M13mp18 and pUC19 vectors. *Gene*, **33**(1): 103–19.

Chapter 14

DNA cloning

> **FYI 14.1**
>
> **CsCl gradients**
>
> In traditional preparations of highly purified plasmid DNA, the lysate containing the plasmid DNA was mixed with the dye, ethidium bromide, that slips between the bases of the double helix (intercalation) and unwinds the DNA. The amount of ethidium bromide that can intercalate depends on the topology of the DNA molecule. Covalently closed supercoiled molecules bind less ethidium bromide than linear DNA because the DNA is constrained. Linear molecules, such as broken chromosomes, can bind more ethidium bromide because they can unwind as the ethidium bromide binds. Linear molecules can be saturated with ethidium bromide to the point of approximately one molecule of ethidium bromide for every two base pairs of DNA. The ethidium saturated DNA is centrifuged in a cesium chloride gradient. The differential binding of the dye leads to different buoyant densities for chromosomal and plasmid DNA in the cesium chloride gradient. This allows the plasmid DNA to be separated from the chromosome. While ethidium bromide–cesium chloride gradients are time consuming, for extremely pure plasmid DNA, they are still very useful.

Cloning is the process of moving a gene from the chromosome it occurs in naturally to an autonomously replicating vector. In the cloning process, the DNA is removed from cells, manipulations of the DNA are carried out in a test-tube, and the DNA is subsequently put back into cells. Because *E. coli* is so well characterized, it is usually the cell of choice for manipulating DNA molecules. Once the appropriate combination of vector and cloned DNA or construct has been made in *E. coli*, the construct can be put into other cell types. This chapter is concerned with the details of the individual steps in the cloning process:

1. How is the DNA removed from the cells?
2. How is the DNA cut into pieces?
3. How are the pieces of DNA put back together?
4. How do we monitor each of these steps?

Isolating DNA from cells

Plasmid DNA isolation

The first step in cloning is to isolate a large amount of the vector and chromosomal DNAs. Isolation of plasmid DNA will be examined first. In the general scheme, cells containing the plasmid are grown to a high cell density, gently lysed, and the plasmid DNA is isolated and concentrated. When the cells are growing, the antibiotic corresponding to the antibiotic resistance determinant on the plasmid is included in the growth media. This ensures that the majority of cells contain plasmid DNA. Without the antibiotic selection, an unstable plasmid (i.e. one without a *par* function) can be lost from the cell population in a few generations.

Cells can be lysed by several different methods depending on the size of the plasmid molecule, the specific strain of *E. coli* the plasmid will be isolated from, and how the plasmid DNA will be purified. Most procedures use EDTA to chelate the Mg^{++} associated the outer membrane and destabilize the outer membrane. Lysozyme is added to digest the peptidoglycan and detergents are frequently used to solubilize the membranes. RNases are added to degrade the large amount of RNA found in actively growing *E. coli* cells. The RNase gains access to the RNA after the EDTA and lysozyme treatments. This mixture is centrifuged to pellet intact cells and large pieces of cell debris. The supernatant contains a mixture of soluble cell components, including the plasmid, and is known as a lysate.

The methods used to purify the plasmid DNA from the cell lysate rely on the small size and abundance of the plasmid DNA relative to the chromosome, and the covalently closed circular nature of plasmid DNA. Most plasmids exist in the cytoplasm of the cell as circular DNA molecules that are highly supercoiled. The lysate is treated with sodium hydroxide to denature all of the DNA, and with detergent, SDS. The pH is then abruptly lowered, causing the SDS to precipitate and bring with it denatured chromosomal DNA, membrane fragments, and other cell debris. Most of the plasmid DNA renneals to form dsDNA because each strand is a covalently closed molecule and the two strands are not physically separated from each other. The small size of the plasmid allows the plasmid molecules to remain in suspension. The supernatant, which contains plasmid DNA, proteins, and other small molecules can be treated in a number of different ways to purify the plasmid. The most common protocol relies on a column resin that binds DNA. A small amount of the resin is mixed with the plasmid-containing supernatant and the plasmid-bound resin is collected in a small column. The remaining cell components are washed away and the plasmid is eluted from the resin. This procedure is quick, simple, and reliable and can be easily carried out on a large number of samples. Many modifications of this procedure have been devised.

Chromosomal DNA isolation

To isolate chromosomal DNA, cells are lysed in much the same way as for plasmid DNA isolation. The cell lysate is extracted with phenol or otherwise treated to remove all of the proteins. The chromosomal DNA is precipitated as long threads. The chromosomal DNA is very fragile and breaks easily. For these reasons, the chromosomal DNA is not usually purified using columns. Rather, the precipitated threads are collected by centrifugation.

Cutting DNA molecules

Once DNA has been purified, it must be cut into pieces before the chromosomal DNA and the plasmid DNA can be joined. The problem is to cut the DNA so that it will be easy to join the cut ends of the chromosomal DNA to the cut ends of the plasmid DNA. A group of enzymes, called **restriction enzymes**, are used for this purpose. Restriction enzymes are isolated from different bacterial species.

Bacteria use restriction enzymes and modification enzymes to identify their own DNA from any foreign DNA that enters their cytoplasm. The restriction part of the system is an enzyme that recognizes a specific DNA sequence or **restriction site** and cleaves the DNA by catalyzing breaks in specific phosphodiester bonds. The cleavage is on both strands of the DNA so that a double-stranded break is made. The modification part of the system is a protein that recognizes the same DNA sequence as the restriction enzyme. The modification enzyme methylates the DNA sequence so that the restriction enzyme no longer recognizes the sequence. Thus, the bacteria can protect its own DNA from the restriction enzyme. Any DNA that enters the bacteria and contains the unmethlyated restriction site is cut and degraded. There are three types of restriction–modification systems (Table 14.1). The types are distinguished based on the

FYI 14.2

The discovery of restriction enzymes

Restriction enzymes were discovered in *E. coli* in the 1950s by scientists studying bacteriophage. Bacteriophage λ can be grown on an *E. coli* K12 strain and titered on *E. coli* K12 to determine the number of phage per milliliter. A high-titer phage lysate will contain approximately 10^{10} plaque-forming units per ml (pfu/ml). If this phage lysate is titered on an *E. coli* B strain, the titer will drop to 10^6 pfu/ml. One of the phage that forms a plaque on *E. coli* B can be used to make a high-titer lysate on *E. coli* B. The lysate grown on *E. coli* B will titer on *E. coli* B at approximately 10^{10} pfu/ml but will titer on *E. coli* K12 at 10^6 pfu/ml. This four-log drop in plating efficiency can be traced to genes encoded by the bacterial chromosome of each strain. The system is known as host restriction and modification. Host restriction is carried out by a restriction endonuclease and modification is carried out by the protein that modifies the restriction site.

Table 14.1 Characteristics of the three types of restriction–modification systems.

Type	No. proteins	Complex	Cofactors	Recognition sequence
I	3	yes	SAM ATP Mg^{++}	3 bp (spacer) 4–5 bp Cleavage occurs 400–700 bp away
II	2	no	SAM	4 to 9 bp palindrome Cleavage is symmetrical within palindrome
III	2	yes	SAM ATP Mg^{++}	Asymmetric 5–6 bp Cleavage on 3′ side 25–27 bp away Cleavage on one strand only

SAM, S-adenosylmethionine.

number of proteins in the system, the cofactors for these proteins, and if the proteins form a complex.

Type I restriction–modification systems

Type I systems are the most intricate and very few of them have been described. Three different proteins form a complex that carries out both restriction and modification of the DNA. The complex must interact with a cofactor, S-adenosylmethionine, before it is capable of recognizing DNA. The S-adenosylmethionine is the methyl donor for the modification reaction and all known Type I systems methylate adenine residues on both strands of the DNA. The restriction reaction requires ATP and Mg^{++} for cleavage of the DNA. The complex also has topoisomerase activity. The DNA sequence recognized by Type I enzymes is also complex. The sequence is asymmetric and split into two parts (Fig. 14.1a). The first part is a 3 bp sequence, next is a 6–8 bp spacer of nonspecific sequence, and finally there is a 4–5 bp sequence. Cleavage of the DNA occurs randomly, usually no closer than 400 bp from the recognition sequence and sometimes as far away as 7000 bp.

Type II restriction–modification systems

Type II systems are composed of two independent proteins. One protein is responsible for modifying the DNA and one for restricting the DNA. Modification of the DNA uses S- adenosylmethionine as the methyl donor. The Type II modification enzymes methylate the DNA at one of three places, with each specific modification enzyme methlyating the same residue every time. The modifications that have been found are 5-methlycytosine, 4-methylcytosine, or 6-methlyadenosine. The DNA sequence recognized by Type II restriction enzymes is symmetric and usually palindromic (Fig. 14.1b). The DNA sequence is between 4 and 8 bp in length, with most restriction enzymes recognizing 4 or 6 bp. Both the cleavage of the DNA and modification of the DNA occur symmetrically on both strands of the DNA within the recognition sequence. Restriction enzymes function as dimers of a single protein so that each protein monomer can interact with one strand of the DNA. Thus, both strands of the DNA are cleaved at the same time, generating a double-stranded break. Several

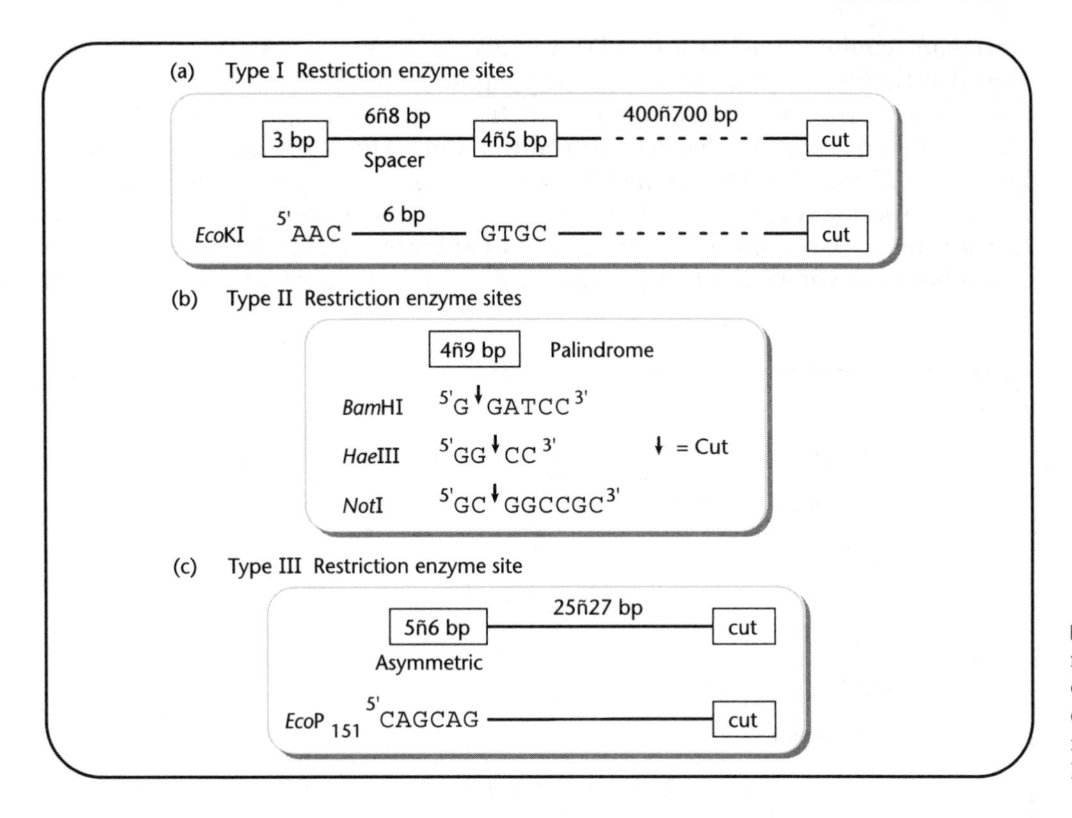

Fig. 14.1 The sequences recognized by restriction enzymes. (a) Type I restriction enzyme sites. (b) Type II restriction enzyme sites. (c) Type III restriction enzyme sites.

thousand Type II systems have been identified. Type II restriction enzymes are the most useful for cloning because they generate DNA molecules with a specific sequence on the ends (Fig. 14.2).

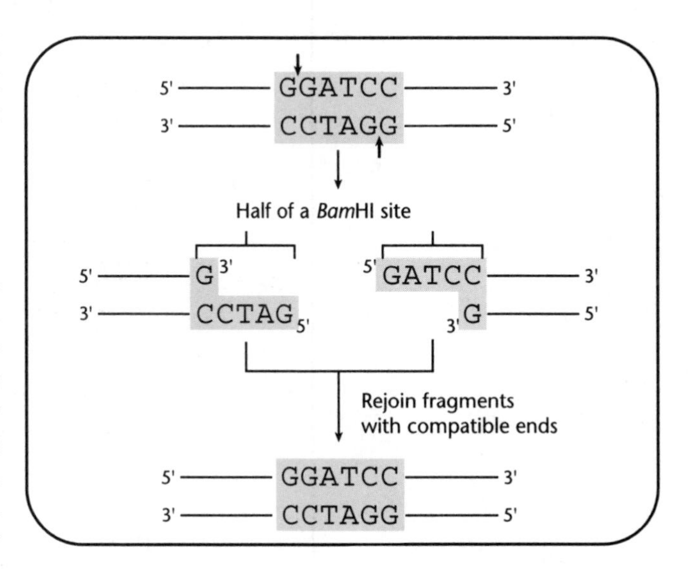

Fig. 14.2 Type II restriction enzymes generate DNA molecules with specific sequences on both ends. These ends can be rejoined to regenerate the restriction site.

Type III restriction–modification systems

Type III systems are composed of two different proteins in a complex. The complex is responsible for both restriction and modification. Modification requires S-adenosylmethionine, is stimulated by ATP and Mg^{++}, and occurs as 6-methyladenine. Type III modification enzymes only modify one strand of the DNA helix. Restriction requires Mg^{++} and is stimulated by ATP and S-adenosylmethionine. The recognition sites for Type III enzymes are asymmetric and 5–6 bp in length. The DNA is cleaved on the 3′ side of the recognition sequence, 25–27 bp away from the recognition sequence. Type III restriction enzymes require two recognition sites in inverted orientation in order to cleave the DNA (Fig. 14.1c).

Restriction–modification as a molecular tool

Type II restriction enzymes have several useful properties that make them suitable for cutting DNA molecules into pieces for cloning experiments. First, most cloning ex-

FYI 14.3
Naming restriction enzymes

Restriction enzymes are named for the species and strain in which they are first identified. For example, *Bam*HI was the first enzyme found in *Bacillus amyloliquefaciens* H. *Cla*I was the first restriction enzyme found in *Caryophanon latum*. Because the enzymes are named after the species they come from, the first three letters in the restriction enzyme are always italicized. Some species encode more than one restriction enzyme in their genome. Hence, the names, *Dra*I and *Dra*III or *Dpn*I and *Dpn*II.

periments require manipulation of DNA molecules in a test-tube. The fact that the Type II restriction enzymes are a single polypeptide aids in the purification of the enzyme for in vitro work. Second, Type II restriction enzymes recognize and cleave DNA at a specific sequence. The cleavage is on both strands of the DNA and results in a double-stranded break. Cleavage of the DNA leaves one of three types of ends, depending upon the specific restriction enzyme (Fig. 14.3a). Some enzymes leave a **5′ overhang**, some a **3′ overhang**, and some leave **blunt ends**. The ends with either a 5′ or 3′ overhang are known as **sticky ends**. Any blunt end can be joined to any other

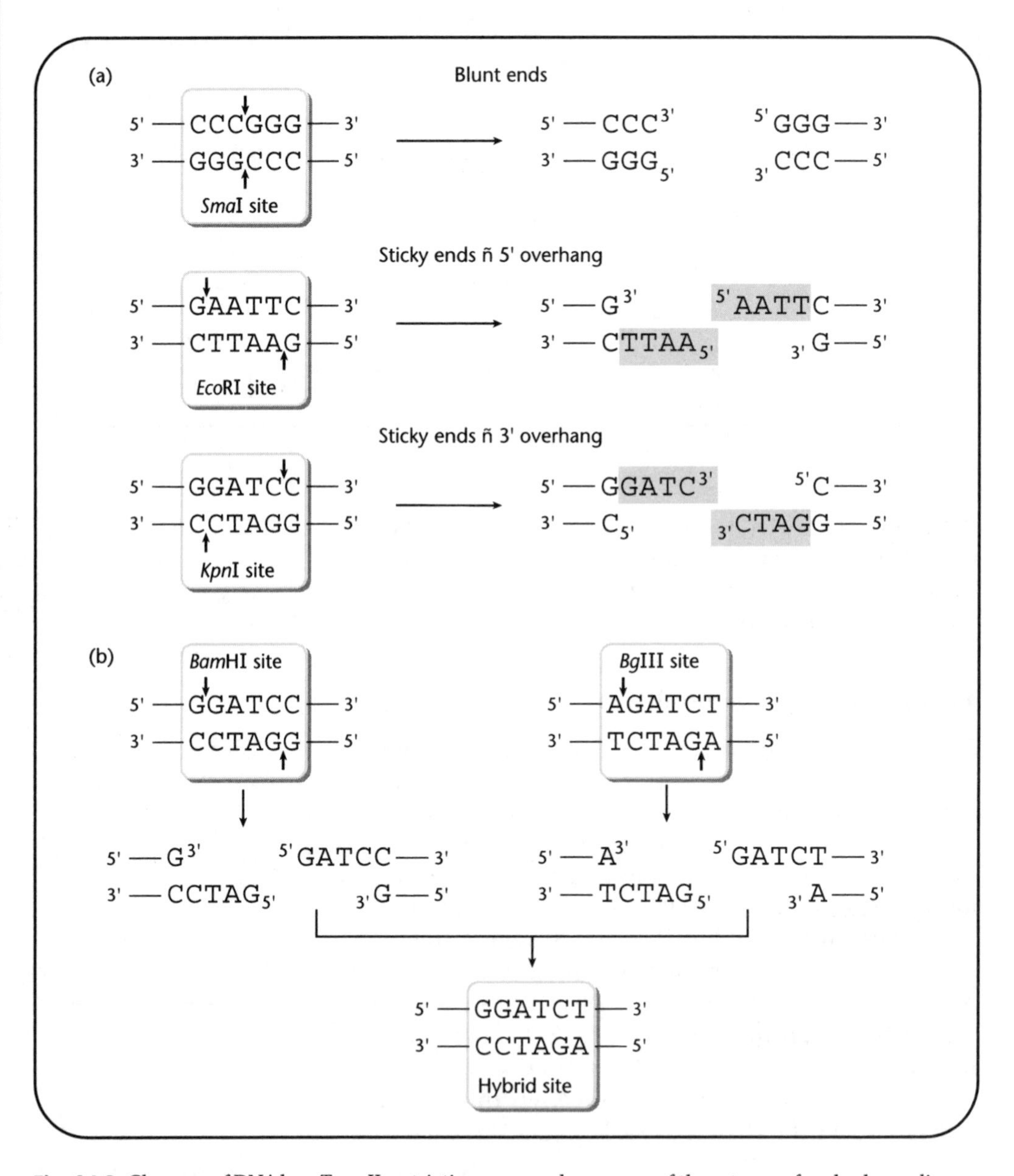

Fig. 14.3 Cleavage of DNA by a Type II restriction enzyme leaves one of three types of ends, depending on the enzyme used. (a) The ends generated can be blunt ends, sticky ends with a 5′ overhang, or sticky ends with a 3′ overhang. (b) The sticky ends from two molecules cut with two different restriction enzymes can be joined if the overhangs can hybridize. In this example, the hybrid site formed is no longer a substrate for either enzyme.

blunt end regardless of how the blunt end was generated. Sticky ends can be joined to other sticky ends, provided that either the same Type II restriction enzyme was used to generate both sticky ends that are to be joined or that the bases in the overhang are identical and have the correct overhang (Fig. 14.3b).

Generate double-stranded breaks in DNA by shearing the DNA

Usually restriction enzymes are used to cut the chromosomal DNA for cloning. The drawback of this approach is uncovered when the gene contains a restriction site for the enzyme being used to construct the library. One way to solve this problem is to only partially digest the chromosomal DNA with the restriction enzyme. Another way is to shear the chromosomal DNA and clone the randomly sheared DNA into a blunt end restriction enzyme site in the vector. One way to shear chromosomal DNA is by passing it quickly through a small needle attached to a syringe.

Joining DNA molecules

As described above, both plasmid and chromosomal DNA can be independently isolated from cells and digested with restriction enzymes. If, however, DNA with double-stranded ends is simply transformed back into *E. coli*, *E. coli* will degrade it. The double-stranded ends must be covalently attached. A version of this reaction is normally carried out in the cell by an enzyme known as **DNA ligase**. During DNA replication, the RNA primers are replaced by DNA (see Chapter 2). At the end of this process, there is a nick in the DNA that is sealed by DNA ligase. The double-stranded break formed by the restriction enzyme can be thought of as two nicks, each of which is a substrate for ligase.

If a plasmid molecule that has been digested with a restriction enzyme is subsequently treated with ligase, the plasmid molecule ends can be covalently closed by ligase (Fig. 14.4). Ligation is an energy-requiring reaction that occurs in three distinct steps. In the first step, the adenylyl group from ATP is covalently attached to ligase and inorganic phosphate is released. Next, the adenylyl group is transferred from ligase to the 5′ phosphate of the DNA in the nick. Lastly, the phosphodiester bond is formed when the 3′ OH in the nick attacks the activated 5′ phosphate. AMP is released in the process.

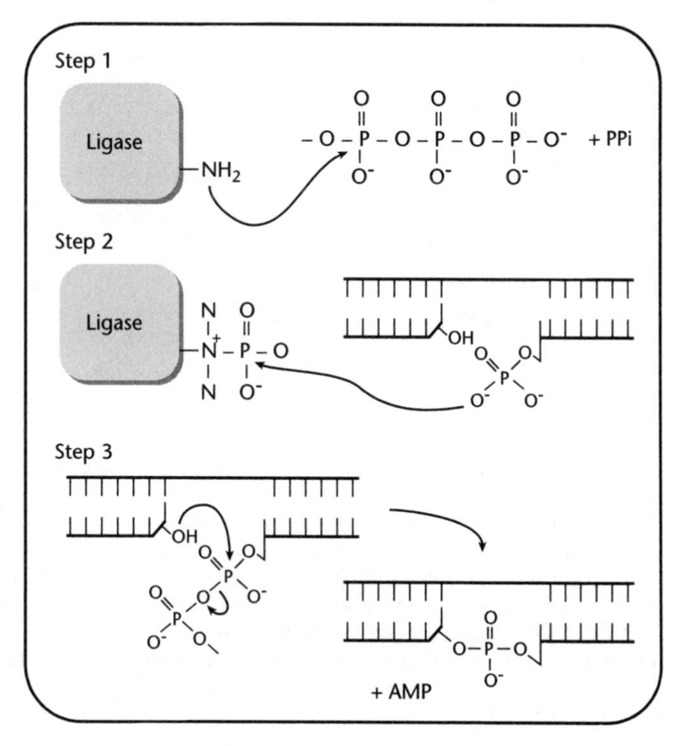

Fig. 14.4 The ends of DNA molecules can be joined and the phosphate backbone of the DNA reformed by an enzyme called DNA ligase. Ligase uses three steps to reform the backbone. (Step 1) A ligase molecule and an ATP molecule interact and the adenylyl group of ATP is covalently attached to the amine group of a specific lysine residue in the ligase protein. (Step 2) The adenylyl group is transferred to the 5′ phosphate in the nicked DNA. (Step 3) The 3′ OH attacks the activated 5′ phosphate, reforming the backbone and releasing AMP.

Because of this mechanism, ligase requires both a 3′ OH and a 5′ phosphate. If the 5′ phosphate is missing (see below), the nick cannot be sealed by ligase.

Manipulating the ends of molecules

While the goal of cloning is to construct a vector carrying the piece of DNA of interest, the reactions of the cloning process are concerned with the DNA ends. How the DNA ends are formed is very important because it dictates if the ends can be ligated to form the desired construct. The physical state of the DNA ends can be manipulated in vitro to influence the ligation reaction (Fig. 14.5a). For example, blunt ends cannot be ligated to sticky ends. The sticky ends can be made into blunt ends by the reaction of sev-

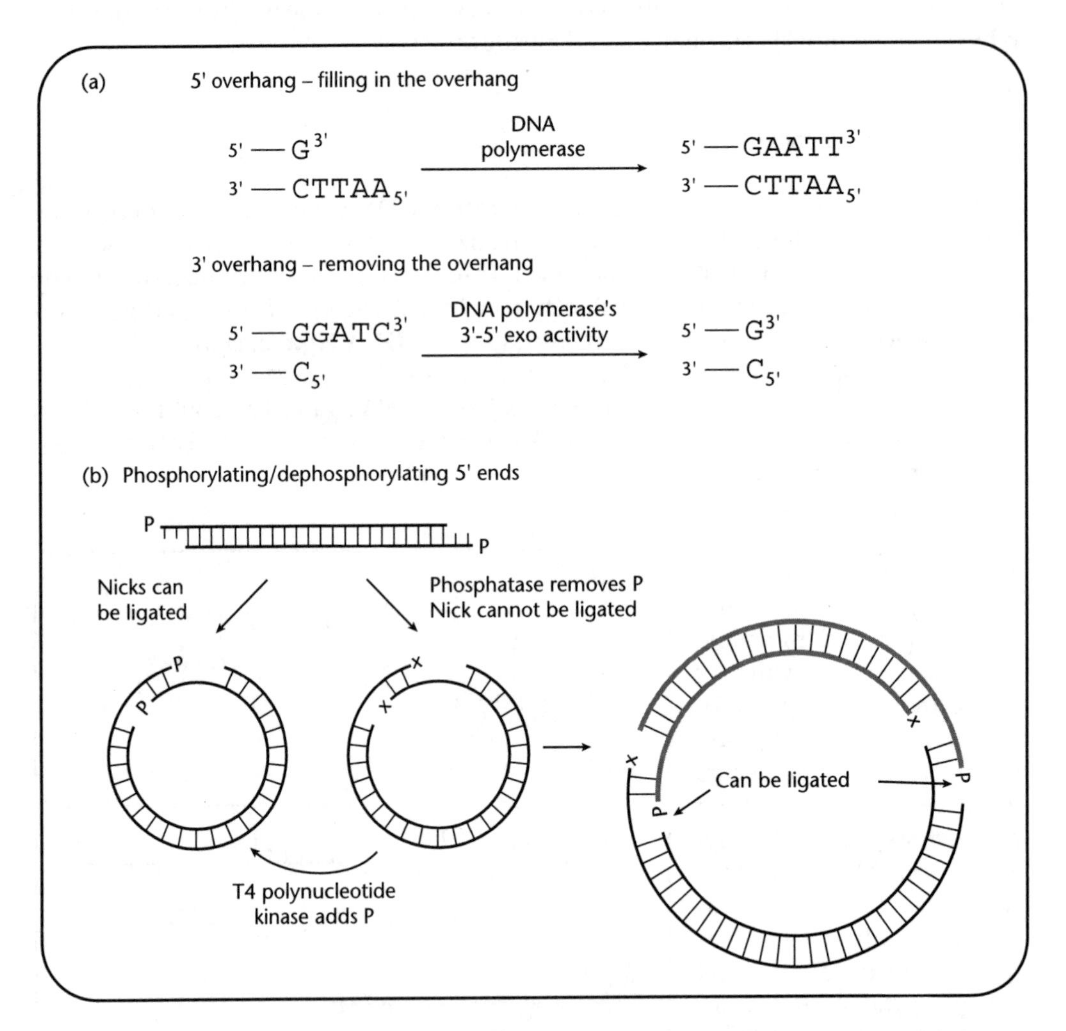

Fig. 14.5 The ends of DNA molecules can be manipulated in vitro to meet the requirements of the experiment. (a) 5′ overhangs can be filled in by DNA polymerase and 3′ overhangs can be removed by the 3′ to 5′ exonuclease activity of DNA polymerase. (b) 5′ phosphates are required for ligase to function. If the 5′ phosphate is missing, ligase cannot seal the nick. Phosphates can be removed by phosphatase and added by T4 polynucleotide kinase. Some cloning strategies take advantage of this by removing the 5′ phosphates from the vector so that it cannot re-ligate without an insert. The insert allows two of the four nicks to be sealed. This molecule is stable enough to be transformed into cells where the other two nicks will be sealed.

eral different enzymes. If the sticky end contains a 5′ overhang, then any one of several different DNA polymerases can be used to add the missing bases to the 3′ OH using the 5′ overhang as a template. A 3′ overhang cannot be filled in, rather the overhang must be removed. Many DNA polymerases have a 3′ to 5′ exonuclease activity and this activity can be used to remove 3′ overhangs.

The 5′ phosphate can also be manipulated (Fig. 14.5b). If DNA molecules are missing the 5′ phosphate, the phosphate can be added by an enzyme called T4 polynucleotide kinase. T4 polynucleotide kinase is an ATP-requiring enzyme that was originally identified in the bacteriophage T4. Molecules that have been phosphorylated by T4 polynucleotide kinase can be ligated to other molecules by ligase.

When pieces of chromosomal DNA are mixed with cut vector DNA, ligation of several different molecules can take place. The vector DNA ends can be ligated to reform the vector, the ends of a piece of chromosomal DNA can be ligated to each other, the ends of several vector or several chromosomal molecules can be ligated, or the ends of a piece of chromosomal DNA can be ligated to the ends of a piece of vector DNA. The ligation mix is usually put back into cells by transformation (see Chapter 11) and the antibiotic marker on the vector is selected for. Only cells transformed by molecules that contain vector DNA will form colonies. Of the molecules in the ligation mix, only religated vector DNA or vector DNA with a chromosomal insert are a possibility in the transformants. To reduce the number of vector molecules that are religated without a chromosomal insert, the 5′ phosphates on the vector can be removed by an enzyme known as a **phosphatase**. The ends of vector molecules that have been dephosphorylated (the 5′ phosphate has been removed) can only be ligated to chromosomal DNA molecules that have 5′ phosphates (Fig. 14.5b). These molecules still have a nick on each strand but this nick can be sealed inside the cell.

FYI14.4

What is agarose?

Agarose is a polysaccharide composed of modified galactose residues that form long chains. These chains form the matrix that the DNA must move through. Smaller molecules pass through the matrix faster and larger molecules get caught up in the matrix. Agarose is particularly useful for making gels because it has a high gel strength at low concentrations of agarose. Practically, this means that the gels are strong enough to be handled even when the agarose is present at less than 1% (weight per volume). Agarose is isolated from algae such as seaweed. Different seaweeds have different modifications on the repeating galactose residues. The biological function of agarose is to protect the seaweed from drying out at low tide. Agarose has been used for many years as a stabilizer in the preparation of ice cream and other foods.

Visualizing the cloning process

At each step of the cloning process, what is happening to the DNA molecules in the test-tube can be monitored using a technique called **gel electrophoresis**. In this technique, a gel (Fig. 14.6) containing small indentations or wells is cast. The DNA is loaded into the wells and the gel is placed in an electric current. Because DNA is negatively charged, it will move in the gel towards the positive pole. The DNA migrates or moves in the electric current based on size and shape. The larger a DNA molecule, the slower it moves. The more compact, or supercoiled a piece of DNA, the faster it moves.

The gel can be made from several different polymers, depending on the specifics of the experiment. Agarose forms a matrix that will separate DNA molecules from ~500 bp up to entire chromosomes (several million base pairs). If an electric current is constantly applied to an agarose gel from only one direction, agarose gels will separate DNA from ~500 bp to ~25,000 bp. If the direction and the timing of the current are varied over the electrophoresis time, then entire chromosomes can be separated in agarose. An alternative polymer, polyacrylamide, can be used to separate molecules a few base pairs in length to approximately 1000 bp.

Once the DNA has been separated in the gel, the gel is immersed in a solution containing ethidium bromide. If ultraviolet light is used to illuminate the gel, the ethidium bromide that is bound to the DNA will fluoresce, indicating the presence of bands of DNA (Fig. 14.6). Each band is composed of DNA molecules that are similar in size and shape. For example, when plasmid DNA is extracted from the cell, the majority of

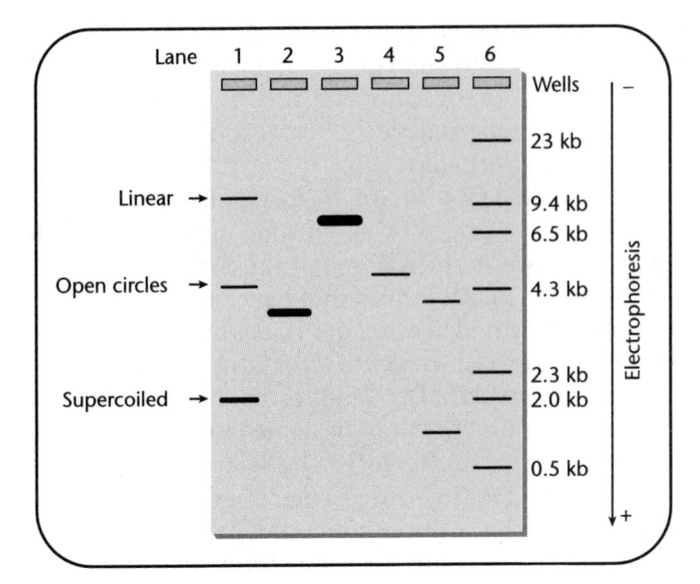

Fig. 14.6 A diagram of an agarose gel after electrophoresis and staining of the DNA with a fluorescent dye such as ethidium bromide. Lane 1 contains supercoiled, open circular and linear DNA forms of the same plasmid. Supercoiled DNA runs faster because of its topology. Lane 2 contains a 5 kb supercoiled plasmid and lane 3 contains a 10 kb supercoiled plasmid. Lanes 4 and 5 contain DNA fragments of different sizes. The band in lane 4 is larger than the bands in lane 5, meaning that the larger band contains more base pairs. Lane 6 contains DNA fragments of known molecular weights (molecular weight standards). Note that the molecular weight standards cannot be used to predict the sizes of supercoiled DNA.

it is supercoiled. Supercoiled DNA migrates very fast in an agarose gel (Fig. 14.6, lane 1). Some of the DNA will get nicked in the process of being extracted. The nick allows all of the supercoils to be removed, resulting in an open circle. Open circles migrate slower than supercoiled DNA. Some of the DNA will have a double-stranded break after isolation and the resulting molecules are linear. Linear DNA migrates the slowest of the three forms. If two different plasmids, one 5 kb and one 10 kb are isolated and run in an agarose gel, the supercoiled 5 kb plasmid will migrate faster than the 10 kb supercoiled plasmid (Fig. 14.6, lanes 2, 3). Likewise, the open circle 5 kb plasmid species migrates faster than open circle 10 kb plasmid species and the 5 kb linear species will migrate faster than the 10 kb linear species.

If a circular plasmid DNA is digested with a restriction enzyme that has two recognition sites in the plasmid, two linear pieces of DNA will result (Fig. 14.7). The shorter piece will migrate faster than the longer piece. If the DNA starts as a linear molecule and is digested with a restriction enzyme that recognizes the DNA in two places, then the DNA will be cut into three pieces (Fig. 14.7). Once all of the molecules are linear, they will migrate in the agarose gel based mainly on size.

Constructing libraries of clones

Sometimes it is desirable to make clones of all of the genes from an organism and subsequently to fish out the clone of interest. A large group of clones that contains all of the pieces of a chromosome on individual vector molecules is known as a library. To construct a library, vector DNA is isolated and digested with a restriction enzyme that only cuts the vector once. Chromosomal DNA is isolated and digested with the same restriction enzyme. The cut chromosomal DNA and the cut vector DNA are mixed together and treated with ligase. This mixture is transformed into *E. coli* and plated on agar containing the antibiotic that corresponds to the antibiotic resistance determinant on the vector. *E. coli* cells that are capable of forming colonies must either contain the vector without a chromosomal DNA insert or a vector with a chromosomal DNA insert. If the correct ratio of chromosomal DNA to vector DNA is used (usually two chromosomal molecules for every one vector molecule), rarely does a vector have two distinct pieces of chromosomal DNA inserted into it. Every clone should carry a unique piece of chromosomal DNA. If enough independent colonies are isolated, the entire sequence of the chromosome should be represented by the population of colonies. Libraries can be made from any kind of DNA, regardless of species.

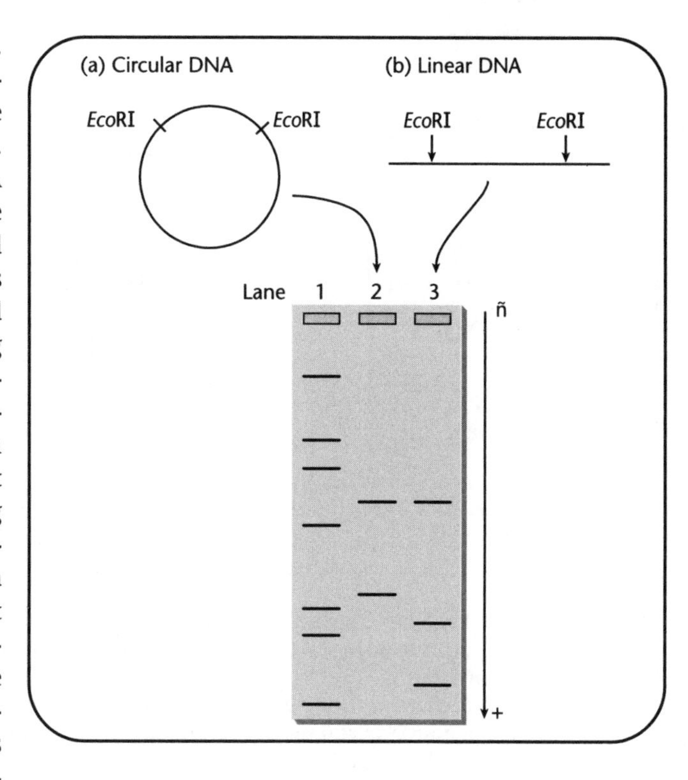

Fig. 14.7 The number of bands a molecule is cut into depends on if the starting molecule is linear or circular. Lane 1, molecular weight standards. Lane 2, bands from a cut circular molecule. Lane 3, bands form the same molecule as in lane 2 except the starting molecule was linear and not circular.

How many independent clones are needed so that the library contains the entire genome? This depends on several factors and can be calculated using the formula:

$$P = 1 - (1 - F)^N \text{ or } N = \frac{\ln(1 - P)}{\ln(1 - F)}$$

where:

P = probability of any unique sequence being present in the library.

N = number of independent clones in the library.

F = fraction of the total genome in each clone (size of average insert/total genome size).

For example, a library with ~10 kb inserts is made from *E. coli*, which has a genome size of 4639 kb. To ensure that any given *E. coli* gene has a 99% probability of being on a clone in the library would require 2302 independent clones.

$$N = \frac{\ln(1 - 0.99)}{\ln(1 - 10/4639)} = 2302$$

DNA detection—Southern blotting

In 1975, E.M. Southern described a technique to detect sequence homology between two molecules, without determining the exact base sequence of the molecules (Fig.

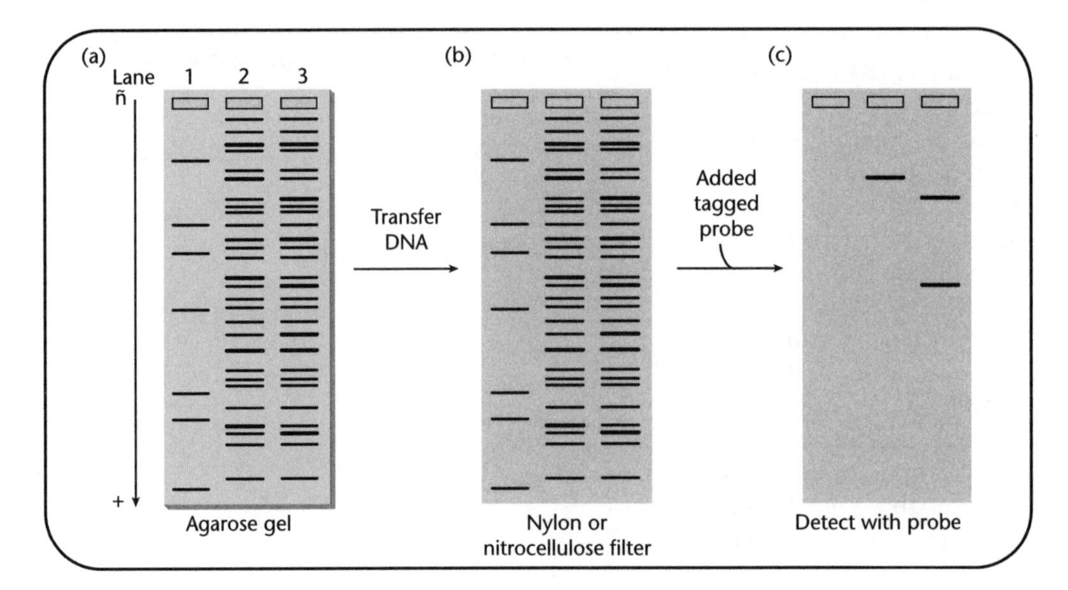

Fig. 14.8 The steps in a Southern blot. (a) DNA (usually chromosomal DNA) is digested with different restriction enzymes and run on a gel. The DNA is cut in many different places and leads to many fragments of different sizes. (b) The DNA is transferred from the gel to a nylon or nitrocellulose filter. (c) A fragment of DNA that contains the gene of interest is tagged and used as the probe. The probe is added to the membrane containing DNA and allowed to hybridize to any DNA fragment on the membrane to which it has homology. Excess probe is washed away and the probe is detected. If any fragments with homology are present, the size of the fragments can be determined based on where probe is detected.

14.8). The technique relies on fractionating the DNA on an agarose gel and denaturing the fractionated DNA in the agarose. The denatured DNA is transferred to a solid support, such as a nylon or nitrocellulose filter. A second DNA, called the **probe**, is labeled with a tag, denatured, and applied to the filter. Probes can be tagged with radioactivity and detected with X-ray film. They can also be labeled with fluorescent nucleotides or enzymes such as alkaline phosphatase or horseradish peroxidase. The enzymes are then detected with special substrate molecules that change color or emit light when cleaved by the enzyme. The probe will hybridize with any DNA on the filter that has complementary base sequences. Once the excess, non-hybridized probe is washed away, the tag attached to the probe can be detected.

Southern blotting can be used in many types of experiments. For example, if you have a clone of your favorite gene and you want to know if your gene exists in other species. You can isolate the chromosomal DNA from all of the species you want to test, digest the DNA with one or several restriction enzymes, and prepare a Southern blot. The clone of your favorite gene is used as the probe. If another species has sequence homology to your gene, then the band corresponding to the fragment containing the homology will be detected. The different species can be as diverse as bacteria, yeast, mice, rats, plants, humans, or as simple as several different types of bacteria. Blots with many different species of DNA included have become known as zoo blots!

A version of Southern blots can be used to identify clones from other species that are related to any gene of interest (Fig. 14.9). This technique is known as **colony blotting**. A population of cells containing a library from another species is plated on several agar plates and the cells are allowed to grow into colonies. The colonies are transferred to either a nylon or nitrocellulose filter and lysed on the membrane. The DNA from the lysed colonies is attached to the membrane by either UV crosslinking for nylon or heat for nitrocellulose. A plasmid carrying the gene of interest is labeled with a tag and used as the probe. The labeled probe will hybridize with DNA from a lysed colony only if the colony carries a clone that contains complementary sequences. Once the appropriate clone is identified, that colony from the agar plate is purified and used as a source of the cloned gene of interest. For example, a cloned yeast gene can be used to fish out of a human library a related human clone.

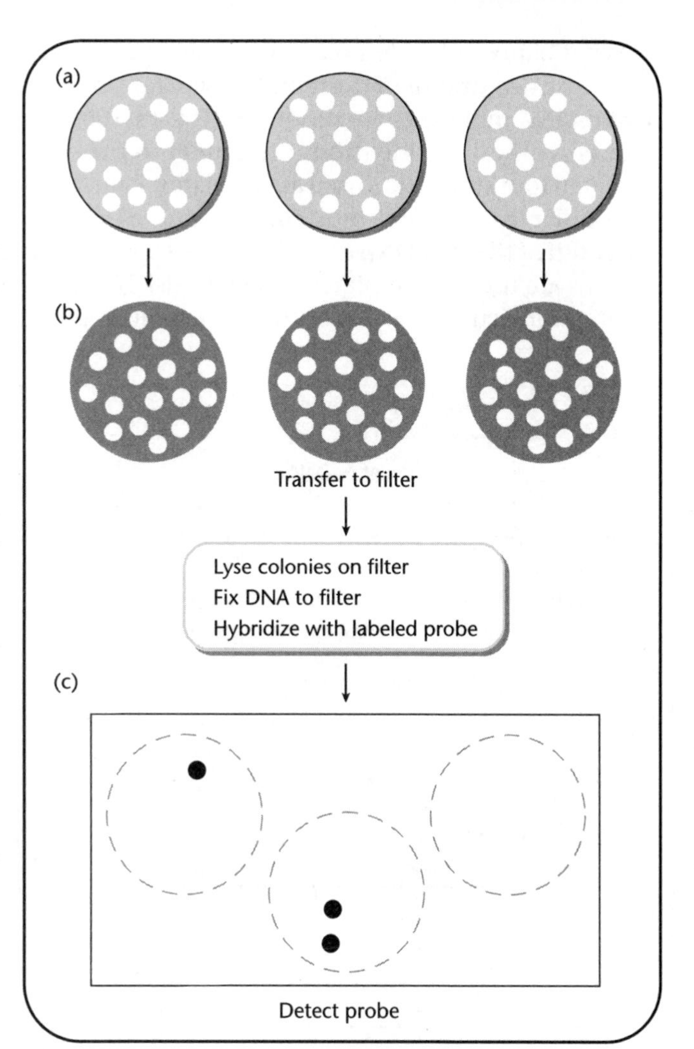

Fig. 14.9 A diagram of a colony blot. (a) The cells to be tested are plated on agar plates and incubated until they form colonies. (b) The colonies are transferred to a filter and lysed, releasing their DNA. The DNA is attached to the filter. (c) A tagged probe is added to the filters and the colonies that carry DNA that is homologous to the probe can be identified.

DNA amplification—polymerase chain reaction

In 1993, the Nobel Prize in Chemistry was awarded for a novel and extremely important development called polymerase chain reaction (PCR, Fig. 14.10a). PCR allows almost any piece of DNA to be amplified *in vitro*. Normally DNA replication requires an RNA primer, a DNA template, and DNA polymerase. PCR uses the DNA template, two DNA primers, and a unique DNA polymerase isolated from a bacterium that grows at 70°C. The unique properties of this polymerase are that, unlike most DNA polymerases, it is capable of synthesizing DNA at 70°C and it is stable at even higher temperatures.

In PCR, many cycles of DNA synthesis are carried out and these cycles are staged by controlling the temperature that the reaction takes place. For example, the DNA template is double-stranded DNA and the two strands must be separated before any DNA synthesis can take place. The PCR reaction mix is first placed at 90–95°C to melt the DNA. The temperature is lowered to ~40–55°C to allow the primer to anneal to the single-stranded template. Finally, the temperature is raised to 70°C to allow the DNA to be synthesized from the primer. This cycle of melting the template, annealing the primers, and synthesizing the DNA is repeated between 25 and 35 times for each PCR reaction.

What happens to a template molecule in each cycle of DNA synthesis? In the first cycle, the two strands of the template separate, one primer anneals to each strand and two dsDNA molecules are produced (Fig. 14.10b). One strand of each dsDNA molecule is synthesized only from the primer to the end of the template. In the second cycle, all four strands are used as templates. Four dsDNA molecules are produced, two are similar to the dsDNA molecules produced in the first cycle and the other two have three out of the four DNA ends delineated by the primers. In the third cycle, the eight strands are used as templates and eight dsDNA molecules are produced. Two of the dsDNA molecules are similar to the products produced in cycle one, four of the dsDNA molecules have three out of four ends delineated by the primers, and the other two molecules now have all four ends delineated by the primer. In the remaining cycles, the number of dsDNA molecules continues to increase linearly. In cycle 4, 8/16 molecules have all four ends delineated by the two primers, in cycle 5, 24/32, cycle 6, 56/64, cycle 7, 120/128, and cycle 8, 248/256. After 35 cycles, the dsDNA molecule with all four ends delineated by the primers is the predominant molecule in the PCR reaction mix.

The DNA primers used in PCR are chosen so that the piece of DNA of interest is amplified. The DNA primers are synthesized in vitro. By carefully choosing primers, the exact base pairs at either end of the amplified fragment can be predetermined. The template DNA can be from any source. Only a small amount of template is needed. Once a fragment of DNA has been amplified, it can be cloned into an appropriate vector, used as a probe, restriction mapped, or used in a number of other techniques.

The DNA replication that

FYI 14.5

Restriction mapping

Different restriction enzymes recognize different DNA sequences. For example, *Bam*HI recognizes GGATCC and *Eco*RI recognizes GAATTC. If a linear DNA fragment is cut once by *Bam*HI, and once by *Eco*RI, what will happen if you cut the DNA fragment with both *Bam*HI and *Eco*RI? In this example, three fragments would be generated. The sizes of the individual fragments depend on where the enzymes cut in relation to each other. Determining where different enzymes cut is known as restriction mapping. The restriction maps in A and D can be distinguished from those in B and C. Digesting the fragment with both *Eco*RI and *Bam*HI will give the relative order of *Bam*HI and *Eco*RI sites.

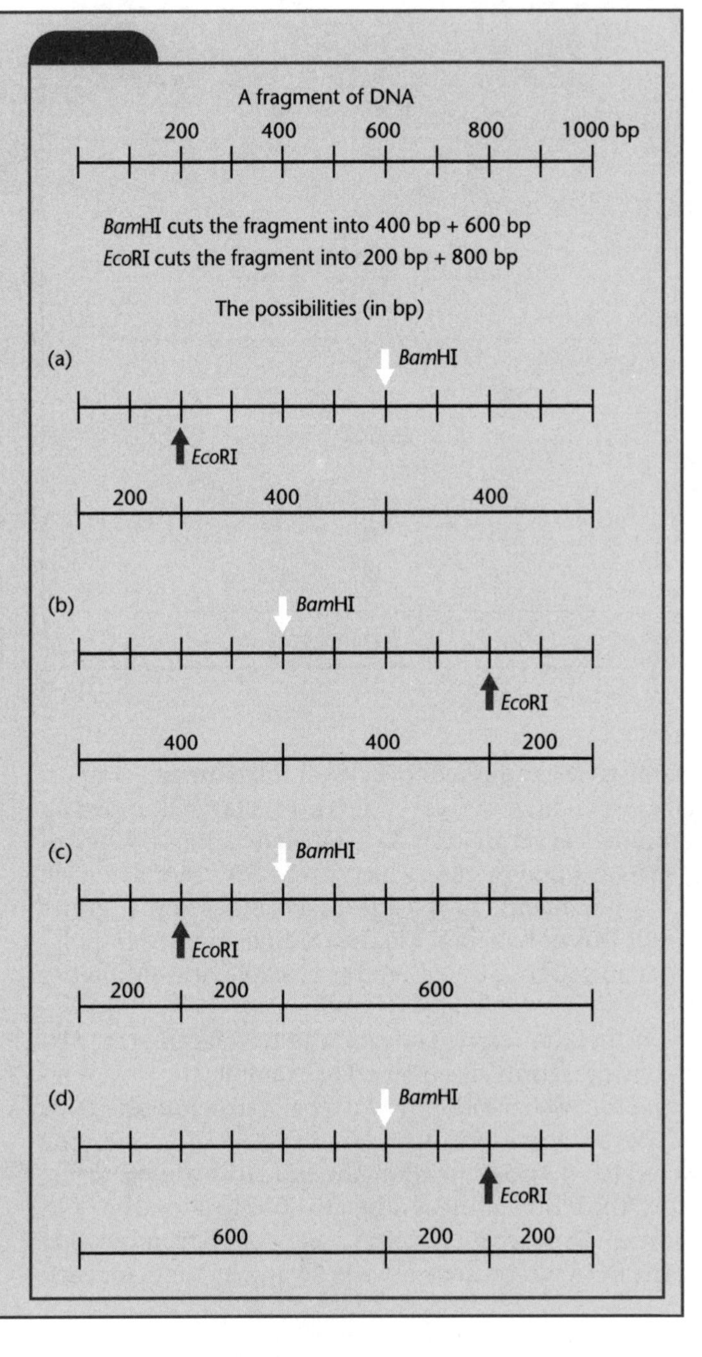

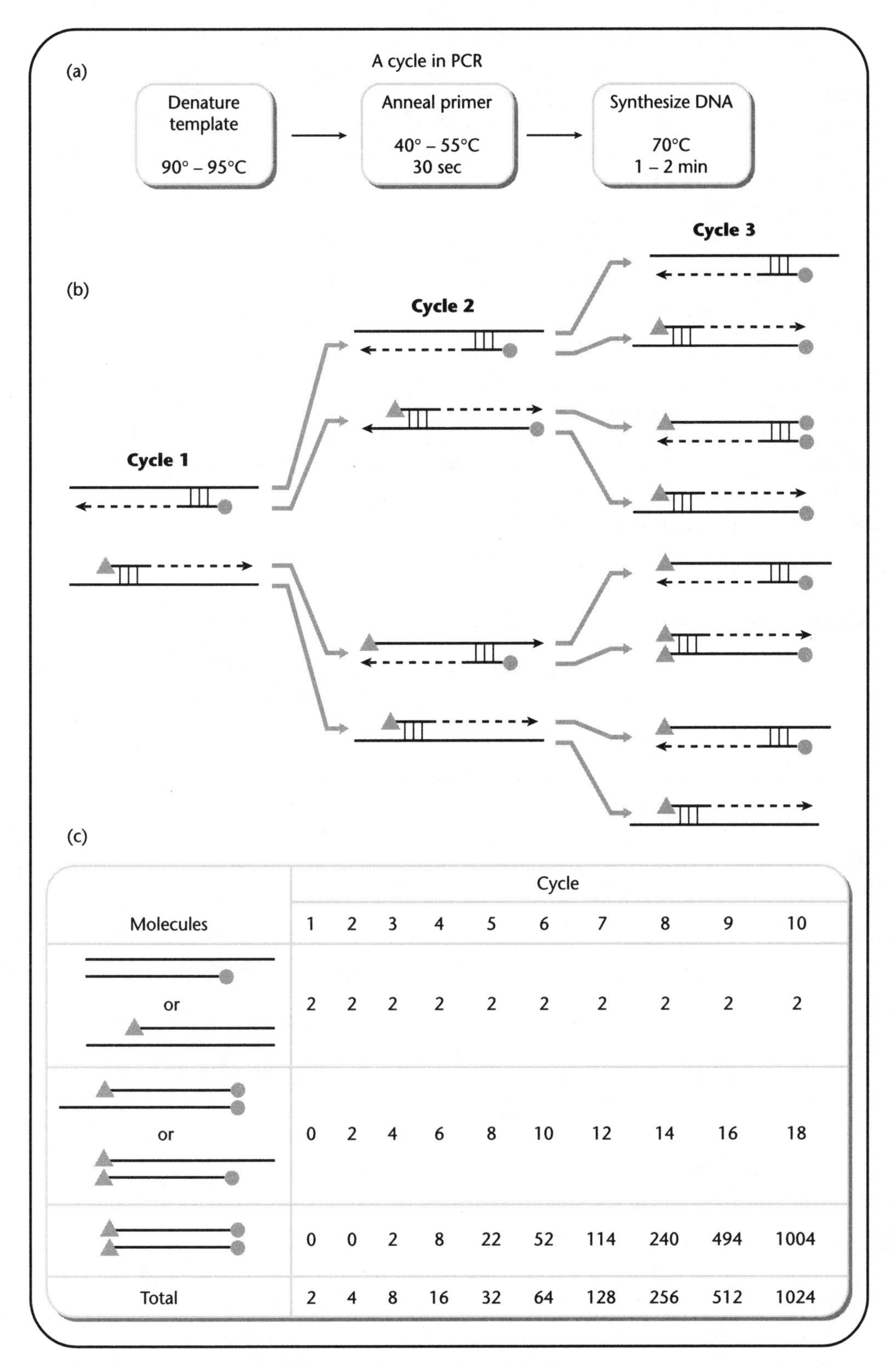

Molecules	Cycle 1	2	3	4	5	6	7	8	9	10
or	2	2	2	2	2	2	2	2	2	2
or	0	2	4	6	8	10	12	14	16	18
	0	0	2	8	22	52	114	240	494	1004
Total	2	4	8	16	32	64	128	256	512	1024

Fig. 14.10 A diagram of the polymerase chain reaction. (a) The general reaction taking place in each cycle. The template must be denatured, the primers annealed, and the DNA synthesized. (b) The fate of the template molecules in the first three cycles. The circle and triangle attached to the ends of the molecules represent the DNA ends that are delineated by the primers. (c) How many molecules of each type are formed in successive cycles. There will always be two molecules that match the ones formed in cycle 1. The number of molecules with three of the four ends matching the primer ends increases by two in each successive cycle. The number of molecules with all four ends determined by the primers is amplified dramatically. At the end of 35 cycles, >99% of the DNA molecules will have all four ends specified by the primers.

FYI 14.6

Amplifying DNA using PCR: functional uses of PCR

At its most basic level, PCR is used to amplify a small amount of DNA into a large amount of DNA. The large amount of DNA is then analyzed by restriction mapping, cloning, sequencing, or some other technique. PCR is not limited to laboratory uses; it has had profound impacts on many other disciplines. PCR is used to amplify DNA from crime scenes. The small amounts of DNA found in unlikely places can now be analyzed to help prove guilt or innocence. DNA extracted from Egyptian mummies can be amplified by PCR. The relationships between individuals buried in the pyramids have been studied using PCR and restriction mapping. Many disease-causing microbes cannot be grown in the laboratory or take a long time to grow. Not knowing the microbes present in a sick person can prevent them from getting the proper course of treatment in time. PCR can identify the presence of specific bacterial species simply by using species-specific primers in a PCR reaction and a sample (sputum, blood, etc.) from the sick person. A PCR test can be conducted in a few hours or in time to drastically alter the course of treatment. In any discipline where the amount of DNA has traditionally been limiting, PCR can be used to circumvent that problem and help provide information.

takes place in PCR, like in vivo DNA replication, is not 100% accurate. Occasionally, a mistake is made. If the amplified fragment is to be cloned, the resulting clones must be sequenced to ensure that they carry a wild-type copy of the gene. If the amplified fragment is to be used as a probe, a few mutant copies in a mixture that contains a large number of wild-types copies will not present a problem. Thus, depending on the use of the PCR fragment, these contaminating mutant copies of the fragment must be accounted for.

Adding novel DNA sequences to the ends of a PCR amplified sequence

The primers used for PCR amplification can be synthesized with unique DNA sequences on the 5′ ends (Fig. 14.11). These unique sequences can contain restriction enzymes sites, transcription factor binding sites, or any other sequence of interest. The unique sequences can be a few bases in length or up to 75–100 bases. If the added sequences are restriction enzyme sites, the resulting PCR fragment can be digested with the corresponding restriction enzyme for cloning into an appropriate vector. By predetermining the use of the PCR fragment, it can be customized in a variety of different ways.

Site-directed mutagenesis using PCR

PCR can be used to introduce a specific mutation into a specific base pair in a cloned

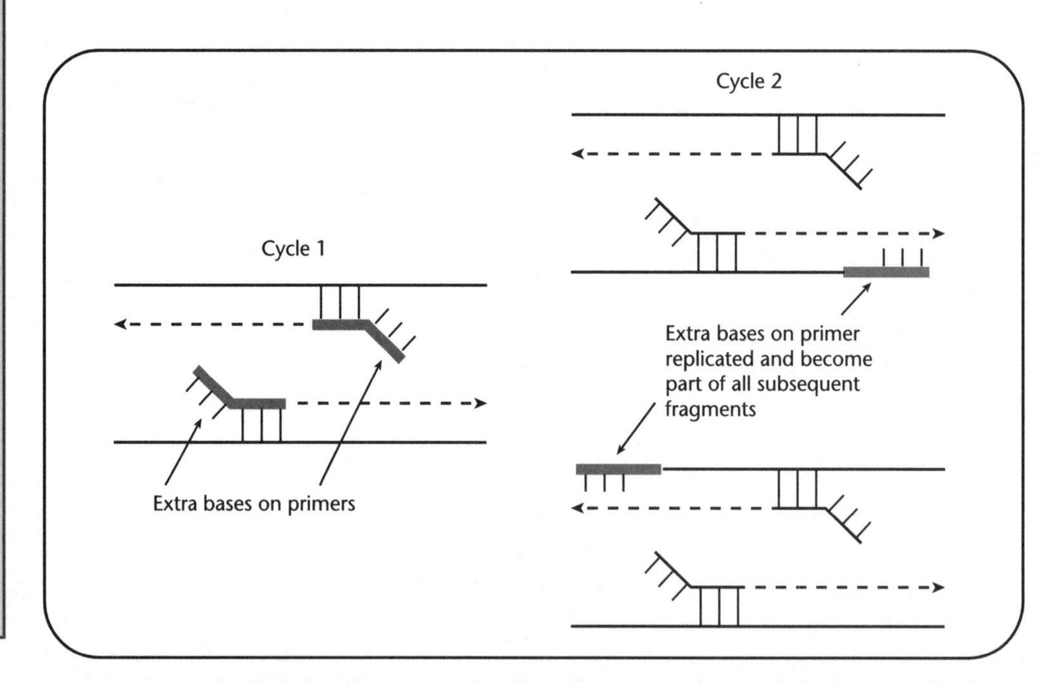

Fig. 14.11 PCR can be used to add specific sequences to the end of a DNA molecule. These specific sequences can be restriction enzymes sites or any other sequence of interest. In cycle 1, the added bases simply do not hybridize to anything. In cycle 2, the added bases are replicated, effectively making them a part of the PCR products. In subsequent cycles, the added sequences will be replicated as part of the DNA molecules.

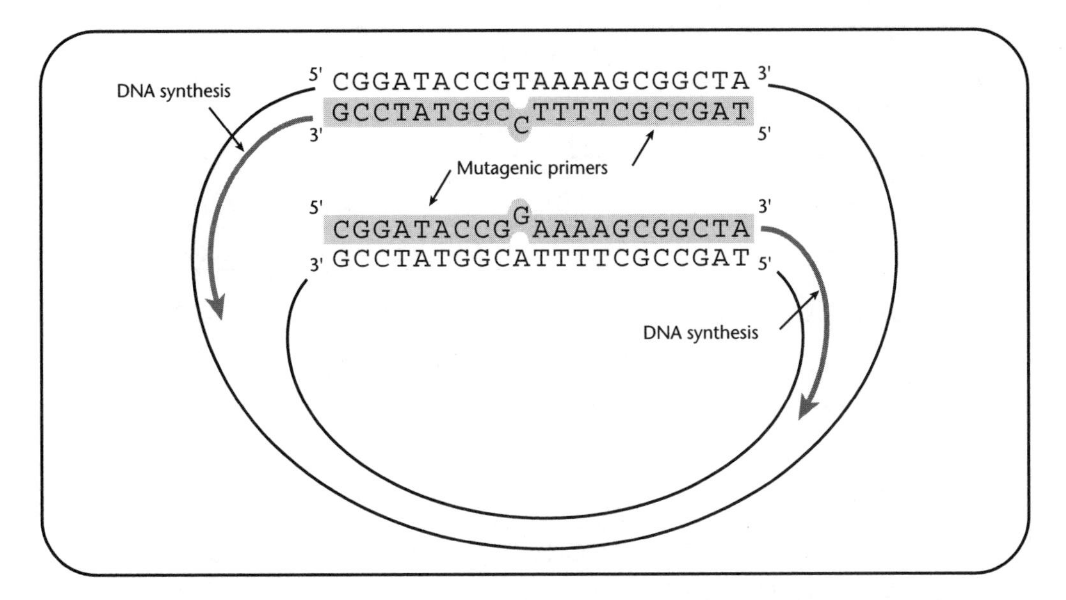

Fig. 14.12 One strategy for site-directed mutagenesis using PCR. Primers are designed to synthesize both strands of the plasmid containing the sequence to be mutagenized. Included in the primers are the base changes to be incorporated into the final mutant product. These primers are used in a PCR reaction to synthesize both strands of the plasmid with the incorporated change.

gene (Fig. 14.12). A primer is designed that contains the mutant base pair in place of the wild-type base pair. This mutant primer is used in a PCR reaction with the plasmid containing the cloned gene. PCR is used to synthesize the entire plasmid and after several rounds of replication, the large majority of plasmid molecules contain the mutation. This mixture of mutant and wild-type molecules is transformed into *E. coli* and the plasmids in individual colonies are tested for the presence of the mutation by determining the DNA sequence of the cloned gene.

Cloning and expressing a gene

In many cases, the goal is to not only clone the gene of interest but it is also to have the cloned gene expressed. When cloning a gene behind a promoter, the spacing of the elements needed for transcription and translation is critical. The closer the spacing of the elements to the optimum spacing for each element, the better the regulation and the better the expression of the cloned gene. The spacing can be manipulated during the cloning process. If a PCR fragment is used as a source of the gene to be cloned, base pairs can be inserted as needed by the design of the primers. For example, many vectors contain a promoter, mRNA start site, ribosome binding site (Shine–Delgarno sequence), and ATG start codon (Fig. 14.13a). Each of these elements are in the correct order with the correct spacing. Following the ATG start codon are the multiple cloning sites (MCS). A gene is cloned into one of the restriction sites in the MCS. In this case, the reading frame must be maintained from the ATG start codon through the MCS and into the cloned gene (Fig. 14.13b). If the cloned gene is out of frame, this can be corrected by changing the primers used to amplify the fragment (Fig. 14.14).

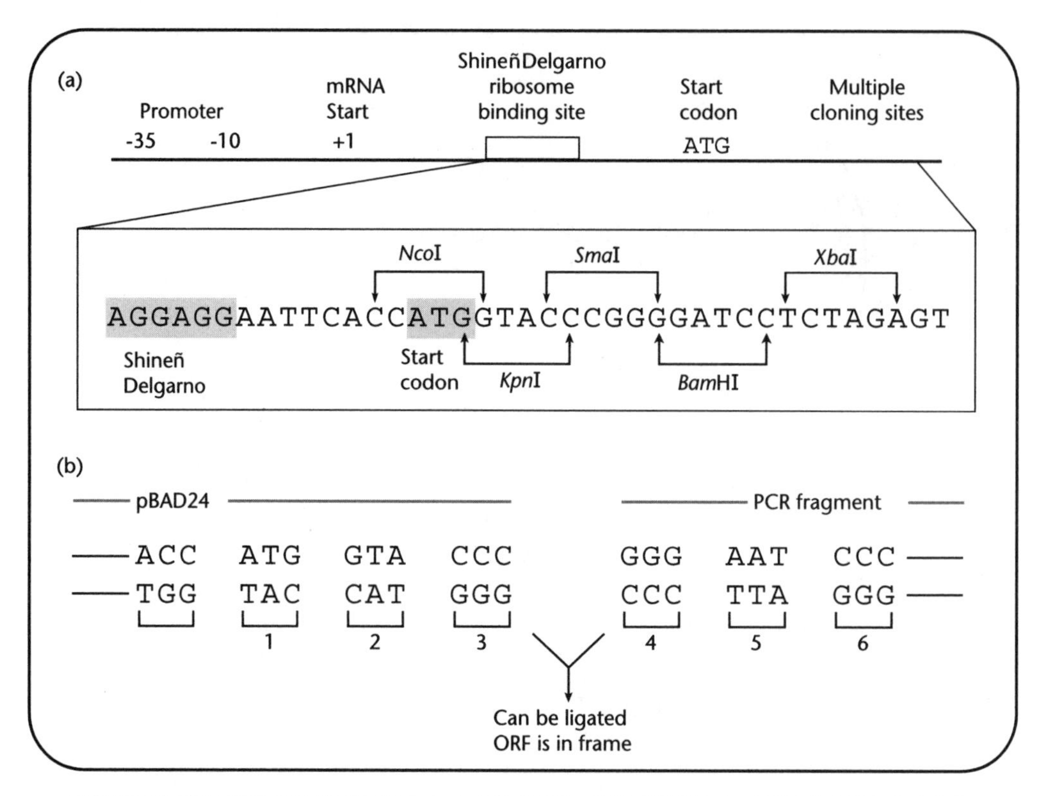

Fig. 14.13 Cloning vectors can incorporate many different features including transcription and translation signals. (a) In this example, the cloning vector pBAD24 is shown. It contains a regulated promoter, mRNA start site, ribosome binding site, and a start codon. The multiple cloning sites are located after the start codon. (b) To clone into this vector and have your protein of interest be expressed from the regulated promoter, your DNA sequence must be cloned so that the reading frame of the gene is maintained.

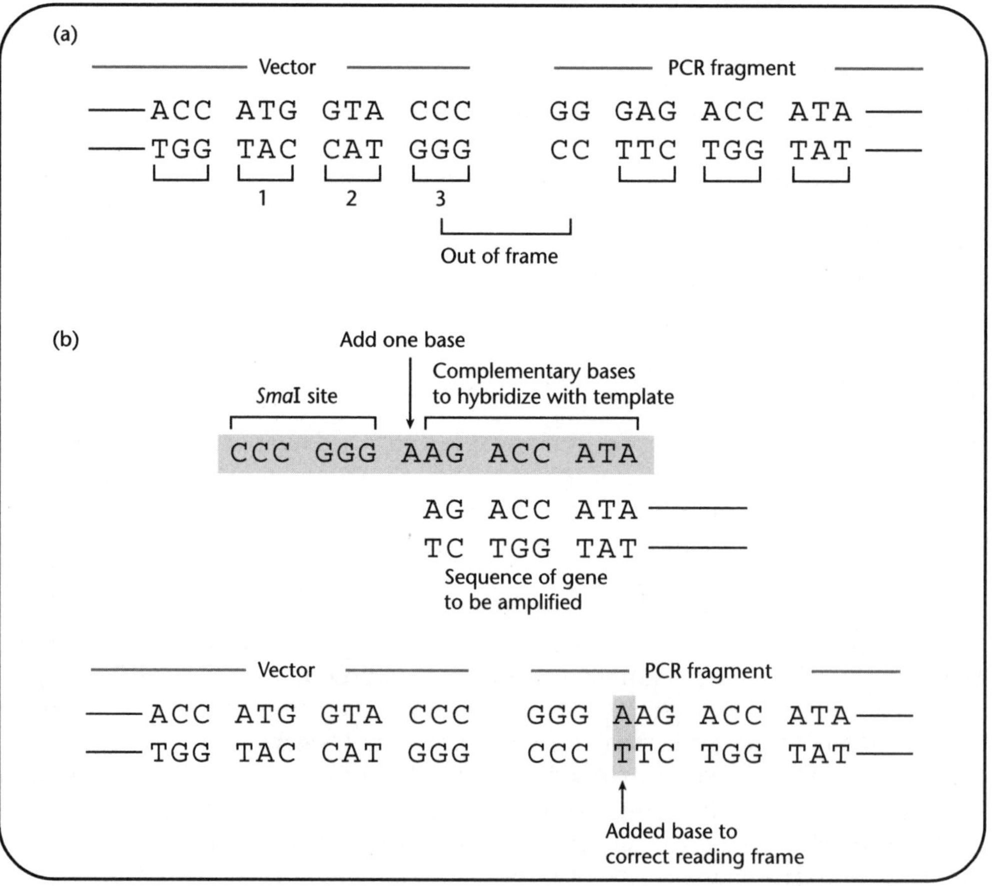

Fig. 14.14 If there are no restriction sites that leave the reading frame intact, then the fragment to be cloned can be amplified by PCR and the correct number of bases added to the primer.

DNA sequencing using dideoxy sequencing

DNA sequencing is the determination of the exact sequence of bases in a given DNA fragment. The start of the DNA sequence is determined by the placement of a DNA primer. Normal deoxyribonucleoside triphosphate precursors (dNTPs or dATP, dTTP, dCTP, and dGTP) and DNA polymerase are added to carry out DNA synthesis (Fig. 14.15a). In addition to the normal dNTPs, four special dideoxyribonucleosides (ddATP, ddTTP, ddCTP, and ddGTP) are included in small amounts. The special ddNTPs have a fluorescent tag attached to them. ddGTP has a fluorescent tag that fluoresces at one wavelength, ddATP a tag that fluoresces at a different wavelength,

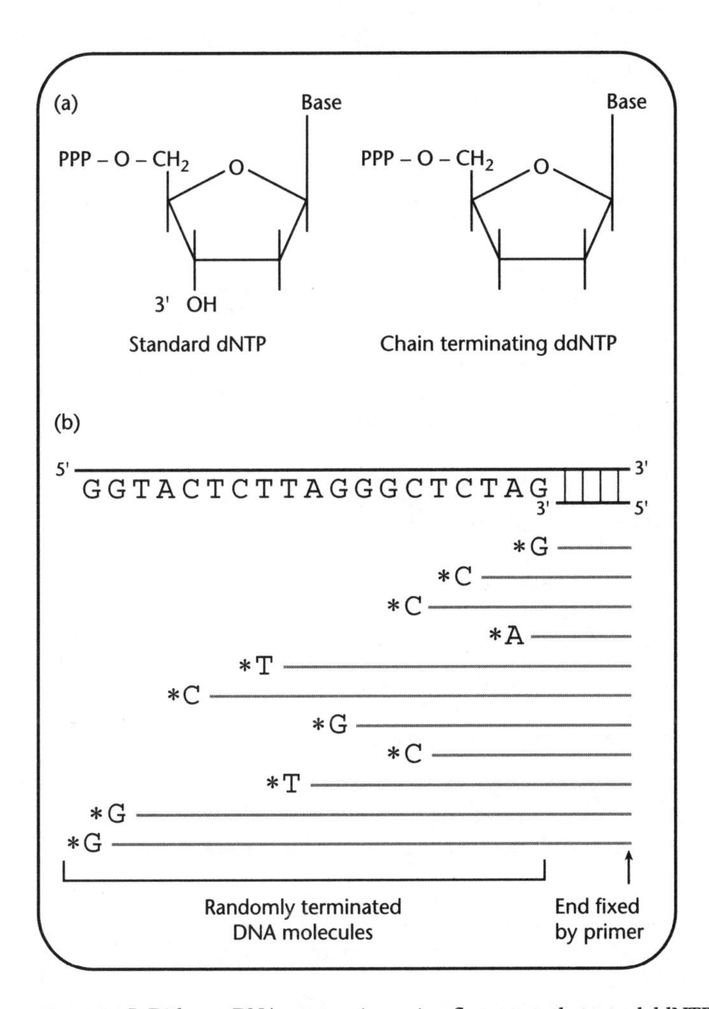

Fig. 14.15 Dideoxy DNA sequencing using fluorescently tagged ddNTPs. (a) A dideoxyribonucleoside triphosphate does not contain a 3′ OH and as such terminates DNA replication. (b) The fluorescently tagged ddNTPs are randomly incorporated into growing DNA molecules. This leads to a collection of molecules and in this population are molecules that are only a few bases long all the way up to about 500 bases. The fragments will differ in size by only one base pair. The population is separated in a gel or column using an electric current. The smallest molecules migrate fastest and will reach the bottom of the gel or column first. A laser is positioned at the bottom of the gel or column and detects the fluorescent bases as they migrate through. The different bases are tagged with different colored dyes. The signals detected by the laser are relayed to a computer that records what color of dye was attached to each sized fragment.

ddTTP a tag with a third wavelength, and ddCTP a tag with a fourth wavelength. ddNTPs do not have a 3′ OH and therefore block further synthesis of DNA. The fluorescently tagged ddNTPs are randomly incorporated into the growing DNA molecules (Fig. 14.15b). The results of DNA synthesis in the presence of tagged ddNTPs are a collection of DNA molecules different from each other by one base.

The fluorescently tagged ddNTP at the end of each molecule is dictated by the sequence of the template DNA. The tagged fragments are subsequently separated on either a polyacrylamide gel or on a very thin column. At the base of the column or polyacrylamide gel is located a laser. As the DNA fragments run off the column or gel, they pass through the laser beam, fluoresce, and the wavelength of the fluorescence is recorded and sent to a computer. The order of the fluorescently tagged molecules coming off the column reflects the sequence of the template DNA.

The automation of DNA sequencing has greatly simplified the process and made it much faster. Approximately 350 to 500 bp of DNA sequence can be read from one

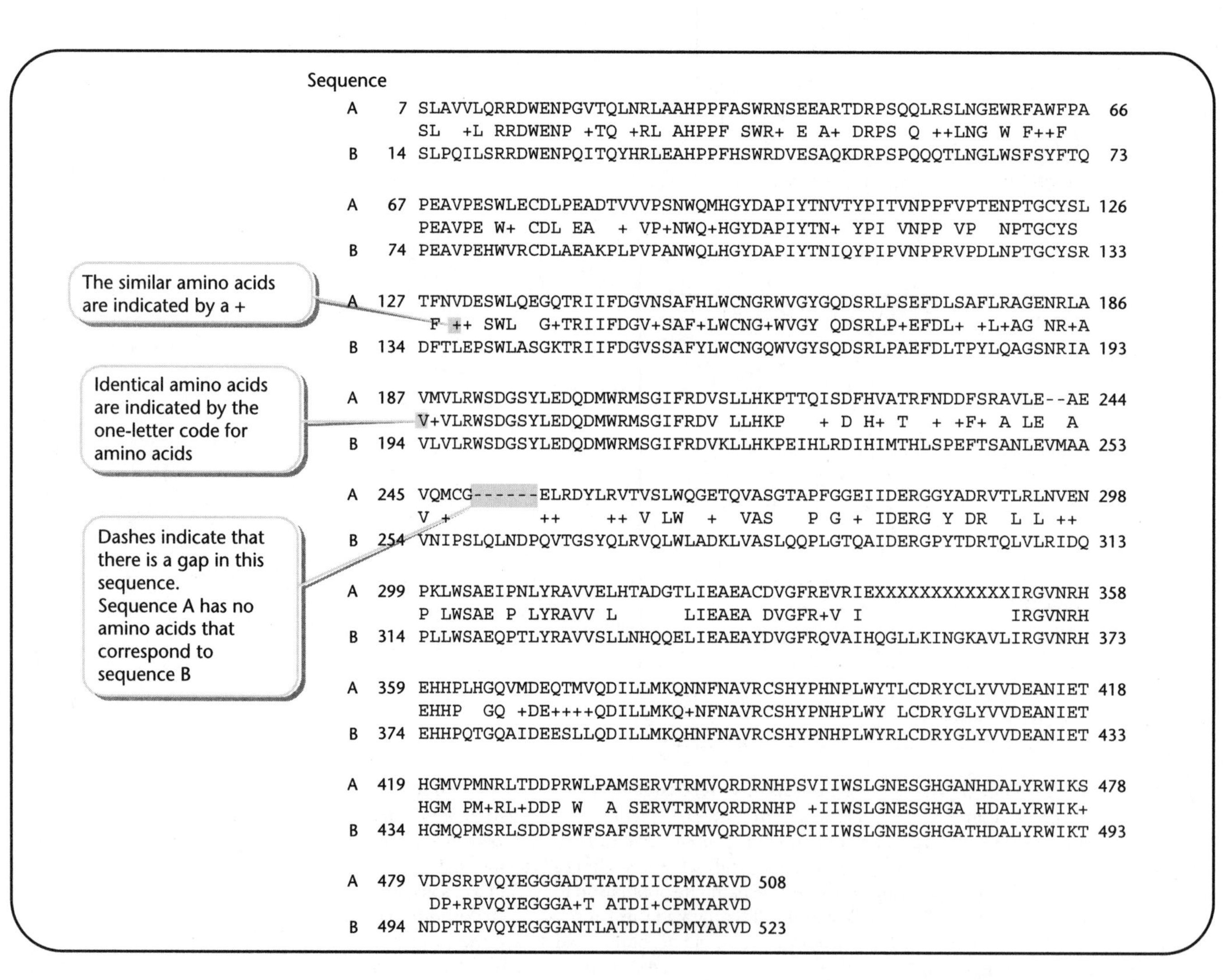

Fig. 14.16 Sequences that are very similar over the entire length of the protein indicate that the two proteins may have a similar function. The greater the similarity, the greater the chance the functions are the same.

primer. By carrying out separate sequencing reactions using primers located every 350 to 450 bp on the template, the sequence of the entire template can be determined.

DNA sequence searches

To date, many millions of base pairs of DNA from many species have been sequenced. For example, the chromosomes of at least 50 bacterial species, several yeasts, and the large majority of all human chromosomes have been determined. These sequences contain an incredible amount of information. So much in fact that special computer programs had to be designed to help interpret just a fraction of the data.

When a DNA sequence is published in a scientific journal, it is also deposited in a computer database known as GenBank. When a sequence is placed in GenBank, the known and predicted features of the sequence are also indicated. These include promoters, open reading frames, and transcription factor binding sites. Just a listing of As, Cs, Gs, and Ts are known as a **raw sequence** and the sequence with all of the features indicated is known as an **annotated sequence**.

It is possible to search the sequences in GenBank using several different programs. You can search by the name of an interesting gene very easily. If you have the sequence of a gene of interest, it is possible to search for related sequences. GenBank can be searched using a DNA sequence or using that DNA sequence translated into the

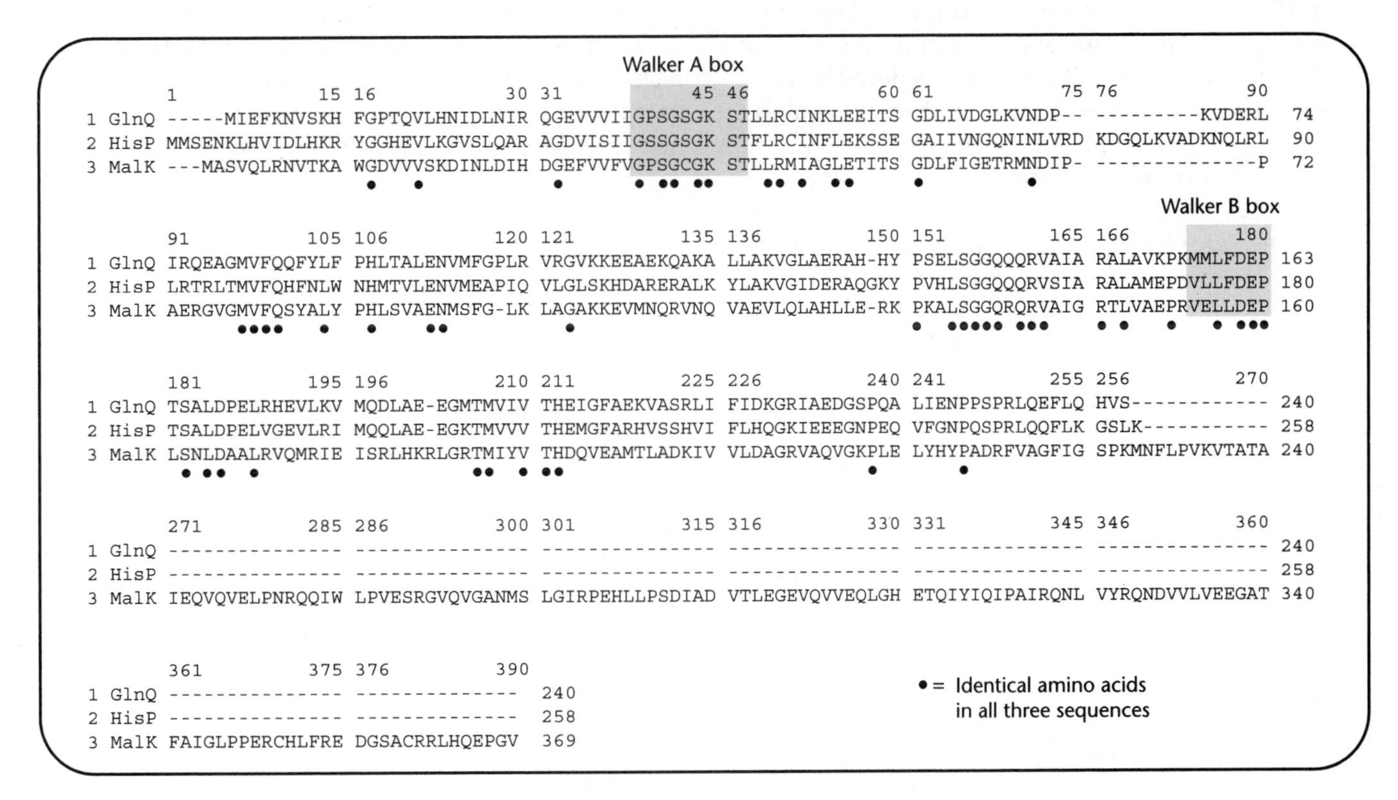

Fig. 14.17 Limited regions of homology can indicate a specific feature such as a nucleotide binding site or a metal binding domain. In this example, the highest regions of homology surround the Walker A and Walker B boxes. These amino acids form a part of the protein that binds ATP.

protein sequence. The programs used for these searches are capable of identifying not only exact matches but also sequences that have differing degrees of similarity.

What can be learned from sequence searches? First, DNA sequence searches are more stringent than protein sequences. Two DNA sequences either have an adenine in the same position or they do not. Protein sequences can have the same amino acid in the same place and are, thus, **identical** at that position. Proteins can also have similar amino acids in one position, such as valine in one protein and alanine in the other. Because both amino acids are hydrophobic, they can frequently carry out the same functions. In this case, the proteins are said to be **similar** in a given position.

If two proteins have similarity over a large segment of their sequences, they may have similar functions (Fig. 14.16). This kind of analysis is especially useful if the function of one of the proteins has been identified. Knowing the function of one of the proteins suggests that the other protein should also be checked for this function. More limited regions of sequence similarity or identity can indicate the presence of a cofactor binding site. An example of this is the Walker box, which is an ATP binding site (Fig. 14.17). Sequence similarities can provide very valuable information about an unknown sequence and dramatically influence the direction of experiments on the novel gene or protein.

Summary

In 1962, the Nobel Prize in Medicine and Physiology was awarded to Watson and Crick for the discovery of the structure of DNA. The technology developed in the 40 years since has revolutionized how biological research is conducted. The ability to manipulate genes in vitro has greatly increased not only the experiments that are now possible but also how scientists think about biological problems. Each of the techniques described in this chapter allows scientists to manipulate a novel gene in many different ways with the goal of uncovering its unique role in the cell.

Study questions

1 How is plasmid DNA purified away from chromosomal DNA?
2 How does ethidium bromide interact with DNA and how does it help in visualizing DNA on an agarose gel?
3 What are the distinguishing features between the three types of restriction enzymes?
4 Which class of enzymes is used for cloning and what characteristic(s) of this class make it suitable for the job?
5 Could you clone the gene for a restriction enzyme into *E. coli* in the absence of the modification enzyme? What would happen to the cellular DNA if you tried this experiment?
6 Can a blunt end be ligated to a sticky end?
7 Can a clone library be constructed from human DNA? Broccoli DNA? *E. coli* proteins? Human membranes? Why or why not?
8 What kind of gel would you use to separate a 100 bp DNA fragment from a 35 bp DNA fragment? A 750 bp fragment of DNA from a 6 kb fragment of DNA?
9 How would you determine if a novel bacterium contains a gene for degrading lactose?
10 In a PCR reaction, the reactions are cycled between 90–95°C, 40–55°C and 70°C. What biochemical reactions are occurring at each temperature?
11 You have the DNA sequence 5′ TGCGCTAGGCTCATGGCCTTATAGACTCAGTCAAACGTCGTAGT 3′ in your gene. Design an 18 bp primer to change the two Cs to As in the sequence: 5′ GGCCTTA 3′.
12 Two protein sequences are 55% identical and 83% similar. What does this mean? Can two sequences have a greater identity than similarity?

Further reading

Grunstein, M. and Hogness, D. 1975. Colony hybridization: a method for the isolation of cloned DNAs that contain a specific gene. *Proceedings of the National Academy of Science USA*, **72**: 3961.
Lobban, P. and Kaiser, A.D. 1973. Enzymatic end-to-end joining of DNA molecules. *Journal of Molecular Biology*, **78**: 453.
Mertz, J. and Davies, R. 1972. Cleavage of DNA: RI restriction enzyme generates cohesive ends. *Proceedings of the National Academy of Science USA*, **69**: 3370.
Sambrook, J., Fritsch, E.F., and Maniatis, T. 1989. *Molecular Cloning: A Laboratory Manual*. Cold Spring Harbor, NY: Cold Spring Harbor Laboratory.

Chapter 15

Bioinformatics and proteomics

In the preceding 14 chapters, individual cellular processes have been examined with the goal of understanding the components of the system, the regulation of the components, and the function of the process. Up until a few years ago this was the focus of biological research. It has given, and continues to reveal, a thorough understanding of how individual or small numbers of cellular components function together. What is missing from this view is a bigger picture of how a cell functions.

As a cell, population of cells, or organism is growing and dividing, many if not most of the processes that have been discussed are happening at the same time. The cell is not only concerned with replicating its DNA, but also that energy sources are being brought into the cell, amino acids are being synthesized, extracellular signals are being intercepted and responded to, and all of the other details that are required for it to function are being addressed. The areas of bioinformatics and proteomics are beginning to address how the cell integrates all of the intracellular and extracellular information so that it develops and thrives in the appropriate manner.

Bioinformatics

Bioinformatics is defined as the study of information content and information flow in biological systems and processes. It is the study of how genomes are organized, structured, maintained, propagated, and expressed. The techniques of bioinformatics are used after the complete base pair sequence of the genome of a cell or organism has been determined. The **genome** of an organism, be it from a bacterium, a virus, or a human, is all of its DNA, including extrachromosomal DNA from plasmids or resident DNA sequences such as integrated phages or viruses.

Bioinformatics involves the use of molecular tools such as DNA sequencing and sophisticated computer-based analysis technologies to gain an understanding of the information encoded by an organism's genome. Computer databases store biological data in various forms, including DNA sequences, protein sequences, three-dimensional protein structures, promoter sequences, and regulatory protein binding sites to name a few. The databases are annotated to reflect the results of biological experiments on the individual components in the database. In this manner, what has been determined experimentally is used to make predictions about uncharacterized sequences. Using comparisons, bioinformatics looks at the molecular organization and structure of a cell's genome including what genes are present and what genes are absent, how or when the genes are expressed, and what other information is present in the genome besides the coding regions. Bioinformatics is not possible without the knowledge gained from studying the individual components of a cell and the sequence of the cell's genome.

Strategies for sequencing genomes

Chapter 14 describes how the sequence of a DNA molecule is determined. These same chemical reactions are used when sequencing one small DNA molecule or a whole genome. However, when sequencing an entire genome, the genome must be broken into smaller pieces so that each piece can be amplified and enough DNA isolated to carry out the sequencing reactions (Fig. 15.1). Many strategies have been developed for breaking the genome into smaller pieces, cloning the pieces, sequencing each piece, and assembling the pieces of the DNA sequence with the aid of a computer.

Each DNA sequencing reaction can yield approximately 500 base pairs of sequence information from one primer. The first primer is usually located in the vector DNA, whose DNA sequence is known. If a clone contains a chromosomal insert that is larger than 500 base pairs, then the sequence determined from one reaction can be used to synthesize another primer for use in a second sequencing reaction (Fig. 15.2). This strategy is used to determine the nucleotide sequence of the entire insert.

Why not just clone the genome in 500 bp segments and sequence each clone from a primer that starts in the vector? While this strategy has been used and does work, it means isolating and sequencing a large number of clones. Usually, random pieces of DNA are cloned and sequenced. The sequence of each clone is put into a computer and the computer looks for overlapping regions of sequence (Fig. 15.3). Once the computer finds an overlapping sequence, it assumes that the two clones are contiguous or are **contigs**. As more and more clones are sequenced, the nucleotide sequence of the entire genome is reconstructed in the computer. Due to the amount of overlap required between two clones to order them, if the genome was divided into 500 bp segments, many more clones would be required to complete the genome sequence than if larger clones were used. Additional complications arise, depending on the features of the genome

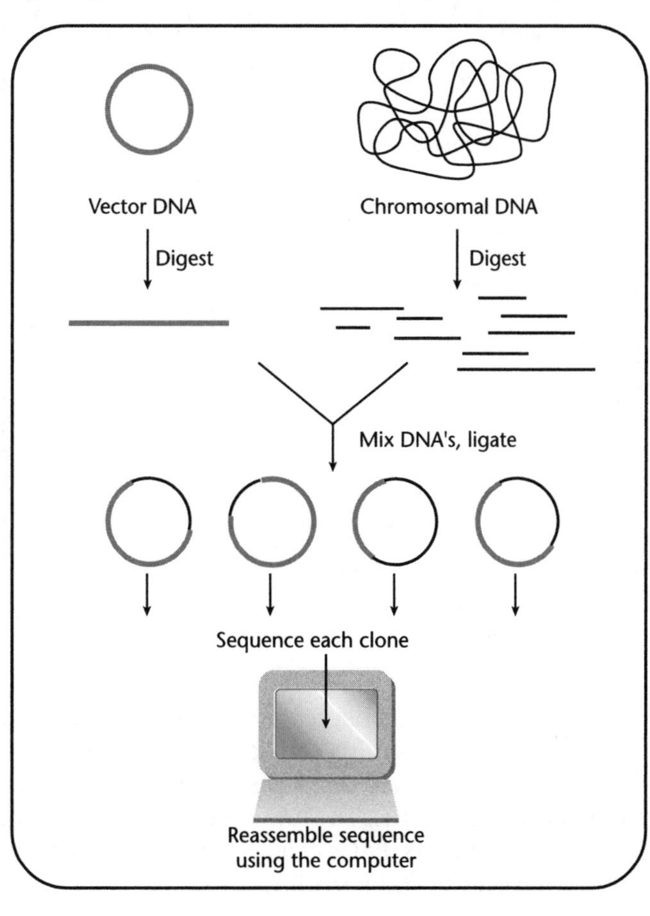

Fig. 15.1 The general strategy for sequencing a genome. Genomic DNA and vector are isolated, digested with an appropriate restriction enzyme, mixed together, ligated, and transformed into *E. coli*. Many independent clones are isolated. The DNA from each clone is isolated and the genomic insert is sequenced. The sequence of each clone is entered into the computer and the once enough clones have been sequenced, the complete sequence of the genome can be assembled in the computer.

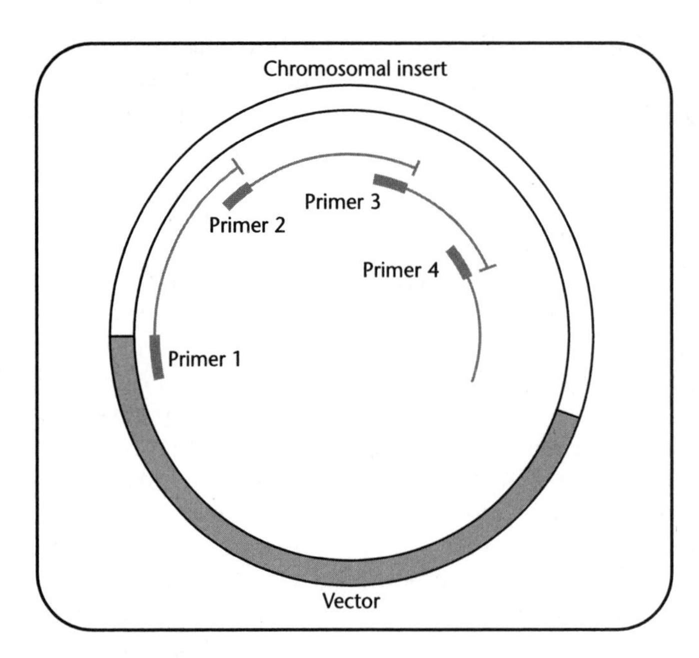

Fig. 15.2 The strategy for sequencing an insert in a clone. Each sequencing reaction can read ~500 base pairs. The first sequencing reaction uses primer 1, which hybridizes to the vector and reads into the insert. The second primer is made using the data obtained from the first sequencing reaction and allows more sequence from the insert to be read. As many primers as needed can be made so that the entire sequence of the insert can be determined.

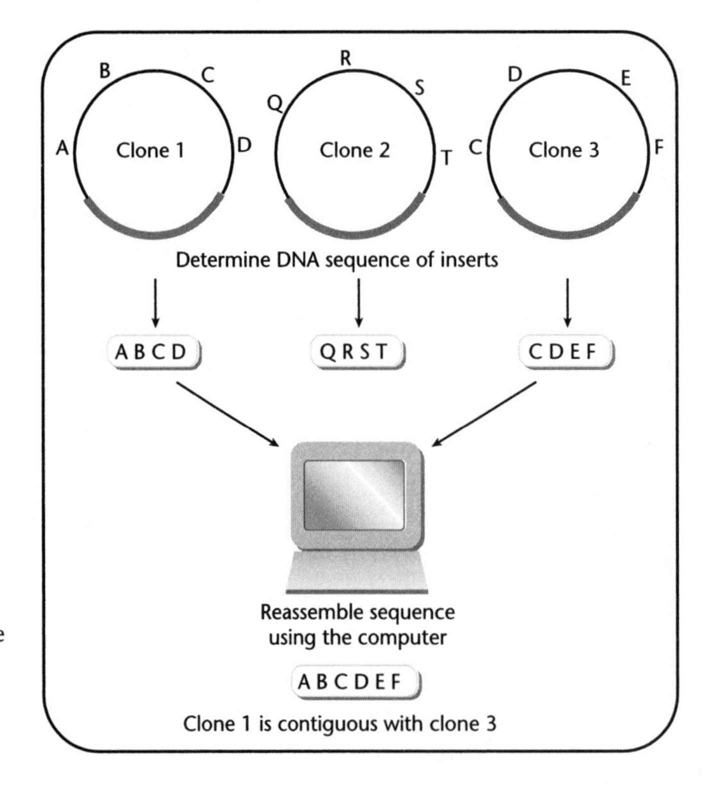

Fig. 15.3 Assembling the sequence of an entire genome from individual clones. Computer programs are used to look for overlaps in the sequence of individual clones. Once an overlap has been found, the sequence from the two clones is assembled into a contig.

being sequenced. Repeated sequences are especially difficult to work with. If a sequence is present in more than one place in a genome, then the smaller the clone, the more likely it is to contain only the repeat and no unique sequence. Reconstructing repeated sequences can frequently lead to the wrong order of clones in contigs.

The plasmid and phage vectors described in Chapter 13 are useful for fragments of DNA in the tens of kilobase pair range. These vectors have been used successfully to sequence many genomes. In the quest to make sequencing genomes and putting contigs together easier, vectors that can accommodate fragments of DNA in the hundreds of kilobase pair range have been developed. One class of vectors that accepts large DNA fragments is the artificial chromosomes. **Artificial chromosomes** contain all of the DNA sequences necessary for them to be replicated as a chromosome. Cosmids, the first artificial chromosome vectors, are plasmids containing a plasmid *ori* such as ColEI *ori* and λ cos sites. These features allow a cosmid vector to replicate like a plasmid, be packaged in λ phage heads and moved from cell to cell like λ. Cosmids can accept DNA fragments in the ~40–45 kb range. PACs rely on P1 sequences to replicate and partition the DNA and as such, are stable, low copy number plasmids. The low copy number usually leads to an increase in stability of cloned fragments. PACs can accept ~90 kb inserts. BACs use sequences from the F factor to replicate and partition the DNA. Like PACs, they are maintained as stable, low copy plasmids but unlike PACs, they can accept several hundred kilobases of DNA. YACs use sequences from yeast chromosomes to replicate and partition the DNA (Fig. 15.4). YACs can accept up to 1000 kb of DNA. All of these vectors are amenable to easy, quick, and economical isolation of their DNA.

Bacterial genomes

The 3569 base genome of the RNA virus, MS2, was the first genome to be sequenced and analyzed. By 1995, the 1,830,137 base pair genome of the bacterial pathogen, *Haemophilus influenzae*, was completely sequenced. Since then, the sequences of many microbial genomes have been published. Table 15.1 lists descriptions of several sequenced microbial genomes. As of 2002, the complete sequence of 70 different microbial genomes has been published and the sequence of 189 microbial genomes is currently in progress.

Once sequencing of an entire genome was well underway, it became obvious that the amount of information generated and thus the amount of information that could be analyzed is enormous. Each base pair of a genome is one bit of information. The *Haemophilus influenzae* genome has 1,830,137 pieces of information. The discipline of bioinformatics, by necessity, grew hand in hand with the ability of desktop computers to process large amounts of information.

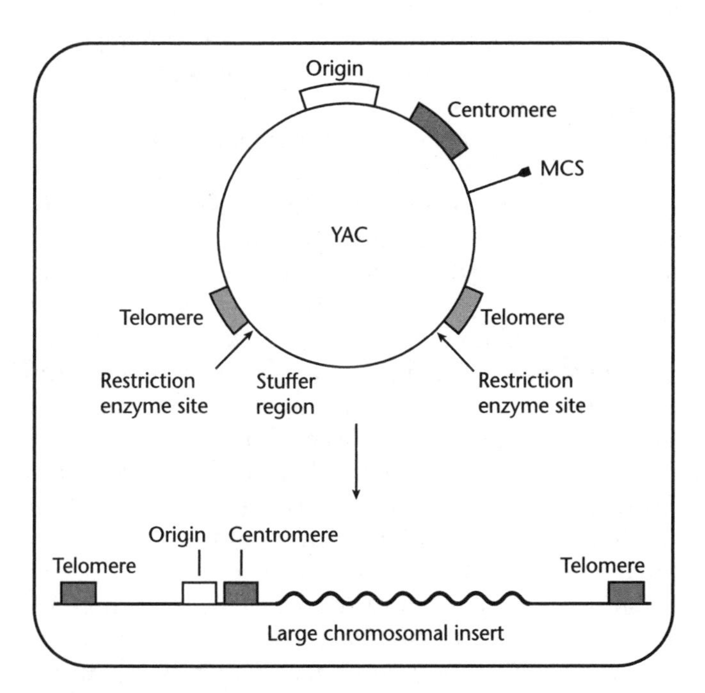

Fig. 15.4 Sequencing whole genomes has required advances in DNA cloning technology. Clones that can carry bigger segments of DNA are more useful. To this end, artificial chromosome technology has been developed. The artificial chromosomes can be for bacteria (BAC), phage P1 (PAC), phage lambda (cosmids), or yeast (YAC). Shown is an example of how YACs are constructed. A YAC contains an origin for DNA replication, a centromere sequence that is used to separate the chromosomes before cell division, and two telomere regions that function as the ends of the YAC when it is replicated as a chromosome. YACs also have a multiple cloning site (MCS) and a stuffer region that is removed to expose the telomere sequences. Thousands of kb of chromosomal DNA can be cloned into the MCS.

Table 15.1 Examples of some completely sequenced genomes and the number of potential proteins (ORFs) that they contain.

Organism	No. of base pairs	ORFs
Escherichia coli K 12	4,639,221	4288
Bacillus subtilis	4,214,810	4100
Lactococcus lactis IL 1403 (dairy cultures)	2,365,589	2310
Pseudomonas aeruginosa (opportunistic pathogen)	6,264,403	5570
Vibrio cholerae (causes cholera)	4,033,460	3885
Clostridium perfringens (causes gas gangrene)	3,031,430	2670
Streptococcus pneumoniae (causes pneumonia)	2,160,837	2236
Streptococcus pyogenes (causes rheumatic fever)	1,852,442	752
Helicobacter pylori (causes peptic ulcers)	1,667,867	1590
Treponema pallidum (causes syphilis)	1,138,006	1041
Mycoplasma pneumoniae (causes pneumonia)	816,394	677
Borrelia burgdorferi (causes Lyme's disease)	910,725	853

Analyzing genomes

Analysis techniques are continually being invented and perfected to determine how much information resides in the genome, what this information is and what it means. The tools of bioinformatics allow scientists to analyze and compare genomes once the DNA sequence has been obtained. For example, after the sequence of a DNA fragment is determined, it is added to a database and compared to DNA sequences already in the database. If this comparative analysis shows similarity between the newly obtained sequence and previously characterized sequences, then predictions can be made about the functions encoded in the newly obtained DNA sequence. This comparative aspect of sequence annotation is also used to determine common genes, regulatory mechanisms and biological pathways among different species, and in evolutionary studies to help determine relationships among different species. Searching databases for new genes or features is often referred to as data mining. Data mining can uncover unique or interesting features of a genome that in turn can be studied experimentally.

The first features that are usually identified in a newly sequenced genome are the open reading frames (ORFs). ORFs are identified by analyzing several properties of known genes from the organism. There are no hard and fast rules as to what is an ORF and what is not an ORF. First, the length of the ORF is considered using known genes as a standard. If the shortest known gene is 100 amino acids in length, then ORFs that are smaller than this may be eliminated. Several different lower limits can be tried in an analysis to determine how many potential ORFs fit the criterion. For example, in yeast, there are ~100,000 potential ORFs of 100 amino acids or less. Because there are so many, they are usually omitted from any analysis. Next, the codon usage in the ORF is examined. Remember that the genetic code is redundant. Different organisms show favoritism with respect to specific codons that they use. Using information from known genes, a codon bias can be determined and used to predict which ORFs match the bias. These are more likely to be ORFs that encode a gene product. If a gene encodes a protein composed of only rare codons, then the protein is not likely to be made anyway.

An analysis of the genes that an organism contains in its genome can provide a significant amount of information about the organism. More information can be garnered by looking at the frequency of combinations of nucleotides such as AT, CC, GATC, etc. The frequencies of specific nucleotide combinations can be compared to what is expected if the genome were composed of random sequences. If a sequence is present or absent at a much greater frequency than expected for a random sequence then it is potentially interesting. Non-random sequences can have very important functions such as a role in gene regulation or DNA replication, the binding site for a specific protein or RNA, or the site for a specific modification such as methylation. For example, in many eukaryotes the dinucleotide CpG is methylated and the methylation influences the expression of genes. The location of CpG in the genome is very important and non-random.

Bacteria can be classified by the overall G + C content of their genome. Some bacterial genomes are G + C rich, some are A + T rich. The sequence of the entire genome allows the investigation of the G + C or A + T content in local regions. In several cases, regions of genomes have been shown to have a G + C content that is very different from the overall G + C content of the genome (Fig. 15.5). These regions with different G + C contents probably did not evolve with the genome. In some cases, the unusual region has been shown to be a remnant of a phage or acquired as a block from another species.

The *E. coli* K12 genome

The genome of *E. coli* K12, which was completely sequenced in 1997, contains 4,639,221 base pairs. Many different kinds of information have been garnered from an analysis of the *E. coli* genome. First, 4288 potential or known open reading frames were identified (Table 15.2). Surprisingly, only one-half of *E.coli*'s ORFs are well characterized. Even in an organism as well studied as *E. coli*, we can only assign functions to approximately half of its genes! The average distance between ORFs is 118 base pairs and only 70 genes are more than 600 base pairs apart. Genes that encode proteins account for 87.8% of the genome, 0.8% of the genome encodes stable RNAs, and 0.7% is repetitive DNA, leaving 11% for regulatory regions and any other function we don't know about yet. This leads to the conclusion that the *E. coli* genome is tightly packed with genes, a situation very different from many eukaryotic genomes where the gene can be spaced kilobases or hundreds of kilobases apart.

The sequence of the genome has also revealed that many genes are oriented in the same direction as the replication forks move. *E. coli* replicated bidirectionally from *oriC*. All seven ribosomal operons, 53 out of 86 tRNA genes, and approximately 55% of protein-encoding genes are oriented so that the replication fork sees the 5′ end of

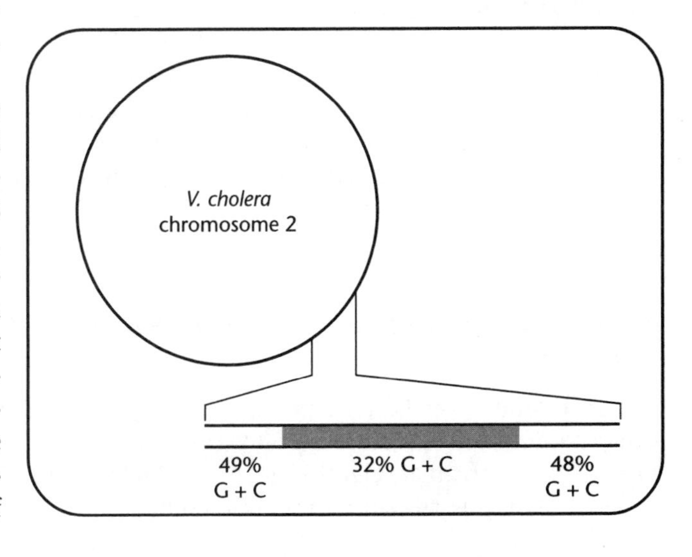

Fig. 15.5 The overall G + C content of a cell is a characteristic of that cell. Regions in its genome that are very different in G + C content are frequently regions that have been acquired from a bacteriophage, plasmid, or from another bacterium.

Table 15.2 The number and distribution of *E. coli* ORFs.

Identified Genes/Proteins	Number	%
Regulatory proteins	45	1.05
Structural proteins	182	4.24
Transport and binding proteins	281	6.55
Energy metabolism	243	5.67
DNA replication, recombination and DNA repair proteins	115	2.68
Transcription, RNA synthesis, RNA metabolism and RNA modification	55	1.28
Translation and post-translational protein modification	182	4.24
Cell adaptation and protection	188	4.38
Biosynthesis of cofactors, prosthetic groups and carriers	103	2.40
Nucleotide biosynthesis and metabolism	58	1.35
Amino acid biosynthesis and metabolism	131	3.06
Fatty acid and phospholipids metabolism	48	1.12
Carbon compound metabolism	130	3.03
Central intermediary metabolism	188	4.38
Other known genes/gene products	26	0.61
Phage, transposons and plasmids	87	2.03
Putative Genes/Proteins		
Regulatory proteins	133	3.10
Membrane proteins	13	0.30
Structural proteins	42	0.98
Transport proteins	146	3.40
Chaperones	9	0.21
Enzymes	251	5.85
Hypothetical, unclassified or unknown	1632	38.06

FYI 15.1

The human genome

In 2000, the 3 billion base pair genome of a haploid human was described in a preliminary format, suggesting that detailed sequence information and the corresponding analysis would be available sooner than the planned 15 years needed to complete the project. Approximately 30,000 genes or potential genes were identified. The draft sequence has revealed that the human genome contains 3164.7 million base pairs. The average gene is 3000 bp in length with the largest gene to date being the dystrophin gene at 2.4 million bp. Only 50% of the human genes have had functions identified. Interestingly, 99% of the nucleotide sequence is exactly the same in all people. Chromosome one has the most genes at 2968, while the Y chromosome has the fewest genes at 231.

the gene first. For these genes replication and transcription both proceed in the same direction, leading to less interference between the replication fork proteins and RNA polymerase.

Proteomics

Proteomics, is the study of the when and where all of the gene products of a genome are expressed. A **proteome** is defined as all the proteins present in a cell at a specific time or under specific conditions. The types and amounts of proteins present in a cell are constantly changing in response to environmental or developmental factors. Proteomics attempts to identify all of the proteins that are expressed under any given condition.

Techniques for examining the proteome—SDS-PAGE and 2-D PAGE

The proteome of a cell can be examined using several techniques. The first technique, called sodium dodecyl sulfate-polyacrylamide gel electrophoresis (SDS-PAGE), separates proteins on an acrylamide gel in an electric field (Fig. 15.6). SDS, a detergent, is added to denature the proteins extracted from the cells and to give most proteins an overall negative charge. Protein then separate based on their molecular weights. The larger the protein, the slower it migrates through the acrylamide gel. This technique is very useful for examining a small number of proteins or proteins that are present in large quantities. It does not give a good separation of all of the cell's proteins, especially those that are present in a small amount or those that are of a similar molecular weight.

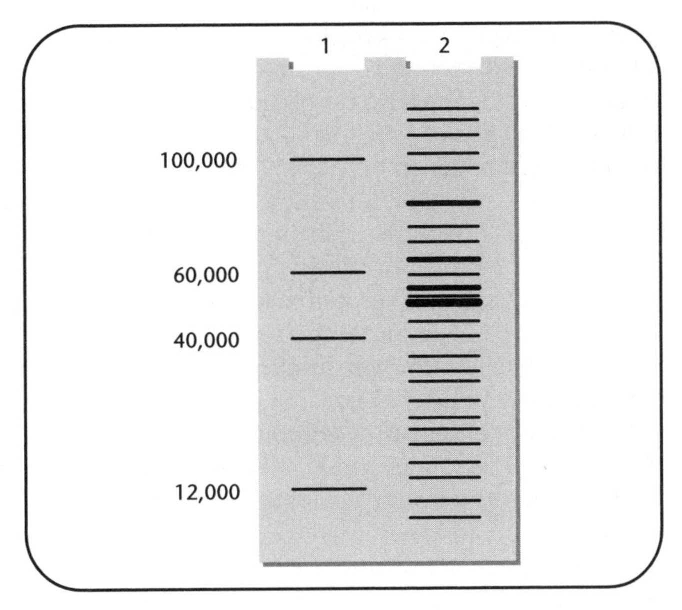

Fig. 15.6 A standard protein gel with samples electrophoresed based on molecular weight can separate only 50–100 different proteins. These gels are prepared using sodium dodecyl sulfate-polyacrylamide gel electrophoresis.

A second way to examine the size and composition of a proteome is through the technique called two-dimensional polyacrylamide gel electrophoresis (2-D PAGE). Total protein from a culture of cells is extracted and fractionated in a cylindrical polyacrylamide gel that separates proteins based on the overall charge of the protein (isoelectric point or pI, Fig. 15.7). Proteins will migrate in the pH gradient of this first dimension until their net charge reaches zero. The separated proteins are then fractionated on a second polyacrylamide gel that separates them in a second dimension based on their size or molecular weight. In the second dimension, the larger the protein, the slower it will migrate. Large proteins will be retained at the top of the gel and small proteins will be found at the bottom of the gel. Because 2-D PAGE separates proteins on the basis of two characteristics, more proteins, especially those of similar molecular weights, can be visualized.

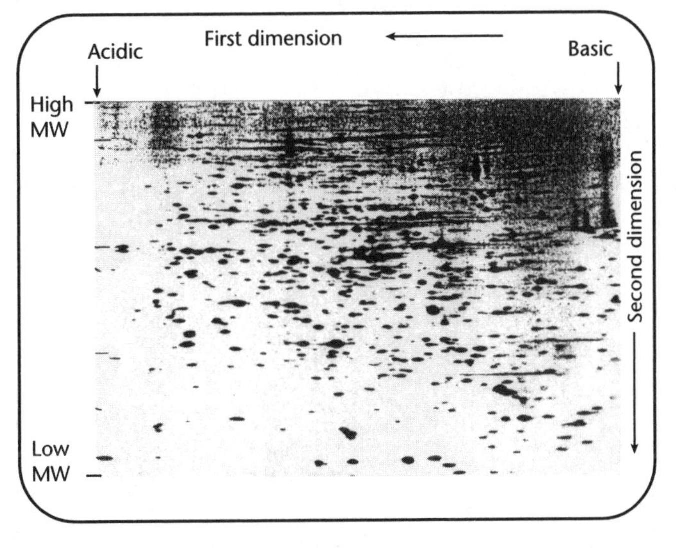

Fig. 15.7 An example of a two-dimensional polyacrylamide gel. Each spot represents a different protein, and the intensity of the spot is measurable and easily comparable to other protein spots. The proteins are first separated on the basis of their isoelectric point or overall charge (pI). The second separation is on the basis of their molecular weight (M_r). The largest proteins are closest to the top of the gel because they migrate the slowest.

2-D PAGE analysis can be used for comparing cultures of the same bacteria that were grown under two different experimental conditions. For example, two cultures grown at different temperatures or with different carbon sources can be examined. This type of comparative analysis requires the use of at least two 2-D gels, one gel displaying the separated proteins of the culture growing in condition one and the other gel displaying the separated proteins of the culture growing in condition two.

The number of proteins that can be detected on 2-D PAGE is limited by the separation of the proteins and the amount of the protein that is present. The amount of protein that is detectable can be increased by adding a radioactive amino acid to the growing cultures. This leads to all of the proteins being radioactively labeled. The 2-D gels are electrophoresed as normal and are subsequently exposed to X-ray film (autoradiography). The amount of radioactivity in each protein spot can then be determined. Radioactive labeling of the proteins amplifies the signal and allows for detection of smaller amounts of protein.

2-D gels identify the proteome as a group of specific spots of protein. The identity of the proteins is not known. Further characterization of the expressed proteins usually includes the removal of specific proteins from the 2-D gel and amino terminal sequencing of individual proteins. Amino terminal sequencing of the protein can help identify it and may indicate the gene that encodes it. While amino terminal sequencing is possible, it is time consuming and not practical to carry out on the hundreds of different proteins in a 2-D gel.

2-D PAGE analysis can be combined with other experimental approaches to extend their capabilities. For example, 2-D gels can be used in western blots where a specific protein is identified by an antibody. The antibody can be made against a protein from one bacterium and then used to determine if other bacterial species have the same protein. 2-D gels allow a more accurate screening of the proteome for the presence or absence of a specific protein. This approach is being used on the bacterial pathogens, *Helicobacter pylori*, *Chlamydia trachomatis*, and *Borrelia garinii* to determine if these species share any proteins that can be used in the development of vaccines.

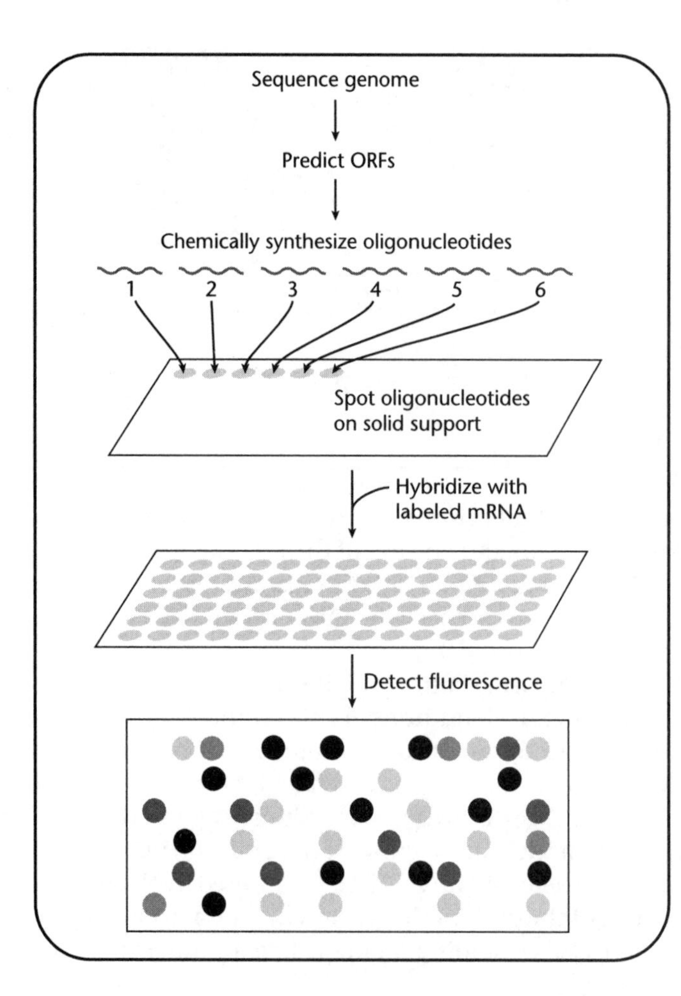

Fig. 15.8 A flowchart for the microarray technique. The microarray contains short single-stranded oligonucleotides corresponding to all of an organism's ORFs. A given single-stranded oligonucleotide is placed in a known and identifiable location on the microarray. Fluorescently tagged mRNA isolated from an actively growing culture of this organism is hybridized to the microarray. The regions that fluoresce as a result of hybridization will reveal which genes were being expressed at the time the mRNA was isolated from the organism.

Techniques for examining the proteome—microarray technology

A unique way to examine proteomes, called microarrays, takes advantage of advances in several fields. In a **microarray**, equal amounts of small pieces of DNA (oligonucleotides) that correspond to every ORF in a genome are attached to a solid support such as a glass slide (Fig. 15.8). Each oligonucleotide is put in a separate spot and the location and contents of each spot is cataloged. Cells are grown under the desired experimental conditions in the presence of a fluorescent molecule that is incorporated into mRNA. The labeled mRNA is isolated from cells and exposed to the arrayed oligonucleotides. The different types and amounts of mRNA present hybridize or bind to the oligonucleotides. Using a laser connected to a computer, the amount of fluorescence in each spot is measured and this information is transmitted to the computer. The fluorescence corresponds to the amounts and types of mRNA present in the original sample. The catalog of the contents of the spots is used to determine which genes and how much mRNA corresponding to each of those genes is made under a specific experimental condition. Thousands of different oligonucleotides can be placed in a 1–2 cm area. Differences in fluorescence as small as a few fold can be detected in a microarray.

Microarrays require a significant amount of technology. The sequence of the genome of the organism in question must be known. The ORFs or potential ORFs in the genome must be identified. Oligonucleotides corresponding to every ORF must be chemically synthesized and attached to the solid support to make the microarray. The oligonucleotides can be attached to a glass slide or a silicone chip. The oligonucleotides can be put on the slide or chip by hand but this is time consuming and not very practical. In many cases, robots are used to spot the oligonucleotides in tightly packed configurations. Once the mRNA has been hybridized to the arrayed oligonucleotides, the microarrays must be exposed to a laser to detect the fluorescent dye. The amount of fluorescence in each spot is detected and this information is relayed to a computer. The computer stores this information and must be used to analyze the large amounts of data generated in each experiment. Microarray technology is very powerful but it is also very expensive.

How is microarray technology used to ascertain the composition of different proteomes? Total mRNA from a culture of cells grown under normal conditions can be applied to a microarray for that cell type. Those genes that are expressed under normal growth conditions can be determined. A second experiment can then be performed using a second microarray. This time, however, mRNA from a culture of cells grown under a stress condition, such as heat shock, can be applied to the microarray. The genes that are expressed under the stress condition can be determined. When the information from the two microarrays are compared, the genes that are preferentially expressed or repressed during heat shock can be identified (Fig. 15.9).

A second approach to comparative microarrays is to label the mRNA from one growth condition with a fluorescent dye of one color and the mRNA from the second growth condition with a different fluorescent dye of a different color. The two mRNA samples are mixed together and applied to one microarray. The computer keeps track of which fluorescent dye is present in which spot and how much is present. The mRNAs that differ between the two samples and how much they differ by is determined.

FYI15.2

The practical uses for comparative microarrays

Microarrays are being used in many areas of research to broaden our understanding of individual cells as well as multicellular organisms. One promising use of microarrays is in the field of diagnostics. It has been known for a very long time that two people with the same disease can react very differently to the same treatment for that disease. Using microarrays containing the human genome, scientists are investigating the proteomes of different people who have the same disease. The idea is to determine if the proteomes of these people are the same or different and how many proteomes can be found for a single disease. It is thought that by understanding proteomes, the reaction to different treatments can be predicted and the appropriate treatment tailored to a specific individual.

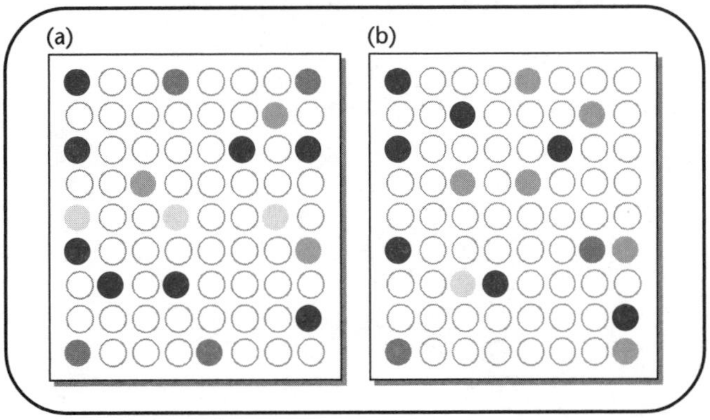

(a) (b)

Fig. 15.9 Comparing different microarray experiments can determine how many and which genes are expressed under different conditions. The two microarrays can be hybridized and examined separately as shown here or two different fluorescent tags can be used for the two mRNA samples and a single microarray can be used. The intensity and identity of each spot can be determined.

Summary

The fields of bioinformatics and proteomics apply state-of-the-art technology to fundamental problems in biology. Bioinformatics takes the information we have learned through experimentation and uses it to predict characteristics of unknown genes and genomes. Proteomics looks at the expression of the entire genome under any imaginable condition. Both of these technologies attempt to understand all of the things that are taking place in a cell any given time. These technologies have the potential to impact how diseases are treated, how drugs are developed, and how vaccine targets are identified. The fields of bioinformatics and proteomics are leading to a better understanding of how species evolve, how a cell ages, and how certain species can persist under harsh and often damaging conditions.

Study questions

1 What is bioinformatics?
2 What is proteomics?
3 Describe an experimental approach that would define the impact of carbon limitation on *E. coli*.
4 What types of genes make up a genome?
5 Can *E. coli* have more than one proteome? Give an example.
6 What are microarrays? How are they made?
7 If you have two strains of the same bacteria, one that is pathogenic and one that is not, how would you predict what genes are needed for the pathogenic phenotype?
8 What features of a genome, besides the actual genes, are important or can be important to the cell?

Further reading

Blattner, F.R. *et al.* 1997. The complete genome sequence of *Escherichia coli* K12. *Science*, **277**: 1453–74.
Heller, M.J. 2002. DNA microarray technology: devices, systems, and applications. *Annual Review Biomedical Engineering*, **4**: 129–53.
Nilsson, C.L. 2002. Bacterial proteomics and vaccine development. *American Journal of Pharmacogenomics*, **2**: 59–65.
TIGR Microbial Database. 2002. http:/www.tigr.org.

Glossary

3′ overhang—Restriction enzymes that cleave the DNA asymmetrically leaving single-stranded bases. If the single-stranded bases end with a 3′ hydroxyl, the enzyme is said to leave a 3′ overhang.

5′ overhang—Restriction enzymes that cleave the DNA asymmetrically leaving single-stranded bases. If the single-stranded bases end with a 5′ phosphate, the enzyme is said to leave a 5′ overhang.

–10 site—A part of the promoter that is ~10 base pairs upstream of the +1 site. The –10 and the –35 site constitute the sites to which RNA polymerase binds.

–35 site—A part of the promoter that is ~35 base pairs upstream of the +1 site.

–1 site—The base at which RNA polymerase starts polymerizing RNA.

3′ to 5′ exonuclease—A subunit of all DNA polymerases capable of removing nucleotides from an exposed 3′ end. This is the editing (proofreading) function used to ensure that the correct nucleotide was added by DNA polymerase III to a growing DNA chain.

α fragment—The first ~60 amino acids of β-galactosidase that can combine with the last ~960 amino acids of β-galactosidase (the ω fragment) to form an active enzyme.

ω fragment—The last ~960 amino acids of β-galactosidase that can be combined with the α fragment of β-galactosidase to form an active enzyme.

Activation—(of a gene, operon, or regulon) A mechanism of gene regulation that requires the induction of the expression of the genes. Frequently, activation involves a regulatory factor binding to the promoter region and activating the promoter.

Alkylating agent—Mutagenic chemicals that attach an alkyl group (methyl or ethyl) to a base and change the base's hydrogen-bonding capabilities. Guanine is the most sensitive to alkylating agents.

Allele—A unique form of a given gene. For example, the wild-type copy of a gene is one allele and a mutant copy of the same gene is also an allele of that gene. Allele is a genetic term used to describe the form of the gene present in the cells.

Allele number—A unique number or letter given to each mutation. Different mutations are distinguished by their allele numbers.

Anabolic—The process or the enzymes that build (i.e. synthesize) a substrate rather than break down a substrate. An example is the amino acid tryptophan, whose synthesis requires the action of four different enzymes encoded by five different genes.

Annotated sequence—The sequence of a piece of DNA that lists, in addition to the order of the bases, other features of the DNA including promoters, mRNA start sites, open reading frames, transcription and translation regulation sites, and other unusual sequences. These features can be either predictions or biologically determined entities.

Antibiotic resistance determinant—A gene whose protein product confers resistance to a specific antibiotic. The mechanism of resistance differs for each antibiotic. Some antibiotics are inactivated by cleaving them, some by modifying them, and some by simply pumping them out of the cell as fast as they come into the cell. The most commonly used antibiotic resistant determinants for *E. coli* are ampicillin, kanamycin, tetracycline, and chloramphenicol resistance.

Antiparallel—A description for the opposite polarities of the two strands of a DNA molecule. One strand is orientated 5′ to 3′ and the other strand is 3′ to 5′.

AP endonuclease—An enzyme that catalyzes the breakage of phosphodiester bonds within a DNA molecule at a site upstream and downstream of an apurinic or apyrimidinic site.

Apurinic site—A site in the DNA where a purine has been removed from the DNA phosphate–sugar backbone. The bond affected is the N-glycosidic bond.

Apyrimidinic site—A site in the DNA where a pyrimidine has been removed from the DNA phosphate–sugar backbone. The bond affected is the N-glycosidic bond.

Assimilation—The part of the homologous recombination process where the RecA–ssDNA filament binds to and unwinds another dsDNA, promoting base pairing of the complementary nucleotides.

ATPase—An enzyme that degrades ATP. Frequently, the ATPase provides the energy released from the degradation of ATP to other proteins that require energy to carry out their function.

Artificial chromosome—A cloning vector designed to accept 40–1000 kilobases of cloned DNA. These vectors rely on the features that make up a chromosome for their replication and partitioning into daughter cells.

Attenuation—A regulatory mechanism of some anabolic operons (i.e. *trp* operon) that controls the efficiency of transcription after transcription has initiated, but before mRNA synthesis of the operon's genes takes place.

Autophosphorylation—The ability of an enzyme to phosphorylate itself. The addition of phosphate groups to an enzyme often results in the activation of the enzyme's function.

Bacteriophage —A virus that infects bacteria. Bacteriophage (or phage for short) are usually specific for a single bacterial species.

Bacteriocidal —Any compound or physical treatment that kills the bacteria.

Bacteriostatic —Any compound or treatment that prevents bacteria from growing but does not kill them.

Basal level of transcription —Refers to the amount of transcription that originates at a given promoter when the promoter is functioning at its lowest level. For a promoter that is activated by a protein, it is the amount of transcription in the absence of the activator protein. For a promoter that is repressed by a protein, it is the amount of transcription in the presence of the repressor protein.

Base analogs —Chemicals with similar properties to the natural occurring bases (A, T, G, C, U). DNA polymerase III will incorporate base analogs into a growing DNA strand as if they were the naturally occurring nucleotide.

Base modifiers —Chemicals that modify the structure of bases so that they no longer base pair in a predictable manner (A with T and G with C).

Base substitution mutations —A mutation where one base has been substituted for another. Base substitutions are also another name for point mutations.

Bidirectional DNA synthesis —DNA replication proceeding outward from the origin in both directions at the same time. The consequences of this are that the leading and the lagging strands are synthesized at the same time.

Bioinformatics —Defined as the "study of information content and information flow in biological systems and processes". Bioinformatics bridges the biological sciences with the computer sciences.

Biopanning —The process of screening a library of clones for a clone with a useful characteristic. Biopanning frequently refers to screening a phage display library for a phage with a useful or interesting insert.

Blue/white screen —The visual screen that is used to tell when a plasmid has a cloned insert. The blue color results from an α fragment of β-galactosidase combining with an ϖ fragment of β-galactosidase to form an active β-galactosidase molecule. When the α fragment is inactivated by a cloned insert, no active β-galactosidase is made.

Blunt ends —A double-stranded DNA end where both strands end at the same base. Some restriction enzymes cleave symmetrically in their recognition sequence. There are no single-stranded bases after the enzyme has cut the DNA. These enzymes are said to leave blunt ends.

Broad host range plasmids —Plasmids that can be stably maintained in more than one species. They must contain an origin that functions in different species or more than one origin.

Building blocks —A group of small molecules that are used to build macromolecules. For example, the four bases and the phosphate sugars used for the backbone are the building blocks of DNA. Amino acids are the building blocks for proteins.

Burst size —The number of phage produced by one phage infecting one bacterium.

Bypass suppressors —Suppressor mutations that bypass the need for the gene containing the primary mutation. Bypass suppressors also work with deletions in the gene containing the primary mutation.

Capsid —The protein coat that surrounds the phage genome in a phage particle.

Capsule or **capsular polysaccharide** —The thick slime layer composed of polysaccharides that surrounds the cell and protects it from dehydration and the immune system. A given bacterial species can produce more than one type of capsule.

Carbohydrates —Composed of simple sugars and used as a source of energy for the cell as well as a component of several cell structures.

Catabolic —The process or enzymes that break down a substrate rather than build a substrate. An example is the enzyme β-galactosidase, which breaks lactose into glucose and galactose.

Catabolite repression —Cells growing in medium containing glucose do not express, at high levels, certain sugar metabolizing operons (i.e. *lac*) even if the inducers of those operons are present. Catabolite repression relies on the levels of cAMP and CAP protein to relay information on the sugars that are present.

Chain initiation, chain elongation, and chain termination — The steps needed to produce RNA molecules. RNA polymerase must bind to a promoter, initiate RNA chain formation, elongate the RNA chain, and then terminate it at the appropriate place.

Chemotaxis —The movement of a bacterial cell towards a favorable environment and away from a harmful environment. The rotation of the flagella is used to move the cell.

Chi site —The specific DNA sequence used in the homologous recombination process. The sequence of chi is 5′ GCTGGTGG 3′. Once RecBCD passes the chi site in the correct orientation, its 3′ to 5′ exonuclease activity is inhibited. This generates ssDNA for RecA to bind and assimilate.

Circular permutation —The genomes of some bacteriophage always contain the same genes but they are not always present on the infecting phage in the same order. For example, one phage may have the order ABCDEFG, another may have CDEFGAB, and another may have DEFGABC. These different phage genomes are circularly permuted with respect to each other.

Cloning —The process of putting a piece of chromosomal or other DNA into a vector using in vitro manipulations of the DNA.

Coding strand —The strand in a dsDNA molecule that matches the base sequence of the RNA that is made from it.

Coinheritance —The ability to inherit two genes in a single genetic cross.

Cointegrate —A structure that contains two copies of a transposon in a single DNA molecule. The transposons are arranged as direct repeats.

Cold sensitive —A secondary phenotype caused by some mutations. The gene product containing the mutation is not functional at low temperatures.

Colony —The group of cells that are formed when bacteria are plated on a solid growth media and incubated for a period of time. If the cell concentration is low enough, each colony is formed from one bacterium. For *E. coli*, the average colony contains approximately one million cells.

Colony blotting —A version of Southern blotting where colonies on an agar plate are transferred to a filter and lysed in situ. The DNA from the colonies is attached to the filter. The filter is exposed to a tagged probe the same as for a standard Southern blot. Colony blots are used to fish specific genes out of libraries of clones.

Competence —A transient physiological state in which the bacteria can take up naked DNA and transport it into their cytoplasm.

Competent —Cells that are in the physiological state where they can take up naked DNA.

Competence factors —Small peptides that act to induce the expression of the genes needed to make the cell able to take up naked DNA. Some competency factors bind to surface receptors and initiate a relay of information across the membrane and into the cytoplasm. Other competency factors are transported into the cell to effect a change in regulation.

Composite transposons—Composed of modular units. They have insertion elements on either end. The central piece of DNA can encode many different functions.

Concatomer—A long DNA molecule that contains multiple copies of the same DNA sequences linked end to end. Concatomers are frequently the result of rolling circle replication. An example is a phage genome that is arranged in a head-to-tail manner (i.e. if the genes in the phage DNA are arranged ABC, then in a concatomer the genes would be ABCABCABCABC).

Conjugation—The process of moving DNA from one cell to another through cell-to-cell contact and using a specialized plasmid that encodes an F pilus such as an F factor or an R factor. The proteins needed for conjugation are encoded by a plasmid.

Consensus sequence—A base sequence generated from closely related sequences with similar functions. For example, many operons are controlled by camp–CAP binding to their promoter regions. The DNA sequence that camp–CAP binds to is not identical in every operon. If all of the sequences are aligned, a most probable binding site or consensus sequence can be derived.

Contigs—A group of clones whose inserts contain DNA sequences that overlap with each other. Contigs are used to reconstruct the sequence of a genome in the computer.

Constitutive—The synthesis of a gene product continuously.

Construct—The molecule that is made by rearranging DNA sequences in vitro using cloning techniques.

Control regions—The sequences that are used in controlling the expression of a gene or genes.

Core RNA polymerase—The core subunits, $\beta\beta'\alpha_2$, of RNA polymerase that synthesizes RNA by adding ribonucleotides to a growing mRNA chain. Core RNA polymerase does not need a primer to begin synthesis. It does not specifically recognize promoters and begins RNA synthesis randomly.

cos—The name describing λ's cohesive end sites. *cos* sites are used to circularize the λ genome after it is injected into the cytoplasm of a cell.

Counterselection—A counterselection prevents cells of a specific genotype from growing. It is used in genetic crosses to ensure that the parental cells cannot grow and only the recombinants grow.

"cut-and-paste" transposition—Another name for nonreplicative transposition.

Cytoplasm—The aqueous compartment surrounded by the inner membrane. It contains the chromosome, ribosomes, and many enzymes used to degrade or build molecules. The cytoplasm is also the location for transcription and translation.

Deamination—The loss of exocyclic amino groups from cytosine, adenine, or guanine.

Deletion—The loss of one or many base pairs from the DNA.

de novo protein synthesis—New protein synthesis needed for certain cellular processes. For example, many chromosomes need newly synthesized proteins to initiate DNA replication. This implies that a protein is used only once and then inactivated or degraded. The next time that the protein is needed, it must be newly synthesized.

Density labeling—A way of tagging specific cellular components by using a heavy isotope of one of the naturally occurring element, for example ^{15}N instead of ^{14}N.

Deoxyribonucleoside—Also called, deoxynucleoside or nucleoside for short. This is the nomenclature used to describe a base attached to a sugar (adenosine, guanosine, thymidine, cytidine, uridine)

Deoxyribonucleotide—Also called deoxynucleotide, or nucleotide for short. This is the nomenclature used to describe a base attached to a sugar containing a phosphate group (nucleoside 5′ monophosphate: AMP, GMP, TMP, CMP, UMP).

Dephosphorylation—The removal of a phosphate from a molecule. Some proteins are activated or inactivated by the additional or removal of a specific phosphate.

Depurination—The removal of a purine from the DNA backbone, leaving an unpaired base on one strand of the DNA.

Diploid—A cell containing two complete copies of the chromosome. Diploidy is the state of a cell that contains two complete chromosomes.

Direct repeats—DNA sequences that are repeated in a head-to-tail fashion. If the sequence is ATTGCC then it will be repeated ATTGCC–ATTGCC–ATTGCC.

DNA—Deoxyribonucleic acid. The molecule that is passed from generation to generation and specifies the physical characteristics of the cell that contains it. DNA is composed of a sugar–phosphate backbone and the four bases: thymine, adenine, guanine, and cytosine.

DNA-dependent DNA polymerase I or DNA Pol I—An enzyme that uses a DNA template to polymerize nucleotides onto a free 3′ OH of an existing RNA oligonucleotide (primer). DNA Pol I has a 3′ to 5′ exonuclease activity that is called an editing or proofreading activity. It also has a 5′ to 3′ exonuclease activity that removes nucleotides from a double-stranded DNA molecule's exposed 5′ end. The polymerizing activity is used to fill in small gaps in a DNA molecule that have arisen due to the removal of mismatched base pairs and the removal of RNA primers during the editing process.

DNA-dependent DNA polymerase II or DNA Pol II—An enzyme that uses a DNA template to polymerize nucleotides onto a free 3′ OH of an existing RNA (primer). DNA Pol II has a 3′ to 5′ exonuclease activity that is called an editing or proofreading activity. This polymerase is primarily involved in DNA repair processes.

DNA-dependent DNA polymerase III or DNA Pol III—An enzyme that uses a DNA template to polymerize nucleotides onto a free 3′ OH of an existing RNA oligonucleotide (primer). DNA Pol III has a 3′ to 5′ exonuclease activity that is called an editing or proofreading activity. It also has a 5′ to 3′ exonuclease activity that can remove nucleotides from single-stranded DNA. DNA Pol III is the major replicating enzyme in *E. coli*.

DNA gyrase—A topoisomerase that can introduce negative supercoils.

DNA ligase—An enzyme that catalyzes the formation of a phosphodiester bond between a free 5′ phosphate and a free 3′ OH group.

DnaA box—A specific base pair sequence, 5′ TTATCCACA 3′ present four times in the *oriC*. Ten to 12 molecules of DnaA protein bind the DnaA boxes to form a complex (primosome) needed for the initiation of DNA replication.

DnaA—Ten to 12 molecules of DnaA bind the DnaA boxes to force the hydrogen bonds of the A–T rich region in the *oriC* to dissociate, thus opening the double-stranded DNA molecule for access by other initiator proteins (i.e. DnaB, DnaC) of the primosome.

DnaB—A helicase that is loaded onto the primosome. Its activity is used to further dissociate (unwind) the double strands of the DNA so that replication can be initiated by DnaG primase.

DnaC—An enzyme that adds DnaB helicase to the primosome.

DnaG—A DNA-dependent RNA primase that synthesizes the short ribonucleotide primer (5–11 ribonucleotides in length). This short RNA primer provides a free 3′ OH group for DNA polymerase III to use for adding on deoxyribonucleotides during replication.

Dominant mutation—A mutation that still exhibits a phenotype when it is in the presence of a wild-type copy of the same gene.

Donor cell—In a bacterial genetic cross, the cell that is donating the DNA to the other cell (the recipient cell). For example, in an Hfr cross, the Hfr is in the donor cell and gets transferred into the recipient cell.

Duplication—A mutation resulting from a stretch of bases being repeated in the DNA sequence.

Early genes—Phage genes that are expressed first after the injection of phage DNA into a bacterium.

Editing—Some DNA polymerases contain 3′ to 5′ exonuclease activity that can be used to monitor if the correct base has been inserted into the growing DNA chain. If the wrong base is polymerized into the chain, the 3′ to 5′ exonuclease activity removes it giving DNA polymerase a second chance to get it right.

Efficiency of plating (EOP)—The efficiency with which a phage forms plaques on different host strains. Restriction and modification enzymes were discovered because the EOP of phage grown on different strains of *E. coli* was altered by up to four logs, depending on the last strain the phage was plated on.

Electroporation—The process of putting DNA into cells using a large jolt of electricity. Electroporation can be used for bacterial, yeast, and mammalian cells.

Endonuclease—An enzyme that catalyzes the breakage of a phosphodiester bond within a DNA molecule.

Environment suppressors—Suppressor mutations that alter the cell environment and allow the gene product containing the primary mutation to function at some noticeable level.

Exconjugants—Cell that have undergone conjugation and exchanged DNA.

Expression vectors—Plasmid vectors that are designed to transcribe and translate a cloned gene.

Exonuclease—An enzyme that catalyzes the breakage of phosphodiester bonds, affecting the nucleotides at the very end of a DNA molecule. Exonucleases are described as having a 3′ to 5′ or a 5′ to 3′ nuclease activity. This describes which end of the DNA strand will have its nucleotides removed first.

Exponential growth—A phase in the growth of bacteria where the number of cells doubles in a set amount of time for a number of generations. Exponential growth requires an excess of nutrients. Once the doubling time is no longer constant, then exponential growth is over.

Extragenic suppressors—Suppressor mutations that occur outside the gene containing the primary mutation.

F^+—A strain containing an F factor. The F factor is located extrachromosomally and contains only F DNA.

F factor—The plasmid that is capable of moving DNA from one cell to another. The F factor encodes all of the proteins necessary for transfer of the DNA.

F pilus—The hair-like structure that is located on the surface of a strain carrying an F factor. The F pilus is used to bring the two cells in close contact so that DNA can be transferred between them. The F pilus is 2–3 microns in length and F-containing cells have between one to three of them.

F prime (F′)—An F factor that is located extrachromosomally but carries chromosomal DNA in addition to F factor DNA.

Female cells—Cells that do not contain an F factor, an Hfr, or an F′.

Filamentous phage—Phage whose infectious particles are long filaments of nucleic acid that are coated by protein.

Fimbriae—Hair-like projections that are attached to the outer surface of the outer membrane of certain bacteria. Fimbriae are made of protein and help bacterial cells adhere to animal cells.

Flagella—Flagella are long whip-like structures that are embedded in the membranes. At the base of each flagella in the inner membrane is a motor that rotates the flagella and propels the cell. Flagella can rotate both clockwise and counterclockwise, resulting in the bacteria moving in different directions.

Flush ends—The ends of a linear dsDNA that do not contain any single-stranded bases. Also known as blunt ends.

Forward mutation—Mutations that change the wild-type nucleotide sequence.

Frameshift mutations—The addition or deletion of one or two bases to a nucleotide sequence that results in the original genetic code being read out of frame during translation.

Gain of function mutation—A mutation that has acquired or exaggerated a specific phenotype. Gain of function mutations are frequently dominant.

Gel electrophoresis—The technique of separating molecules in a gel matrix in an electric field. Gels are frequently made of agarose or polyacrylamide.

Gene—A region of the DNA that specifies a protein, a tRNA, or a rRNA.

Gene conversion—In a recombinational cross, both alleles in the cross are usually recovered. In a gene conversion event, one allele is recovered twice and one allele is lost. Gene conversion is caused when the mismatched base pairs in the heteroduplex DNA created during recombination are repaired to the mutant base pairs.

Gene linkage—Tendency of genes located close together on a chromosome to be inherited together in a recombinational cross. The higher the frequency of being inherited together, the closer the two genes are.

Gene product—The protein product that is translated from mRNA. If the RNA is not translated (tRNA, rRNA, or stable RNA), then the RNA is the gene product.

Gene regulation—The mechanism(s) used to control the expression of a gene into its corresponding gene product. Gene regulation can occur transcriptionally, translationally, post-transcriptionally, and/or post-translationally.

Generalized transducing phage—A phage that can incorporate chromosomal DNA into its phage particle in place of phage DNA. Many different segments of the chromosome can be incorporated into different phage particles.

Generalized transduction—Movement of any piece of chromosomal DNA from one bacterium to another using a bacteriophage to carry the DNA.

Genetic recombination—The exchange of DNA sequences between two homologous DNA molecules. Genetic recombination is detected by the formation of new combinations of the sequence.

Genome—An organism's entire set of genes, including extrachromosomal genes from plasmids or resident DNA sequences such as from integrated phage.

Genotype—The information in the DNA that specifies the characteristics of a cell.

Green fluorescent protein (GFP)—A protein isolated from jellyfish that absorbs light at a specific wavelength and emits it at a different wavelength. The emitted light can be seen in living cells using low-light photography. Fusing GFP to a protein of interest allows the movement of the protein of interest to be monitored in vivo.

GST-fusion protein—GST is glutathione S-transferase, a protein that binds glutathione. GST, when fused to a protein of interest allows the fusion protein to be purified in a single step.

Haploid—A cell that contains only a single copy of the chromosome. Haploidy is the state of a cell that contains only one chromosome.

Head-full packaging—Inserting (packaging) DNA into phage heads until the head is completely full of DNA and no more DNA will fit inside.

Helicase—An enzyme that unwinds (i.e. separates) the strands of a DNA molecule by disassociating the hydrogen bonds between the paired bases. This enzymatic activity would be found at a replication fork when the two parental strands need to be separated so that they may be used as templates by the replication machinery.

Heteroduplex DNA—A segment on a double-stranded DNA molecule in which the two DNA strands are of different origin and thus do not have perfectly complementary nucleotide sequences.

Hfr—A strain containing an F factor that has integrated into the chromosome of a bacterium.

High copy number plasmids—Plasmids that are present in the cell in greater than approximately 20 to 30 copies per cell.

Histidine tag—At least six histidine residues in a row that can be added to a protein of interest. The histidines bind to metal ions such as nickel or zinc ions. This allows fusion proteins that contain the his-tag to be purified in a single step.

Holoenzyme—A term used to describe an enzyme containing all of its subunits (both catalytic and regulatory). An example would be RNA polymerase, when this enzyme contains both its core catalytic subunits ($\beta\beta'\alpha_2$) and its regulatory subunit (sigma factor).

Homologous recombination—Recombination between two DNA molecules that have identical or nearly identical DNA sequences.

Homology—In the recombination process, homology refers to DNA sequences that are identical or nearly identical.

Host range—The bacterial species that can be infected by a specific phage.

Hotspots—Sites in the DNA where a given event occurs more frequently than is predicted for a random event. For example, some transposons insert preferentially at specific sites. Some transposons have hotspots, others do not. A hotspot for one transposon will not necessarily be a hotspot for a different transposon.

Housekeeping genes—Those genes whose products are required at some level for the cells to grow. Housekeeping genes are always expressed.

Identical sequences—When comparing two or more sequences, the same base or amino acid is found in all sequences at a given place. Identical sequences can be either DNA or protein sequences.

Illegitimate recombination—Recombination that does not rely on homologous sequences to occur.

Incompatibility—The inability to maintain in the same cell two plasmids that utilize the same origin of replication. An example of this is the inability of closely related F factors and R factors to be maintained in the same cell.

Inducer—A small molecule that can cause an operon to express its genes. An example of an inducer is allolactose. The addition of allolactose results in the inactivation of the Lac repressor and the induction of the operon thus allowing the *lac* genes to be expressed.

Informational suppressors—Suppressors that affect a general cellular process such as transcription, translation, or DNA replication to fix the primary mutation.

Inner membrane—The inner-most membrane of *E. coli*. It forms the major barrier between the cytoplasm and the outside of the cell.

Insertion—Mutations that are the result of addition of base pairs to the genome.

Insertion elements—The simplest type of transposon. They are small (~750 to ~2000 base pairs) and encode only transposase, the protein needed to move them.

Interactive suppressors—Suppressors that compensate for the defect in the gene containing the primary mutation by mutating another gene whose product interacts with the primary gene product.

Intercalation—Certain chemicals that are large flat molecules are capable of slipping between the base pairs of the DNA helix. This mode of interacting with DNA is known as intercalation.

Intermolecular—Events or processes that happen between two distinct molecules.

Intragenic suppressors—Suppressor mutations that occur in the same gene as the primary mutation.

Intramolecular—Events or processes that happen within the same molecule.

Inversion—A mutation that results from a flipping of several base pairs in the DNA sequence.

Inverted repeats—DNA sequences that are arranged such that the sequence is inverted and repeated within the same molecule. If the sequence is ATTGCC then it will be repeated CCGTTA in the same molecule.

in vitro—Reactions that take place in a test-tube and not inside a cell.

Jackpot—A mutation that occurs early in the growth of a culture. Jackpots affect the mutation frequency but not the mutation rate.

Kinase—An enzyme that adds phosphates to a specific molecule or substrate.

Lagging strand—During DNA replication, the lagging strand of DNA is synthesized in a discontinuous fashion using short DNA segments (Okazaki fragments) and relying on many initiating events.

Late genes—Genes that are expressed from phage DNA late in an infection. These usually include the genes that encode capsid and tail proteins.

Leader sequence—A short sequence of mRNA that is used to regulate specific operons. For example in the *trp* operon, the leader sequence contains two *trp* codons that are used to monitor tryptophan levels in the cell.

Leading strand—During DNA replication, the leading strand of DNA is synthesized in a continuous fashion, relying on only one initiation event.

Leaky mutations—Mutations that cause only a partial loss of function of the gene product.
Lesions—A term used to describe types of DNA damage.
Lethal selection—A selection where the cells that do not grow are actively killed by the selective agent. Most antibiotics can be used as lethal selection agents.
Library—A population of clones (usually tens of thousands) that each has one random pieces of chromosomal DNA cloned into a vector. Within the population, every piece of the chromosome can be found on at least one vector. The population is known a library. Libraries can be made for any large piece of DNA such as mouse, yeast, bacterial, or human chromosomes.
Linkage—A measure of how close two genes are on the chromosome.
Lipids—Molecules composed of long chains of carbons. Their major property is that they are greasy, hydrophobic molecules. One type of lipid, known as a fatty acid, is the major component of all membranes.
Lipopolysaccharides (LPS)—Specialized lipids that contain large carbohydrate side chains. They are embedded in the outer leaflet of the outer membrane of certain bacteria. LPS is very densely packed and helps protect bacteria from toxic compounds.
Localized mutagenesis—A technique used to limit mutagenesis to a specific and small region of the chromosome. For example, cells are mutagenized, P1 is grown on the cells, and a specific region of the chromosome is selected in a transductional cross. Only the region of DNA that was incorporated into the transducing particle was mutagenized.
Loss of function mutation—A mutation that results in the inability of a gene product to carry out the same functions as the wild-type gene product.
Low copy number plasmids—Plasmids that are maintained in cells in approximately 1 to 10 copies. Low copy number plasmids require an active partition system to ensure that they are transmitted from mother to daughter cells.
Lysate—After a cell has been broken open, the content of the cell is referred to as a cell lysate. The cell lysate is the starting material for purification of individual cell components.
Lysogeny—The growth phase of phage whereby the phage DNA is incorporated into the bacterial cell and remains in a stable, silent state.
Lysogen—A bacterium that contains a stable, silent phage genome. The phage genome is inherited by the daughter cells of the bacterium.
Lysogenization—The process by which a phage DNA is incorporated stably into the bacterial cell and replicated and inherited by daughter cells.
Lysozyme—An enzyme, usually isolated from egg whites, that is capable of degrading the peptidoglycan cell wall of Gram-negative bacteria.
Lytic—The growth phase of phage that is designed to produce many offspring as quickly as possible.
Macrolesions—Mutations that involve more than one base pair.
Macromolecules—Large chemical structures in cells that are composed of smaller molecules known as building blocks. For example, DNA is a macromolecule that is composed of a sugar–phosphate backbone and the bases A, T, C, and G. Proteins are macromolecules that are composed of amino acids.
Male cells—Cells that contain an F^+, F′, or Hfr.
Maltose binding protein (MBP)—An *E. coli* protein that binds to maltose. The gene encoding MBP can be fused to another gene so that a fusion protein is made. The fusion protein will bind to maltose, allowing purification of the fusion protein in a single step.
Merodiploid—A cell that contains two copies of some chromosomal DNA but not two complete chromosomes.
Messenger RNA (mRNA)—The RNA that is made by transcribing specific parts of the DNA molecule. mRNA encodes protein(s).
Methyl-directed mismatch repair—A mechanism that identifies and corrects mismatched base pairs in a hemimethylated DNA molecule (the newly synthesized daughter strand is not methylated whereas the parental template strand is). The enzymes involved in this repair, MutHLS, identify the mismatched base pair and excise the improperly placed base from the nonmethylated daughter strand. DNA polymerase I fills in the gap spanning the area that used to contain the mismatched base pair.
Methyltransferases—DNA repair enzymes with the capability of removing methyl groups from a base, such as guanine.
Microarray—A small solid support that carries DNA oligonucleotides representing every gene in a specific cell's genome. Microarrays can be used to monitor expression of all of the genes under a given condition.
Microhomology—Small regions of homology that can be used to make duplications and inversions.
Microlesions—Mutations that involve only one base pair.
Middle genes—The genes that are expressed from the phage genome between the early and late genes.
Mismatched base pair—A pair of nucleotides that do not hydrogen bond correctly and thus are not complementary.
Missense mutation—A base pair substitution that changes the nucleotide sequence encoding a polypeptide and also changes the corresponding amino acid sequence of the protein.
Mobilizable plasmids—Plasmids that do not contain all of the genes necessary for moving themselves from one cell to another but which can be moved from cell to cell using another plasmid's conjugation system.
Mobilization—The process of moving a plasmid from one cell to another using plasmid-encoded proteins.
Modification—The process of modifying specific sites in the DNA to prevent restriction enzymes from digesting the DNA. If a bacteria contains a restriction enzyme it must contain a modification enzyme to protect its DNA.
Modification enzyme—The enzyme that modifies the site that a restriction enzyme cuts. Modification enzymes prevent restriction enzymes from digesting the DNA.
Mu—A bacteriophage that uses transposition as an integral part of its lifestyle.
Multiple cloning site cassettes (MCS)—A sequence of DNA usually 50 to 75 base pairs in length that specify the recognition sites for restriction endonucleases.
Muropeptides—Strands of peptidoglycan that are added to a growing cell to build the cell wall.
Mutagens—Chemicals or physical treatments that alter DNA and increase the mutation rate.
Mutation—A change in the base sequence of a DNA molecule.
Mutation frequency—A simple but potentially inaccurate way to measure mutagenic potential. The frequency is how many mutations occurred in a given culture.

Mutation rate—An accurate way to measure mutagenic potential. The mutation rate is the probability that a specific gene will be mutated in one generation.
Negative regulation—A scheme for controlling the amount of a gene product that is made. This scheme relies on limiting production of a gene product by inhibiting its expression. Frequently, a protein (the negative regulator) binds to the promoter of the gene and prevents expression of the gene.
Neutral mutations—Mutations that have no phenotype.
Nitrosoguanidine—A powerful mutagen that tends to produce mainly transition mutations.
Non-composite transposons—Complex transposons that are not formed from modular units. They do not have IS elements at their ends.
Non-lethal selection—Selections where the cells that do not grow are not actively killed.
Non-replicative transposition—A mechanism for transposition where the transposon is cleaved from its original site and ligated into a new site. A minimal amount of DNA synthesis is carried out by mismatch repair enzymes to restore the duplex DNA.
Nonsense mutation—A base pair substitution, in which the nucleotide sequence of a codon encoding an amino acid has been changed to a codon encoding one of three termination codons (UGA, UAA, UAG).
Nucleic acids—Macromolecules composed of a sugar–phosphate backbone and four nucleotide bases. The two types of nucleic acids in living systems are DNA and RNA. DNA contains the bases adenine, cytosine, guanine, and thymine. RNA contains the bases adenine, cytosine, guanine, and uracil.
Nucleoside 5′ triphosphate—The precursor to the nucleotide, it consists of a base, sugar, and three phosphate groups (ATP, GTP, TTP, CTP, UTP).
Nucleotide excision repair—A DNA repair process that uses nuclease activity to remove damaged or mismatched nucleotides.
Null—A mutation that results in a nonfunctional gene product. When a transposon inserts into a gene, it disrupts the genes and no functional gene product can be made from that insertion, except in very rare instances.
Okazaki fragments—During DNA replication, one of the template strands is replicated in a discontinuous fashion (the lagging strand) using a mechanism that requires the synthesis of short stretches of nucleotides called Okazaki fragments.
Oligonucleotide—A short stretch of DNA, usually less than 100 base pairs. Oligonucleotides are short enough to be synthesized in vitro.
Operator—The original nomenclature to describe a regulatory region in front of a group of coordinately regulated structural genes that interacts with a specific repressor protein to control the transcription of these genes.
Operon—A group of linked genes whose expression is coordinately regulated.
Operon fusions—A fusion of a gene of interest to a reporter gene in which the transcriptional signals come from the gene of interest and the translational signals come from the reporter gene.
Optical density—The measurement used to monitor the growth of a culture of cells. Light at a specific wavelength is passed through the culture. The amount of light that is scattered by the culture is recorded. Light scatter is proportional to the number of particles or cells.
Origin for DNA replication (ori)—A DNA sequence that must be present in cis for replication of a DNA molecule to take place. The sequence is usually the binding site for a number of proteins that are required for DNA replication.
Origin of transfer (oriT)—The place in the F factor where transfer of DNA to the recipient strain is initiated.
Outer membrane—The outer membrane is composed of a lipid bilayer, proteins, and lipopolysaccharides. It is the outer-most barrier of Gram-negative cells and acts to protect them from toxic substances.
Overproduction suppressors—Suppressor mutations that increase the expression of either the mutant gene product or the rate-limiting component of the affected pathway.
Oxidative stress response regulon—A group of operons whose genes are coordinately regulated in response to oxidative stress.
***pac* site**—A site on the phage DNA where proteins bind and begin inserting the phage DNA into the phage head.
Pathogenic (pathogenicity)—The ability of an organism to cause damage or death to its host.
Peptidoglycan—The macromolecule that is used to build the rigid structure that surrounds the cell and maintains the cell's shape. Peptidoglycan is composed of several repeating sugar molecules.
Periplasm or periplasmic space—The aqueous compartment between the inner and outer membranes of *E. coli*. It contains proteins that transport nutrients into the cell, waste out of the cell, sense the environment, and convey this information to the cytoplasm. The periplasm resembles the outside of the cell in osmolarity and solute concentration.
Permissive temperature—The temperature at which a cold-sensitive or a temperature-sensitive mutation is able to function.
Phage or bacteriophage—Viruses that infect bacteria.
Phage particles—Phage infect bacteria and can produce offspring. The offspring produced from an infection are called phage particles. Most of the phage particles are infective and can go on to produce more offspring. Sometimes something will be wrong with a phage particle and it will not produce offspring. While all phage are infective, not all phage particles are infective.
Phage genome—The nucleic acid enclosed in the phage particle that specifies the production of more phage.
Phenotypes—The physical characteristics of a cell that are specified by that cell's DNA.
Phosphatase—An enzyme that removes phosphates from a specific molecule or substrate.
Phosphodiester bond—A covalent bond between a free 5′ phosphate of one nucleotide and a free 3′ hydroxyl group of an adjacent nucleotide, formed by the catalytic activity of ligase. This type of bond links nucleotides together to form the phosphate–sugar backbone of DNA.
Phosphorylation—The addition of phosphates to a molecule or substrate.
Photoreactivation—A DNA repair process that reverses the effects of ultraviolet light through the use of visible light.
Photolyase—The enzyme used in photoreactivation to split apart pyrimidine dimers formed by exposure to ultraviolet light. The enzyme consists of two subunits and requires visible light to be active.
Pilus—A filamentous (hair-like) structure attached to the surface of the bacterial cell (singular: pilus; plural: pili).
Plaques—Circular clear areas in a lawn of bacteria growing on an

agar plate. Plaques contain phage but have few or no bacteria growing in them.
Plasmid—A piece of DNA that exists outside the chromosome. Plasmids must contain an origin of DNA replication (*ori*) so that they can be replicated and inherited by daughter cells.
Point mutation—A microlesion involving the addition, deletion, or substitution of one base pair.
Polar mutation—A mutation that affects the expression of downstream genes that are in the same operon.
Polycistronic mRNA—An mRNA molecule that yields two or more proteins.
Polymerase chain reaction—A technique for amplifying any specific segment of DNA in vitro. A piece of DNA is used as the template. Specific DNA primers, one for each strand, are used to initiate DNA synthesis and DNA polymerase from a bacterium that grows at 70°C. The DNA is synthesized by cycling the reaction through three temperatures. 94°C is used to denature double-stranded DNA, a temperature between 40°C and 55°C is used to anneal the primers to the template, and 70°C is used for DNA synthesis. This cycle can be repeated 20 to 30 times so that a significant quantity of DNA can be synthesized.
Polypeptide or protein—A string of amino acids each linked together by a peptide bond.
Post-replicative repair (PRR)—A repair mechanism that does not repair damaged DNA, but rather tolerates the lesions so that the damaged DNA can still be replicated. PRR uses homologous recombination to fill in gaps that form when DNA Pol III hits a lesion in the template and does not fill in a nucleotide, but rather skips ahead. PRR is RecA dependent.
Post-transcriptional regulation—Regulation of the gene product at any step after the RNA has been transcribed.
Post-translational regulation—Regulation of the protein's activity after it has been translated.
Primary mutation—The mutation that leads to the defect that is reversed by a suppressor or pseudorevertant.
Primer—A small stretch of linked nucleotides. In the case of DNA synthesis, this small stretch consists of ribonucleotides that DNA polymerase can attach deoxyribonucleotides to.
Primosome—The initiator complex consisting of DnaA, DnaB, and DnaC, whose activities at *oriC* (specifically at the DnaA boxes of *oriC*) result in the separation of the double-stranded DNA molecule so that DnaG can prime DNA replication.
Probe—A molecule that is tagged in some manner so that it can be easily detected and is used to identify a molecule of interest. For example, a gene of interest can be tagged and used in a colony blot to detect the plasmids that carry the gene of interest. Probes can be tagged with radioactivity, fluorescent dyes, or enzymes, to name a few.
Promoters—A sequence of DNA found adjacent to the coding region of a gene that RNA polymerase containing a sigma factor recognizes and binds to, subsequently initiating mRNA transcription. The promoter sequence contains two highly conserved motifs at −10 and −35.
Promoter recognition—The first step in the synthesis of RNA. RNA polymerase recognizes the promoter and binds to it.
Proofreading—DNA polymerases contain enzymatic activities that allow them to recognize a mispaired nucleotide and remove it before moving on and adding more bases. See editing.
Prophage—The phage DNA that is maintained in a silent state inside a bacterium.
Protease—An enzyme that degrades proteins by catalyzing the breakage of a peptide bond between the amino (NH_2) group of one amino acid and the carboxyl (COOH) group of another amino acid.
Protein or polypeptide—Macromolecules composed of a chain of amino acids that are used for many different purposes in the cell. The amino acid sequence of a protein is specified by the cell's DNA. Proteins are used for building and tearing down molecules, transporting nutrients into the cell, all major processes such as transcription and translation, and generating energy.
Protein fusion—Fusion of a gene of interest to a reporter gene such that all signals needed for transcription and translation come from the gene of interest. The protein that is produced contains a stretch of amino acids from the gene of interest linked to a stretch of amino acids from the reporter gene.
Proteome—All the proteins present in a cell at any given time or under some specific condition.
Proteomics—The study of which of the proteins in a cell are, at what time, and under what condition.
Pseudo-*pac* sites—Sites in the chromosomal DNA molecule that are used to package chromosomal DNA into the heads of generalized transducing phage.
Pseudorevertant—A mutation that reverses the phenotype of another mutation.
Pyrimidine dimer—A type of lesion that forms when the DNA molecule is exposed to ultraviolet light. Examples of dimers include 6–4 pyrimidine-pyrimidone and cyclobutane dipyrimidine. Pyrimidine dimers can be repaired by photoreactivation, UvrABC endonuclease excision repair, or tolerated by post-replicative repair.
Quorum-sensing system—A system that detects a change in the number of cells in a given volume and relays this information to the inside of the cell. Once the signal has been internalized, it affects the regulation of genes whose products allow the cell to adapt to the environment.
R factors—Plasmids that can carry many antibiotic resistance genes as well as the genes needed for conjugation.
Raw sequence—The base sequence of a piece of DNA that simply lists the order of the As, Cs, Gs, and Ts.
Rearrangements—Mutations that result from the scrambling of DNA bases without loss of any base pairs.
Recipient cells—Cells in a genetic cross that are receiving the DNA from the donor cell. For example, in a transduction, the transducing phage is grown on the donor and used to infect the recipient cells.
Recessive—A mutation that does not exhibit any phenotypes when it is placed in a cell with a wild-type copy of the same gene.
Reciprocal—In homologous recombination, DNA is exchanged between two DNA sequences. In a reciprocal cross, the information from one DNA sequence is moved to a second DNA sequence concomitantly with the DNA sequences from the second molecule being moved to the first DNA sequence. No DNA sequences are lost in a reciprocal cross, they are simply rearranged.
Recombinase—The enzyme responsible for the exchange of DNA in a site-specific recombination event.
Regulatory protein—A protein whose function is to regulate or control expression of genes. Expression can be controlled negatively through the repression of gene expression or positively through the activation of gene expression.

Regulon—A group of operons or a subset of genes in a group of operons whose expression is coordinately regulated by a global regulatory mechanism.
Replication fork—The region in a replicating molecule where the replication machinery is adding nucleotides to a growing DNA chain.
Replicative transposition—A mechanism for transposition where one strand of the transposon is cleaved and ligated to one strand of the new insertion site. Extensive DNA replication across the entire transposon by DNA replication enzymes restores the duplex DNA.
Replicon—An independently replicating DNA molecule that is stably maintained in a population of cells. Plasmids are referred to as replicons.
Reporter gene—A well-characterized gene that is used to monitor the expression level of another less well-characterized gene. The reporter gene is placed 3′ to the DNA of the promoter to be studied and expression of that promoter is examined by following the amount and timing of appearance of the protein from the reporter gene.
Repression—Control of gene expression by preventing a gene from being expressed.
Resolvase—The enzyme encoded by some transposons to resolve a cointegrate structure into two DNA molecules that each contain one copy of the transposon. Resolvase carries out a site-specific recombination reaction to resolve the cointegrate.
Response regulator protein—One of two proteins that make up a two-component regulatory system. The response regulator protein is activated, usually by phosphorylation, in response to signals received from a membrane-bound sensor kinase protein (the second component of this system). A response regulator protein is often a transcriptional activator, whose ability to activate gene expression is controlled by the signals it receives from the corresponding sensor kinase protein.
Restriction—The process by which DNA is cut at a specific sequence by an enzyme known as a restriction endonuclease.
Restriction endonucleases or **restriction enzymes**—Proteins produced by many different species of bacteria that recognize foreign DNA sequences and cleave the foreign DNA on both strands at a DNA sequence specific for each restriction enzyme. Bacterial restriction systems are a primitive immune system in that they protect bacteria from foreign DNA coming into the cytoplasm. There are three types of restriction enzymes (Type I, II, and III). The three categories are based on the number of proteins that make up the enzyme, the cofactor requirements, if the proteins form a complex- and the nature of the sequence that is recognized by the enzyme. Commercially useful restriction endonucleases (Type II) recognize and cleave the short, specific sequence of DNA.
Restriction site—The site in the DNA that is cleaved by a restriction enzyme.
Revertant—A mutant cell that regains its wild-type phenotype by a second mutation. Revertants can be either a reversal of the mutant base pair back to the wild-type base pair (true revertant) or a second mutation that compensated for the first mutation (pseudorevertant or suppressor).
Reversion—The process of reversing the phenotype of a mutant bacterium.
Ribosomal RNA—RNA molecules that are used to build a ribosome. Ribosomes are the large complex of RNA and proteins that translate mRNA into proteins.
Ribosome binding site or **Shine–Delgarno sequence**—The bases in the mRNA where the ribosome binds to begin translating the mRNA into protein.
RNA—Ribonucleic acid is a macromolecule that is composed of a sugar–phosphate backbone and the four bases: uracil, adenine, thymine, and guanine.
RNA polymerase—The enzymes responsible for transcribing DNA into RNA. The holoenzyme consists of the subunits: $\beta\beta'\alpha_2\sigma$, with the σ factor acting as the specificity factor, directing RNA polymerase to initiate transcription from specific promoter sequences.
Rolling circle replication—The process by which phage make one long piece of DNA that contains multiple phage genomes. Phage use rolling circle replication to make DNA molecules that they can package into phage heads.
Same-sense or silent mutation—A base pair substitution that changes the nucleotide sequence encoding a polypeptide but fails to result in a change in the amino acid sequence.
Selectable marker—A gene in a vector that makes a gene product that confers a positive or selectable phenotype on cells that contain the vector. The selectable marker is used to select out of a mixed population cells that contain the vector.
Selection—The environmental conditions used to allow only some cells in a population to grow.
Selective agar—Growth media used to identify specific cells and exclude the growth of other cells. For example, selective agar is used to select transductants from a mixture of bacteria and transducing phage particles.
Semiconservative replication—During DNA replication, both strands of the DNA molecule serve as templates for the synthesis of new complementary strands. The newly synthesized complementary strand (also called daughter strand) will hydrogen bond with its corresponding template strand (also called parental strand). Thus, each DNA molecule will consist of a parental and a daughter strand.
Sensor-kinase protein—One of two proteins that make up a two-component regulatory system. The sensor protein is a membrane-bound kinase that autophosphorylates in response to a specific environmental signal. It transfers this phosphoryl group to a response regulator protein so that the regulator may activate or repress express of the appropriate genes.
Septum—The structure that is built between two dividing cells to wall them off from each other. The septum is mainly composed of membranes and peptidoglycan.
Shine–Delgarno sequence—Another name for a bacterial ribosome binding site. The site in an mRNA where the ribosome binds to initiate translation.
Short patch repair—Repair of a uv light induced dimer by excision of the dimer from the DNA by UvrABC and synthesis across the gap by DNA polymerase I.
Short sequence repeats (SSR)—Repeated DNA sequences that can be used for illegitimate recombination. The SSR can be as short as 3–8 base pairs.
Shuttle vectors—Vectors that can replicate and be maintained in two different species. One of the species is usually *E. coli*, where it is easy to manipulate DNA, and the other can be from bacteria, yeast, or even mammals.
Siblings—Cells in a culture that are derived from each other and all contain the same mutation(s).
Sigma factor—The specificity determinant of RNA polymerase.

This subunit directs core RNA polymerase to promoters during initiation of transcription.

Signal transduction—The mechanism by which two-component regulatory systems (sensor–regulator pairs) function. It involves the transferring of signals, usually in the form of phosphoryl groups so that environmental signals outside the cell can influence gene expression inside the cell.

Silent mutation—A mutation that does not confer a phenotype on the cell that carries it.

Similar sequences—When comparing two or more sequences, similar amino acids are found at a given position. Similarity usually refers to amino acid sequences because some amino acids have similar properties and are interchangeable. For example, valine and alanine both have hydrophobic characteristics.

Site-specific recombination—Recombination between two specific sequences in DNA that requires a unique set of proteins to carry out the cleavage and joining reactions. For example, recombination between *attP* in λ and *attB* in the chromosome is a site-specific recombination reaction.

SOS mutagenic repair—A DNA repair regulon that responds to extensive DNA damage. SOS genes are repressed by the transcriptional repressor, LexA. Upon extensive DNA damage, LexA is cleaved, fully activating the expression of the SOS genes whose gene products are needed to repair the damaged DNA.

Southern blotting—A techniques first described by E.M. Southern to detect homologies between two DNA molecules when the sequence of the two DNA molecules is not known. One of the DNAs is electrophoresed on a gel and transferred to a filter. The DNA strands are separated after the DNA has been transferred on the filter. The second sequence is tagged and incubated with the filter. The tagged DNA will bind to the filter only where a homologous DNA sequence has been attached to the filter.

Specialized transducing phage—A phage that can carry a specific segment of chromosomal DNA as part of its genome. The segment of chromosomal DNA is the same in every phage particle.

Specialized transduction—Movement of a specific fragment of chromosomal DNA from one bacterium to another. The fragment of DNA is usually incorporated and replicated as part of bacteriophage DNA. Because of this, the specific fragment of chromosomal DNA can be amplified by growing a large number of the phage.

Spore—The storage form of some bacteria that are made in the developmental pathway called sporulation. Spores contain a bacterial genome surrounded by a tough protective coating. In some species, spores can survive for decades.

Sporulation—A process that results in the formation of a dormant spore that is resistant to chemicals, heat, and physical agents, and which contains an entire copy of the bacterial chromosome. Only some bacteria are capable of sporulation.

Stem loop structures—Found in nucleic acid molecules when single-stranded regions find intramolecular complementary bases to pair with to form regions of double strandedness. Also called hairpin structures or secondary structures. The base pairing that forms is a result of noncovalent, hydrogen bonding between complementary nucleotides within the same single-stranded molecule. Stem loop structures can be found at the ends of genes to signal where transcription stops.

Sticky ends or staggered ends—The linear ends of a dsDNA molecule that contain a few single-stranded nucleotides. Restriction enzymes that cleave the DNA asymmetrically at their recognition site produce sticky ends on the DNA. The sticky end can have either a 5′ overhang or a 3′ overhang.

Suicide vectors—Phage vectors that can infect a specific bacterium but not replicate or produce offspring.

Supercoiling –DNA that is highly twisted and compacted.

Superinfection—The process of a phage infecting a cell that is already lysogenic for the phage. In most cases, the incoming phage cannot replicate.

Suppressor—A mutation that reverses the phenotype of another mutation.

Surface exclusion—The process of a cell containing an F preventing mating with another cell that contains an F. Cells that contain an F produce at least two proteins to indicate the presence of the F. TraS inhibits DNA transfer and TraT inhibits mating pair formation.

Target DNA—The DNA that the transposon is moving into.

Target immunity—The inability of a transposon to transpose near a copy of itself. Only some transposons have this property. Immunity can extend over at least 175 kilobases for some transposons.

Tautomer—An alternative form (a structural isomer) of a base in which a spontaneous transient rearrangement of bonds results in the electrons being distributed differently among the atoms.

T-DNA—The piece of the Ti plasmid that is moved into the plant cells. The T-DNA enters the plant nucleus where it is integrated into the plant nuclear genome.

Temperate phage—A phage that is capable of both lytic and lysogenic growth.

Template—A chain of nucleotides that is used in a polymerization reaction to produce a complementary DNA or RNA strand.

Temperature sensitive—A secondary phenotype of a mutation that causes the mutant gene product to not function at increased temperatures.

Terminal redundancy—The ends of the DNA of certain phage genomes that have overlapping sequence homology. If the sequence of the genes is ABCDE, terminally redundant molecules would have the sequence ABCDEAB. The overlapping sequence homology, or terminal redundancy, allows the ends of the phage DNA to circularize by homologous recombination.

Terminator—The sequence at the end of a gene that instructs RNA polymerase to stop transcribing. Terminators have step loop structures.

Theta replication—The bidirectional replication of a circular DNA molecule where initiation of replication starts at a single *ori*. Structures that look like the Greek letter theta are produced.

Three-factor crosses—Examining the behavior of three genes in a single genetic cross.

Ti plasmids—Plasmids that are found in *Agrobacterium tumifaciens* and are responsible for the formation of crown gall tumors in plants. There are many different versions of the Ti plasmid, each having subtle differences. All are capable of inducing tumors in plants.

Topoisomerase—An enzyme that eliminates or introduces either underwinding or overwinding of a double-stranded DNA molecule. It catalyzes single (type I) and double (type II) strand breaks changing the relative positions of the DNA strands, and then ligating the breaks. It differs from a nuclease in the way it creates these breaks or reseals them. Topoisomerses use the process of transesterification, rather than hydrolysis, to break or reseal a phosphodiester bond between two nucleotides.

Transcription—The synthesis of RNA using a DNA template, RNA polymerase, and the precursor, ribonucleoside 5′-triphosphates.

Transcription fusion or operon fusion—A fusion of the DNA from a promoter that is under study to a reporter gene. The signals needed for transcription come from the promoter under study and the signals needed for translation are provided by the reporter gene.
Transdimer synthesis—A tolerating mechanism used to replicate damaged DNA. The specificity of the polymerizing DNA polymerase III is relaxed so that any nucleotide can be placed across from a lesion in the template strand.
Transducing particles—Bacteriophage particles that contain bacterial chromosomal DNA instead of (generalized transducing phage) or in addition to (specialized transducing phage) phage DNA.
Transductants—Bacteria that have received a piece of chromosomal DNA from a transducing phage.
Transduction—The process of using a bacteriophage to move pieces of chromosomal DNA from one bacterial cell to another.
Transfer RNA (tRNA)—A macromolecule that is used in the process of translating a messenger RNA into a protein. tRNA serves as a bridge between the mRNA and the amino acid specified by the tRNA.
Transformants—Bacterial cells that have taken up free DNA. The transformed DNA, if it contains an origin of replication (*ori*), can exist independently from the chromosome. If it does not have an *ori*, yet contains sequences homologous with the chromosome, it can undergo homologous recombination.
Transformasome—A membrane-enclosed compartment that contains a 30–50 kilobase dsDNA fragment. Transformasomes are used by some bacteria to move free DNA from the environment into the cytoplasm.
Transformation—A process by which bacterial cells acquire free (naked) DNA from their environment. Some bacteria are naturally competent and others can be physically manipulated to take up free DNA.
Transition—A base substitution mutation where a purine replaces a purine (an A for a G or vice versa) or a pyrimidine replaces a pyrimidine (a C for a T or vice versa).
Translation—The synthesis of protein from an mRNA template.
Translational fusion or protein fusion—A fusion of the DNA from a promoter that is under study to a reporter gene. The signals needed for both transcription and translation are provided by the promoter under study.
Translational regulation—Regulation of transcription of an mRNA into a protein.
Translocation—A mutation that results from base pairs being moved to a new site on the chromosome.
Transposable element—Another name for a transposon.
Transposase—A protein encoded by a transposon that is required for movement of the transposon. Transposase actually breaks the DNA backbone at the donor site between the transposon and the adjacent DNA and joins it to the DNA at the new insertion site.
Transposon—A segment of DNA that is capable of moving itself from one piece of DNA to another piece of DNA.
Transversion—A base substitution mutation where a purine replaces a pyrimidine (a G for a T) or a pyrimidine replaces a purine (C for a G).
True revertant—A reversal of the phenotypes of a mutation by changing the mutated base back to the wild-type base.
Two-component signal transduction—A process that consists of at least two proteins, the sensor kinase and the response regulator protein. The sensor kinase senses specific changes in the environment and then communicates the change biochemically (usually in the form of a phosphate) to the response regulator. The response regulator alters the expression of the genes whose products respond to the environmental change.
Two-factor crosses—Examining the behavior of two genes in a single genetic cross.
Uptake signal sequence (USS)—A short and specific sequence of nucleotides that must be present in free DNA in order for it to be taken up by certain bacteria during natural transformation.
Unidirectional DNA synthesis—DNA synthesis from the origin that proceeds in only one direction at a time.
Vector—A molecule that has been manipulated in vitro for cloning genes. Vectors must contain an origin of replication and frequently contain multiple cloning sites and a selectable marker. Many different features can be included, depending on the final species the vector will be used in and the uses it will be put to.
Very short patch repair—A repair mechanism that recognizes and repairs T–G mismatched bases in a specific sequence. VSP is a three-step process by which Vsr endonuclease recognizes the mismatch and cleaves the phosphodiester backbone next to the mismatch and removes the mismatch. DNA Pol I adds back the correct base.
Viable cell count—A way to measure the number of cells in a culture. The culture is diluted and a known amount of it is plated on solid growth media known as an agar plate. The agar plate is incubated until colonies are formed and the number of colonies are counted. Each colony represents one cell in the original culture.
Virulence factors -Factors that allow pathogenic (disease-causing) bacteria to invade or colonize other, usually eukaryotic, cells.
Virus—A protein-coated nucleic acid molecule that can infect sensitive cells, transporting the nucleic acid into the cytoplasm. Once there, the nucleic acid is used to make more of the virus. The progeny viruses leave the cell and go on to infect other cells. Viruses that infect bacteria are called bacteriophage.
Wild type—Refers to a gene or a protein sequence that contains no known mutations and is present in most members of a species.
XGal (5-bromo-4-chloro-3-indoyl-β-D-galactoside)—A chemically synthesized derivative of lactose that can be used to indicate when cells are producing active β-galactosidase. When intact XGal is colorless and when cleaved by β-galactosidase, a blue-colored indigo dye is produced.

Further reading

Ausubel, F.M. 1987. *Current Protocols in Molecular Biology*. New York: Greene Publishing Associates and Wiley Interscience.

Bennett, R.J. and West, S.C. 1995. RuvC protein resolves Holliday junctions via cleavage of the continuous (non-crossover) strains. *Proceedings of the National Academy of Science USA*, **92**: 5635–9.

Bender, J. and Kleckner, N. 1986. Genetic evidence that Tn*10* transposes by a nonreplicative mechanism. *Cell*, **45**: 801–15.

Bertrand, K., Squires, C., and Yanofsky, C. 1976. Transcription termination in vitro in the leader region of the tryptophan operon on *E. coli*. *Journal of Molecular Biology*, **103**: 319–37.

Brock, T.D. 1990. *The Emergence of Bacterial Genetics*. Cold Spring Harbor, NY: Cold Spring Harbor Laboratory Press.

Cairns, J. 1966. The bacterial chromosome. *Scientific American*, January.

Casadaban, M.J. and Cohen, S.N. 1979. Lactose genes fused to exogenous promoters in one step using a Mu-*lac* bacteriophage: in vivo probe for transcritpional control sequences. *Proceedings of the National Academy of Science USA*, **76**: 4530–3.

Chandhury, A.M. and Smith, G.R. 1984. A new class of *Escherichia coli recBC* mutants: implications for the role of RecBC enzyme in homologous recombination. *Proceedings of the National Academy of Science USA*, **81**: 7850–4.

Chauthaiwale, V., Therwath, A., and Deshpande, V. 1992. Bacteriophage lambda as a cloning vector. *Microbiological Reviews*, **56**: 577.

Clark, A.J. 1973. Recombinant deficient mutants of *E. coli* and other bacteria. *Annual Review of Genetics*, **7**: 67–86.

Clewell, D.B. ed. 1993. *Bacterial Conjugation*. New York: Plenum Press.

Clowes, R.C. 1973. The molecule of infectious drug resistance. *Scientific American*. April: 18.

Cohen, S.N., Chang, A.C.Y., Boyer, H.W., and Helling, R.B. 1973. Construction of biologically functional bacterial plasmids in vitro. *Proceedings of the National Academy of Science USA*, **70**: 3240–4.

Cox, M.M. 1991. The RecA protein as a recombinational repair system. *Molecular Microbiology*, **5**: 1295.

Craig, E.A. and Gross, C.A. 1991. Is Hsp70 the cellular thermometer? *Trends in Biochemical Science*, **16**: 135–40.

Craig, E.A., Gambill, B.D., and Nelson, R.J. 1993. Heat shock proteins: molecular chaperones of protein biogenesis. *Microbiological Reviews*, **57**: 402–14.

Craig, N.L. and Nash, H.A. 1983. The mechanism of phage λ site-specific recombination: site-specific breakage of DNA by Int topoisomerase. *Cell*, **35**: 795–803.

Crick, R.H.C., Barnett, L., Brenner, S., and Watts-Tobin, R.J. 1961. General nature of the genetic code for proteins. *Nature*, **192**: 1227–32.

Demerec, M., Adelberg, E.A., Clark, A.J., and Hartman, P.E. 1966. A proposal for a uniform nomenclature in bacterial genetics. *Genetics*, **54**: 61–76.

Drlica, K. and Riley, M. 1990. A historical introduction to the bacterial chromosome. In *The Bacterial Chromosome*, eds. K. Drlica and M. Riley. Washington, DC: ASM Press.

Dubnau, D. 1997. Binding and transport of transforming DNA by *Bacillus subtilis*: the role of type-IV pilin-like proteins—a review. *Gene*, **192**: 191–8.

Dubnau, D. and Provvedi, R. 2000. Internalizing DNA. *Research in Microbiology*, **151**: 475–80.

Ebel, W., Vaughn, G.J., Peters, H.K., and Trempy, J.E. 1997. Inactivation of *mdoH* leads to increased expression of colanic acid capsular polysaccharide in *Escherichia coli*. *Journal of Bacteriology, 179*: 6858–61.

Fields, S. and Song, O. 1989. A novel genetic system to detect protein–protein interactions. *Nature*, **340**: 245–6.

Friedberg, E.C., Walker, G.C., and Siede, W. 1995. *DNA Repair and Mutagenesis*. Washington, DC: ASM Press.

Gilbert, W., and Maxam, A. 1973. The nucleotide sequence of the *lac* operator. *Proceedings of the National Academy of Science USA*, **70**: 3581–4.

Gottesman, S. 1995. Regulation of capsule synthesis: modification of the two-component paradigm by an accessory unstable regulator. In *Two-Component Signal Transduction*, eds. J. Hoch and T.J. Silhavy, pp. 253–62. Washington, DC: ASM Press.

Gralla, J.D. 1996. Activation and repression of *E. coli* promoters. *Current Opinion in Genetics and Development*, **6**: 526–30.

Griffith, F. 1928. Significance of pneumococcal types. *Journal of Hygiene*, **27**: 113–59.

Grossman A.D., Erickson, J.W., and Gross, C.A. 1984. The *htpR* gene product of *E. coli* is a sigma factor for heat-shock promoters, *Cell*, **38**(2): 383–90.

Hartman, P. and Roth, J. 1973. Mechanisms of suppression. *Advances in Genetics*, **17**: 1–104.

Hinneburshch, J. and Tilly, K. 1993. Linear plasmids and chromosomes in bacteria. *Molecular Microbiology*, **10**: 917.

Holliday, R. 1964. A mechanism for gene conversion in fungi. *Genetics Research*, **5**: 292–304.

Howard-Flanders, P. and Theriot, L. 1966. Mutants of *Escherichia coli* defective in DNA repair and in genetic recombination. *Genetics*, **53**: 1137–50.

Jacob, F. and Monod, J. 1961. Genetic regulatory mechanisms in the synthesis of proteins. *Journal of Molecular Biology*, **3**: 318–56.

Kitts, P.A. and Nash, H.A. 1987. Homology dependent interactions in phage λ site-specific recombination. *Nature* (London), **329**: 346–8.

Kleckner, N. 1990. Regulating Tn*10* and IS*10* transposition. *Genetics*, **124**: 449–54.

Kohara, Y., Akiyama, K., and Isono, K. 1987. The physical map of the whole *E. coli* chromosome: application of a new strategy for rapid analysis and sorting of a large genomic library. *Cell*, **50**: 495–508.

Kornberg, A. and Baker, T.A. 1992. *DNA Replication*, 2nd edn. San Francisco: W.H. Freeman.

Kroll, J.S., Wilks, K.E., Farrant, J.L., and Langford, P.R. 1998. Natural genetic exchange between *Haemophilus* and *Neisseria*: intergeneric transfer of chromosomal genes between major human pathogens. *Proceedings of the National Academy of Science USA*, **95**: 12381–5.

Lederberg, J. 1986. Forty years of genetic recombination in bacteria: a fortieth anniversary reminiscence. *Nature*, **324**: 627–8.

Lederberg, J. and Lederberg, E.M. 1952. Replica plating and the indirect selection of bacterial mutants. *Journal of Bacteriology*, **63**: 399.

Liberek, K., Galitski, T.P., Zyliez, M., and Georgopoulos, C. 1992. The DnaK chaperon modulates the heat shock response of *E. coli* by binding to the s32 transcription factor. *Proceedings of the National Academy of Science USA*, **89**: 3516–20.

Lin, J. and Sancar, A. 1992. (A)BC exonuclease: The *Escherichia coli* nucleotide excision repair enzyme. *Molecular Microbiology*, **6**: 2219.

Little, J.W. 1993. LexA cleavage and other self-processing reactions. *Journal of Bacteriology*, **175**: 4943–50.

Little, J.W. and Mount, D.W. 1982. The SOS regulatory system of *Escherichia coli*. *Cell*, **29**: 11–22.

Luria, S. and Delbruck, M. 1943. Mutations of bacteria from virus sensitivity to virus resistance. *Genetics*, **28**: 491–511.

Maruyama, I.N., Maruyama, H.I., and Brenner, S. 1994. λfoo: A λ phage vector for the expression of foreign proteins. *Proceedings of the National Academy of Science USA*, **91**: 8273–7.

Mathews, C.K. and van Holde, K.E. 1998. *Biochemistry*, 2nd edn. California: Benjamin/Cummings.

McCann, J. and Ames, B.N. 1976. Detection of carcinogens as mutagens in the *Salmonella*/microsome test assay of 300 chemicals: discussion. *Proceedings of the National Academy of Science USA*, **73**: 950–4.

McCarty, M. 1985. *The Transforming Principle: Discovering that Genes are Made of DNA*. New York: Norton.

McCarty, M. and Avery, O.T. 1946. Studies on the chemical nature of the substance inducing transformation of pneumococcal types. II. Effect of desoxyribonuclease on the biological activity of the transforming substance. *Journal of Experimental Medicine*, **83**: 89–96.

Mermod, N., Ramos, J.L., Lehrbach, P.R., and Timmis, K.N. 1986. Vector for regulated expression of cloned genes in a wide range of Gram-negative bacteria. *Journal of Bacteriology*, **167**: 447–54.

Meselson, M.S. and Radding, C.M. 1975. A general model for genetic recombination. *Proceedings of the National Academy of Science USA*, **2**: 358–61.

Meselson, M. and Stahl, F.W. 1958. The replication of DNA in *Escherichia coli*. *Proceedings of the National Academy of Science USA*, **44**: 671.

Meselson, M. and Weigle, J. 1961. Chromosome breakage accompanying genetic recombination in bacteriophage. *Proceedings of the National Academy of Science USA*, **47**: 857–68.

Mikawa, Y.G., Maruyama, I.N., and Brenner, S. 1996. Surface display of proteins on bacteriophage lambda heads. *Journal of Molecular Biology*, **262**: 21–30.

Miller, R.V. 1998. Bacterial gene swapping in nature. *Scientific American*, **278**: 67–71.

Mullis, K.B. 1990. The unusual origin of the polymerase chain reaction. *Scientific American*, April, 56–65.

Murray, N. 1983. Lambda vectors. In *Lambda II*, ed. R.W. Hendrix, et al., p. 677. Cold Spring Harbor, NY: Cold Spring Harbor Laboratory Press.

Nash H.A., Bauer, C.E. and Gardner, J.F. 1987. Role of homology in site-specific recombination of bacteriophage λ: evidence against joining of cohesive ends. *Proceedings of the National Academy of Science USA*, **84**: 4049–53.

Neidhardt, F.C. 1987. Multigene systems and regulons. In *Escherichia coli* and *Salmonella typhimurium: Cellular and Molecular Biology*, 2nd edn., eds. F.C. Neidhardt, R. Curtiss III, J.L. Ingraham, E.C.C. Lin, K.B. Low, B. Hagasanik, W.S. Rexnikoff, M. Riley, M. Schaechter, and H.E. Umbarger. Washington, DC: ASM Press.

Ogawa, T. and Okazaki, T. 1980. Discontinuous DNA synthesis. *Annual Reviews of Biochemistry*, **49**: 421.

Oxender, D.L., Zurawski, G., and Yanofsky, C. 1979. Attenuation in the *Escherichia coli* tryptophan operon: role of RNA secondary structure involving the tryptophan codon region. *Proceedings of the National Academy of Science USA*, **76**: 5524–8.

Pardee, A.B., Jacob, F., and Monod, J. 1959. The genetic control and cytoplasmic expression of inducibility in the synthesis of B-galactosidase by *E. coli*. *Journal of Molecular Biology*, **1**: 165.

Peterson, K., Ossanna, N., Thliveris, A., Ennis, D., and Mount, D. 1988. De-repression of specific genes promotes DNA repair and mutagenesis in *Escherichia coli*. *Journal of Bacteriology*, **170**: 1–4.

Peterson K.R., Ossanna, N., and Mount, D.W. 1988. The *Escherichia coli* K12 *lexA2* gene encodes a hypocleavable repressor. *Journal of Bacteriology*, **170**(4): 1975–7.

Ptashne, M. 1992. *A Genetic Switch*, 2nd edn. Cambridge, MA: Blackwell Scientific.

Radding, C.M. 1978. Genetic recombination: strand transfer and mismatch repair. *Annual Reviews of Biochemistry*, **47**: 847–80.

Reznikoff, W. 1992. Catabolite gene activator protein activation of *lac* transcription. *Journal of Bacteriology*, **174**: 655–8.

Ross, W., Shulman, M., and Landy, A. 1982. Biochemical analysis of *att*-defective mutants of the phage lambda site-specific recombination system. *Molecular Biology*, **156**: 505–22.

Rudd, K.E. 1993. Maps, genes, sequences, and computers: an *Escherichia coli* case study. *ASM News*, **59**: 335–41.

Rupp, W.D. 1996. DNA repair mechanisms. In *Escherichia coli* and *Salmonella typhimurium: Cellular and Molecular Biology*, 2nd edn., eds. F.C. Neidhardt, R. Curtiss III, J.L. Ingraham, E.C.C. Lin, K.B. Low, B. Hagasanik, W.S. Rexnikoff, M. Riley, M. Schaechter, and H.E. Umbarger, pp. 2277–94. Washington, DC: ASM Press.

Russel, M. 1991. Filamentous phage assembly. *Molecular Microbiology*, **5**: 1607.

Schwartz, D. and Beckwith, J.R. 1970. Mutants missing a factor necessary for the expression of catabolite sensitive operons in *E. coli*. In *The Lactose Operon*, eds. J.R. Beckwith and D. Zipser, pp. 417–22. Cold Spring Harbor, NY: Cold Spring Harbor Laboratory Press.

Smith, G.R. 1991. Conjugational recombination in *E. coli*: myths and mechanisms. *Cell*, **64**: 19–27.

Smith, G.P. 1985. Filamentous fusion phage: novel expression vectors that display cloned antigens on the virion surface. *Science*, **228**: 1315–17.

Smith, H.O., Gwinn, M.L., and Salzberg, S.L. 1999. DNA uptake signal sequences in naturally transformable bacteria. *Research in Microbiology*, **150**: 603–16.

Snyder, L. and Champness, W. 1997. *Molecular Genetics of Bacteria*. Washington, DC: ASM Press.

Solomon, J.M. and Grossman, A.D. 1996. Who's competent and when: regulation of natural genetic competence in bacteria. *Trends in Genetics*, **12**: 150–5.

Stahl, F.W. 1987. Genetic recombination. *Scientific American*, **256**: 91–101.

Stahl, F.W. and Stahl, M.M. 1985. Recombination pathway specificity of Chi. *Genetics*, **86**: 715–25.

Stahl, F.W., Kobayashi, I. and Stahl, M.M. 1985. In phage λ, *cos* is a recombinator in the Red pathway. *Journal of Molecular Biology*, **181**: 199–209.

Sternberg, N. 1987. The production of generalized transducing phage by bacteriophage lambda. *Gene*, **50**: 69–85.

Sternberg, N. and Hoess, R.H. 1995. Display of peptides and proteins on the surface of bacteriophage λ. *Proceedings of the National Academy of Science USA*, **92**: 1509–613.

Stout, V. and Gottesman, S. 1990. RcsB and RcsC: a two-component regulator of capsule synthesis in *Escherichia coli*. *Journal of Bacteriology*, **172**: 659–69.

Studier, F.W., Rosenberg, A.H., Dunn, J.J., and Dubendoff, J.W. 1990. Use of T7 RNA polymerase to direct expression of cloned genes. *Methods in Enzymology*, **185**: 60–89.

Tabor, S. and Richardson, C.C. 1985. A bacteriophage T7 RNA polymerase/promoter system for controlled exclusive expression of specific genes. *Proceedings of the National Academy of Science USA*, **82**: 1074–8.

Taylor, A.E. and Smith, G.R. 1992. RecBCD enzyme is altered upon cutting DNA at a recombination hotspot. *Proceedings of the National Academy of Science USA*, **89**: 5226–30.

Tijan, R. 1995. Molecular machines that control genes. *Scientific American*, February.

Umbarger, H.E. 1978. Amino acid biosynthesis and its regulation. *Annual Reviews of Biochemistry*, **47**: 533–606.

Walker G.C. 2001. To cleave or not to cleave? Insights from the LexA crystal structure. *Molecular Cell*, **8**(3): 486–7.

Walker, G.C. 1984. Mutagenesis and inducible responses to deoxyribonucleic acid damage in *Escherichia coli*. *Microbiological Reviews*, **4**: 60–93.

Walker, G.C. 1996. SOS regulon. In *Escherichia coli* and *Salmonella typhimurium: Cellular and Molecular Biology*, 2nd edn., eds. F.C. Neidhardt, R. Curtiss III, J.L. Ingraham, E.C.C. Lin, K.B. Low, B. Hagasanik, W.S. Rexnikoff, M. Riley, M. Schaechter, and H.E. Umbarger, pp. 2277–94. Washington, DC: ASM Press.

Wang, J. C. 1985. DNA topoisomerases. *Annual Reviews of Biochemistry*, **54**: 665.

Watson, J.D. 1968. *The Double Helix*. New York: Atheneum.

Weigle, J. 1953. Induction of mutations in a bacterial virus. *Proceedings of the National Academy of Science USA*, **39**: 628–36.

Weinberg, R.A. 1985. The molecules of life. *Scientific American*, October, 48–57.

Weisberg, R.A., Gottesman, S., and Gottesman, M.E. 1977. Bacteriophage λ: the lysogenic pathway. In *Comprehensive Virology, 8: Regulation and Genetics*, eds. H. Fraenkel-Conrat and R.R. Wagner, pp. 197–258. New York: Plenum Press.

West, S.C. 1992. Enzymes and molecular mechanisms of genetic recombination. *Annual Reviews of Biochemistry*, **61**: 603–40.

Whitehouse, H.L.K. 1982. *Genetic Recombination: Understanding Mechanisms*. New York: Wiley.

Wilson, G.G. 1991. Organization of restriction-modification systems. *Nucleic Acid Research*, **19**: 2539–66.

Yanisch-Perron, Vieira, C.J., and Messing, J. 1985. Improved M13 phage cloning vectors and host strains: nucleotide sequences of the M13mp18 and pUC19 vectors. *Gene*, **33**: 103–19.

Yanofsky, C. 1981. Attenuation in the control of expression of bacterial operons. *Nature*, **289**: 751–8.

Young, R. 1992. Bacteriophage lysis: mechanism and regulation. *Microbiological Reviews*, **56**: 430–81.

Zaman, G., Smetsers, A., Kaan, A., Schoenmaters, J., and Konings, R. 1991. Regulation of expression of the genome of bacteriophage M13. Gene V protein regulated translation of the mRNAs encoded by genes I, II, V and X. *Biochimica et Biophysica Acta*, **1089**: 183–92.

Zhou, Y.N., Kusukawa, N., Erickson, J.W., Gross, C.A., and Yura, T. 1988. Isolation and characterization of *Escherichia coli* mutants that lack the heat shock sigma factor ρ32. *Journal of Bacteriology*, **170**: 3640–9.

Zyskind, J.W. and Smith, D.W. 1993. DNA replication, the bacterial cell cycle and cell growth. *Cell*, **69**: 5.

Index

Page numbers in *italics* refer to figures and in **bold** refer to tables/boxes.

activation, 196, 267
agarose, 241–2, **241**, *242*
Agrobacterium tumefasciens
  conjugation to eukaryotes, 173
  Ti plasmids, 150–2, *151*
alkylation
  alkylating agents, 50–1, 62, 267
  removal of alkylated bases by DNA glycosylase, 68
allele, 267
allele number, 39–40, 267
Ames test, 54–6, *56*
amino acids, 2, *3*
anabolic processes, 201, 267
annotated sequence, 253, 267
antibiotic resistance determinants, 89, 90, 142, 149–50, 267
  as molecular tools, 100, *102*, 213–15
AP endonuclease, 67, 267
apurinic site, 46, 267
apyrimidinic site, 46, 267
ara promoter, 218
artificial chromosome vectors, 225, 258–9, 267
assimilation, 75, 267
ATPase, 267
attB site, 84, 113, *113*, 134–7
attenuation, 202–5, *204*, 267
attP site, 84, 113, *113*, 135–8
autocleavage, 209
autophosphorylation, 180, 211, 267

Bacillus subtilis, 176–83, *178–83*, 184–6
bacterial genomes, 259, **260**
bacteriocidal, 268
bacteriocidal selection, 129
bacteriophage, 105, 126, 268
  burst size, 107, 268
  as cloning vectors, 215
    phage display vectors, 228–9, *228*, *229*
    suicide vectors, 227–8, *227*, 276
    vector construction, 225–7, *226*
  discovery of, **105**
  filamentous phage, 106, 270
  host range, 107
  isolation of, **127**
  lifecycle of, 106–7
  lytic–lysogenic options, 107
  phage genome, 105, 273
  phage particles, 273
  structure of, 105–6, *106*
  temperate phage, 107, 276
  see also transduction
bacteriophage lambda, 107–18
  adsorption, 107
  assembly, 114–16, *115*
  DNA injection, 107–9
  DNA restriction and modification, 117–18
  DNA structure, *108*, 109
  genome protection in the bacterial cytoplasm, 109
  induction by SOS system, 117
  lysogenic pathway, 112–13
  lytic pathway, 113–14, 116
    DNA replication during, 114
  lytic–lysogenic decision, 110–12, *111*
  as model for site-specific recombination, 84, *85*, 113
  progeny release, 116
  promoters, 109, *109*, 112, 219
  specialized transduction, 134–40
    merodiploid strain construction, 135–8, *136–8*
    mutation transfer, 138–9, *139–40*
  superinfection, 117, 276
  transcription and translation, 109–10, *110*
  vector construction, 226, *226*
    phage display vectors, 228–9, *228*
    suicide vectors, 227–8
    transposon fusion vectors, 231–2, *231*
bacteriophage M13, 118–19
  adsorption and injection, 118
  DNA replication, 118–19, *119*
  genome protection, 118
  origin, **218**
  phage production and release from cell, 119, *120*
  vector construction, 226–7
    phage display vectors, 228, *228*
bacteriophage MS2, **124**
bacteriophage Mu, 90–1, *91*, 272
  transposon fusion vector construction, 231–2, *231*
bacteriophage P1, 119–21
  adsorption and injection, 119
  DNA replication, 119–20
  generalized transduction, 126–30, *130*
    carrying out transduction, 129–30
    identifying transduced bacteria, 129
    packaging the chromosome, 127–8, *128*
    transfer of chromosomal DNA, 128–9, *128*
  genome protection, 119
  phage assembly, 120
  prophage location in a lysogen, 120
  transducing particles, 120–1
bacteriophage T4, 121–4
  adsorption and injection, 121–2, *122*
  DNA replication, 122, *123*
  gene expression, 122
  T4rII mutations, 123–4, *124*
bacteriostatic, 268
bacteriostatic selection, 129
base analogs, 49, *49*, 268
base modifiers, 49–51, 268
base substitution, 40, 268
β-galactosidase, 197, *198*, 217, *222*
bidirectional DNA replication, 27, *28*, **31**, 143, *146*, 268
bioinformatics, 256–62, 268
  genome analysis, 260–1
  genome sequencing strategies, 257–9, *257–9*
biopanning, 229, 268
biotechnology, 36
blunt ends, 238–9, *238*, 268
  manipulation of, 240–1
Borrelia burgdorferi, 152, **152**, *153*
Breakage and Rejoining Model of recombination, 78
burst size, 107, 268
bypass suppressors, 53, 268

capsid, 105, 116, 268
 assembly, 114
capsular polysaccharide, 5, 8, 268
carbohydrates, 4–5, *5*, 268
catabolic processes, 197, 268
catabolite activator protein (CAP), 200, *201*
catabolite repression, 200, 268
cell growth, 13–15, *13*
 growth curve, **14**
 measurement of, **13**
cell membranes *see* membranes
cell wall, 5, 10
 growth of, 14
chaperone proteins, 205
 heat shock response regulation, 205, 207
Chargaff's rule, **23**
chemotaxis, 8, 268
Chi sequences, **76**, 77, 268
chiasmatype theory of recombination, 78
cholera, **118**
chromosomes, 11
 artificial chromosomes, 225, 258–9, 267
 circular structure, **30**
 DNA isolation, 235
 replication, 14
  see also DNA replication
CI protein, 110–12, *111*, 117
CII protein, 112, *112*
circular permutation, 120, *120*, 268
Citrobacter freundii, 86
cloning, 213, *214*, 234–50, 268
 cutting DNA molecules, 235–9
  restriction–modification as a molecular tool, 237–9
  shearing, 239
  type I restriction–modification systems, 236, **236**, *237*
  type II restriction–modification systems, 236–7, **236**, *237*, *238*
  type III restriction–modification systems, **236**, 237, *237*
 DNA amplification, 245–9, *247–8*, **248**
 DNA detection, 243–5
  colony blotting, 244–5, *245*
  Southern blotting, 243–4, *244*, 276
 DNA isolation from cells, 234–5
  chromosomal DNA, 235
  plasmid DNA, 234–5
 expressing the cloned gene, 249, *250*
 joining DNA molecules, 239–41, *239*
  manipulating the ends of molecules, 240–1, *240*
 library construction, 242–3, 272
 visualization of the cloning process, 241–2
cloning vectors, 213–15, *214*, 277
 artificial chromosome vectors, 225, 258–9, 267
 components included, **218**
 copy number importance, 215
 expression vectors, 217–19, *218*, 270
  for purifying the cloned gene product, 219–20, *219–21*
 gene expression analysis, 221–3
 gene product localization, 221
 identification of vectors containing a chromosomal insert, 217, *217*, 230–1
 multiple cloning sites (MCS), 216, *216*, 249, 272
 pBR322 vector, 215–16, *216*
 phage display vectors, 228–9, *228*, *229*
 phage vector construction, 225–7, *226*
 pUC vectors, 217, *217*
 shuttle vectors, 153, 224–5, *225*, 275
 suicide vectors, 227–8, *227*, 276
CMV promoter, **218**
Cockayne's syndrome (CS), **66**
coding strand, 192
coinheritance, 131, 268
cointegrate, 93, 96, *96*, 268
 formation mechanisms, 97, *97*
colanic acid capsule, 210–11
cold sensitive (Cs) gene products, 42, 268
ColE1 plasmids, 85, 145–6, *145*, 149
colonization of human cell surfaces, **8**
colony, 268
colony blotting, 244–5, *245*, 268
com genes, 177
competence, 175–83, 268
 artificial induction, 187
 regulation, 182–3, *183*
competence factors (Com), 177–85, *178–80*, 268–9
composite transposons, 89, *90*, *99*, 269
concatomers, 114, 116, 269
conjugation, 8, 156–73, 269
 demonstration of, **156**
 DNA transfer, 158–9, *160*
 F factor, 156, *157*, 160
 F-prime, 161, 270
  formation of, 163–4, *164*
  genetic uses of, 165–8
  transfer of, 164–5, *165*
 from prokaryotes to eukaryotes, 173
 Hfr, 160–1
  DNA transfer from, 162–3, *162*
  formation of, 161–2, *161*
  use in gene mapping, 167–72
 machinery, 158
 plasmid transfer, 153, *154*
 R factors, 156–8, *157*
 surface exclusion, 159
conjugative transposons, **100**
consensus sequence, 193–4, 269
contigs, 257, 269
control regions, 196, 269
Copy Choice Model of recombination, 78–9
copy number, 145–8, *147*, 272
 high copy number plasmids, 145
 importance of for cloning vectors, 215
 low copy number plasmids, 145
 setting, 148, *149*
core RNA polymerase, 192, 194–5, 269
cosmids, 258–9
counterselection, 269
cps regulon, 209–11, *210*
Cre protein, **218**
Cro protein, 110–12, *111*
CsCl gradients, **234**
CSF (competence and sporulation factor), 182, *182*
cyclic AMP, *lac* operon activation, 200, *201*
cytoplasm, 5, 11, 14–15, 269

de novo protein synthesis, 149, 269
deamination, 46–7, *47*, 269
 deaminated base removal by DNA glycosylase, 68
deletions, 40, 124
density labeling, 26, 269
deoxyribonucleoside, 17–19, 269
deoxyribonucleotide, 17–19, 269
2-deoxyribose, 17, *19*
dephosphorylation, 178–80, 269
depurination, 45–6, *47*, 269
dideoxy sequencing, 251–3, *251*
diploid cells, 38, 269
direct repeats, 89, 269
DNA (deoxyribonucleic acid), 1–2, 17, 269
 double-stranded, 21–2, *23*
  antiparallel orientation, 22, *23*, 267
  supercoiling, 23–6, *24*
 isolation from cells, 234–5
  chromosomal DNA, 235
  plasmid DNA, 234–5
 lesions, 58–60, *59*, 272
  causative agents, **59**
  double-stranded breaks, 59–60, *59*
  macrolesions, 40, *41*, 272
  microlesions, 40, *41*, 272
  single-stranded breaks, 59
  see also DNA repair; mutations
 major and minor grooves, 22, *24*
 modification, 117–18, 235
  see also restriction–modification systems
 polymerization, 19–21, *21*
 restriction, 117–18
  see also restriction enzymes
DNA amplification, 245–9, *247–8*, **248**
 adding novel DNA sequences to the ends of amplified sequences, 248, *248*
DNA cloning *see* cloning
DNA gyrase, 26, *26*, 269
DNA ligase, 32–4, 239–40, *239*, 269
DNA polymerases, 21, 28–30, *29*, **29**, 269
DNA repair, 44–5, 58–72, **61**
 defects, **60**
  cancer and, **69**
 excision of damaged DNA, 62–8
  glycosylases, 67–8
  MutHLS methyl directed mismatch repair, 65, *66*, 272
  UvrABC directed nucleotide excision repair, 62–5
  very short patch (VSP) repair, 65–7, *67*, 277
 inducibility, 72

mechanisms that tolerate DNA damage, 69–71
post replication/recombinational repair (PRR), 69–70, *71*, 274
transdimer synthesis, 69, *70*, 277
reversal of DNA damage, 60–2
methyltransferases, 61–2, 272
photoreactivation, 60–1, *62*, 273
DNA replication, 26–36
bidirectional replication, 27, *28*, **31**, 143, *146*, 268
of both strands, 32–4
constraints, 28
errors, 34–6
incorporation errors, 44–5
minimization of, 34–6
see also mutations
origin of replication (*ori*), 27, *28*, *33*, 142–3, *144*, 273
plasmid DNA, 143, **143**, *145*, *146*, 153
replication machinery, 28–31, **31**, *35*
biotechnology and, 36
DNA polymerases, 28–30, *29*
DnaG primase, 30–1, *32*
rolling circle replication, 114, *114*, 275
semiconservative versus conservative replication, 26,*27*, 275
theta mode replication, 34, 114, *114*, 276
DNA sequence searches, *252*, 253–4, *253*
DNA sequencing, 251–3, *251*
genome sequencing strategies, 257–9, *257–9*
DnaA, 31–2, 269
DnaA box, 31–2, *33*, 269
DnaB, 31, 32, 270
DnaC, 31, 32, 270
DnaG primase, 28, 30–1, *32*, 270
dominant mutation, 39, 166, 270
donor cell, 270
double-strand invasion model of recombination, 79, *80*
duplication, 40, 270

early genes, 122, 270
editing, 34, *36*, 69, 270
efficiency of plating (EOP), 117, 270
electroporation, 188, 270
endonuclease, 60, *60*, 270
see also restriction enzymes
environment suppressors, 54, 270
Escherichia coli, 5–12
cell structure, 5–12
cytoplasm, 11
F pilus, 8, *9*
fimbriae, 6–8, *8*
flagella, 8, *9*
inner membrane, 10–11, *12*
outer membrane, 6–10, *7*
peptidoglycan cell wall, 10, *11*
periplasmic space, 10
cellular locations, 5, *6*
chi sequences, **76**
chromosomal structure, **30**, **170**
supercoiling, 23
composition, **4**
conjugation, 8, **156**, 158–73
DNA transfer, 158–9, *160*
F-prime, 161, 163–7
Hfr strains, 160–3, 167–72
to *Saccharomyces cerevisiae*, 173
surface exclusion, 159
DNA lesions, 58
DNA polymerases, 21, 29
DNA repair, **60**, **61**, 62–70
DNA replication, 26–34, *27*, *28*, **29**, **31**
gene mapping, 167–72
gene regulation
heat shock response, 205–8
lac operon, 197–200, *199*, **199**, *200*, *201*
regulons, **206**
SOS system, **208**
tryptophan biosynthesis (*trp*) operon, 201–5, *202–4*
two-component systems, 210–11, **210**
genome sequence, 261–2
open reading frames, 261, **262**
genotype, 39–40, **40**
growth, 13–15
insertion sequences, *90*
plasmids, 148, 152, 153
recombination deficiencies, 74–8
restriction enzyme discovery, **235**
shuttle vectors, 224–5
transformation, 188–9
transposition accessory proteins, 94
ethyl methanesulfonate (EMS), 51
exonuclease, 60, *60*, 270
exopolysaccharides, 210
exponential growth, 270
expression vectors, 217–19, *218*, 270
for purifying the cloned gene product, 219–20, *219–21*
extragenic suppressors, 53, 270

F factor, 156, *157*, 160–1, *161*, 270
maintenance of in population, **157**
mobilization of non-conjugatible plasmids, 172, *173*
transfer in conjugation, 158–9, *159*
see also F prime; Hfr strains
F pilus, 8, *9*, 158, *159*, 270
F prime
formation of, 163–4, *164*
genetic uses of, 165–8
transfer of, 164–5, *165*
Fanconi's anemia, **71**
fatty acids, 2–4
filamentous phage, 106, 270
fimbriae, 6–8, *8*, 270
flagella, 8, *9*, 270
flow cytometry, **13**
fluctuation tests, **45**
flushed ends, 59, 270
forward mutation, 270
frameshift mutations, 40, 270
fusion proteins, 219–20

G418 resistance, **218**
gain of function mutation, 38, 270
gel electrophoresis, 241–2, *242*, *243*, 270
2-D PAGE, 263–4, *263*
SDS-PAGE, 262–3, *263*
gene, 17, 123–4, 213, 270
gene conversion, 82–3, 270
gene expression
bacteriophage T4, 122
cloned genes, 249, *250*
see also expression vectors
gene fusion, 222, *222*
using transposition, 229–32, **230**, *231*
gene linkage *see* linkage
gene mapping, 131–3, **131**, *132*, **133**, 188
time of entry mapping, 168, **170**
using Hfr strains, 167–72, *169*
50% rule, 171, *171*
use of several strains to cover the chromosome, 171–2
gene product, 270
gene regulation, 191–2, 270
mutational analysis, *196*
negative regulation, 273
operons, 196–7, *197*, 273
lac operon, 197–200, *199*, **199**, *200*, *201*
tryptophan biosynthesis (*trp*) operon, 200–5, *202–4*
players in, 192–6
regulons, 196–7, *197*, **206**
cps regulon, 209–11, *210*
heat shock response regulation, 205–8
SOS regulon, 208–9
two-component regulatory systems, 209–11, **210**
generalized transduction, 126–30, *127*, 270
genetic code, *43*
genetic recombination *see* recombination
genome, 256, 271
analysis, 260–1
bacterial, 259, **260**
E. coli K12, 261–2, **262**
human genome, **262**
phage genome, 105, 273
sequencing strategies, 257–9, *257–9*
genotype, 15, 39–40, *39*, 271
writing, **40**, **41**
glycogen, 4
glycosylases, 67–8
alkylated bases removed by DNA glycosylase, 68
deaminated bases removed by DNA glycosylase, 68
glycosylases specific for pyrimidine dimers, 68
MutM/MutY and oxidative damage, 68
uracil-N-glycosylase coupled with AP excision repair, 67, *68*
Gram stain, **6**

green fluorescent protein (GFP), 221, 271
GST-fusion protein, 219, 271

Haemophilus influenzae, 183, 185–6
  genome sequence, 259
hairpins, 152, *153*
haploidy, 38, 271
head-full packaging, 120, 271
heat shock proteins (HSPs), 205, **206**
heat shock response regulation, 205–7
helicase activity, 32, 76, 271
Helicobacter pylori, 185–6
heteroduplex DNA, 79, 82, 271
Hfr strains, 160–1, 271
  DNA transfer from, 162–3, *162*
  formation of, 161–2, *161*
  use in gene mapping, 167–72
    50% rule, 171, *171*
    using several strains to cover the chromosome, 171–2
histidine tag, 219, 271
Holliday model of recombination, 79, *80*
holoenzyme, 192, 194, 271
homologous recombination *see* recombination
homology, 74, 271
  F factor insertion and, 161–2, *161*
  microhomology, 48, *48*, 272
host range, 271
  bacteriophage, 107
  broad host range plasmids, 152–3, 268
hotspots, **45**, 93, 271
housekeeping genes, 191, 271
human genome, **262**
hydrogen bonding, 21, *22*
hydroxylamine, 50, *50*

illegitimate recombination, 86, 271
incompatibility, 148–9, *150*, 271
incorporation errors, 44–5
inducer, 197, 271
informational suppressors, 53, *55*, 271
initiator protein, 143, *145*, 148
inner membrane, 10–11, *12*, 271
insertion, 40, 271
insertion sequences (IS), 89, *90*
interactive suppressors, 54, *55*, 271
intercalators, 51, 271
intermolecular recombination, 83
intragenic suppressors, 53, 271
intramolecular recombination, 83
inversion, 40–2, 85–6, 271
inverted repeats (IR), 89, 195, 271
ionizing radiation, 51

jackpot mutation, 271

KHO antigens, **10**

lac genes, 130, *137*, 163–4, 166, 197–200
  gene mapping, 168–70
  in pUC vectors, 217, *217*
  as reporter genes, 223, *223–4*
    transposon fusion vector construction, 229–32, **230**, *231*
lac operon
  activation by cAMP and the CAP protein, 200, *201*
  promoter, 218
  repression, 197–200, *199*, **199**, *200*
lactose metabolism, 197–9, *198*
lagging DNA strand, 32, *34*, 271
LamB, 107
lambda *see* bacteriophage lambda
late genes, 122, 271
leader sequence, 202, 271
leading DNA strand, 32, *34*, 271
leaky mutations, 272
lesions, DNA *see* DNA; mutations
lethal mutations, 166, *167*
lethal selection, 129, 272
LexA repressor protein, 72, 117, 208–9
  autocleavage, 209
library construction, 242–3, 272
linkage, 131, 188, 270, 272
  determination by two-factor crosses, 131–2, *131*
  see also gene mapping
lipids, 2–4, *4*, 272
lipopolysaccharides (LPS), 6, *7*, 272
localization signals, **218**
localized mutagenesis, 133–4, *135*, 272
loss of function mutation, 38, 272
lox sites, **218**
lysate, 272
lysogen, 107, 272
lysogeny, 107, 272
  bacteriophage lambda, 110–13
lysozyme, 272
lytic infection, 107, 272
  bacteriophage lambda, 110–12, 113–14, 116

M13 *see* bacteriophage M13
macrolesions, 40, *41*, 272
macromolecules, 1
maltose binding protein (MBP), 219–20, *219*, 272
membrane-derived oligosaccharides (MDOs), 211
membranes, 2–4, 5–11
  growth of, 14, *14*
  inner membrane, 10–11, *12*, 271
  outer membrane, 6–10, *7*, 273
  pores, 9
merodiploid cells, 39, 166, *166*, 272
  construction with specialized transducing phage, 135–8, *136–8*
Meselson and Radding model of recombination, 81–2, *81*
messenger RNA (mRNA), 2, 17, 192, 272
  polycistronic mRNA, 197, *197*
methyl directed mismatch repair, 65, *66*, 272
methylation, transposition and, 94, *94*
methylmethane sulfonate (MMS), 50–1
methyltransferases, 61–2, 272
microarray technology, 264–5, *264*, *265*, **265**, 272
microhomology, 48, *48*, 272
microlesions, 40, *41*, 272
middle genes, 122, 272
mismatched base pairs, 35–6, 58–9, 82, 272
  methyl directed mismatch repair, 65, *66*, 272
  recombination and, 82
missense mutations, 43, *44*, 124, 272
mobilization, 154, 272
  of non-conjugatible plasmids, 172, *173*
modification, 117–18, 235, 272
  see also restriction–modification systems
multidrug resistance, **158**
  see also antibiotic resistance determinant
multiple cloning sites (MCS), 216, *216*, 249, 272
muropeptides, 14, 272
mutagens, 44, 272
  Ames test, 54–6, *56*
  ionizing radiation, 51
  localized mutagenesis, 133–4, *135*, 272
  specificity of, **45**
  ultraviolet (UV) light, 51, *52*
mutants, **43**
  exploitation of, 56
mutations, 15, 38–56, **43**, 272
  classes of, 40–3
  forward mutation, 270
  frameshift mutations, 40, 270
  frequency of, 44, 272
  gain of function mutation, 38, 270
  gene regulation and, 196, *196*
  induced mutations, 44, 48–52
    base analogs, 49, *49*
    base modifiers, 49–51
    intercalators, 51
    ionizing radiation, 51
    site-directed mutagenesis using PCR, 248–9, *249*
    ultraviolet (UV) light, 51, *52*
  leaky mutations, 272
  lethal mutations, 166, *167*
  loss of function mutation, 38, 272
  mutator strains, 52
  null mutations, 100, 273
  point mutations, 40, 42–3, 124, 274
    missense mutations, 43, *44*, 124, 272
    nonsense mutations, 43, *44*, 124, 273
    silent mutations, 42–3, *44*, 276
  polar mutations, 42, *42*, 48, 274
  primary mutation, 53, 274
  rate of, 44, 273
    fluctuation tests, **45**
    hotspots, **45**
  reverting mutations, 52–4
    pseudorevertant, 53
    reversion, 52–3, *53*
    suppression, 53–4, *54*
    true revertant, 53

spontaneous mutations, 44–8
deamination, 46–7, *47*
depurination, 45–6, *47*
genetic rearrangement, 48
incorporation errors, 44–5
tautomerism, 45, *46*
transposition, 48
T4rII mutations, 123–4
transfer of, 167
MutHLS methyl directed mismatch repair, 65, *66*, 272
MutM, 68
MutY, 68

N protein, 109–10
negative regulation, 273
Neisseria meningitidis, 185
nitrosoguanidine (NTG), 50, *50*, 273
non-composite transposons, 89–90, *91*, 273
non-replicative transposition, 93, *93*, 95, *95*, 273
nonlethal selection, 129, 273
nonsense mutations, 43, *44*, 124, 273
nucleic acids, 1–2, 17, *18*, 273
nucleoside 5′ triphosphate, 19–21, 273
nucleosides, 17–19, **19**
synthesis, *20*
nucleotide excision repair (NER), 62–3, *64*, 273
nucleotides, 19, **19**, *21*
synthesis, *20*
null mutations, 100, 273

O4-methylguanine methyltransferase, 61–2
O6-methylguanine methyltransferase, *63*
O6-methylthymine methyltransferase, 61–2
Okazaki fragments, 32, 34, 273
oligonucleotide, *21*, 30, 273
open reading frames (ORFs), 260
E. coli, 261, **262**
operators, 111–12, 196, 273
operon fusion, 222, *222*, 273
operons, 196–7, *197*, 273
lac operon regulation, 197–200, *199*, **199**, *200*
activation by cAMP and the CAP protein, 200, *201*
repression, 197–200
tryptophan biosynthesis (*trp*) operon regulation, 200–5, *202–4*
optical density (OD), **13**, 273
origin of replication (*ori*), 27, *28*, *33*, 142–3, *144*, 273
origin of transfer (*oriT*), 158, *160*, 273
outer membrane, 6–10, *7*, 273
overproduction suppressors, 53–4, 273
oxidative damage, 68
oxidative stress response regulon, 68, 273

P1 *see* bacteriophage P1
pac site, 120, *121*, 126–7, 273
partitioning, 145–8, *147*
pathogenicity, 273
pBR322 cloning vector, 215–16, *216*
peptidoglycan, 10, *11*, 273
growth of, 14, *15*
periplasm, 5, 10, 273
permissive temperature, 42, *42*, 273
phage *see* bacteriophage
phage display vectors, 228–9, *228*, *229*
phenotype, 15, 39–40, *39*, 273
writing, **41**
phosphatase, 241, 273
phosphodiester bond, 17–19, 273
phosphorylation, 178–80, 273
autophosphorylation, 180, 211, 267
photolyase, 60, 273
photoreactivation, 60–1, *62*, 273
pilus, 184, 273
F pilus, 8, *9*, 158, *159*, 270
plaques, 107, 273–4
plasmids, 120, 142–54, *143*, 274
amplification, 149
broad host range plasmids, 152–3, 268
ColE1 plasmids, 85, 145–6, *145*, 149
copy number, 145–8, *147*, 272
high copy number plasmids, 145
importance of for cloning vectors, 215
low copy number plasmids, 145
setting, 148, *149*
DNA isolation, 234–5
DNA replication, 143, **143**, *145*, *146*, 153
F factor, 156, *157*
maintenance in population, **157**
genes carried, 149–52
incompatibility, 148–9, *150*, 271
linear plasmids, 152, *152*, *153*
mobilization of non-conjugatible plasmids, 172, *173*
moving from cell to cell, 153–4, *154*
naming of, **142**
origins of replication, 142–3, *144*
partitioning, 145–8, *147*
Pseudomonas aeruginosa camphor-resistant plasmid, **150**
R factors, 156–8, *157*
discovery of, **158**
Ti plasmids, 150–2, *151*, 276
transposons in, **143**
as vectors, 215
copy number importance, 215
multiple cloning sites (MCS), 216, *216*
see also cloning vectors
point mutations, 40, 42–3, 124, 274
missense mutations, 43, *44*, 124, 272
nonsense mutations, 43, *44*, 124, 273
silent mutations, 42–3, *44*, 276
polar mutations, 42, *42*, 48, 274
poly A sites, **218**
polycistronic mRNA, 197, *197*, 274
polymerase chain reaction (PCR), 245–9, *247–8*, **248**, 274
adding novel DNA sequences to the ends of a PCR amplified sequence, 248, *248*
site-directed mutagenesis, 248–9, *249*
porins, *10*
post replication/recombinational repair (PRR), 69–70, *71*, 274
post-transcriptional regulation, 191, 274
post-translational regulation, 191, 274
primary mutation, 53, 274
prime, 161
primer, 274
primosome, 274
probe, 244, 274
promoters, 109, *109*, 112, 193–4, 274
ara promoter, 218
bacteriophage lambda, 109, *109*, 112, 219
CMV promoter, **218**
expression vectors, 217–19
heat shock promoters, 207
lac operon, 218
promoter recognition, 274
T7 promoter, 219, **219**
proofreading, 34, 69, 274
prophage, 107, 274
P1 prophage location in a lysogen, 120
Vibrio cholera, **118**
proteases, 189, 205, 274
heat shock response regulation, 205, 207
protein fusion, 274
proteins, 2, 274
de novo protein synthesis, 149, 269
membrane proteins, 9, 10–11, *10*
proteome, 262, 274
analysis techniques, 262–5
2-D PAGE, 263–4, *263*
microarray technology, 264–5, *264*, *265*
SDS-PAGE, 262–3, *263*
proteomics, 262–5, 274
pseudo-pac sites, 121, 127, *128*, 274
Pseudomonas aeruginosa plasmids, **150**, 153
pseudorevertant, 53, 274
pUC vectors, 217, *217*
pyrimidine dimers, 274
repair, 62–3, 68
post replication/recombinational repair (PRR), 69–70, *71*
transdimer synthesis, 69, *70*

Q protein, 109, 113, *113*
quorum sensing system, 180–1, 274

R factors, 156–8, *157*, 274
discovery of, **158**
mobilization of non-conjugatible plasmids, 172
R protein, 116
RapC, 182
raw sequence, 253, 274
RcsC (regulator of capsule synthesis), 210–11
rearrangements, 40–2, 48, 274
RecA, 70, **74**, 75–8, *76*, 82, 117
RecBCD, 70, 76–7, *77*, 82
recessive mutation, 39, 166, 274

recipient cells, 274
recombinase, 83, 274
recombination, 270–1
homologous, 70, 74–83, 271
enzymes involved, **75**
models of, 78–82, *78*, *80*, *81*
reciprocal nature, 83, 274
transduction and, 128–9, *128*
illegitimate, 86, 271
site-specific, 83–6, *83*, 276
bacteriophage lambda model, 84, *85*, 113
regulatory protein, 274
regulons, 196–7, *197*, **206**, 275
cps regulon, 209–11, *210*
heat shock response regulation, 205–7
SOS regulon, 71–2, 208
mutagenic repair, 69, 276
regulation, 208–9
replication *see* DNA replication
replicative transposition, 93, *93*, 95–6, *95*, 275
replicon, 143, 275
reporter genes, 221–3, 275
transposon fusion vector construction, 229–32, *231*
repression, 196, 275
res site, 96
resolvase, 96, 275
response-regulator protein, 178–80, 210, 275
restriction, 117–18, 275
restriction enzymes, 117–18, 215, 235, 275
discovery of, **235**
as molecular tools, 237–8
naming, **238**
see also restriction–modification systems
restriction mapping, **246**
restriction sites, 215, 275
frequency of, **213**
restriction–modification systems, 235–8, **236**
as molecular tools, 237–8
type I, 236, *237*
type II, 236–7, *237*, *238*
type III, 237, *237*
reversion, 52–3, *53*, 275
pseudorevertant, 53
true revertant, 53
Rho protein, 195
ribose, 17, *19*
ribosomal RNA (rRNA), 2, 17, 192, 275
ribosome binding site (RBS), 195, 275
RK2 plasmid, 153
RNA (ribonucleic acid), 1–2, 17, 192, 275
synthesis, 192–5, *193*, 268
see also transcription
RNA phage MS2, **124**
RNA polymerase, 28, 192–5, *193*, 275
core RNA polymerase, 192, 194–5, 269
rolling circle replication, 114, *114*, 275
RSF1010 plasmid, 153–4

S protein, 116
S-adenosylmethionine, 236, 237
Saccharomyces cerevisiae, **41**, 173
Salmonella, 54–5, *56*, 86, 126
screening, 129, 268
SDS-PAGE, 262–3, *263*
selectable markers, 213–15, 275
selection, 129, 275
selective media, 129, 275
semiconservative replication, 26, *27*, 275
sensor-kinase protein, 178–80, 210, 275
septum, 13, 275
sequence searches, *252*, 253–4, *253*
sequencing *see* DNA sequencing
shearing DNA, 239
Shine–Delgarno sequence, 195, 275
short patch repair, 63, 275
short sequence repeats (SSR), 86, 275
shuttle vectors, 153, 224–5, *225*, 275
sigma factors, 192–3, 275–6
heat shock response regulation, 207
signal transduction, 209–11, 276
silent mutations, 42–3, *44*, 276
single-strand invasion model of recombination, 81–2, *81*
site-directed mutagenesis using PCR, 248–9, *249*
site-specific recombination, 83–6, *83*, 276
bacteriophage lambda model, 84, *86*, 113
SOS regulon, 71–2, 208
mutagenic repair, 69, 276
regulation, 208–9
SOS response, 117, *117*, 208–9, **208**, *209*
Southern blotting, 243–4, *244*, 276
colony blotting, 244–5, *245*, 268
SpoOK, 182
spores, 181–2, 276
sporulation, 181–2, *181*, 186, 276
regulation, 182–3, *183*
staggered ends *see* sticky ends
stem loop structure, 195, 276
sticky ends, 59–60, 238–9, *238*, 276
manipulation of, 240–1
strain construction, 133
merodiploid strains, 135–8, *136–8*, 166
Streptococcus pneumonia, **175**, 183–6, **187**
Streptomyces griseus, 60
suicide vectors, 227–8, *227*, 276
supercoiling, 23–6, *24*, 276
negative supercoiling, 23
superinfection, 117, 276
suppression, 53–4, *54*
suppressor, 53
bypass suppressors, 53, 268
environment suppressors, 54, 270
extragenic suppressors, 53, 270
informational suppressors, 53, *55*, 271
interactive suppressors, 54, *55*, 271
intragenic suppressors, 53, 271
overproduction suppressors, 53–4, 273
surface exclusion, 159, 276
SV40 origin, **218**

T4 *see* bacteriophage T4
T4 polynucleotide kinase, 241
T7 promoter, 219, **219**
T-DNA, 150–1, *151*, 173, 276
target DNA, 93
target immunity, 99–100, *100*, 276
tautomerism, 45, *46*, 276
temperate phage, 107, 276
temperature sensitive (Ts) gene products, 42, 276
template strand, 192, 276
terminal redundancy, 119, *120*, 276
terminase, 116
terminator, 195, *195*, 276
theta mode replication, 34, 114, *114*, 276
three-factor crosses, 132–3, *132*, **133**, 276
thrombin, 220
Ti plasmids, 150–2, *151*, 276
time of entry mapping, 168, **170**
topoisomerases, 23–5, *25*, 276
transcription, 11, 192–5, *192*, *194*, 276
basal level, 268
coupled transcription and translation, **196**
termination, 194–5
transcriptional fusion, 222, *222*, 277
transcriptional regulation, 191–2, 196
transdimer synthesis, 69, *70*, 277
transducing particles, 120–1, 127, 276
transduction, 126–40, 277
generalized versus specialized transduction, 126, *127*, 270
P1 as a model for generalized transduction, 126–30, *130*
carrying out transduction, 129–30
identification of transduced bacteria, 129
packaging the chromosome, 127–8
transfer of chromosomal DNA, 128–9, *128*
specialized transducing phage, 134–40, 276
merodiploid strain construction, 135–8, *136–8*
mutation transfer, 138–9, *139–40*
uses for, 130–4
gene mapping, 132–3, *132*, **133**
linkage determine, 131–2, *131*
localized mutagenesis, 133–4, *135*
strain construction, 133
transfer RNA (tRNA), 2, 17, 192, 277
transformants, 175, 277
transformasome, 186, *186*, 277
transformation, 175–89, 277
artificial, 187–8
electroporation, 188
influencing factors, 188–9
as a genetic tool, 188
machinery, 184–7
as a molecular tool, 188–9
natural competency, 175–83, **177**
as a pathogenic problem, **187**
process of, 183–4, *184*, *185*
rationale for, 187
transforming principle, 175, **175**
chemical nature of, **176**
transition, 40, 277
translation, 11, 277
coupled transcription and translation, **196**

translational fusion, 222, *222*, 277
translational regulation, 191, 277
translocation, 42, 277
transposable elements *see* transposons
transposase, 93–4, 277
transposition, 48, 89–103
  fate of donor site, 97–8, *98*
  frequency of, 91, *94*
  fusion vector construction, 229–32, **230**, *231*
  machinery, 93–5, *93*
    accessory proteins encoded by the host, 94–5
    accessory proteins encoded by the transposon, 94
  measurement of, **92**
  non-replicative, 93, *93*, 95, *95*, 273
  replicative, 93, *93*, 95–6, *95*, 275
  target immunity, 99–100, *100*, 276
  see also transposons
transposons, 48, *49*, 85, 277
  composite, 89, *90*, *99*, 269
  conjugative, **100**
  eukaryotic, **103**
  fusion vector construction, 229–32, **230**, *231*
  genetically engineered derivatives, *103*
  as molecular tools, 100–1, *101–3*
  non-composite, 89–90, *91*, 273
  in plasmids, **143**
  structure of, 89–91
  see also transposition
transversion, 40, 277
traS gene, 159
traT gene, 159
trichothiodystrophy (TTD), 65
true revertant, 53
tryptophan biosynthesis (*trp*) operon regulation, 200–5, *202–4*
2-D PAGE, 263–4, *263*
two-component signal transduction system, 178, *179*, *180*, 277
two-factor crosses, 131–2, 277

ultraviolet (UV) light, 51, *52*, 58
  SOS response to, 117, 208, **208**
uptake signal sequence (USS), 185–6, *186*, 277
uracil-N-glycosylase coupled with AP excision repair, 67, *68*
UvrABC directed nucleotide excision repair, 62–3, *64*

vectors *see* cloning vectors
very short patch (VSP) repair, 65–7, *67*, 277
viable cell count (VCC), **13**, 277
Vibrio cholera, **118**
virulence factors, 89, 90, 277
viruses, 105, 277

Weigle reactivation, **208**
wild-type cells, 15, 277

xeroderma pigmentosum, **64**
XGal, 277

zoo blots, 244